Ben-Guang Rong (Ed.)
Process Synthesis and Process Intensification
De Gruyter Graduate

Also of Interest

Product and Process Design.
Driving Innovation
Harmsen, de Haan, Swinkels; 2018
ISBN 978-3-11-046772-7, e-ISBN 978-3-11-046774-1

Advanced Process Engineering Control.
Agachi, Christea, Csavdári, Szilagyi; 2016
ISBN 978-3-11-030662-0, e-ISBN 978-3-11-030663-7

Process Technology.
An Introduction
De Haan; 2015
ISBN 978-3-11-033671-9, e-ISBN 978-3-11-033672-6

Chemical Reaction Technology.
Murzin; 2015
ISBN 978-3-11-033643-6, e-ISBN 978-3-11-033644-3

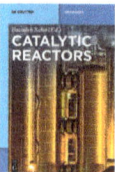

Catalytic Reactors.
Saha (Ed.); 2015
ISBN 978-3-11-033296-4, e-ISBN 978-3-11-033298-8

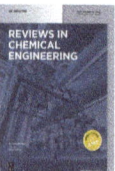

Reviews in Chemical Engineering.
Luss, Brauner (Editors-in-Chief)
ISSN 0167-8299, e-ISSN 2191-0235

Process Synthesis and Process Intensification

Methodological Approaches

Edited by
Ben-Guang Rong

DE GRUYTER

Editor
Dr. Ben-Guang Rong
University of Southern Denmark
Department of Chemical Engineering
Biotechnology & Environmental Technology
Campusvej 55
5230 Odense
Denmark
E-mail: bgr@kbm.sdu.dk

ISBN 978-3-11-046505-1
e-ISBN (PDF) 978-3-11-046506-8
e-ISBN (EPUB) 978-3-11-046515-0

Library of Congress Cataloging-in-Publication Data
A CIP catalog record for this book has been applied for at the Library of Congress.

Bibliographic information published by the Deutsche Nationalbibliothek
The Deutsche Nationalbibliothek lists this publication in the Deutsche Nationalbibliografie;
detailed bibliographic data are available on the Internet at http://dnb.dnb.de.

© 2017 Walter de Gruyter GmbH, Berlin/Boston
Cover image: tuachanwatthana/iStock/Getty Images plus
Typesetting: le-tex publishing services GmbH, Leipzig
Printing and binding: CPI books GmbH, Leck
♾ Printed on acid-free paper
Printed in Germany

www.degruyter.com

Preface

As the world is speeding up on the track to global industrialization, the landscapes of worldwide economy, energy, natural resources, environment as well as products and markets are changing rapidly. At the same time, the world is exploring the track to pursue global sustainable development in terms of resources, environment, climate, economy and society which call for efforts in every aspect including advanced science and technology. In chemical and process engineering discipline, we commit to develop and design manufacturing processes to produce products from both fossil-based and renewable biomass-based feedstocks. Process synthesis and process intensification contribute to their advanced manufacturing technologies through optimal design of the processes, equipment and units. Many manufacturing processes are for the products with massive production to serve global consumption which have high impact to the resource utilization, environment and climate change, and which directly relate to the sustainable development of the economy and society. Therefore, process synthesis and process intensification are among the paramount research subjects towards sustainable products and manufacturing processes.

The contributed works from process synthesis (PS) and process intensification (PI) are reflected by the subject investigator's thinking, designing and creating process which are manifested by testing and embedding novel ideas, concepts, principles, mechanisms, and decisions into the object system and the objective materials processing operation(s). This working process of the interactions between the subject investigator and the object system can be properly represented as the method and approach of the corresponding PS and PI works, which is characterized by the domain topic and problem scope. Considering the common elements in PS and PI works, this book intends to introduce the methodological approaches of PS and PI by focusing on three basic elements: domain topics and problems – methods and approaches – cases and examples, as such 10 chapters are covered and discussed for this purpose.

The several common basic concepts and elements in PS and PI works are first discussed in Chapter 1 "Introduction to basic concepts and elements in process synthesis and process intensification". In Chapter 2, "Process synthesis and process intensification for multicomponent distillation systems – the systematic methodology", six distinct subspaces of multicomponent distillation systems and their systematic generation procedures are presented. Therein three subspaces and their systematic procedures are developed in the context of process synthesis, while the other three subspaces and their systematic procedures are developed in the context of process intensification. Chapter 3 "Reaction intensification by microwave and ultrasound techniques in chemical multiphase systems" presents microwave and ultrasound techniques as the attractive methods for intensification of reaction processes, where cases of fatty acid epoxidation, enantioselective hydrogenation, sugar alcohol production and starch oxidation are illustrated. The approach of process intensification by combining the reaction and separation into the single equipment is introduced in Chapter 4 "Process

https://doi.org/10.1515/9783110465068-201

intensification by reactive distillation", three examples of biodiesel production, fatty esters synthesis and methyl acetate productions have shown the applications of the reactive distillation technology. Chapter 5 "Process synthesis and intensification of hybrid separations" introduces the hybrid method for separation process synthesis and intensification, where two examples of hybrid distillation-pervaporation and hybrid liquid-liquid extraction-distillation are introduced for bioethanol and biobutanol purification. Process intensification based on equipment miniaturization approach is illustrated in Chapter 6 "Process intensification for microdistillation using the equipment miniaturization approach", where some elegant configurations of microdistillation are illustrated including 3D-print technique to produce the functional components like packings. Chapter 7 "Integrated biofuels process synthesis: integration between bioethanol and biodiesel processes" presents a two-stage methodology for synthesis and intensification of integrated biofuel processes, where superstructure is constructed in the first stage including mass and energy integrations, the optimization solution of the superstructure through the mixed integer non-linear programming (MINLP) is a network flowsheet which is then going to the second stage for further intensification. Case study of the integrated lignocellulosic bioethanol and biodiesel (FAEE) process is illustrated with the two-stage methodology. Process synthesis based on process analytical technology (PAT) is presented in Chapter 8 "Process synthesis for natural products from plants based on PAT methodology", where it has shown that different analytical instruments are necessary to obtain the streams information during plants processing operations, case study of conceptual process synthesis for recovery of artemisinin from *Artemisia annua* is illustrated. Chapter 9 "Process synthesis for energy efficiency based on the pinch analysis approach" summarizes the situations to use pinch analysis approach for heat integration during process synthesis, where retrofit of the heat exchanger network for an industrial case study is presented. Integration of process design and control is discussed in Chapter 10 "Process synthesis and intensification by integration between process design and control", specifically closed-loop analysis procedure is introduced and two examples for the intensified processes are studied.

Considering the extensive research in both PS and PI, there are many more topics and cases to be covered from both PS and PI works. It is certainly the most enjoyable moment whenever impressive process technologies are presented and whatever new methods and approaches are developed. As PS and PI will be the generation-to-generation subjects towards sustainable products and process technologies, meanwhile, PS and PI are focusing on addressing creativity and innovation in process and equipment technologies, one thing is sure that the future generations will develop more and more new methods and approaches for both PS and PI which will significantly expand the landscapes of future products and their manufacturing processes.

My deepest gratitude goes to the authors of each chapter for their efforts to introduce the presented topics and methods. I hope that the covered topics and cases in this book will be helpful to serve as the examples to introduce the methods and approaches for PS and PI works.

Ben-Guang Rong

Contents

Preface — V
List of Contributors — XIV
Editor Biography — XVI

Ben-Guang Rong

1 **Introduction to basic concepts and elements in process synthesis and process intensification — 1**
1.1 Introduction — 1
1.2 Basic elements in synthesis and intensification of distillation systems — 3
1.3 Basic facets of process synthesis — 4
1.3.1 Introduction to process synthesis — 4
1.3.2 Two approaches for finding alternatives in process synthesis: pure synthesis versus practical synthesis — 7
1.4 Basic facets of process intensification — 9
1.5 Evaluation indicators for PS and PI works — 11
1.6 Concepts of process synthesis work and process intensification work — 12
1.7 Conceptual design of process and equipment — 13
1.8 Concepts of the system and technical system — 14
1.8.1 System concept — 15
1.8.2 Technical system concept — 16
1.9 Software and hardware systems for a technical system — 18
1.9.1 Software elements — 18
1.9.2 Hardware elements — 19
1.9.3 Emergence of the technical system — 21
1.10 Triple parties in the working process of the methodology: the subject, object, and scientific method — 21
1.11 Synthesis and analysis as basic scientific methods — 23
1.11.1 Synthesis-dominated versus analysis-dominated methods — 23
1.11.2 Synthesis and analysis as basic methods in PS and PI works — 25
1.11.3 Process synthesis versus process analysis — 26
1.12 Systematic procedure as an element — 28
1.12.1 General procedures for process synthesis — 29
1.12.2 Specific procedures for PS and PI — 29
1.13 Concept of the technical system as a technology whole — 30
1.14 Example: the basic elements addressed in distillation intensification for the dividing-wall column subspace — 34
1.15 Conclusions — 39
1.16 Bibliography — 39

Ben-Guang Rong

2   **Process synthesis and process intensification for multicomponent distillation systems – systematic methodology —— 41**
2.1   Introduction —— 41
2.2   Problem definition —— 44
2.2.1   Multicomponent mixtures —— 44
2.2.2   Product specifications —— 45
2.3   Dominant criteria for evaluation of a distillation system —— 45
2.4   Basic concepts for multicomponent distillations —— 45
2.4.1   Sharp and nonsharp splits —— 45
2.4.2   Simple distillation column —— 46
2.5   Basic software and hardware elements for a distillation system —— 47
2.5.1   Software elements —— 47
2.5.2   Hardware elements —— 49
2.6   Distinct separation sequences —— 50
2.7   Subspace of sharp sequence configurations —— 52
2.8   Subspace of nonsharp sequence configurations —— 54
2.9   Subspace of the original thermally coupled configurations —— 62
2.10   Subspace of the thermodynamically equivalent structures —— 68
2.11   Intensified distillation systems with fewer columns —— 74
2.11.1   Subspace of the intensified simple column configurations —— 75
2.11.2   Subspace of intensified distillation systems —— 82
2.12   Subspace of dividing-wall columns —— 90
2.12.1   DWCs from thermodynamically equivalent structures with side columns —— 91
2.12.2   DWCs from prefractionation columns —— 97
2.13   Summary and remarks —— 103
2.13.1   Primary results —— 103
2.13.2   Distinct subspaces for multicomponent distillation —— 104
2.13.3   About the subspaces and their relationships —— 104
2.13.4   About the method scopes of the subspaces for process synthesis and process intensification —— 105
2.13.5   About the mechanisms and systematic procedures —— 105
2.13.6   Common elements of the methods for distillation subspaces —— 105
2.13.7   About further studies and applications —— 106
2.14   Conclusions —— 107
2.15   Nomenclature —— 107
2.16   Bibliography —— 108

Adriana Freites Aguilera, Pasi Tolvanen, Victor Sifontes Herrera, Jean-Noël Tourvielle, Sébastien Leveneur, and Tapio Salmi

3   **Reaction intensification by microwave and ultrasound techniques in chemical multiphase systems —— 111**
3.1   Microwave irradiation —— 111
3.1.1   Background —— 111
3.1.2   Heating mechanisms —— 113
3.1.3   Dielectric loss and permittivity —— 114
3.1.4   Selective heating —— 115
3.1.5   Case 1: conventional heating versus microwaves for epoxidation of vegetable oils —— 117
3.1.6   Case 2: effect of resonant microwave fields on temperature distribution in time and space —— 119
3.2   Process intensification by ultrasound: sonochemistry —— 124
3.2.1   What is ultrasound? —— 124
3.2.2   Ultrasonification techniques —— 126
3.2.3   Case 1: catalyst activation by ultrasonification —— 127
3.2.4   Ultrasonification of starch —— 135
3.3   Conclusions —— 138
3.4   Bibliography —— 138

Anton A. Kiss

4   **Process intensification by reactive distillation —— 143**
4.1   Introduction —— 143
4.2   Principles of reactive distillation —— 144
4.3   Modeling reactive distillation —— 147
4.3.1   Residue curve map —— 149
4.4   Design and control —— 152
4.5   Reactive distillation equipment —— 161
4.6   Applications of reactive distillation —— 163
4.7   Case study: biodiesel production by heat-integrated reactive distillation —— 163
4.8   Case study: fatty esters synthesis by dual reactive distillation —— 169
4.9   Case study: industrial reactive distillation process for methyl acetate production —— 173
4.10   Concluding remarks —— 176
4.11   Bibliography —— 177

Massimiliano Errico

5   **Process synthesis and intensification of hybrid separations —— 182**
5.1   Introduction —— 182
5.2   Pervaporation-assisted distillation —— 185

5.2.1   Hybrid distillation/pervaporation processes for bioethanol
         purification — 189
5.2.2   Hybrid distillation/pervaporation processes for biobutanol
         purification — 192
5.2.3   Hybrid distillation/pervaporation processes: final remarks — 194
5.3   Liquid–liquid extraction-assisted distillation — 195
5.3.1   Hybrid liquid–liquid extraction/distillation processes
         for bioethanol purification — 197
5.3.2   Hybrid liquid–liquid extraction/distillation processes
         for biobutanol purification — 199
5.3.3   Hybrid liquid–liquid extraction/distillation processes: final
         remarks — 201
5.4   Synthesis, design, and optimization of alternative hybrid configurations
         for biobutanol separation — 202
5.5   Conclusions — 207
5.6   Bibliography — 208

Petri Uusi-Kyyny, Saeed Mardani, and Ville Alopaeus
6   **Process intensification for microdistillation using the equipment
         miniaturization approach — 213**
6.1   Introduction — 213
6.2   Development of small-scale distillation units — 217
6.2.1   Reflux ratio control — 219
6.2.2   Reboiler types — 220
6.3   Distillation column structures — 221
6.3.1   Brass column with heat pipe type of operation — 221
6.3.2   Stainless steel plate type of column — 222
6.3.3   Modular copper column — 223
6.3.4   Laser-welded square column — 224
6.3.5   3D-printed coiled compact distillation column — 225
6.3.6   3D-printed modular coiled distillation column — 226
6.3.7   Conclusion of distillation column structure review — 228
6.4   Metal foam as a packing material — 229
6.5   3D-printed packings — 231
6.6   Application of microscale distillation for small-scale piloting — 231
6.6.1   Distillation model — 233
6.6.2   Process model — 233
6.6.3   Apparatus and instrumentation — 234
6.6.4   Impurity accumulation test — 236
6.6.5   Conclusions from the small-scale pilot test runs — 236
6.7   Conclusions — 238
6.8   Bibliography — 239

Carlo Edgar Torres-Ortega and Ben-Guang Rong
7   **Integrated biofuels process synthesis: integration between bioethanol and biodiesel processes —— 241**
7.1   Introduction —— 241
7.1.1   Energy world consumption projections —— 241
7.1.2   Worldwide transportation sector —— 242
7.1.3   Biofuel potential —— 243
7.1.4   Rural and industrial market and development —— 244
7.1.5   Environmental situation —— 245
7.1.6   Fuel properties of bioethanol and biodiesel —— 246
7.2   Lignocellulosic bioethanol production process —— 247
7.2.1   Biomass handling —— 248
7.2.2   Pretreatment of lignocellulosic materials —— 250
7.2.3   Hydrolysis of cellulose and hemicellulose, and fermentation strategies —— 250
7.2.4   Separation and dehydration of bioethanol —— 251
7.3   Fatty acid ethyl esters: biodiesel production process —— 253
7.3.1   Extraction and conversion of oils —— 255
7.3.2   Conversion of oil into alkyl esters —— 255
7.3.3   Separation and purification of biodiesel —— 257
7.3.4   New uses for glycerol —— 258
7.4   Integration between bioethanol and biodiesel processes —— 258
7.4.1   Mass integration —— 260
7.4.2   Energy integration —— 261
7.4.3   Integration between units → intensification —— 262
7.5   Methodological framework for synthesis and intensification —— 263
7.5.1   First stage: formulating and solving the superstructure synthesis problem through MINLP —— 265
7.5.2   Second stage: intensification through recombination of column sections —— 267
7.6   Case study: integrated lignocellulosic bioethanol and biodiesel process synthesis —— 269
7.6.1   Problem description for superstructure optimization —— 270
7.6.2   Superstructure setting and MINLP solution —— 271
7.6.3   Synthesis-intensification: column section methodology —— 276
7.6.4   Evaluation with the process simulator —— 279
7.7   Discussions —— 283
7.8   Conclusions —— 284
7.9   Bibliography —— 285

Chandrakant R. Malwade, Haiyan Qu, Ben-Guang Rong, and Lars P. Christensen
8   **Process synthesis for natural products from plants based on PAT methodology** —— **290**
8.1   Introduction —— 290
8.1.1   Natural products from plants —— 290
8.1.2   Need for recovery of natural products from plants —— 294
8.1.3   Challenges in recovery of natural products from plants —— 295
8.1.4   Process synthesis for separation of multicomponent mixtures —— 297
8.2   Process synthesis for recovery of natural products from plants —— 299
8.2.1   Process analytical technology —— 300
8.2.2   PAT-based methodology for recovery of natural products from plants —— 304
8.3   Recovery of artemisinin from *Artemisia annua* – a case study —— 309
8.3.1   Generation of initial process flowsheet —— 310
8.3.2   Evaluation of initial process flowsheet —— 316
8.3.3   Measurement of solid–liquid equilibrium of artemisinin and impact of impurities —— 317
8.3.4   Generation of improved process flowsheet —— 319
8.4   Conclusions —— 320
8.5   Bibliography —— 322

Yufei Wang and Xiao Feng
9   **Process synthesis for energy efficiency based on the pinch analysis approach** —— **325**
9.1   The hierarchy of process synthesis —— 325
9.2   Heat exchanger networks —— 326
9.2.1   Pinch and energy targets —— 326
9.2.2   Capital cost-related targets —— 330
9.2.3   Synthesis of HENs —— 331
9.2.4   HEN synthesis example —— 333
9.2.5   Retrofit of HENs —— 336
9.3   Utility selection —— 340
9.3.1   Grand composite curve —— 340
9.3.2   Utility selection —— 342
9.3.3   Combined heat and power generation —— 344
9.3.4   Integration of heat pumps —— 346
9.4   Heat integration of reactors and distillation columns —— 348
9.4.1   Appropriate placement of reactors —— 348
9.4.2   Heat integration characteristics of a single distillation column —— 350
9.4.3   Heat integration of a distillation system —— 350
9.4.4   Appropriate placement of distillation column —— 352

9.4.5   Use of the grand composite curve for heat integration of distillation
          column — **354**
9.5   Heat integration across plants — **356**
9.5.1   Total site profiles — **357**
9.5.2   Direct and indirect heat integration — **358**
9.5.3   Direct heat integration — **359**
9.5.4   Indirect heat integration — **359**
9.6   An industrial case study — **361**
9.6.1   Procedure for energy integration for total site system retrofit — **361**
9.6.2   Basic data for the case study — **362**
9.6.3   HEN subsystem integration — **364**
9.6.4   Separation subsystem integration and its coordination with HEN
          subsystem — **365**
9.6.5   Coordination of HEN subsystem and utility subsystem — **366**
9.6.6   Steam subsystem retrofit — **367**
9.7   Concluding remarks — **368**
9.8   Bibliography — **369**

Juan Gabriel Segovia-Hernández, Fernando Israel Gómez-Castro, José Antonio
Vázquez-Castillo, Gabriel Contreras-Zarazúa, and Claudia Gutiérrez Antonio

10   **Process synthesis and intensification by integration between process
          design and control — 370**
10.1   Introduction — **370**
10.2   Process synthesis and integration of process design and control — **372**
10.3   Closed-loop analysis — **376**
10.3.1   Case study 1: reactive distillation to produce diphenyl carbonate — **377**
10.3.2   Case study 2: thermally coupled distillation columns for the separation
          of a multicomponent mixture — **392**
10.4   Conclusions — **400**
10.5   Bibliography — **400**

Index — **405**

List of Contributors

A. F. Aguilera
Laboratory of Industrial Chemistry and Reaction
Engineering
Johan Gadolin Process Chemistry Centre
Åbo Akademi University
Biskopsgatan 8
20500 Åbo/Turku, Finland

V. Alopaeus
Department of Chemical and Metallurgical
Engineering
School of Chemical Engineering
Aalto University
00076 Aalto, Finland

L. P. Christensen
Department of Chemical Engineering
Biotechnology and Environmental Technology
University of Southern Denmark
Campusvej 55
5230 Odense, Denmark

G. Contreras-Zarazúa
Departamento de Ingeniería Química
División de Ciencias Naturales y Exactas
Universidad de Guanajuato
Campus Guanajuato
Gto., 36005, México

M. Errico
Department of Chemical Engineering
Biotechnology and Environmental Technology
University of Southern Denmark
Campusvej 55
5230 Odense, Denmark

X. Feng
School of Chemical Engineering & Technology
Xi'an Jiaotong University
Xi'an 710049, China

F. I. Gómez-Castro
Departamento de Ingeniería Química
División de Ciencias Naturales y Exactas
Universidad de Guanajuato
Campus Guanajuato
Gto., 36005, México

C. Gutiérrez Antonio
Facultad de Química
Universidad Autónoma de Querétaro
Qro., 76010, México

V. S. Herrera
Laboratory of Industrial Chemistry and Reaction
Engineering
Johan Gadolin Process Chemistry Centre
Åbo Akademi University
Biskopsgatan 8
20500 Åbo/Turku, Finland

A. A. Kiss[1,2]
[1]AkzoNobel – Research Development & Innovation
Process Technology SRG
Zutphenseweg 10
7418 AJ Deventer, The Netherlands.
[2]Sustainable Process Technology Group
Faculty of Science and Technology
University of Twente
PO Box 217
7500 AE Enschede, The Netherlands

S. Leveneur
Laboratoire de Sécurité des Procédés Chimiques
(LSPC)
EA4704
INSA-Rouen 685 Avenue de l'université
BP 08
76801 Saint-Etienne-du-Rouvray, France

C. R. Malwade
Department of Chemical Engineering
Biotechnology and Environmental Technology
University of Southern Denmark
Campusvej 55
5230 Odense, Denmark

https://doi.org/10.1515/9783110465068-202

S. Mardani
Department of Chemical and Metallurgical
Engineering
School of Chemical Engineering
Aalto University
00076 Aalto, Finland

H. Qu
Department of Chemical Engineering
Biotechnology and Environmental Technology
University of Southern Denmark
Campusvej 55
5230 Odense, Denmark

B.-G. Rong
Department of Chemical Engineering
Biotechnology and Environmental Technology
University of Southern Denmark
Campusvej 55
5230 Odense, Denmark

T. Salmi
Laboratory of Industrial Chemistry and Reaction
Engineering
Johan Gadolin Process Chemistry Centre
Åbo Akademi University
Biskopsgatan 8
20500 Åbo/Turku, Finland

J. G. Segovia-Hernández
Departamento de Ingeniería Química
División de Ciencias Naturales y Exactas
Universidad de Guanajuato
Campus Guanajuato
Gto., 36005, México

P. Tolvanen
Laboratory of Industrial Chemistry and Reaction
Engineering
Johan Gadolin Process Chemistry Centre
Åbo Akademi University
Biskopsgatan 8
20500 Åbo/Turku, Finland

C. E. Torres-Ortega
Department of Chemical Engineering
Biotechnology and Environmental Technology
University of Southern Denmark
Campusvej 55
5230 Odense, Denmark

J.-N. Tourvielle
Laboratory of Industrial Chemistry and Reaction
Engineering
Johan Gadolin Process Chemistry Centre
Åbo Akademi University
Biskopsgatan 8
20500 Åbo/Turku, Finland

P. Uusi-Kyyny
Department of Chemical and Metallurgical
Engineering
School of Chemical Engineering
Aalto University
00076 Aalto, Finland

J. A. Vázquez-Castillo
Departamento de Ingeniería Química
División de Ciencias Naturales y Exactas
Universidad de Guanajuato
Campus Guanajuato
Gto., 36005, México

Y. F. Wang
State Key Laboratory of Heavy Oil Processing
China University of Petroleum
Beijing 102249, China

Editor Biography

Dr. Ben-Guang Rong is an Associate Professor in Chemical Engineering at University of Southern Denmark (SDU, Odense). He has over 20 years teaching and research experience in Chemical Engineering and Process Systems Engineering. His research areas focus on chemical & biochemical process synthesis and design; biofuels separations; distillation technology; process modeling, simulation and optimization. He has authored and coauthored more than 108 papers in peer reviewed Journals and Proceedings. He is currently teaching the undergraduate courses of Chemical Engineering Thermodynamics, Separation Process, Chemical Process Design, and the graduate course of Industrial Separation Technologies at SDU. Since 1999–2008, he worked as senior researcher and Docent at Lappeenranta University of Technology (LUT, Finland). He was a member of the team who developed the Process Simulator ECSS (Engineering Chemistry Simulation System) at Qingdao University of Science and Technology (1989–1998, Qingdao, China). ECSS simulator has been commercialized and widely used in Industries, Universities and Research Institutes in China, where he also co-authored the book *Steady-State Process Simulation Techniques*.

https://doi.org/10.1515/9783110465068-203

Ben-Guang Rong

1 Introduction to basic concepts and elements in process synthesis and process intensification

Abstract: In this chapter, we introduce some basic concepts and elements for process synthesis (PS) and process intensification (PI). These basic concepts and elements are taken partially from distillation systems synthesis and intensification, as described in Chapter 2. In Chapter 2, we present six distillation subspaces and their systematic procedures. Three of these subspaces and their procedures were developed in the context of PS and the other three in the context of PI. However, both synthesis and intensification involve some common basic concepts and elements. In this chapter, we discuss these common concepts and elements in the context of general PS and PI works. We hope that at least some of the concepts and elements can be examined and addressed in the methodological approaches for PS and PI.

Keywords: Basic concept and element, process synthesis work, process intensification work, software and hardware, technical system, scientific method, methodology

1.1 Introduction

In chemical and process engineering, both process synthesis (PS) and process intensification (PI) aim to achieve technical systems with desired functionality and utility. These technical systems are applied in manufacturing processes to change and transform materials via certain steps to the desired products. Depending on the tasks and scope of the problem, a technical system can refer to a whole production process (e.g., complete manufacturing process), section of a production process (e.g., reaction section, separation section), equipment in a production process (e.g., microstructured equipment, reactive distillation equipment), or unit of a production process (e.g., operation units such as reactor, separator, mixer, heat exchanger, compressor, turbine, and pump). No matter the size and problem scope of a technical system, intellectual efforts in both PS and PI are required to realize its functionality and utility and to pursue global optimum performance.

Both the problem scope and performance indicators are important in pursuing a concrete technical system for either a specific synthesis work or a specific intensification work. That is, to pursue a concrete technical system, one has to define the task and scope of the problem; to pursue the global optimum performance of a technical system, one has to specify indicators to evaluate and measure performance.

In this chapter, we add the word "work" to PS and PI. A PS or PI work is related to the synthesis or intensification process and must relate to the methodology in which

https://doi.org/10.1515/9783110465068-001

all the major activities and decisions are managed and all relevant basic elements are examined and addressed.

To perform a specific synthesis or intensification work, one has to conceptualize the technical system. To conceptualize the technical system, one has to clarify its functionality and utility. To clarify its functionality and utility, one has to define the problem scope and task. At the same time, for global optimum performance, one has to achieve some novelty and creativity in the technical system. However, to achieve global optimum through novelty and creativity, one has to approach new concepts, principles, and mechanisms. To approach new concepts, principles, and mechanisms, one has to characterize the unique features and phenomena of the problem and task. To characterize the unique features and phenomena of the problem and task, one has to rely on deep analysis and sufficient information, which may need either measurement and data compilation or observation using special tools for characterization and representation.

Normally, a technical system consists of software and hardware systems. To achieve a novel and creative technical system, one needs to design an advanced software system or an advanced hardware system, commonly both. After deep analysis and acquisition of sufficient information on the problem and task, the unique features and phenomena of the problem can be described. After clarifying the unique features and phenomena, the new concepts, principles, and mechanisms emerge. After the new concepts, principles, and mechanisms emerge, the advanced software system is conceptualized. After conceptualizing the advanced software system, the advanced hardware system is configured. Thus, a novel and creative technical system is achieved that incorporates advanced software and hardware systems and whose global optimum performance can be evaluated and measured using defined performance indicators.

Therefore, during a PS or PI work, the workflow concerns the method and procedure for organizing the basic elements that emerge during the working process, either explicitly or implicitly. These 12 basic elements can be summarized as follows:
1. The defined problem scope and task
2. defined functionality and utility
3. defined performance indicators
4. analyzed unique features and phenomena
5. emerged new concepts, principles, and mechanisms
6. emerged advanced software elements
7. emerged advanced hardware elements
8. formulated assembly method and procedure between the software and hardware systems
9. emerged technical system (conceptual design)
10. observed novel and distinct features of the technical system

11. achieved optimum performance, as evaluated and measured by the predefined indicators for the technical system
12. proven novelty of the technical system (qualitatively or quantitatively)

In this chapter, we discuss the basic elements and concepts as an introduction to PS and PI.

1.2 Basic elements in synthesis and intensification of distillation systems

To examine how the basic elements are addressed in some specific PS and PI works, we consider multicomponent distillation synthesis and intensification as an example. In Chapter 2, the systematic methodology for synthesis and intensification of multicomponent distillation systems is presented. PS focuses on the systematic synthesis procedure for generating feasible distillation alternatives, whereas PI focuses on intensifying the distillation system by using fewer columns. In total, six distinct distillation subspaces are obtained.

The methodology for both synthesis and intensification of distillation systems is the systematic synthesis procedure for conceptual design of the systems' structures. However, in either a specific subspace from PS or a specific subspace from PI, the methodology must account for the synthesis and intensification process from the concrete concept and mechanism to the concrete technical systems. The basic elements of the synthesis and intensification of distillation systems are identified as follows (see Chapter 2):
– specific mechanisms (for synthesis or intensification of each subspace)
– systematic synthesis procedure (for both synthesis and intensification)
– generated subspace (the distinct synthesized or intensified systems)
– distinct features of the generated systems (for each subspace)

In addition, the following basic elements have been defined at the beginning or during the synthesis and intensification for multicomponent distillation problems:
– defined problem scope and task (multicomponent separation)
– defined functionality and utility (distillation systems)
– defined performance indicators (three dominant criteria)
– analyzed unique features and phenomena (e.g., remixing and irreversibility, movable column sections, thermodynamically equivalent structures, special side columns, etc.)

By generalizing the basic elements of distillation systems to any process technical systems (e.g., reaction systems, other separation systems, heat transfer systems), for any concrete PS or PI work, we can generalize the following basic elements of the methodology:

– process or equipment with targeted functionality and utility (problem scope)
– concept, mechanism, and principle (should emerge or be given)
– evaluation criteria of the technical system (performance, goals, indicators)
– software aspects (should be given or emerge)
– hardware aspects (should be given or emerge)
– procedure and method (should be formulated to assemble the software and hardware systems)
– technical system(s) obtained (should present or show its state)
– distinct features of the obtained system(s) (should analyze and highlight)
– evaluation of the specified evaluation indicators (qualitative and quantitative)
– other specific analyses to check the feasibility, performance, and efficiency of the obtained system(s), depending on the systems and methods used

Therefore, to perform a PS or PI work, some basic elements need to be covered and addressed in the methodology. Depending on the problem scope and task, the methodology may address some or all of the basic elements.

In Chapter 2, both synthesis and intensification of distillation systems are included within the scope of the *conceptual process design*. The results of synthesis and intensification are subspaces with distinct distillation systems. The methods of synthesis and intensification are the systematic procedures that define the physical structures of all distillation systems step-by-step. Thus, all the subspaces consist of distinct distillation systems, which are feasible conceptual designs so that it is not necessary to check their feasibility. Therefore, conceptual process design is concerned with developing systematic methods and procedures to define the physical structures of technical systems. In doing so, the conceptual designs represent the feasible structural designs of the technical systems that fulfill the functionality and utility of the given problem and task.

In this chapter, we first briefly introduce the basic facets of PS and PI. Then, we introduce some basic elements that are concerned with the methodology for PS and PI works.

1.3 Basic facets of process synthesis

1.3.1 Introduction to process synthesis

Generally, PS addresses conceptual process design problems. The overall objective is to generate global optimal process alternatives for the given process task. Process synthesis is a relative new research area in chemical engineering [1–4], and is also an active research area in process systems engineering. It is a major theme in international conferences on process systems engineering and computer-aided process engineering. The number of publications on PS in the last three decades is overwhelming and a review of the literature is out of the scope of this chapter; however, we can highlight various facets of PS research and its applications. The four facets listed relate to PS in general; they are not about a specific PS work in which the basic elements can be examined.

Facet 1: Process synthesis problems and tasks
- heat exchanger networks
- multicomponent distillations
- separation and hybrid processes
- reaction paths and reactor networks
- mass exchanger networks
- utility and cogeneration systems
- entire process synthesis
- integration of process design and control
- biofuels and renewable energy processes
- biomass-based production processes
- problems from specific industrial sectors

PS problems can concern the complete production process, sections of a total process, the utility system, and mass and heat exchange networks, etc. However, they seldom refer to only one unit operation, unless new or improved unit alternatives are being proposed and compared.

Facet 2: Process synthesis methods and techniques
- heuristics-based approach
- systematic procedure
- algorithmic method
- thermodynamic and pinch analysis method
- evolutionary method
- case-based reasoning method
- conflict-based approach
- phenomena-based approach
- graphical tool-based method
- mixed-integer nonlinear programing (MINLP) optimization method
- hybrid and integrated approach
- process analytical technology (PAT)-based method

Depending on the problems and tasks, the available PS methods and techniques can focus on the following:
- addressing the structural design of the process system
- addressing both structural and parametric designs of the process system
- representing and characterizing the process unique features and phenomena

Facet 3: Process synthesis applications
–   fossil-based manufacturing processes
–   biomass-based manufacturing processes
–   continuous manufacturing processes
–   batch manufacturing processes
–   refinery industry
–   chemical industry
–   petrochemical industry
–   energy industry
–   pharmaceutical industry
–   fine chemical industry
–   food industry
–   biorefinery industry
–   biofuels industry
–   biochemical industry
–   other emerging product and process industry

In the manufacturing industries, where processes involve the technologies used to transform materials into final products, PS can be widely used to design and develop new processes or to improve and retrofit existing processes.

Facet 4: Process synthesis evaluation indicators
–   technical indicators (yield, conversion, selectivity, purity, etc.)
–   number of units applied in the process
–   total energy consumption
–   total capital investment
–   total production cost
–   energy cost savings
–   capital cost reduction
–   total annual cost
–   profitability indicators
–   other economic indicators (e.g., product minimum selling price)
–   operability and controllability
–   safety and hazardous reduction
– CO_2 reduction
–   total greenhouse-gas (GHG) emission reduction
–   waste water reduction
–   total minimum emissions
–   lifecycle analysis (LCA) indicators
–   other environmental indicators
–   sustainable indicators (e.g., ecological products, green engineering)

Undoubtedly, evaluation and comparison are also important facets of PS problems. On the one hand, different synthesis problems call for different evaluation indicators; on the other hand, one specific synthesis work seldom calls for a single evaluation indicator. Several evaluation indicators are usually used for evaluation and comparison of system alternatives. The evaluation indicators concern all the properties and aspects of the process systems. They can generally be classified into four types: technical indicators, economic indicators, environmental indicators, and sustainable indicators.

In summary, PS concerns study of the theories, methods, and support tools used to synthesize various processes in the form of conceptual designs for the technical systems. Such technical systems can be total process systems, process subsystems and equipment, or process units. Such technical systems are evaluated using specified performance indicators.

1.3.2 Two approaches for finding alternatives in process synthesis: pure synthesis versus practical synthesis

The major purpose of PS methodology is to generate and evaluate *process alternatives*. When doing a specific PS work, there are two approaches for finding process alternatives, i.e., convergence and divergence. The convergence approach ends up with a single or a few promising alternatives, whereas the divergence approach ends up with many more alternatives, which might be either a new subspace of alternatives or an expansion of the existing alternative space.

To distinguish the different synthesis problems in terms of alternative space, we can classify the synthesis problems into two categories.

Category I: Convergence approach for process alternatives (practical synthesis)

Process synthesis in category I is concerned with how to search the known alternative space. When the target of a PS problem is within the scope of process design to produce a final process design, the PS problem is to pursue a single optimal process alternative or a few competitive alternatives for a predefined design project. For example, PS for base case design is typically such a category I synthesis problem. We call this category of PS *practical synthesis*, which is to follow the convergence approach to obtain a single or several alternatives, as shown in Figure 1.1.

For a known space with predefined alternatives, PS in category I is to search for an optimal alternative for the given problem and is usually application-oriented PS.

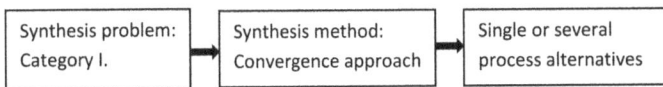

Synthesis problem: Category I.	→	Synthesis method: Convergence approach	→	Single or several process alternatives

Fig. 1.1: Convergence approach for alternatives in process synthesis: practical synthesis.

Category II: Divergence approach for process alternatives (pure synthesis)

Every PS problem is concerned with a space of alternatives. PS in category II is concerned with how to create a larger or more complete alternative space. Such a PS problem is to create the space of the alternatives or the space of potential new alternatives in which an optimal alternative can be searched for when considering a specific case and application. We call this category of PS *pure synthesis*, which is to follow the divergence approach to generate more alternatives or a more complete space of alternatives, as shown in Figure 1.2.

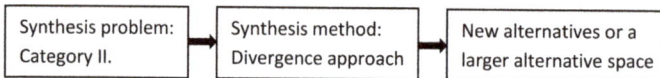

Synthesis problem: Category II.	→	Synthesis method: Divergence approach	→	New alternatives or a larger alternative space

Fig. 1.2: Divergence approach for alternatives in process synthesis: pure synthesis.

Beyond the known alternative space, PS in category II is to generate competitive alternatives or to formulate a larger and more complete alternative space. For example, synthesis of multicomponent distillation configurations is such a category II PS problem.

Figure 1.3 illustrates the alternative space evolution during PS. The subspaces may be isolated without clear connections between their alternatives; searching for the optimal alternative should cover the whole space.

For example, the convergence approach can search the original space of both subspace 1 and 2 to find an optimal alternative based on given evaluation indicators. The

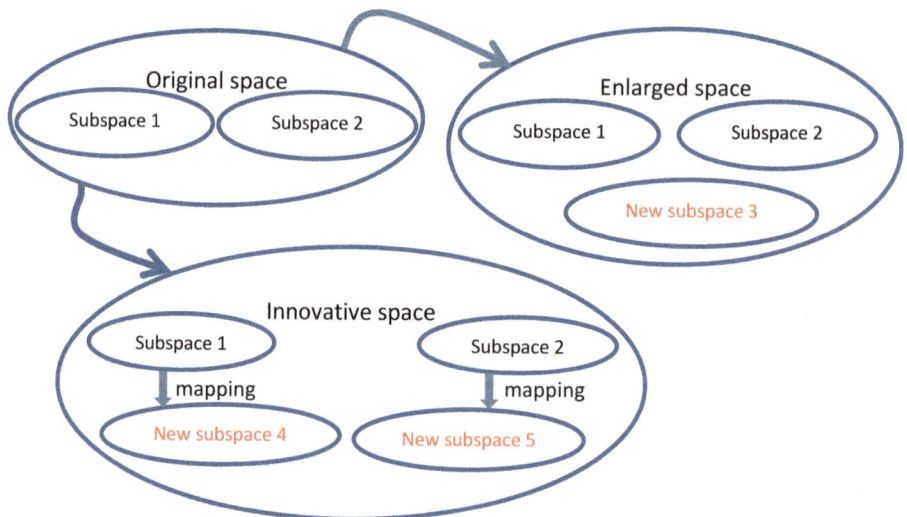

Fig. 1.3: Process synthesis for evolution of the space and subspaces for the alternatives.

divergence approach can enlarge the alternative space with new subspace 3, or can transform subspaces 1 and 2 to new subspaces 4 and 5 in terms of generated alternatives. Many PS works belong to category I for practical synthesis with a given problem and task.

In summary, in terms of PS methodology for generation of process alternatives, four heuristics can be summarized as follows:

Heuristics 1: Divergence approach for PS can very often reveal new alternatives.

Heuristics 2: Structural change is the driving force for achieving innovative and creative alternatives.

Heuristics 3: Systematic synthesis procedure is necessary to achieve a more complete alternative space.

Heuristics 4: Corresponding principle is a useful principle for obtaining innovative alternatives from known alternatives.

1.4 Basic facets of process intensification

Generally, PI addresses process and equipment design and development with the purpose of significantly improving technological performance and efficiency. A specific PI work usually also focuses on conceptual design, emphasizing novel concepts and ideas, nonconventional methods and principles for approaching nonconventional units and devices, as well as novel processes and equipment. Therefore, PI is often a set of radically innovative principles in process design, which can bring significant benefits in terms of process efficiency, energy savings, reduced capital and operating expenses, waste reduction, process safety, etc.

Stankiewicz and Moulijn [5] defined process intensification: "Process intensification consists of the development of novel apparatus and techniques that, compared to those commonly used today, are expected to bring dramatic improvements in manufacturing and processing, substantially decreasing equipment-size/production-capacity ratio, energy consumption, or waste production, and ultimately resulting in cheaper, sustainable technology."

On the other hand, in an academic context, PI has similar scope to PS in terms of conceptual design. PI is to study the theories, methods, and support tools with the aim of intensifying various processes in the form of conceptual designs of the technical systems. Such technical systems can be process units, process subsystems and equipment, total process systems, and are evaluated using specified performance indicators.

Similar to PS, we can highlight four facets of PI research and application.

Facet 1: Process intensification problems and tasks [6]

- reaction processes and reactors
- separation processes and separation equipment
- mass transfer processes

- heat transfer processes and heat exchangers
- mixing processes and mixing devices

Mass and energy transfer processes, especially mass and energy transfer modes and efficiency are the most important problems and tasks studied in PI. Specifically, non-conventional concepts and mechanisms to pursue mass and energy transfer efficiency are emphasized. Thus, novel equipment and devices for reactions, heat transfers, separations, and mixings are the major highlights of PI.

Facet 2: Process intensification methods and techniques [6, 7]
- multifunctional reactors
- multifunctional separation equipment
- hybrid separations
- alternative energy sources
- mechanisms for intensified heat transfer
- mechanisms for intensified mass transfer
- mechanisms for electrically enhanced process
- microfluidics
- microstructure
- miniaturization

PI sets the target of pursuing novel equipment and devices. The methods and techniques of PI call for creative ideas and novel principles and mechanisms to implement the processing tasks, mostly at the subsystem and equipment level. Some methods in PS, such as the heuristic, algorithmic, and optimization methods, may not be suitable for PI. On the other hand, general creative and innovative methods are not available in both PS and PI. Therefore, a systematic procedure incorporating novel ideas and mechanisms to determine the conceptual designs of the process and equipment is probably the most promising PI method. For example, we have shown that such a systematic procedure can be developed for systematic synthesis of the intensified distillation systems in Chapter 2.

Facet 3: Process intensification applications [7]
- bulk chemical industry
- fine chemical and pharmaceutical industry
- offshore processing
- nuclear industry
- fossil and biofuel energy industry
- food and drink industry
- textile industry
- metals industry
- polymer processing
- glass and ceramics industry
- aerospace

- effluent treatment
- refrigeration/heat pumps
- microelectronics industry

As the major results of PI are intensified units and equipment, the application scope of PI is even wider than that of PS. Reay et al. [6] summarized PI application areas, including petrochemical and fine chemicals, offshore processing, miscellaneous process industries, and described the extended application areas of the built environment, electronics, and the home.

Facet 4: Process intensification evaluation indicators

The major objectives to be pursued in PI works are a significant improvement in performance and efficiency, achieving plants that are smaller, safer, cleaner, cheaper, more energy efficient, and use less to do more. These objectives are also the major evaluation indicators for the results for PI works. Evaluation indicators can be classified into the following four types:

- technical indicators
- economic indicators
- environmental indicators
- sustainable indicators

These four types cover all the indicators listed for PS. A specific PI work seldom calls for a single evaluation indicator; instead, several evaluation indicators are usually used for evaluation and comparison of the intensified systems.

1.5 Evaluation indicators for PS and PI works

As mentioned, evaluation of a PS or PI work is very important. Both PS and PI works need to be evaluated and judged during the working process as well as after the work has been accomplished. The evaluation always contains some uncertainty because different problems may specify different evaluation indicators. Due to the diversity of problems and tasks, and the different scopes of study for a technical system, there are many evaluation indicators to be selected and applied for a specific PS or PI work.

Evaluation of process alternatives or intensified systems involves specifying or selecting different indicators from the four types given in Section 1.4. It does not necessarily follow the sequence of the four types of indicator; however, numerical calculation of the indicators is usually needed to relate and translate the data and information between different type indicators. Figure 1.4 illustrates the possible data and information translation between the four types of indicators. In the innermost layer are the technical indicators, which directly relate to the design and operating parameters and performance variables of the technical system. This means that the evaluation indicators are not independent. Generally, the outer layer indicators are directly or indirectly related to the data and information of the inner layer indicators. For example, a technical indicator can measure the energy savings of the technical system. These energy savings can be related and translated into an economic indicator of the reduction in operating costs, into an environmental indicator of the equivalent CO_2 reduction, and so on.

1.6 Concepts of process synthesis work and process intensification work

The facets summarized in Sections 1.3 and 1.4 give a brief overview of PS and PI as research areas. There are comprehensive books and literature that introduce PS and PI. However, we focus on the basic elements for an individual PS or PI work for the purpose of identifying and introducing these basic elements, and relating them to the methodology of PS and PI problems and tasks.

Basically, a PS or PI work is a specific work that is presented in a form that can be evaluated and reviewed. A PS work relates to a specific synthesis task and to the working process that addresses how the synthesis was done. Therefore, a PS work relates to the methodology in which all the major activities and decisions are presented. Similarly, a PI work relates to a specific intensification task and its methodology. For example, a PS work can be seen as an article within a PS area that addresses the synthesis of a specific technical system and its methodology. Similarly, an article within a PI area that addresses the intensification of a specific technical system and its methodology is seen as a PI work.

In terms of methodology, we are concerned with the following questions:
– How to create a specific PS work or a specific PI work?
– How to appreciate and evaluate a specific PS work or a specific PI work?

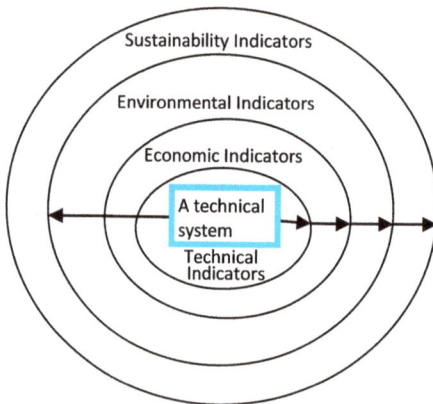

Fig. 1.4: Calculation and comparison information translated into different performance indicators for evaluating a technical system (*arrows* represent data and information translation between the indicators).

By focusing on a specific PS or PI work, we can identify and examine its basic elements. We can then examine how these basic elements are addressed in the methodology of the work.

For example, for a specific PS work on a distillation system, the following typical questions can be examined concerning its basic elements:
- What distillation system has been synthesized?
- What is the method?
- What is the result?
- What is the evaluation and justification?
- What is the novelty?
- How can the synthesized distillation system be implemented in applications?

Similarly, for a specific PI work on a reaction system, the following typical questions can be examined concerning its basic elements:
- What reaction system has been intensified?
- What is the method?
- What is the result?
- What is the evaluation and justification?
- What is the novelty?
- How can the intensified reaction system be implemented in applications?

Both PS and PI works are about work done for a specific task or problem. A work is composed of partial art in terms of the concept and endeavor of researchers and engineers (like artworks from artists) and partial technology in terms of scientific principles and laws. The designer and researcher must participate in the creation process for a specific PS or PI work, which is a major part of PS and PI methodology.

1.7 Conceptual design of process and equipment

When considering the common elements of PS and PI, one symbiosis element is that PS and PI converge to conceptual process design for a technical system, as illustrated in Figure 1.5.

For both PS and PI, the conceptual design is an essential element. From a methodology point of view, we emphasize that conceptual design is both the design stage in terms of the design process and the design result for the process and equipment. That is, the *conceptual design* represents a feasible technical system that fulfills the required functionality and utility and can be evaluated with the given performance indicators.

A striking example of conceptual design is the methyl acetate process, which is a favorite example in both PS and PI and has been introduced in many PS and PI works [7–11]. The reaction and separation are synergetically intensified into the re-

```
┌─────────────────────────┐   ┌─────────────────────────┐
│    Process Synthesis    │   │  Process Intensification │
└─────────────────────────┘   └─────────────────────────┘
            │                               │
            ▼                               ▼
┌───────────────────────────────────────────────────────┐
│      Conceptual Design of Process and Equipment         │
└───────────────────────────────────────────────────────┘
                        │
                        ▼
        ┌───────────────────────────────────────┐
        │      Optimal process alternatives      │
        │      Novel devices and equipment       │
        └───────────────────────────────────────┘
```

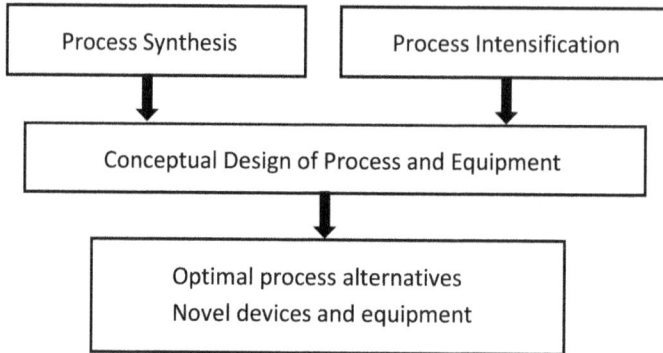

Fig. 1.5: Process synthesis and process intensification converge at conceptual design.

active distillation equipment, which circumvents both the reaction equilibrium limit and the azeotropes of the mixture.

The conceptual design of a technical system typically has the following features:
– determined concept(s) and principle(s) for the technical system
– structural design of the technical system
– can be represented in a certain form (either a process or an equipment)
– may have different alternatives
– can be evaluated with the concerned indicators (qualitative or quantitative)

Within the scope of conceptual design for a PS or PI work, there are two basic observations that can be highlighted: the obtained technical system, which represents the objective result of the PS or PI work (the highlight on the object) and the idea or concept of the method, which represents the subjective endeavor of the PS or PI work (the highlight on the subject). The two following basic observations are also the answers to the two basic questions that arise with any specific PS or PI work:
– What is the designed technical system?
– Who designed the technical system and how?

1.8 Concepts of the system and technical system

The conceptual design is a feasible process alternative or a physical structure of the technical system. Therefore, we need both the system concept and the technical system concept to represent the conceptual design.

1.8.1 System concept

"System" is a ubiquitous concept not only in modern science and technology texts, but also in texts concerning societies, natural and social environments, and any disciplines humankind concerns. It can refer to anything that we are interested in and that we would like to pay attention to. In a scientific context, depending on the subject and discipline, we can generally distinguish between existing systems and systems to be designed and created. The former may refer to any existing things in nature and society, while the latter may refer to any artificial things that man brings into the world. This division primarily determines the scientific methodology employed or developed to study a system, namely the analysis-oriented method or the synthesis-oriented method. For existing things, analysis or systems analysis is the fundamental scientific method, whereas for artificial things, synthesis or systems synthesis is the fundamental scientific method.

A striking example of system as a fundamental concept is in chemical engineering thermodynamics. There, system is defined as a quantity of matter or a region of space chosen for study [12]. The things of the system are divide from their surroundings by a boundary. Chemical engineering thermodynamics is mostly concerned with study of the various changes of the system and the various interactions between the system and its surroundings. Figure 1.6 illustrates the thermodynamic system and its surroundings, with the boundary.

Essentially, study of a thermodynamic system uses analysis-based methods to predict the change in thermodynamic properties between the thermodynamic states of the system. Without considering any devices or hardware system, we can predict and estimate the thermodynamic property changes of the system for various processes, like the isothermal process, isobaric process, adiabatic process, constant volume process, compression process, and expansion process. In chemical engineering, such processes mostly occur in an open system and, therefore, without specific designation, we mean here that the thermodynamic system is an open system that can have both energy and mass exchange with the surroundings.

As a consequence, the system concept in thermodynamics is primarily concerned with software aspects (change in the state and properties) without considering hard-

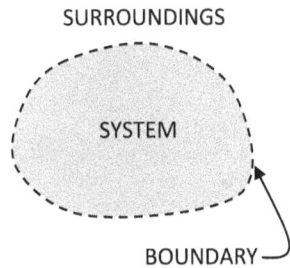

Fig. 1.6: A thermodynamic system and its surroundings: exchange of work, heat, and matter across the boundary.

Reversible process: Q_{rev}, $W_{rev}=W_{ideal}$

System: Initial state
T_1, P_1, V_1, U_1, H_1, S_1

System: Final state
T_2, P_2, V_2, U_2, H_2, S_2

Irreversible process: $Q_{irrev} \neq Q_{rev}$, $W_{irrev} \neq W_{ideal}$

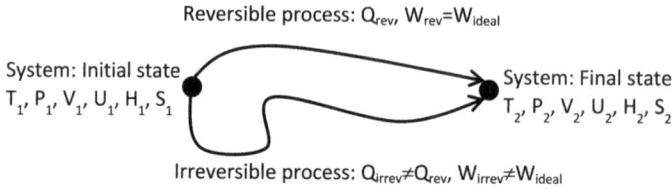

Fig. 1.7: Changes in states and property changes of a thermodynamic system during different processes without considering hardware system. (T, P, V, U, H, and S represent state properties of the system, i.e., temperature, pressure, volume, internal energy, enthalpy, and entropy, respectively; Q_{rev}, Q_{irrev} the heat of reversible and irreversible process; and W_{rev}, W_{irrev} the work of reversible (ideal) and irreversible process).

ware aspects such as how to make such process changes and with what devices. The change is simply related to the surroundings, as shown in Figure 1.7.

Obviously, such process and related state and properties changes in the thermodynamic system, without considering any hardware aspects, are software aspects. One important software aspect is the intended process. To consider both software and hardware aspects simultaneously, we need the technical system concept.

1.8.2 Technical system concept

When we want to implement a process for an intended change in the thermodynamic system, we inevitably need to introduce the hardware system for implementation of the process and realization of the intended change. Then, we need to design various devices to implement and realize various processes and changes, as shown in Figure 1.8. Such configured devices that combine both software and hardware aspects are called technical systems. A technical system is defined as having a certain software system and a certain hardware system, combined in such a way that the system can properly implement and realize the predefined process and intended change(s).

The simplest and most widely used devices as technical systems include valves, mixers, compressors, pumps, heat exchangers, and turbines. Yet, they are designed through the combination of both software and hardware systems. The mention of these simple devices is to emphasize that it is only through simultaneously combin-

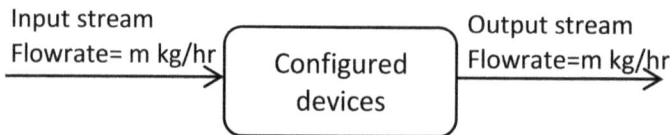

Input stream
Flowrate= m kg/hr

Configured devices

Output stream
Flowrate=m kg/hr

Fig. 1.8: Configured devices as technical systems with both software and hardware to implement various processes.

ing the software system and the hardware system that a technical system can provide the functionality and utility for the intended process changes.

Figure 1.9 shows that a technical system is the combination of its software system and its hardware system, and that the combination mechanism of the software and the hardware implements the process with intended change(s).

Obviously, when such a predefined process and intended change is designated as functionality and utility, then such technical systems can broadly include any process systems. Essentially, nearly every process technical system involves processing mass and energy. Depending on its software and hardware elements, and depending on the specified functionality and intended process changes, the performance and efficiency of the technical system can be measured by the material utilization and conversion efficiency, energy efficiency, environmental and sustainable indicators, hardware costs, as well as the safety, operability, and controllability of the technical system. Therefore, it is natural to pursue different design alternatives by PS, and it is natural to pursue the highest performance and efficiency by PI. It is important to realize that both PS and PI inevitably concern the design of software and hardware systems, any changes in these systems, and their assembly and combination.

In general, we can summarize the features of a technical system. All or some of the following features may be involved during a specific PS or PI work for conceptual design of the technical system:

- functionality and utility
- working principle and mechanism
- software system
- hardware system
- conceptual design structure
- decomposable functional parts and components

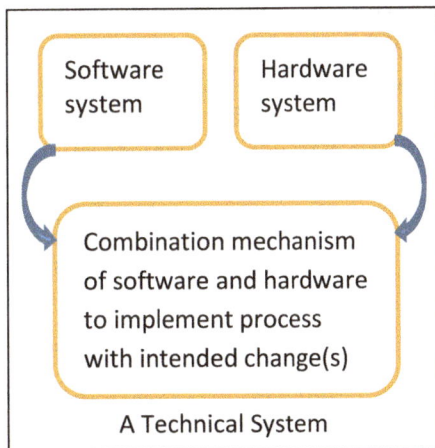

Software system

Hardware system

Combination mechanism of software and hardware to implement process with intended change(s)

A Technical System

Fig. 1.9: A technical system is the combination of its software and hardware systems.

- characteristic parameters (variables)
- quantifiable performance indicators

1.9 Software and hardware systems for a technical system

As discussed, a technical system is an artificial process system that consists of a certain number of decomposable functional parts and components. By design, the functional parts and components are coordinated and work together to fulfill the expected functionality and utility. For PS and PI, it is safe to say that any technical system consists of its software and hardware systems. To this point, a theorem can be put as follows:

Theorem. *To configure a technical system, it is inevitable to consider both the software system and the hardware system, as well as the combination of software and hardware systems.*

Corollary. *To obtain a new technical system, we need to conceive or change either the software elements or the hardware elements, or both. At the same time, we need to configure or change the combination of software and hardware elements.*

1.9.1 Software elements

The software system is primarily determined by the working principle and mechanism of the technical system. The software system primarily determines the main hardware elements of the system and their arrangement.

In general, any intangible elements that are significant and relevant to the design of a technical system are software elements. If we consider all the facets of PS and PI, as discussed in Sections 1.3 and 1.4, there are probably infinite software elements and hardware elements to be considered. However, when we deal with a specific PS or PI work, both the software and hardware elements should be finally determined in order to approach the technical system. This means that, for a specific PS or PI work, both the software elements and the hardware elements are finite. These basic elements should emerge and be determined during the working process to synthesize or intensify the technical system.

For process technical systems, we can list the following basic software elements:
- intended process and change
- paths of the process and change
- concepts of the material processing tasks
- mass transfer mechanisms
- heat transfer mechanisms
- energy forms and force field
- thermodynamics of the process

– kinetics of the process
– flow pattern and hydrodynamics
– unique features and phenomena of the process
– new concepts, principles, and mechanisms that emerge during the design process
– operating mode (static, moving, rotating, spinning, swing, etc.)
– assembly method and procedure between the software and hardware systems

For a specific PS or PI work, the possible software elements are limited by the problem and task, as well as the specified functionality and utility. This means that it is always a finite task to design and finalize the software elements for a specific technical system.

Taking a distillation system as an example, the essential working principle is based on the difference in volatility of the components in the mixture. The distillation separation is typically realized through mass and heat transfers between vapor and liquid phases of the flowing fluids. Therefore, managing the distribution of components between vapor phase and liquid phase is the central consideration of its software system. Thermodynamics, phase equilibrium, thermal profile, and fluid hydrodynamics are the basic elements of its software system.

For a distillation system (without reaction), we can list the following basic software elements:
– relative volatility
– thermodynamic behavior (azeotropes or no)
– phase equilibria
– mass and heat transfers
– mass and energy conservation
– hydrodynamics
– gravity or high-gravity
– temperature and pressure profiles
– vapor and liquid flows and hydraulics
– feasible individual splits
– feasible separation sequences

It is possible to list other elements, but there are a finite number of basic elements for the software system of a distillation system when dealing with a specific PS or PI work.

1.9.2 Hardware elements

Any technical system, like its software system, must have a hardware system. The hardware system must match its software system and fulfill the functionality and utility of the technical system.

In general, any tangible elements that are significant and relevant for the design of a technical system are hardware elements. Hardware elements include all the physical parts needed to construct the technical system. All the physical parts should be combined or assembled according to their functions. Therefore, the physical parts are the basic hardware elements of the functional parts of the technical system.

On the other hand, the materials being processed are the initial tangible things in a technical system. These materials mostly serve as joint agents and form the interface between the software elements and the hardware elements in the technical system.

Like a software system, if we consider all the facets of PS and PI, there are infinite hardware elements that can be listed. However, when we deal with a specific PS or PI work, the hardware elements should be determined along with the software elements. This means that, for a specific PS or PI work, the hardware elements are finite. These basic hardware elements should emerge and be determined to match the software elements during the working process to synthesize or intensify the technical system.

For process technical systems, we can list the following basic hardware elements:
- operation units
- standard functional parts and components
- geometrized functional parts and components
- internals of equipment
- contact surface materials and geometrized surfaces
- facility agents (catalyst, mass separation agent, solvent, additive, membrane, other media)
- auxiliary devices to manipulate the various profiles (e.g., flow rate, temperature, pressure, concentration, velocity, density, viscosity, and particle size)

For example, for a distillation system, the hardware elements are typically for generating the vapor and liquid flows, for providing physical geometric space to process the bulk vapor and liquid flows, and for maintaining sufficient contact of vapor and liquid phases for mass and heat transfers.

For a distillation system (column type), we can list the following basic hardware elements:
- condenser for condensing vapor into liquid
- reboiler for vaporizing liquid into vapor
- column shell for physical space to the bulk vapor and liquid flows
- internals (trays and packings) for vapor and liquid phase contacting for mass and heat transfers
- column sections for functional section with a certain number of stages or a certain height of packings
- pump for changing the pressure of liquid streams
- compressor for changing the pressure of vapor streams
- reflux drum for containing the condensed liquid or both vapor and liquid
- decanter for heterogeneous azeotropic systems
- instruments for parameter measurements or control purposes

- controllers for the control system
- auxiliary devices such as valves and containers

Similarly, there other elements can be listed but there are a finite number of basic elements for the hardware system of a distillation system when dealing with a specific PS or PI work.

1.9.3 Emergence of the technical system

Having determined the software and hardware elements, we next need to consider the following questions: What is the method for combining the software and hardware systems? Can we articulate the mechanism to combine the elements between the software and hardware system?

The combination of software and hardware elements gives concrete conceptual designs for the technical system. Therefore, methods and procedures to combine the software and hardware elements are also crucial elements of the methodology of PS and PI. Such methods and procedures very often rely on the mechanisms for simultaneously coordinating the software and hardware elements so that both the software and hardware systems are finalized in the technical system. Consequently, combination of the software and hardware systems gives concrete conceptual designs for the technical system, which may present as the results of synthesis or intensification in the following forms:

- configuration of the technical system
- geometrized equipment of the technical system
- physical structure of the technical system
- feasible structural alternatives of the technical system

These concrete conceptual designs for the technical system can then be evaluated and compared on the basis of specified indicators.

1.10 Triple parties in the working process of the methodology: the subject, object, and scientific method

For a given PS or PI task, to produce a specific PS or PI work, the following questions must be asked:

- Who is doing the task (a task needs someone as the subject?)
- What is to be done (a technical system needs to be generated as the object)?
- How to do the task (scientific method needs to be applied or developed)?
- What is the judgement (judgement and evaluation of the work)?

A hypothesis about PS or PI work can be stated as follows:

Assuming that a true creative and innovative work has been achieved for the given synthesis or intensification task (a process or equipment), what dimensions can we identify that have been involved in and have resulted in the true creative and innovative work?

We would like to indicate the following dimensions:
- the subject designer/investigator and his/her role in the task
- the object system and its specific features and characteristics
- the scientific method of synthesis and analysis
- the judgement and evaluation of the result of the work

The subject designer/investigator, the object system, and the scientific method are the three main parties of the design (working) process in terms of the methodology, as illustrated in Figure 1.10. Regarding the scientific method, both synthesis and analysis are the basic methods in applications.

As introduced in Section 1.6, the concept of a PS or PI work is used to connect the work to its methodology. In other word, when we develop or formulate a methodology, it is for a specific PS or PI work concerning the object system, as shown in Figure 1.10.

The subject designer is also a necessary party in the working process of the methodology. This is because the role of the subject designer in a PS or PI work is similar to the role of the artist in an artwork. An artwork is exclusively the creation and contribution of the subject artist. A PS or PI work has these artistic attributes due

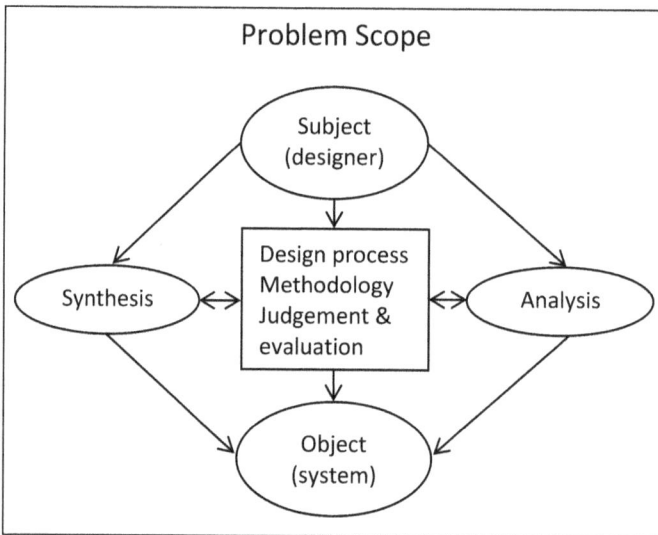

Fig. 1.10: The triple parties in the methodology of the working process for process synthesis and intensification: the subject, object, and scientific method.

to the similar role of the subject designer in creation. The subject designer must participate in the creation process for a specific PS or PI work. However, a PS or PI work has the aim of creating a technical system that must fulfill the required functionality and utility with the desired optimal performance, which is evaluated using specified performance indicators. Therefore, a PS or PI work must have scientific attributes that follow scientific laws, principles, and mechanisms.

The scientific method is another necessary party of the methodology. For either a PS or PI work, synthesis and analysis must be used simultaneously during the working process. Therefore, both the synthetic and analytic methods are fundamental scientific methods for carrying out a PS or PI work.

1.11 Synthesis and analysis as basic scientific methods

According to Ritchey [13], G. W. Leibniz (1646–1716) was among the first to define analysis and synthesis as modern methodological concepts: "*Synthesis* ... is *the process* in which we begin from principles and build up theorems and problems, ... while *analysis* is *the process* in which we begin with a given conclusion or proposed problem and seek the principles by which we may demonstrate the conclusion or solve the problem."

Also according to Ritchey [13], Bernhard Riemann (1826–1866) had once discussed the wider applications of the analytic and the synthetic methods for scientific investigation in general in his introduction "On the method which is applicable to the study of the physiology of the finer sense organs." Riemann used a specific case study and presented 13 paragraphs in his methodological introduction to "The mechanism of the ear," where analytic and synthetic methods were the basic scientific methods applied.

Today, we generally accept that synthesis and analysis are the basic scientific methods that are ubiquitously applied in scientific inquiry.

1.11.1 Synthesis-dominated versus analysis-dominated methods

In engineering and technology, it is impossible to find a problem that is to be solved by pure synthesis or by pure analysis. An engineering problem is usually solved by complementary application of synthesis and analysis. During problem solving, there is a certain mode that represents the interplay between the subject investigator and the object system and presents the cause–effect result to the object system.

Depending on the cause–effect result on the object system, when the subject investigator gives the ideas, concepts, principles, and mechanisms to *create the object system*, the problem-solving mode is the synthesis-dominated method. Essentially, a synthesis-dominated method inputs information to the object system as ideas, concepts, principles, and mechanisms, as shown in Figure 1.11. In such a case, the subject investigator still needs analysis of the problem to get the ideas, concepts, principles, and mechanisms to be input into the object system.

When the object is an existing system (e.g., an already created object system) that calls the subject investigator to *produce information and knowledge* to understand its properties and behaviors and to evaluate its performance and efficiency with the specified indicators, the problem solving mode is analysis-dominated. Essentially, such analysis-dominated method outputs information from the object system as properties, behaviors, evaluated performance, and efficiency (Figure 1.12). In such a case, the subject investigator still needs synthesis to define the method and tools to be used in analysis of the object system.

In both synthesis-dominated and analysis-dominated methods, there is interplay between the subject and object, which is manifested by the working process. However, in both cases, problem solving still needs the complementary application of synthesis and analysis. Moreover, for some tasks, problem solving is manifested by such cause–effect results that both the object system and its properties, behaviors, performance, and efficiency emerge during the working process. In these cases, the problem solving mode relies almost equally on synthesis and analysis and is named here the *synalysis method* (SYNthesis+anALYSIS). Figure 1.13 illustrates such the synalysis method.

Pure analysis methods focus on obtaining precise numerical solutions for variables and parameters and cannot reach the creativity and innovation of the technical system. By contrast, pure synthesis methods focus on the conceptual structure to pursue novelty and nonconventional structures and cannot reach the numerical precision required for application and operation of the technical system. In the practical application of a technical system, synthesis and analysis methods must apply simultaneously, which call for the synalysis method.

Essentially, pure synthesis embeds the concepts and ideas from the subject into the object. This inevitably change the software or hardware, or both, of the technical system. Such a change is manifested by qualitative or quantitative indicators, or by

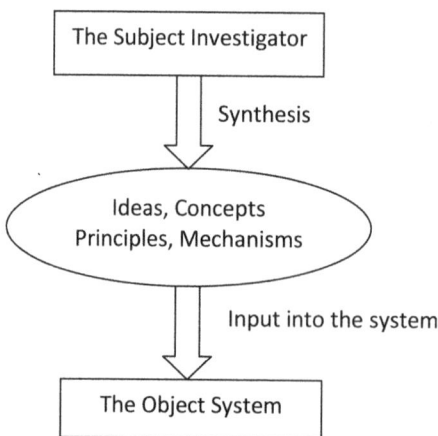

Fig. 1.11: Synthesis-dominated method: input information into the system.

The Object System

Analysis

The Subject Investigator

Output of the system

Properties, behaviors, performance and efficiency indicators

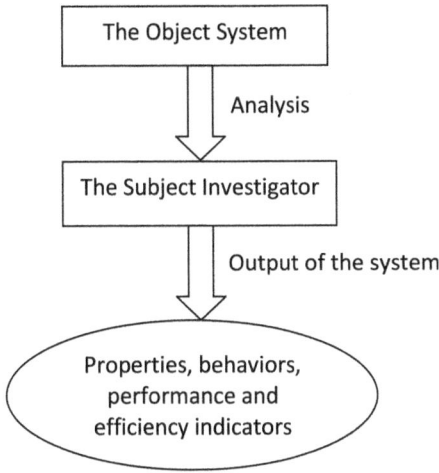

Fig. 1.12: Analysis-dominated method: output information from the system.

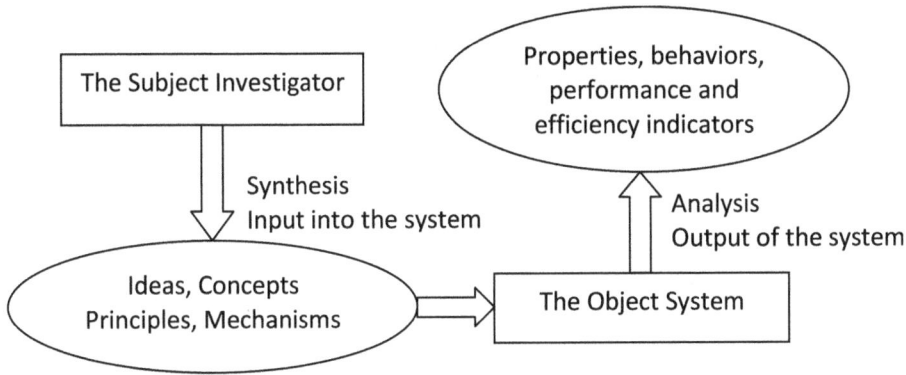

Properties, behaviors, performance and efficiency indicators

The Subject Investigator

Synthesis
Input into the system

Analysis
Output of the system

Ideas, Concepts Principles, Mechanisms

The Object System

Fig. 1.13: Synalysis method as a combination of synthesis and analysis.

pure analysis. On the other hand, pure analysis digs out information from the object by the subject. It may rely on an existing analysis method or develop a new method. The extracted information is valuable for understanding and implementing the technical system.

1.11.2 Synthesis and analysis as basic methods in PS and PI works

For both PS and PI, synthesis and analysis are the basic methods. Figure 1.10 shows that the working process of the methodology is to apply simultaneously the synthesis and analysis methods. The objective of both PS and PI works is to create a technical system that fulfills the required functionality and utility and can be evaluated with the specified indicators for its performance. The synthesis and analysis methods are applied to address the basic elements of the PS or PI work. During the working process of the methodology, the following can be highlighted in terms of the methods:

- For the problem and task, it is primarily the analysis method that gives the features and special phenomena of the problem.
- For the software and hardware system, both analysis and synthesis methods are needed to approach the basic elements and then the software and hardware systems.
- For the technical system, it is primarily the synthesis method that combines the software and hardware systems to produce the technical system.
- For the performance, it is primarily the analysis method that evaluates the performance using specified indicators.

Due to the diversity of the scopes of the technical systems, there are important situations in which one method is more suitable than the other. This concerns the question of which method is most appropriate as the primary point of departure for study of a technical system. Figure 1.14 presents the diverse scopes of study for technical systems based on the synthesis and analysis methods.

Whenever we do a specific PS or PI work, the object system is always the central element of the work and the results are directly or indirectly related to the technical system. However, depending on the tasks and intentions, the works have different scopes of study, which are briefly classified as follows:

- invented system: new concept and idea, new principle
- improved system: modification, retrofit, improved for a certain situation and case
- designed system: application purpose, case problem is defined, parameters designed to fulfill the specifications
- searched system: application purpose, case problem is defined, compare and select among the alternatives
- examined system: development of heuristics, steady-state and dynamic properties, safety
- controlled system: dynamic and control structure of the system, controller tuning

All studies concerning technical systems involve application of the synthesis and analysis methods. Invented and improved systems rely on synthesis-dominated methods, whereas designed and examined systems rely on analysis-dominated methods. Searched and controlled systems rely more evenly on both methods (synalysis method).

1.11.3 Process synthesis versus process analysis

Both synthesis and analysis are basic scientific methods and are generally applied in scientific inquiry. On the other hand, PS and process analysis are concerned with the study of process systems. PS is a synthesis-oriented method to synthesize and design

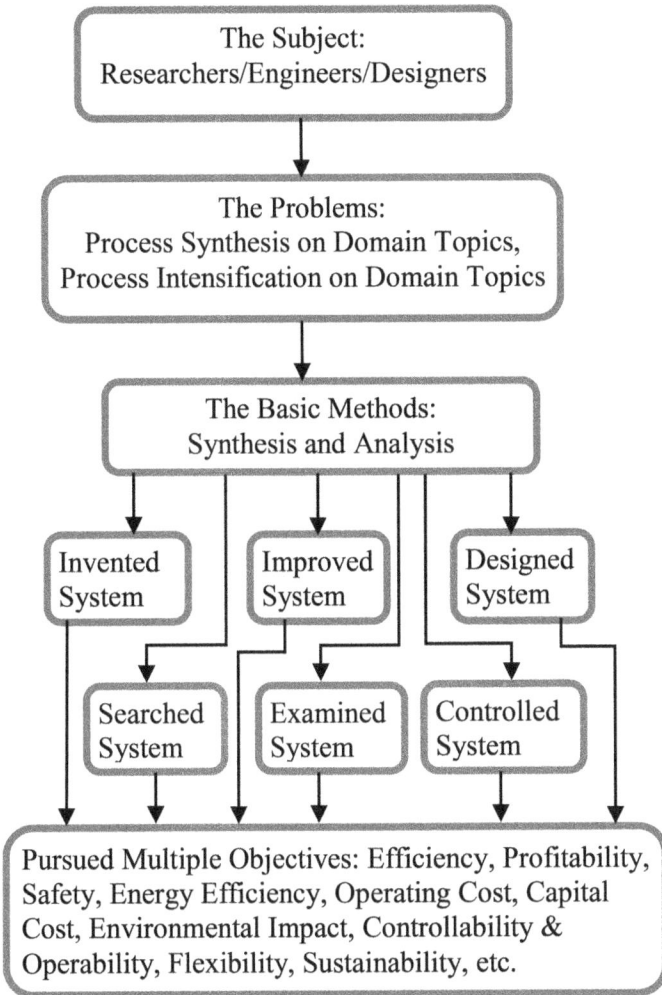

Fig. 1.14: Synthesis and analysis as basic methods for studying technical systems with diverse scopes.

process systems, whereas process analysis is an analysis-oriented method to understand and implement the process systems.

In process systems engineering, PS and process analysis have developed into distinct research areas. The methods for PS were briefly discussed in Section 1.3. Methods for process analysis mostly rely on process modeling, process simulation, and process optimization with numerical and algorithmic methods. PS concerns the structure of the process system, where the system's software, hardware, and their combination are finally determined. The software system, the hardware system, and their combination and arrangement are the key tasks and problems in PS. Changes and modifications to the software and hardware systems are essential for the synthesis of new alternatives. Process analysis is concerned with rigorous design, equipment construction, application, control, and operations needed to implement the process system. Other important tasks in process analysis are analyzing both steady-state and dynamic properties and behaviors and evaluating performance using specified indicators.

The experience of engineering research and practice has established the general principle that, on the one hand, without PS to explore the possible ideas and concepts to be applied in the process technology, it is impossible to generate all the feasible and novel structural alternatives. On the other hand, without process analysis to determine numerical values for the system's parameters, it is impossible to reach the final implementation and operation of the process system. The two methods are complementary; PS focuses more on the qualitative decisions for structural design of the process system and process analysis focuses more on the quantitative decisions for parametric design of the process system.

The relationship between PS and process analysis is perhaps like the relationship between a seed and the earth. Without the seed of PS, the earth of process analysis will become barren and bald land; On the other hand, without the earth of process analysis, the seed of PS cannot grow into flourishing and harvestable crops.

1.12 Systematic procedure as an element

When considering conceptual design as the result of a PS or PI work, the design process is a complex decision-making process in which synthesis and analysis are applied to deal with each of the basic elements, and decisions need to be made to determine all the elements. Therefore, the working process to deal with the elements and make decisions always involves a procedure to determine the technical system (i.e., the conceptual design). Depending on the problem and task, the procedure consists of different steps to reach a conceptual design of the technical system. Moreover, the steps are not always explicit *a priori*, but are related to the elements considered for the problem. One needs to identify the steps of the working process in order to formulate the procedure. Therefore, formulating the systematic procedure by identifying the critical steps is the key element in developing the methodology for a PS or PI work. It is the systematic procedure that produces alternatives for the conceptual design and, correspondingly, it is the quality of the generated alternatives that determines the performance of the technical system.

There are many studies that explicitly or implicitly describe the procedures, methods, and approaches, especially for PS works [11, 14–18]. The procedures usually consist of a limited number of steps to address either general issues or specific issues for the given problems and tasks. Depending on the scope of the problem, they can be considered as general procedures or specific procedures.

1.12.1 General procedures for process synthesis

A widely applied general procedure is the hierarchical procedure from Douglas [14] for the conceptual design of chemical processes. The hierarchical procedure is a general procedure that consists of the steps needed to address the major issues and tasks involved in determining the process structures, as listed in Table 1.1. Details of the steps in the hierarchical procedure are illustrated by a case study of the HDA (hydrodealkylation) process. The alternative conceptual designs were finally presented as the heat-integrated HDA process alternatives. A similar procedure is illustrated by the onion model [15, 16] in which the layers and steps address major tasks such as the reaction system, separation system, heat exchanger network, utility plant, and waste treatment plant (Table 1.1).

The main steps in such general procedures aim to determine the most suitable operation units, their interconnections, and their integration, which are normally at the total process level. Some major tasks in the general procedure can be listed as follows:
– select suitable operation units
– determine the subsystems for reaction and separation
– determine the interconnection between the units
– determine heat integration among the units
– determine mass integration among the units

Such general procedures can be applied to total process systems or process subsystems to generate conceptual design alternatives.

1.12.2 Specific procedures for PS and PI

There are also systematic procedures that have been developed for synthesis of process subsystems. For example, for separation processes, Wibowo and Ng [19, 20] developed different systematic procedures for synthesis of solid-processing and crystallization processes. We have developed systematic procedures for synthesis and inten-

Tab. 1.1: General procedures for conceptual process design.

Steps	Hierarchical procedure [14]	Onion model [15, 16]
1	Batch versus continuous	Reactor system
2	Input/output structure of the flowsheet	Separation system
3	Recycle structure of the flowsheet	Heat recovery system
4	Separation system	Utility system
5	Heat exchanger network	Waste treatment system

sification of multicomponent distillation systems [21, 22]. Doherty and Malone [11] developed systematic procedures for synthesis of azeotropic distillation systems.

Specific procedures can address elements such as mechanisms for novel equipment, concepts and principles for hybrid processes and nonconventional devices and methods. Such specific procedures aim to intensify the technical systems at equipment and subsystem levels, which call for specific steps to determine the elements of the novel structures of the equipment and subsystems. The procedures also generate feasible conceptual design alternatives, which are based on the unique features and phenomena that emerge during problem-solving processes (e.g., special mass and heat transfer phenomena and mechanisms).

Some major steps in specific procedures (equipment and subsystem level) can be listed as follows:
-   identify unique features and phenomena
-   emerge novel concepts and mechanisms
-   determine software elements
-   determine hardware elements
-   determine new mechanisms to combine software and hardware system
-   emerge novel alternatives for the technical system

In summary, it is feasible and practical to develop and present the systematic procedure for a PS or PI work, with either implicit steps or explicit steps, as the methodology. For future works, it is worth paying more attention to the systematic procedure when doing PS and PI for specific system synthesis and intensification. The synthesized or intensified technical system and the systematic procedure are the two fundamental elements for a specific PS or PI work.

1.13 Concept of the technical system as a technology whole

When we consider the technical system as the result of a PS or PI work, this technical system is usually presented as a conceptual design. If we want to deliver this PS or PI work for scientific publication, then there are normally two situations regarding evaluation of the technical system:
-   The technical system was evaluated in the work with the specified performance indicators of the problem scope. The work normally also includes the methods for generating and analyzing the technical system.
-   The PS or PI work was evaluated by area experts as reviewers; thus, the technical system was evaluated by the reviewers.

Statistically, the evaluations, reviews, and comments on the presented technical system (either PS or PI) concern and include the following questions:
- What is new about the technical system? What is the novelty?
- What is the method? Is there a new synthesis or intensification method?
- Is there a systematic procedure for the methodology?
- What is the method for designing parameters of the technical system?
- Are the designed parameters optimized?
- What is the optimization method? Is the solution local or global optimization?
- If there are different alternatives, what are the heuristics for choosing and applying the alternative systems?
- What are the dynamic and control properties analyzed for the technical system?
- What is the controllability and operability of the technical system?
- Are there any safety problems with the technical system?
- What are the applications and implementation situations and cases of the technical system?
- How complete is the evaluation and comparison with other existing technical systems?
- Are the evaluation methods and indicators sufficient?
- What are the equipment design, sizing, and costing methods?
- Are there any other issues to be evaluated?

All the concerned issues and aspects are relevant to the technical system. Therefore, for a technical system delivered from a PS or PI work, there is usually a third situation where the technical system is evaluated. The technical system delivered from a specific PS or PI work is subsequently studied and evaluated by other research works. This produces more knowledge and information about this technical system, which will improve its application and implementation as a mature technology for different cases.

It is reasonable to ask the following questions concerning the state of a technical system as a technology whole:
- What is the perfectness of the technical system?
- Is there an absolute optimized technical system?

It is difficult to define a perfect technical system, and any optimization is a relative optimization within a certain scope of the specified objective(s). When considering a technical system for its functionality and utility, the system is in a specific technical state. In a thermodynamic system, where every thermodynamic state is a unique equilibrium between the system and its surroundings, a change of state is accomplished by a specific process. The thermodynamic state of a thermodynamic system or the technical state of a technical system means that the system has certain properties and attributes, and a change in the state means changes in the properties and attributes of the system.

A technical system is an artificial system that goes through different states during its evolution from invention to application. The possible states of a technical system can be summarized as follows:
-   emerging state as concepts, principles, and mechanisms during synthesis, intensification, invention, or innovation
-   synthesis state to change and determine its structure (including application cases, system specifications, and constraints)
-   conceptual state as a conceptual design for the technical system
-   experimenting test state with laboratory- or pilot-scale models (if this is a necessary state)
-   design state to change and determine its parameters (including application cases, system specifications, and constraints)
-   analysis state to study its steady-state and dynamic properties and performance (including application cases, system specifications, and constraints)
-   control state to design its control structure and tune controllers
-   simulation state as a virtual system with modeling and algorithms
-   installation state including its construction, installation, and commissioning
-   operation state running at the designed and tuned set-point values
-   retrofit state to partly repeat the design, analysis, installation, and operation
-   life end state to decommission its operation and application

For further extension of the relevant points and aspects of a process technical system, we can summarize all the relevant knowledge and information on the process technical system to help comprehend the concept of the process technical system as a technology whole, which is illustrated in Figure 1.15.

For a process technical system, statistically, all the related knowledge and information will eventually merge as a technology whole. Probabilistically, different knowledge and information comes from different works by addressing different problem scopes and tasks.

For the concept of the technical system as a technology whole, we can make the following five observations, which are related to the states of a technical system:

Observation 1: A basic phenomenon is that a technical system has no gestalt or eternal form and state. For synthesis and design, different designers can have different designs for the technical system. The reason is that one can change any aspects of both its software elements and its hardware elements, and also change the combination and assembly of its software and hardware systems. For analysis and optimization, depending on the cases, analysis and optimization scopes, and constraints imposed, different analyzers can produce different analyses and optimized results for the technical system. The reason is that one can use different analysis methods and introduce different evaluation indicators and their combinations to analyze and optimize the technical system.

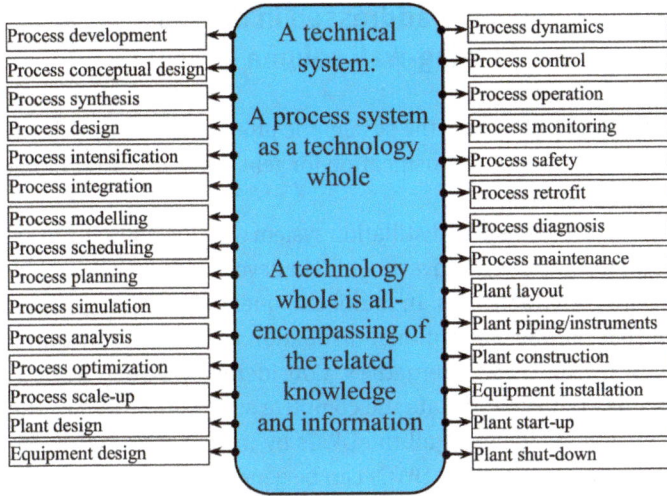

Fig. 1.15: The concept of a process technical system as a technology whole.

Observation 2: There are always uncertainty and randomness in selecting and determining the software and hardware elements for a technical system. The emerged software and hardware systems also have some of this uncertainty and randomness, so there can always be different conceptual designs produced as different alternatives for a technical system.

Observation 3: From an application point of view, when the specific application case problem is clearly defined and when the evaluation indicators are specified, one alternative system can have obvious advantages over any alternative systems for the defined application case.

Observation 4: From a research point of view, there is no one single methodology that can cover all of the different research aspects and scopes of research for a technical system. For a process system, research can address any relevant topical aspect. They all contribute to knowledge and information about the process technical system as a technology whole.

Observation 5: The intrinsic arbitrary and uncertainty features for PS and PI dictate that there is no absolute optimization of a technical system. Optimization is always relative to certain conditions and constraints imposed on a technical system.

1.14 Example: the basic elements addressed in distillation intensification for the dividing-wall column subspace

In this section, as an example, we illustrate the systematic procedure and associated basic elements addressed in distillation intensification for generation of the dividing-wall column (DWC) subspace.

The DWC is considered as an intensified distillation system with the potential to reduce both energy consumption and capital investment. The several known DWCs were developed through some inventive activities. In multicomponent distillation, we can generate a complete subspace of the conventional distillation configurations, which include a large number of conventional alternatives. Considering the attractive potential of DWCs, it is very desirable to generate the complete subspace of DWCs. The problem is that it is impossible to generate all the DWCs by inventive activities, because experience has shown that only a few DWCs can be generated by such chaotic inventive activities. Also, the invented DWCs do not include the methods for systems design of their physical structures.

If we want to generate the complete subspace of DWCs, the following questions arise: How many possible DWCs are there? What basic elements do we need to consider? Is there any systematic procedure to generate DWCs?

In Chapter 2, we present a systematic procedure to address these questions. The systematic procedure can produce DWCs by defining their physical structures step by step. In total, nine steps formulate the systematic procedure, which also serves to address the emerged basic elements for the DWC subspace. We briefly summarize the emerged and addressed basic elements during DWC intensification. Details of the development of the systematic procedure and its application to generate the DWC subspace are presented in Section 2.12 of Chapter 2.

The following general elements are addressed for this DWC intensification problem:
- the problem scope and task: multicomponent separation
- the functionality and utility: distillation systems with N nearly pure products for a mixture with N components
- the performance indicators: fewer columns than conventional configurations

The following specific elements are addressed for synthesis and intensification to generate the DWC subspace:
- The element of the unique features and phenomena: Some column sections are movable. Thermodynamically equivalent structures can flexibly rearrange and recombine the column sections.
- The element of mechanism: Two mechanisms emerged as mechanisms 7 and 8. Some columns can be incorporated into other columns as internal dividing walls.

- The element of the software system: Thermal couplings can simultaneously change the mass and heat transports between columns. Thermal couplings can be implemented as external thermal couplings or internal thermal couplings.
- The element of the hardware system: The number of condensers and reboilers can be changed, the number of column sections in a column can be changed, a column can be incorporated into another column as an internal dividing wall, and the number of columns in a distillation system can be changed.
- The element of the systematic procedure: The systematic procedure consists of nine steps to change and combine the software and hardware elements to generate the DWC subspace.
- The element of the intensified technical system: A complete subspace of intensified DWCs is generated.
- The element of the distinct features of the generated DWCs: All the generated DWCs use fewer columns than the conventional configurations.

A case example is presented to examine the elements addressed for DWC column generation. The case illustrates the change in the states of the technical system during synthesis and intensification in terms of the software and hardware elements. The technical systems at states I–IV are described next.

The technical system at state I: a feasible conceptual design named as a simple column configuration (Figure 1.16).

The technical system in this state is a conventional simple column configuration with three simple columns for the three sharp splits. Each simple column is a standard binary distillation column with one feed, one rectifying section with an overhead con-

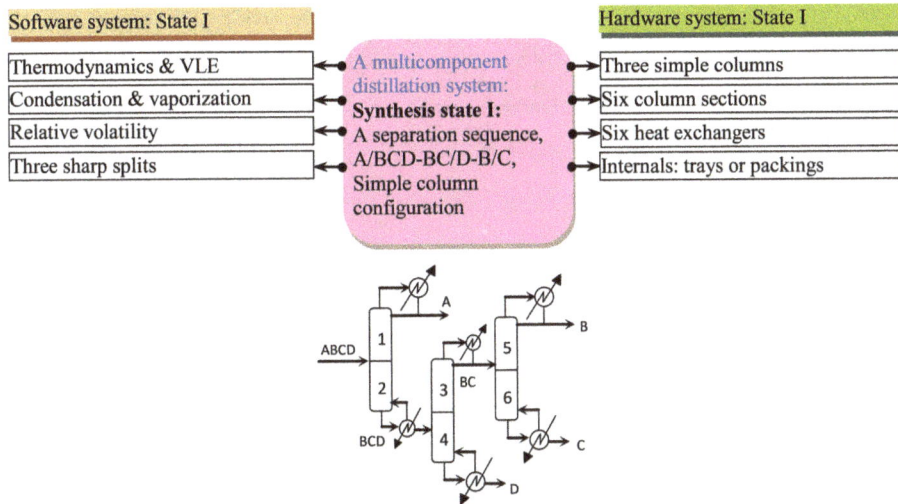

Fig. 1.16: The technical system at state I: a simple column configuration.

denser, and one stripping section with a bottom reboiler. In total, six heat exchangers and six column sections are applied. Either trays or packings can be considered for the column sections. Thermodynamics, phase equilibria, partial condensation and vaporization are the basic software elements for distillation. The mixture does not involve azeotropes and the relative volatilities are suitable for ordinary distillations.

The technical system at state II: a feasible conceptual design named as a thermally coupled configuration (Figure 1.17).

The technical system in this state is a thermally coupled configuration with three columns. Each column keeps its rectifying section and stripping section. A thermal coupling mechanism is introduced into the system, which dramatically changes both the software and hardware elements. The thermal coupling streams serve as simultaneous mass and heat transports between the columns. At the same time, one reboiler and one condenser were eliminated. Thus, only four heat exchangers are applied in the thermally coupled configuration.

The technical system at state III: a feasible conceptual design named as a thermodynamically equivalent structure (Figure 1.18).

The technical system in this state is a thermodynamically equivalent structure with three columns. The mechanism of rearrangement of the column sections is introduced, which can recombine the movable column sections (sections 4 and 5) among the columns. Both the software and hardware elements are dramatically changed. The separation tasks are changed in the columns, and the columns have a different number of sections. The first column has three sections to produce products A and D, the second column has two sections as a rectifier to produce B, and the third column has only one section as a stripper to produce C.

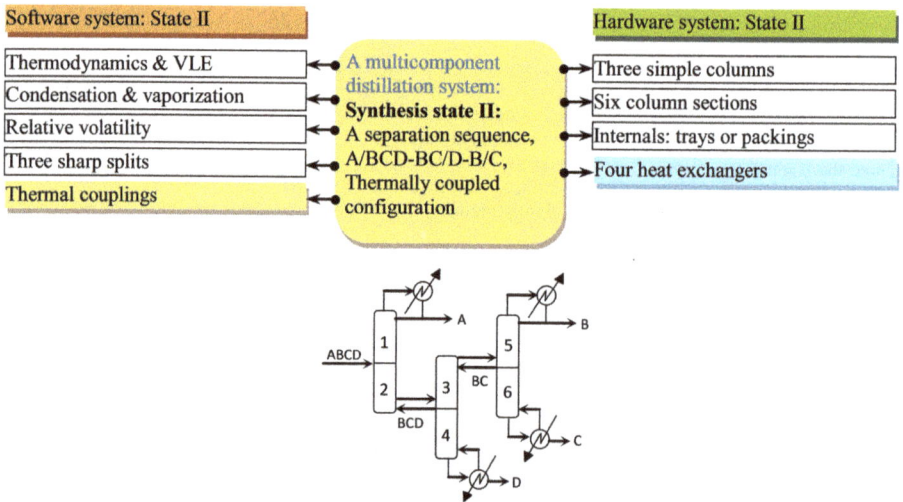

Fig. 1.17: The technical system at state II: a thermally coupled configuration.

Software system: State III		Hardware system: State III
Thermodynamics & VLE	A multicomponent distillation system: Synthesis state III: A separation sequence, A/BCD-BC/D-B/C, Thermodynamically equivalent structure	Six column sections
Condensation & vaporization		Four heat exchangers
Relative volatility		Internals: trays or packings
Three sharp splits		Two movable column sections
Thermal couplings		Three different columns
Re-assigned separation tasks		

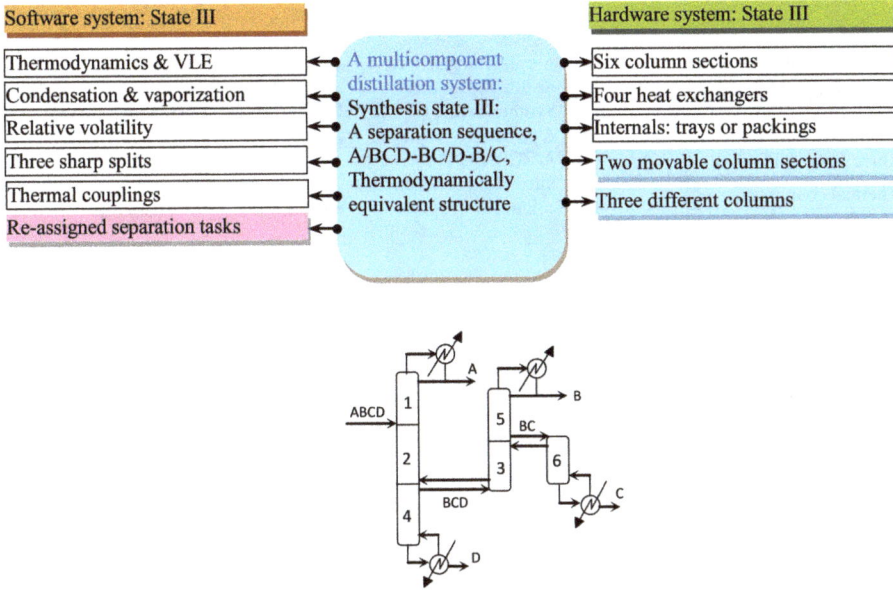

Fig. 1.18: The technical system at state III: a thermodynamically equivalent structure.

The technical system at state IV: a feasible conceptual design named as a DWC (Figure 1.19).

The technical system in this state is an intensified DWC with one column. The mechanism of incorporating the side column as an internal dividing wall is introduced. The two side columns are incorporated into the thermally linked column at the same time as two dividing walls, which results in the intensified distillation system with one DWC. Both the software and hardware elements are dramatically changed. The thermal couplings are implemented inside the column at the ends of the dividing walls. All three separation tasks are accomplished in the single column through the corresponding functional zones divided by the internal dividing walls. The six functional zones in the DWC correspond to the six column sections in the earlier thermodynamically equivalent structure.

From the DWC subspace and the case illustration, we observe the following:

– It is possible to identify the basic elements for PS or PI works involved in this DWC subspace case.
– The key step in the working process for DWC intensification is separation of the software and hardware systems and examination of their respective changes.
– The major element of the PS and PI works is explicit articulation of the changes in the software and hardware systems. It is possible to articulate the changes in software and hardware elements in this DWC intensification case.

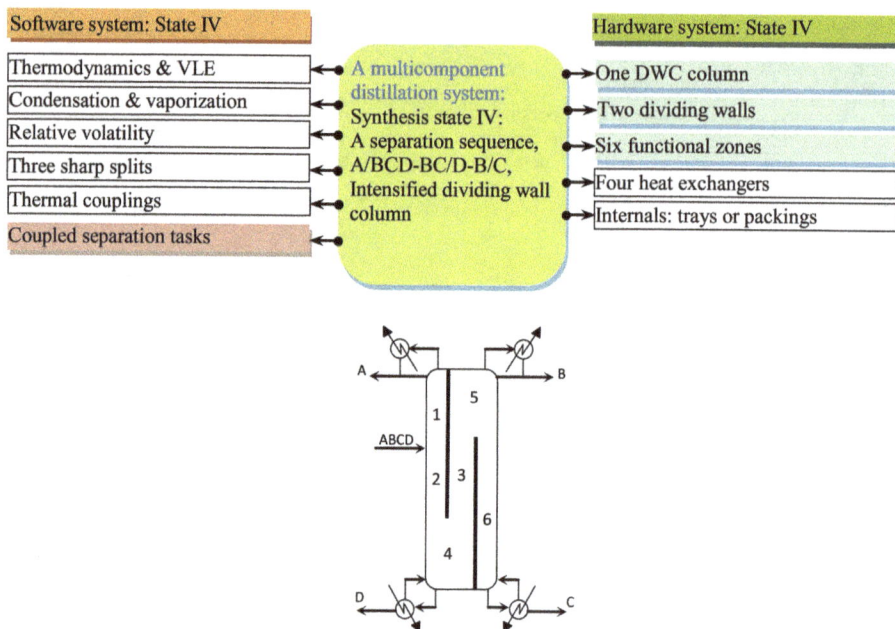

Software system: State IV

| Thermodynamics & VLE |
| Condensation & vaporization |
| Relative volatility |
| Three sharp splits |
| Thermal couplings |
| Coupled separation tasks |

A multicomponent distillation system:
Synthesis state IV:
A separation sequence,
A/BCD-BC/D-B/C,
Intensified dividing wall column

Hardware system: State IV

One DWC column
Two dividing walls
Six functional zones
Four heat exchangers
Internals: trays or packings

Fig. 1.19: The technical system at state IV: a dividing wall column.

- The changes and evolution of the software and hardware systems correspond to the states of the technical system with different alternatives. Each state is a feasible conceptual design of the distillation system in this DWC case.
- For PS or PI works in general, the scope of the problem and tasks determine whether it is possible to articulate the software and hardware changes, whether in an explicit or implicit way, and how to examine the effects of the introduced software and hardware elements.
- This case shows that distillation intensification for DWCs is a working process, which is to determine a conceptual design for the technical system from the initial state to the final state. The possible software and hardware elements to be introduced and changed eventually emerge and are articulated in different steps. The identified and emerged steps are formulated in the systematic procedure, which represents the major element of the methodology for multicomponent distillation intensification.
- For PS or PI works in general, the working process to determine a conceptual design for the technical system often starts from an initial state, where there are infinite and uncertain software and hardware elements to be chosen and conceived, and ends in the final state where there are finite and certain software and hardware elements chosen and applied in the technical system.

1.15 Conclusions

In this chapter, we have discussed some basic concepts and elements for PS and PI. These basic concepts and elements are considered as components when addressing the working process in terms of methodology for a specific PS or PI work. The discussed basic concepts and elements have focused on the concept of a PS or PI work, the conceptual design, the technical system, the software system, the hardware system, the systematic procedure, synthesis and analysis as basic methods, and the triple parties of the methodology. The basic elements addressed in distillation intensification for the generation of the DWC subspace have been discussed as an example.

Instead of closing the chapter with conclusive points, we would like to raise the following 15 questions to open discussion on the basic concepts and elements involved in PS and PI works, as well as on the methodology of carrying out PS and PI works:

1. What key elements constitute a PS work?
2. What key elements constitute a PI work?
3. How do we clearly articulate the software elements for a PS or PI work?
4. How do we clearly articulate the hardware elements for a PS or PI work?
5. How do we formulate the procedure to assemble the software and hardware systems?
6. What are the precise criteria for evaluating the technical system?
7. Do we have the complete list of design parameters for the technical system?
8. Do we have the design method for the technical system?
9. Do we have the optimization method for the technical system? Is it single objective or multiple objectives optimization?
10. How do we precisely predict the steady-state properties and performance of the technical system?
11. How do we precisely predict the dynamic properties and performance of the technical system?
12. What is the controllability and operability of the technical system?
13. What is the safety of the technical system?
14. How do we precisely predict the environmental impact?
15. How do we precisely predict the sustainability?

1.16 Bibliography

[1] Hendry JE, Rudd DF, Seader JD. Synthesis in the design of chemical processes. AIChE Journal, 1973, 19, 1–15.
[2] Rudd DF, Powers GJ, Siirola JJ. Process Synthesis. Prentice-Hall Inc.Englewood Cliffs, New Jersey, USA. 1973.
[3] Nishida N, Stephanopoulos G, Westerberg AW. A review of process synthesis. AIChE Journal, 1981, 27, 321–351.

[4]   Anderson JL. Process synthesis, Advances in Chemical Engineering Vol. 23. Academic Press Inc., London, 1996.

[5]   Stankiewicz A, Moulijn JA. Process intensification: transforming chemical engineering. Chemical Engineering Progress, 2000, 22–34.

[6]   Reay D, Ramshaw C, Harvey A. Process Intensification-Engineering for Efficiency, Sustainability and Flexibility, 2nd edition, Butterworth-Heinemann, UK, 2013.

[7]   Stankiewicz A, Moulijin JA. Re-engineering the Chemical Processing Plant: Process Intensification, Marcel Dekker, New York, USA, 2004.

[8]   Agreda VH, Partin LR. Reactive distillation process for the production of methyl acetate. U.S. Patent 4,435,595 assigned to Eastman Kodak Company, 1984.

[9]   Agreda VH, Partin LR, Heise WH. High purity methyl acetate via reactive distillation, Chem. Eng. Progr. 1990, 86, 40–46.

[10] Sirrola JJ. Industrial Applications of Chemical Process Synthesis. In: Anderson JL (ed). Process synthesis, Advances in Chemical Engineering Vol. 23, Academic Press Inc., London, 1996, 2–62.

[11] Doherty MF, Malone MF. Conceptual Design of Distillation Systems. McGraw-Hill Companies, Inc. New York, USA, 2001.

[12] Smith JM, Van Ness HC, Abbott MM. Introduction to Chemical Engineering Thermodynamics, 7th edn. McGraw-Hill High Education (Asia), International Edition 2005.

[13] Ritchey T. Analysis and synthesis: on scientific method based on a study by Bernhard Riemann, Systems Research, 1991, 8(4), 21–41.

[14] Douglas JM. Conceptual design of chemical processes, McGraw-Hill, New York, USA, 1988.

[15] Linnhoff B, Townsend DW, Boland D, Hewitt GF, Thomas B, Guy AR, Marsland RH. User guide on process integration for the efficient use of energy. The Institution of Chemical Engineers.

[16] Smith R. Chemical Process Design and Integration, Wiley, Chichester, UK 2005.

[17] Biegler LT, Grossmann I, Westerberg A. Systematic methods of chemical process design. Wiley, New York, USA. 1998.

[18] Seider WD, Seader JD, Lewin DR, Widagdo S. Product and Process Design Principles: Synthesis, Analysis and Design. 3rd ed(n. Wiley, Hoboken NJ, USA, 2009.

[19] Wibowo C, Ng KM. Workflow for process synthesis and development: Crystallization and solids processing, Ind Chem Eng Res, 2002, 41, 3839–3848.

[20] Wibowo C, Ng KM. Unified approach for synthesizing crystallization-based separation processes, AIChE Journal, 2000, 46, 1400–1421.

[21] Rong BG. Synthesis of dividing wall columns (DWC) for multicomponent distillations – A systematic approach. Chem Eng Res Des 2011, 89(8), 1281–1294.

[22] Rong BG. A systematic procedure for synthesis of intensified nonsharp distillation systems with fewer columns, Chemical Engineering Research and Design, 2014, 92, 1955–1968.

Ben-Guang Rong

2 Process synthesis and process intensification for multicomponent distillation systems – systematic methodology

Abstract: This chapter presents process synthesis (PS) and process intensification (PI) for multicomponent distillations. PS focuses on defining all the possible distillation configurations, whereas PI focuses on defining distillation systems with fewer columns. In total, seven subspaces are obtained for multicomponent distillations, each containing distinct distillation systems. A systematic synthesis procedure was developed for each subspace. Each procedure consists of distinct mechanisms and explicitly defines the physical structures of the distillation systems step by step. The systematic procedures are generalized procedures and applicable for mixtures with any number of components.

The relationships of the distinct subspaces and their systematic procedures are discussed. Systematic procedures can be developed for both PS and PI in the generation of distinct distillation systems. The methodology is a strong interplay between PS and PI for multicomponent distillation system synthesis and intensification. Therefore, a systematic synthesis procedure that integrates PS and PI is a highly efficient methodology for intensifying process systems.

Keywords: Process synthesis, process intensification, multicomponent distillation, distillation system, systematic procedure, subspace, methodology

2.1 Introduction

Multicomponent distillation is both a classical subject in process synthesis (PS) and an ever-growing subject in both PS and process intensification (PI). In this chapter, we introduce PS and PI for conceptual design of multicomponent distillation systems. Due to its wide use in process industries, synthesis of optimal distillation systems has always been one of the most important topics in PS [1, 2]. Early works made significant progress in distillation schemes for ternary mixtures [3–10]; however, the various configurations were developed gradually in a somewhat chaotic evolutionary process [11].

The driving force for distillation system synthesis and intensification is the need to reduce energy consumption and capital expenditure. Distillation uses energy as the separating agent so energy consumption is significant. Distillation is usually applied for very large capacity bulk production of products such as fuels and chemicals. In these cases, distillation features high energy consumption and large capital investment. Distillation is also the major industrial separation technology used for renewable biomass-based manufacturing of products such as biofuels and biochemicals.

https://doi.org/10.1515/9783110465068-002

Therefore, pursuing energy savings and capital reductions is important and the main aim of distillation synthesis and intensification, and of subsequent equipment design, control, and operation.

For the conceptual design of distillation configurations during synthesis and intensification, the numbers of columns and heat exchangers are the dominant criteria in evaluating a distillation configuration. Because sharp separation sequences need the minimum number of simple columns, traditional distillation configurations with only sharp splits have been the preferred alternatives [3, 12–22]. Nevertheless, taking into account all the application cases and situations, studies have shown that it is not obvious whether some structures are less expensive than others.

In industrial processes, the mixtures to be separated often contain four or more components. Industrial experience shows that if the optimum alternative is not predefined, it will not be found [23]. Thus, the optimal design of distillation configurations for multicomponent mixtures depends on the ability to define all of the possible alternatives *a priori*. However, unlike ternary mixtures, there are a large number of possible distillation configurations for mixtures of four or more components. When nonsharp separation sequences are considered, it is impossible to obtain all the possible configurations by inventive activities; instead, systematic synthesis procedures are needed.

The synthesis of multicomponent distillation configurations has made significant progress in the last decade. Different methods are reported in the literature for synthesis of multicomponent distillation configurations (four or more components). Agrawal [24] presented a method for obtaining distillation configurations with $N - 1$ columns through modifying a network superstructure for a multicomponent mixture. The principles of the method are simple, but the details are somewhat overwhelming [11]. Moreover, the derived substructures are not always feasible. A matrix method was later presented for synthesis of multicomponent distillation sequences [25, 26], the major limitation being that the schemes generated from the matrices need to be checked for feasibility, which is difficult when generating schemes with fewer columns [26]. Neither the network-superstructure method nor the matrix method give any hints for designing parameters for the schemes. Caballero and Grossmann [27–30] presented a state-task network superstructure method for mathematically formulating the alternatives space of multicomponent distillation configurations. The alternatives also need to be checked for feasibility; however, the power of the superstructure method is that it can screen the predefined alternative space for a given mixture and an objetive function.

We have investigated the synthesis of complex distillation flowsheets with thermal couplings for five-component separations [31, 32]. Initially, we focused on development of heuristics and parametric designs [31, 32]. During development of the design procedure for these complex distillation flowsheets, two mechanisms were discovered for transforming five-component distillation configurations [32, 33]. First, a five-component simple column configuration can be directly transformed into its corresponding thermally coupled configuration. Second, the column sections in a thermally coupled configuration can be recombined among the columns with the designated movable sections. The two mechanisms were first discovered for the five-component sharp distillation configurations and constituted the methodology for synthesis of the subspace of the thermally coupled configurations from the subspace of the sharp separation sequences [33, 34]. Subsequently, we developed a systematic method that can synthesize from the traditional distillation configurations to the thermally coupled configurations and, finally, to the thermodynamically equivalent structures [35]. This method was based on formulating all the feasible separation sequences *a priori*, including both sharp and nonsharp sequences. The method was demonstrated for quaternary distillation and is valid for an N-component mixture to obtain configurations with $N − 1$ or more than $N − 1$ columns [35].

Starting from the sharp and nonsharp separation sequences, we recently focused on applying PI principles to the synthesis of intensified distillation systems. The target was to synthesize distillation systems with fewer columns. Specifically, we considered distillation systems with less than $N − 1$ columns for an N-component mixture. Such distillation intensification is also based on a systematic synthesis procedure to define the subspaces of intensified distillation structures. The systematic procedures for intensificiation are based on different mechanisms and strategies to coordinate the multiple separation tasks, which can explicitly define the intensified distillation systems [36, 37].

In this chapter, we formulate the systematic methodology for multicomponent distillation on both synthesis and intensification. By doing so, we try to answer the following questions:
- What are the different kinds of distinct configurations for a multicomponent distillation?
- What are the subspaces with distinct alternatives?
- What are the systematic methods and procedures for approach the distinct subspaces?
- What are the relationships of the distinct subspaces for a multicomponent distillation?

To answer these questions, we focus on the following objectives:
- Both sharp separation sequences and nonsharp separations sequences are considered.
- Process synthesis focuses on defining all the feasible distillation alternatives.
- Process intensification focuses on defining distillation alternatives using fewer columns.
- The obtained feasible distillation alternatives are classified into distinct subspaces.
- Systematic procedures are developed as methods to obtain the subspaces for both distillation synthesis and distillation intensification.

– Systematic procedures give the design steps for both structural and parametric design of the distinct distillation systems.
– The relationships of the distinct distillation subspaces and their systematic procedures are defined.

Recently, thermally coupled distillation configurations and intensified equipment such as dividing-wall columns (DWCs) have been studied for their application in a number of different cases and separation problems. It is beyond the scope of this chapter to review these studied and applied cases.

2.2 Problem definition

The separation problem is that if the number of components in a mixture is N, at least N nearly pure products are obtained by ordinary distillation. This means that the mixture is ideal or nearly ideal without azeotropes. For such zeotropic mixtures, the separation between any adjacent components of the mixture is feasible without introducing an external mass-separating agent (MSA) to enhance the relative volatility. Thus, all splits are feasible and all feasible separation sequences can be formulated by only considering their relative volatilities. Such feasible separation sequences are the volatility-dominated distillation configurations.

For mixtures that form azeotropes under the examined thermodynamic conditions, the feasible splits should first be determined by introducing an external MSA or other means to circumvent the azeotropes. Methods for determining the feasible separation sequences for such problems are available in the literature, e.g., Doherty and Malone [38]. Such feasible separation sequences are the azeotrope-dominated distillation configurations.

2.2.1 Multicomponent mixtures

For a multicomponent mixture, the components are represented in this chapter as a certain number of adjacent capital letters, e.g., ABCD for a four-component mixture. For a mixture with N components, except the lightest and heaviest components, the other $N - 2$ components are called middle components. For example, for a five-component mixture ABCDE, with its components ordered according to decreasing volatility, A is the lightest component, E is the heaviest component, and B, C, D are the middle components.

2.2.2 Product specifications

For the separation of a multicomponent mixture by distillation, the possible products can be classified into three different types, as follows:

Product type 1: The number of products is equal to the number of components in the feed mixture. Each component is separated as a final product with specified purity.

Product type 2: The number of products is less than the number of components in the feed mixture. More than one component in the feed mixture can be present in the same product.

Product type 3: The number of products is more than the number of components in the feed mixture. The same component can be produced from different locations as different products with different purities or specifications.

2.3 Dominant criteria for evaluation of a distillation system

As distillation uses an energy-separating agent (ESA) and is usually applied in industry for large capacity production, from a design point of view, there are three dominant criteria for evaluating a distillation system, as follows:

– energy cost: pursuing high energy efficiency with minimum energy consumption
– capital cost: pursuing minimum capital investment
– equipment design and operation: preferring easy design, control, and operation

2.4 Basic concepts for multicomponent distillations

2.4.1 Sharp and nonsharp splits

For a multicomponent mixture, three types of individual splits can be used to perform a split of the mixture. Figure 2.1 shows a four-component mixture as an example. The split between two adjacent components B and C sharply distributes the two key components into the two products, as shown in Figure 2.1(a). A sharp split performs a separation in such a way that each component is distributed primarily in one product stream. However, recovery of the key component in the product is not necessarily 100%. Depending on the specification, the product stream AB can contain component C as an impurity or even component D. Similarly, the product stream CD can contain component B as an impurity or even component A.

A nonsharp split is a separation between two nonadjacent key components; in other words, there must be a middle component(s) between the two key components. Depending on the number of middle components in the mixture, there are two types of nonsharp splits. An asymmetric nonsharp split is shown in Figure 2.1(b), whereby one

AB/CD ⟨ AB / CD   ABCD ⟨ AB / BCD   ABCD ⟨ ABC / BCD

(a)   (b)   (c)

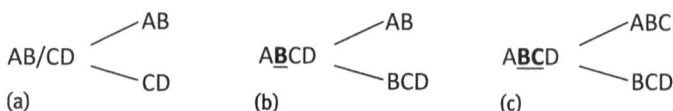

Fig. 2.1: Three types of individual splits: (a) sharp split, (b) partially sloppy split, (c) fully sloppy split.

or more middle components are distributed in the two products. If the middle components of a mixture are not distributed simultaneously, the asymmetric nonsharp split is called a *partially sloppy split*. A symmetric nonsharp split is shown in Figure 2.1(c), whereby all middle components of the mixture are distributed in the two products. A symmetric nonsharp split must distribute all the middle components of a mixture simultaneously, which is called *fully sloppy split* [39, 40]. For a mixture with four or more components, one can introduce different sharp splits or different partially sloppy splits to perform a separation; however, there is only one fully sloppy split possible.

As shown in Figure 2.1, a nonsharp split for a mixture is represented by underlining the distributed middle components; thus, ABCD means that a nonsharp split is performed for the mixture ABCD with the two middle components B, C distributed in both top product ABC and bottom product BCD. A sharp split is represented by an oblique line between the two adjacent keys; therefore, AB/CD means that a sharp split is introduced between B and C for the mixture ABCD, with top product being AB and bottom product CD.

In theory, to introduce nonsharp splits into multicomponent distillation separation, the following points should be considered:

- Nonsharp splits in distillation can intrinsically improve thermodynamic efficiency by reducing irrevesibility. Theoretically, nonsharp splits can improve separation efficiency and reduce energy consumption.
- There are many more feasible separation sequences with nonsharp splits than with only sharp splits.
- To predefine a complete space for a multicomponent distillation, we must include all the nonsharp separation sequences.

2.4.2 Simple distillation column

A simple distillation column receives one feed stream and produces an overhead product with a condenser and a bottom product with a reboiler. For simplicity, the condenser is specified as a total condenser that condenses all the vapor from the top of the column into liquid, and the reboiler is a partial reboiler that boils part of the bottom liquid into vapor. One can use a simple column to perform any intended individual splits with the three types of sharp and nonsharp splits, as shown in Figure 2.2.

Each simple column has a rectifying section with an overhead condenser and a stripping section with a bottom reboiler.

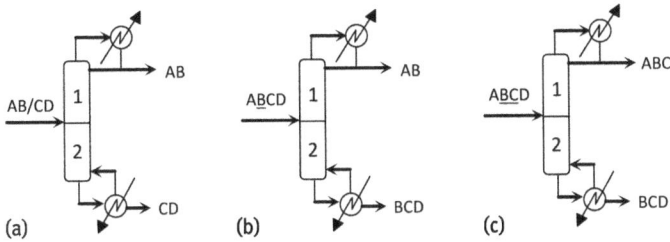

Fig. 2.2: Simple columns corresponding to the three types of individual splits in Figure 2.1.

2.5 Basic software and hardware elements for a distillation system

As discussed in Chapter 1, a technical system is considered to be an artificial system that consists of a certain number of individual functional parts and components. By design, the individual functional parts and components are coordinated and work together produce the expected functionality and utility. For PS and PI, it is fair to say that any technical system consists of its software and hardware systems.

It is necessary to discuss briefly the software and hardware elements of a distillation system.

2.5.1 Software elements

Any technical system, depending on its functionality and utility, must have its own software system. The software system is primarily based on its working principle and mechanism, and primarily determines its main hardware elements and their arrangement.

Distillation is a technical system with the functionality and utility to separate a fluid mixture into fractions or constituents of the required purity and containing the specified components. The essential working principle of a distillation system is based on the difference in volatilities of the components in the mixture to be separated. The distillation separation is typically realized through mass and heat transfers between vapor and liquid phases of the flowing fluids. Therefore, managing the distribution of components between vapor and liquid phases is the central consideration of its software system. Thermodynamics, phase equilibrium, thermal profile, and fluid hydrodynamics are the basic elements of its software system.

For distillation systems, it is also convenient to categorize the software aspects either as basic or fundamental aspects or as aspects imposed by the constraints and limits of the specific case.

2.5.1.1 Basic software aspects of a distillation system

The fundamental principle of a distillation system is thermal refining of the individual components in a fluid. For any distillation operation, the basic design consideration is determination of the proper temperature and pressure profiles for the phase behavior of the mixture, so that the desired products with specified purities can be obtained. This corresponds to the phase equilibrium behavior of the mixture, which is primarily related to the relative volatilities of the components in the mixture. When a mixture has normal phase equilibria behavior, which corresponds to the ideal and nearly ideal situations (i.e., no azeotropes appear), the thermal profile of the distillation system is only determined by the components of the feed mixture.

For a distillation system (without reaction), we can list the following basic software elements:
- relative volatility
- thermodynamics
- phase equilibria
- mass and heat transfers
- mass and energy conservation
- hydrodynamics
- gravity or high-gravity
- temperature and pressure profiles
- vapor and liquid flows and hydraulics
- feasible individual splits
- feasible separation sequences

2.5.1.2 Constraints and limits related to software aspects

Azeotropes are an additional constraint in determining the thermal profiles of the distillation mixture. Temperature and pressure profiles should take into consideration the presence of azeotropes. Very often, external mass-separating agents need to be introduced into the mixture to modify the thermodynamic and phase equilibria behaviors. In such situations, the feasible splits and the actual temperature and pressure profiles of the distillation system should simultaneously take into account the constraints imposed by the azeotropes and the phase behaviors resulting from use of external separating agents. Therefore, the constraints affect the following software elements:
- thermodynamics
- phase equilibria
- feasible individual splits
- feasible separation sequences

Doherty and Malone [38] introduced a comprehensive method and tool to determine the feasible splits and separation sequences to deal with the constraints of azeotropic mixtures.

Hardware elements remain the same for nearly all distillation situations, except that a decanter might be needed for heterogeneous azeotropic systems.

2.5.2 Hardware elements

Any technical system must include a hardware system. The hardware system must match the software system and fulfill the functionality and utility requirements of the technical system.

For a distillation system, hardware elements are typically used for generating the vapor and liquid flows, providing physical geometric space to process the bulk vapor and liquid flows, and maintaining sufficient contact of vapor and liquid phases for mass and heat transfers.

For a distillation system (column type), we can list the following basic hardware elements:
-   condenser, for condensing vapor into liquid
-   reboiler, for vaporizing liquid into vapor
-   column shell, for physical space for the bulk vapor and liquid flows
-   internals (trays and packings), for vapor and liquid phase contact for mass and heat transfers
-   column sections, for functional sections with a certain number of stages or a certain height of packings
-   pump, for changing the pressure of liquid streams
-   compressor, for change the pressure of vapor streams
-   reflux drum, for containing the condensed liquid or both vapor and liquid
-   decanter, for heterogeneous azeotropic systems
-   instruments, for parameter measurements or control purposes
-   controllers, for the control system
-   auxiliary devices, such as valves and containers

We show that both distillation synthesis and intensification are concerned with changes in the software and hardware elements and with the combinations of software and hardware elements to obtain distinct distillation systems. Special distillation schemes other than column types are not considered here.

2.6 Distinct separation sequences

We know that, in a forest, every tree grows from its own root. If every feasible multicomponent distillation structure is a tree, then its root is the distinct separation sequence (DSS), which is composed of the intended individual splits for the feed mixture. A DSS is defined as a separation sequence that has at least one different intended individual split from any other feasible separation sequences [35, 41].

For a multicomponent distillation, the first synthesis problem is to determine how many DSS alternatives can be generated for a multicomponent mixture. Given the three types of individual splits for a multicomponent mixture, as shown in Figure 2.1, one can introduce a set of intended individual splits to formulate a separation sequence to obtain the individual components as final pure products. When all three types of individual splits are used to formulate the separation sequences, three types of DSSs can be formulated as follows:

Type 1: DSSs with all sharp splits

Type 2: DSS with all fully sloppy splits (for mixtures with three or more components)

Type 3: DSSs involving partially sloppy splits

For an N-component mixture, the number of distinct sharp sequences (SSC_N) of type 1 can be predicted by the following equation from Thompson and King [13]:

$$SSC_N = \frac{[2(N-1)]!}{N!(N-1)!}$$ (2.1)

For any N-component mixture ($N \geq 3$), there is only one DSS of type 2 [40]. The number of DSSs of type 3 can also be calculated by the method introduced by Rong et al. [41, 42]. Table 2.1 illustrates the number of DSSs for each of the three types of sequences for three-, four- and five-component mixtures. Compared with DSSs with only sharp splits, there are many more DSSs with nonsharp splits.

As an example, Table 2.2 presents the formulated DSSs for each of the three types for a four-component mixture. In total, there are 5 sharp sequences and 17 nonsharp sequences.

Tab. 2.1: Number of distinct separation sequences of each type for a multicomponent mixture.

Number of components	Number of DSSs of type 1	Number of DSSs of type 2	Number of DSSs of type 3	Total number of DSSs
3	2	1	0	3
4	5	1	16	22
5	14	1	554	569
Number of splits	$N-1$ (minimum)	$N(N-1)/2$ (maximum)	Between $N-1$ and $N(N-1)/2$	

Tab. 2.2: The three types of distinct separation sequences for a four-component mixture.

Sequence order	Distinct separation sequences	Number of splits
Type 1 distinct separation sequences		
a	A/BCD → B/CD → C/D	3
b	A/BCD → BC/D → B/C	3
c	AB/CD → A/B → C/D	3
d	ABC/D → AB/C → A/B	3
e	ABC/D → A/BC → B/C	3
Type 2 distinct separation sequence		
f	ABCD → ABC → BCD → A/B → B/C → C/D	6
Type 3 distinct separation sequences		
g	A/BCD → BCD → B/C → C/D	4
h	ABC/D → ABC → A/B → B/C	4
i	ABCD → B/CD → A/B → C/D	4
j	ABCD → BC/D → A/B → B/C	4
k	ABCD → BCD → A/B → B/C → C/D	5
l	ABCD → A/BC → B/C → C/D	4
m	ABCD → AB/C → A/B → C/D	4
n	ABCD → ABC → A/B → B/C → C/D	5
o	ABCD → A/BC → BCD → B/C → C/D	5
p	ABCD → A/BC → BC/D → B/C	4
q	ABCD → AB/C → B/CD → A/B → C/D	5
r	ABCD → ABC → BC/D → A/B → B/C	5
s	ABCD → A/BC → B/CD → B/C → C/D	5
t	ABCD → ABC → B/CD → A/B → B/C → C/D	6
u	ABCD → AB/C → BC/D → A/B → B/C	5
v	ABCD → AB/C → BCD → A/B → B/C → C/D	6

We consider these DSSs as the first level synthesis for a multicomponent distillation. They serve as the starting point for generation of distinct subspaces using different synthesis and intensification methods. We introduce the synthesis and intensification procedures for generation of the following distinct subspaces for a multicomponent distillation:

- the subspace of the sharp sequence configurations (subspace SSC)
- the subspace of the nonsharp sequence configurations (subspace NSSC)
- the subspace of the original thermally coupled configurations (subspace OTC)
- the subspace of the thermodynamically equivalent structures (subspace TES)
- the subspace of the intensified simple column configurations (subspace ISC)
- the subspace of the intensified distillation systems (subspace IDS)
- the subspace of the intensified dividing-wall columns (subspace DWC)

These distinct subspaces are all for multicomponent distillations, as illustrated in Figure 2.3.

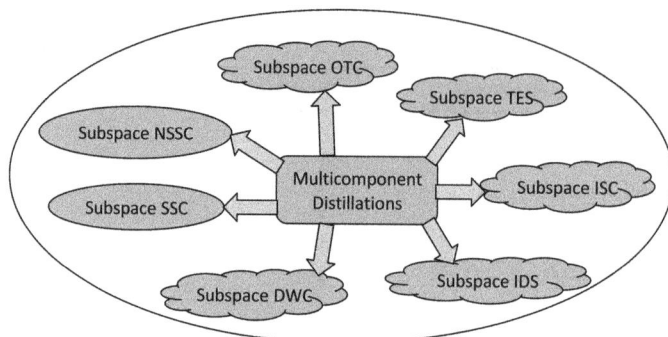

Fig. 2.3: Different subspaces of distinct distillation systems for a multicomponent distillation.

Some subspaces are well described in the literature (e.g., SSC subspace), whereas others have not yet been fully defined regarding all their alternatives (e.g., IDS and DWC subspaces).

2.7 Subspace of sharp sequence configurations

The alternatives in this subspace are straightforward to generate. The simple column configurations are directly generated from the DSSs with all sharp splits (type 1). By introducing a simple column for each of the sharp splits in the separation sequence, the simple column configuration is obtained. Therefore, the number of simple column configurations is equal to the number of sharp separation sequences calculated using Equation (2.1). Figure 2.4 shows that this subspace for a four-component mixture consists of five simple column configurations.

The distinct features of the simple column configurations can be summarized as follows (for an N-component mixture):
- minimum number of $N - 1$ simple columns
- $N - 1$ condensers and $N - 1$ reboilers, giving a total of $2(N - 1)$ heat exchangers
- $N - 1$ rectifying column sections and $N - 1$ stripping sections, giving a total of $2(N - 1)$ sections

Because the numbers of columns and heat exchangers are the dominant criteria for evaluating a distillation configuration, earlier distillation syntheses primarily focused on searching for the optimal simple column configuration for a given mixture. Specifically, a heuristic-based approach was the primary method [15–19].

Essentially, the synthesis of distillation sequences for this subspace is application oriented. There is no need for a special synthesis method to define the alternative structures because all the sharp sequences are known *a priori* with the number of alternatives predicted by Equation (2.1). Given a specific mixutre, the synthesis in-

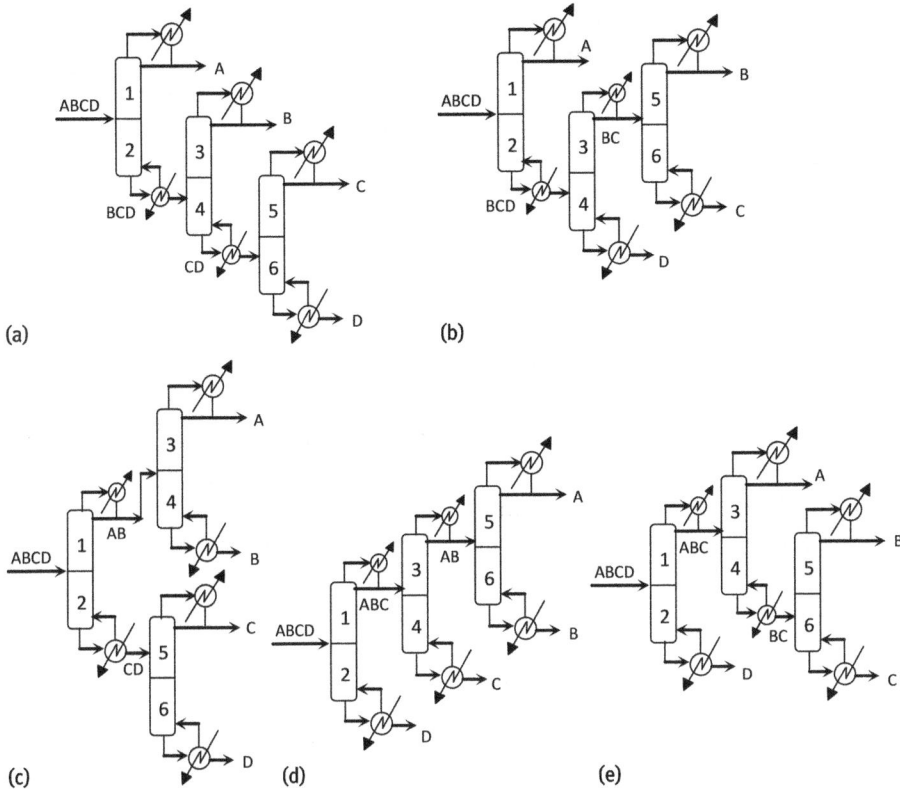

Fig. 2.4: SSC subspace of the sharp sequence configurations for a four-component mixture. (a)–(e) are corresponding to the DSSs of (a)–(e) in Table 2.2.

volves searching the subspace to determine which separation sequence is the optimal sequence.

Due to its simplicity, such simple column configuration is attractive to both academia and industry; thus, this subspace was extensively studied in earlier works. The minimum number of $N-1$ columns has served as the benchmark for synthesis of other possible multicomponent distillation schemes for an N-component distillation. Another benchmark is the number of condensers and reboilers used in simple column configurations, which is $2(N-1)$.

However, there is intrinsic thermodynamic inefficiency in distillation configurations with only sharp splits, which incurs high energy consumption. One way to reduce the thermodynamic inefficiency is to apply nonsharp splits. However, with nonsharp splits, a distillation configuration uses more simple columns and heat exchangers than one with only sharp splits. This is a big disadvantage in real applications. However, for ternary nonsharp sequences, Petlyuk et al. [43] showed that the number of columns and heat exchangers in the fully thermally coupled configuration can be the same as in the sharp configurations. The fully thermally coupled configuration from the nonsharp sequence has been proved to have the minimum energy requirement for ternary distillation [44]. Therefore, it is important to consider nonsharp sequences in the synthesis of multicomponent distillation configurations.

We will use the simple column configurations as the benchmark for synthesis of other distillations configurations. Specifically, the benchmark number of columns is $N - 1$ and the benchmark number of heat exchangers is $2(N - 1)$ for an N-component mixture.

2.8 Subspace of nonsharp sequence configurations

This subspace is for synthesis of distillation configurations from nonsharp separation sequences. The alternatives generated are called nonsharp sequence configurations (NSSCs). When nonsharp splits are introduced for mixtures of four or more components, the situation is much more complex. First, there are larger number of possible nonsharp sequences than sharp sequences. For example, for a four-component mixture, there are only 5 sharp sequences, but 17 nonsharp sequences, as shown in Table 2.1 [35]. Second, a separation sequence can have more than one nonsharp split, which gives many more degrees of freedom to arrange the distillation configurations. As a consequence, different types of distillation configurations with nonsharp splits can be obtained, which can have different number of columns and heat exchangers.

For such nonsharp sequences, for an N-component mixture, we present the synthesis method for generating nonsharp sequence configurations with $N - 1$ columns, which is equal to the number of columns needed for the sharp sequences. The synthesis procedure consists of two mechanisms to coordinate the simple columns defined by the distinct nonsharp separation sequences.

The starting point is to draw the simple column configuration (SCC) for the nonsharp sequence. Here, the simple column configuration is defined as that it is obtained by employing a simple column for each of the intended individual splits in the nonsharp sequence. A simple column can perform either a sharp or nonsharp split for a mixture with an overhead condender and a bottom reboiler, as shown in Figure 2.2. Thus, the number of simple columns is equal to the number of individual splits in the nonsharp sequence, and the number of heat exchangers is twice the number of individual splits. For example, the SCC for the nonsharp separation sequence of A<u>BC</u>D → A/BC → B/CD → B/C → C/D, is illustrated in Figure 2.5.

The number of intended individual splits in any nonsharp sequence is higher than that of the sharp sequences. As a consequence, for an N-component mixture, the number of columns in an SCC for any nonsharp sequence is higher than that of the sharp sequences (i.e., more than $N - 1$ columns). To achieve nonsharp sequence configurations with $N - 1$ columns, we introduce two mechanisms to reduce the number of simple columns.

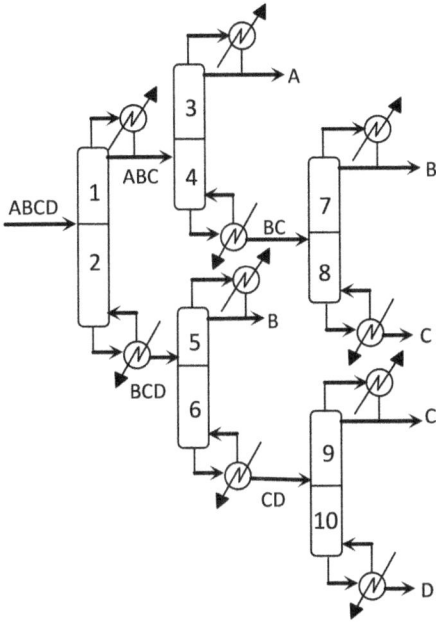

Fig. 2.5: SCC representation of the distinct separation sequence with nonsharp splits.

Mechanism 1: Open heat integration (OHI) is a mechanism to combine columns to reduce both the number of columns and the number of heat exchangers in the SCC.

This mechanism combines two columns through heat integration, which is implemented between a condenser and a reboiler associated with the same product (either the same single middle component or the same mixture of middle components). Mass communication is allowed at the heat-integrated location along the combined column. The same product is withdrawn at a single location of the combined column; therefore, it has the effect of simultaneous mass and heat integration. This mechanism can reduce both the number of columns and the number of heat exchangers of the SCC. For example, Figure 2.6 is obtained from Figure 2.5 through OHI of the two columns coproducing the middle component C. The number of columns is reduced from five to four.

Mechanism 2: Closed heat integration (CHI) is a mechanism to combine columns to reduce both the number of columns and the number of heat exchangers in the SCC.

This mechanism combines two columns through heat integration, which is implemented between a condenser and a reboiler associated with two different products. The product can be a single middle component or a mixture of middle components. Mass communication is not allowed at the heat-integrated location between the combined columns. The two different products are withdrawn at two different locations of the combined column. For example, Figure 2.7(a) illustrates the combination of two columns through CHI between the reboiler associated with the submixture BC and the condenser of the single product B. The submixture BC is withdrawn above the CHI

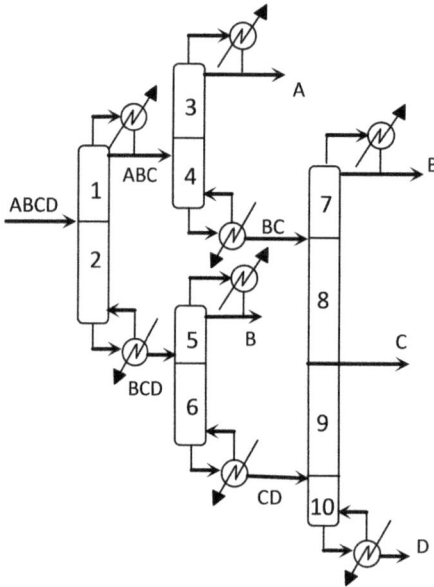

Fig. 2.6: Open-heat-integration mechanism for combining simple columns.

location and product B is withdrawn below the CHI location. CHI without mass communication prevents remixing of the two different products around the CHI location in the combined column. If such remixing can be dealt with by proper locations for withdrawal of the two products, CHI can be replaced with OHI. For example, Figure 2.7(b) illustrates the replacement of CHI with OHI for combining the two columns. Similar to the CHI in Figure 2.7(a), BC and B must be withdrawn at two different locations of the combined column in Figure 2.7(b). The two different withdrawing locations for BC

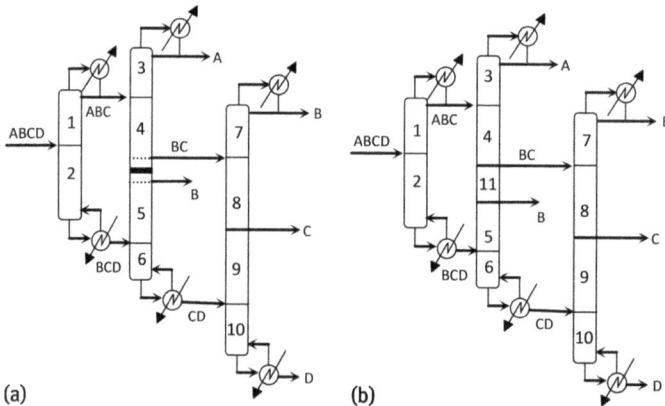

Fig. 2.7: Heat-integrated distillation configurations obtained by mechanisms 1 and 2. (a) Mechanism 2: closed heat integration (CHI), (b) Mechanism 1: open heat integration (OHI).

and B designate a small number of stages (or packings) between them, as shown by section 11 in Figure 2.7(b).

Like mechanism 1, mechanism 2 can also reduce both the number of columns and the number of heat exchangers in the SCC. Mechanisms 1 and 2 have been studied earlier in the synthesis of heat-integrated thermally coupled configurations [45, 46]. Few configurations with less than $N - 1$ columns were obtained for an N-component distillation. For example, for a four-component mixture, only one configuration was obtained with two columns (less than $N - 1$) through mechanisms 1 and 2.

Mechanisms 1 and 2 are universal mechanisms for coordinating simple columns in an SCC; thus, a systematic procedure has been developed for the synthesis of nonsharp sequence configurations (NSSC subspace), which is presented in Figure 2.8.

Implementing the systematic procedure for each of the nonsharp sequences in Table 2.2 (types 2 and 3), the complete subspace of the NSSCs for a four-component mixture is obtained, as shown in Figure 2.9.

When both CHI and OHI mechanisms can be used to achieve an NSSC with $N - 1$ columns, we prefer to use OHI than CHI to avoid remixing. In other words, to combine a condenser and a reboiler with the same product, the OHI mechanism is performed first, then, consider combining a condenser and a reboiler with the different products through the CHI mechanism. Note that a CHI can always be replaced with an OHI, as illustrated in Figure 2.9(s') for Figure 2.9(s).

The distinct features of the configurations in the NSSC subspace, compared with the SSC subspace, are as follows:

- The number of feasible NSSCs is much higher than that of the SSC subspace.
- The same number of $N - 1$ columns as the SSC subspace.
- The same number of $2(N - 1)$ heat exchangers as the SSC subspace.
- Each column produces an overhead product with a condenser and a bottom product with a reboiler, which is similar to the SSC subspace.
- At least one column has more than two column sections in the NSSC, whereas each column in the SSC subspace has only two sections.
- A product can be obtained from the intermediate location of a column in the NSSC, which is different from the SSC subspace where a product is obtained from either the overhead or the bottom of a column.
- The submixture BC is transported in one direction between the intermediate locations of two columns.
- In the SSC subspace, the number of condensers and reboilers associated with submixtures is $N - 2$. In the NSSC subspace, the number of condensers and reboilers associated with submixtures is more than $N - 2$.

In summary, the NSSCs for nonsharp sequences have similar structures as the SSCs for sharp sequences (i.e., the same number of $N - 1$ columns and each has an overhead condenser and a bottom reboiler). However, due to the nonsharp splits introduced,

Step 0. Given a zeotropic mixture with N components.

↓

Step 1. Formulate a distinct separation sequence (DSS) by introducing a set of intended individual splits for the given mixture with both sharp and nonsharp splits.

↓

Step 2. Represent the DSS with a simple column configuration (SCC) with one simple column for each of the individual splits in the DSS.

↓

Step 3. Identify the pairs of a condenser and a reboiler associated with the same product of the middle component(s), apply mechanism 1 to combine the columns.

↓

Step 4. Identify the pairs of a condenser and a reboiler associated with two different products of the middle component(s), apply mechanism 2 to combine the columns.

↓

Step 5. Do you want to examine another distinct nonsharp separation sequence (DSS)? If yes go to step 1, if no go to step 6. Repeat this cycle until all the feasible nonsharp sequences have been examined.

↓

Step 6. Store all the generated NSSC configurations in the NSSC subspace and stop.

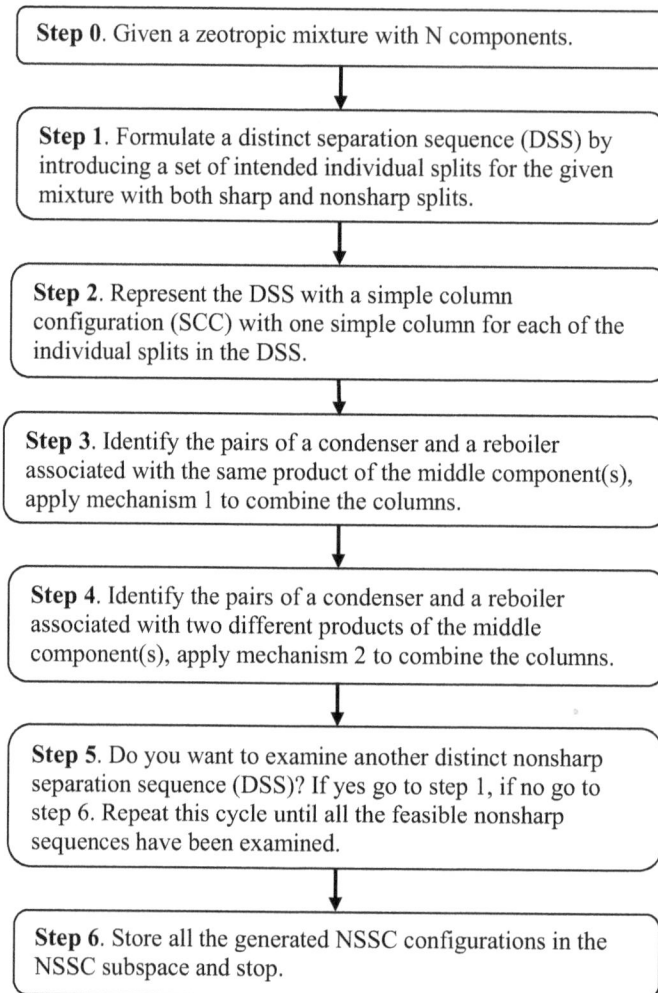

Fig. 2.8: Systematic procedure for synthesis of the NSSC subspace from nonsharp separation sequences.

NSSCs are thermodynamically more efficient and thus have the potential to reduce energy consumption.

It is clear that mechanisms 1 and 2, both based on the heat integration principle, are universal mechanisms for coordinating individual simple columns into a multicomponent distillation configuration. When dealing with simple columns for nonsharp sequences, it has the dramatic effect of reducing the number of simple columns to approach $N-1$.

Note that by combining columns with heat integration, the intermediate stream BC is transported in one-way communication between the columns. Moreover, pro-

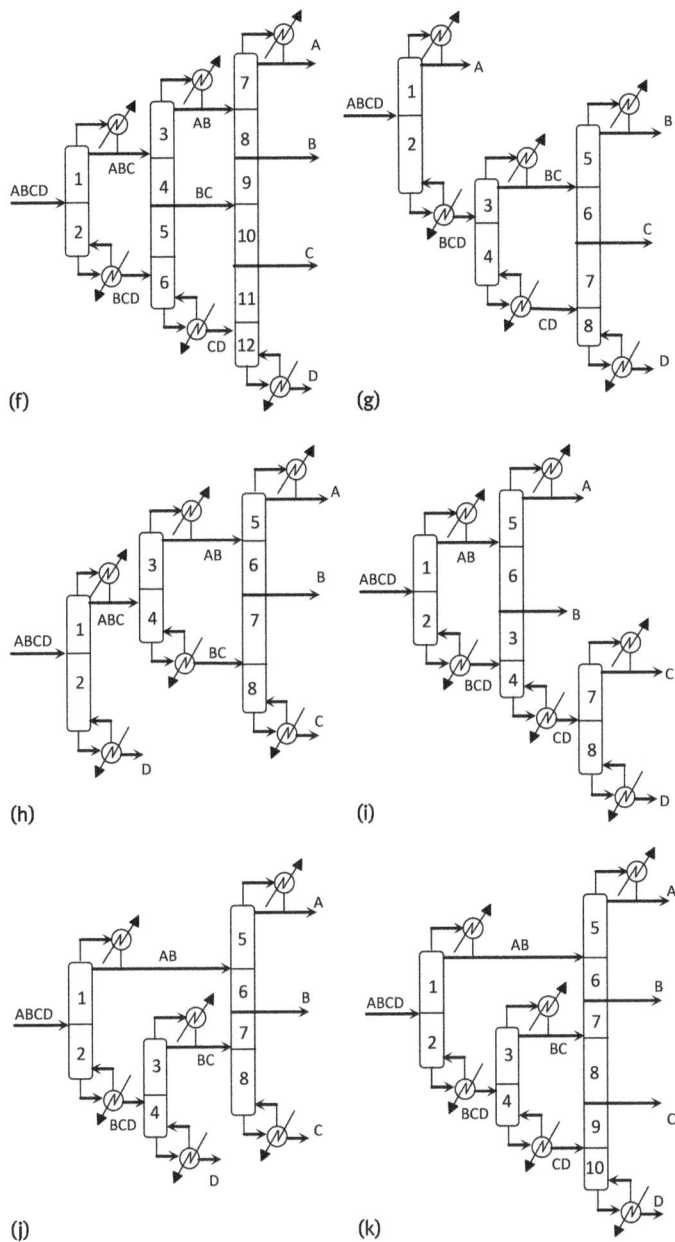

Fig. 2.9: NSSC subspace for the nonsharp separation sequences for a four-component mixture. (f)–(k) are corresponding to the DSSs of (f)–(k) in Table 2.2.

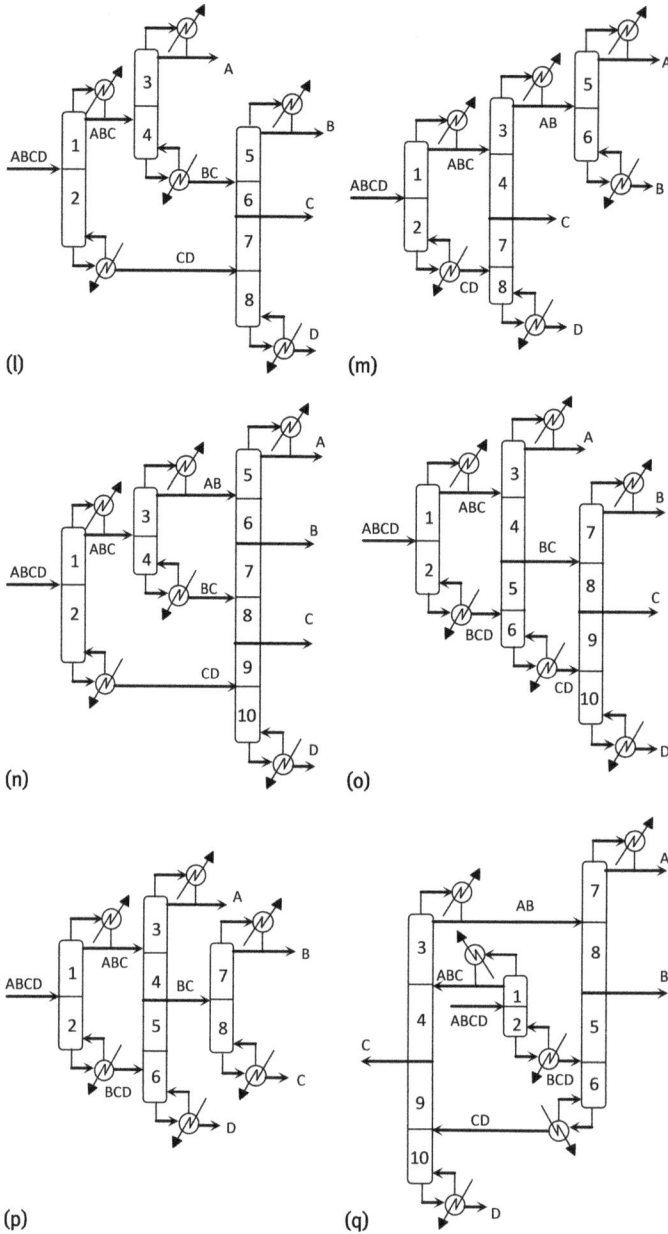

Fig. 2.9 (cont.): NSSC subspace for the nonsharp separation sequences for a four-component mixture. (l)–(q) are corresponding to the DSSs of (l)–(q) in Table 2.2

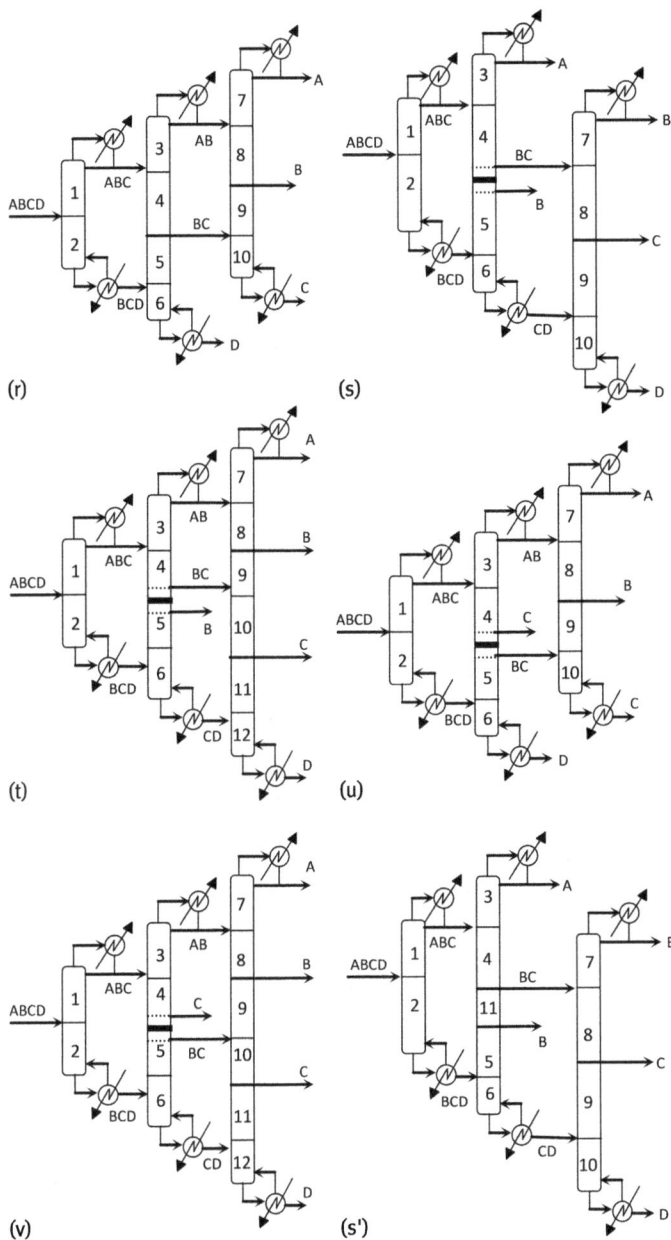

Fig. 2.9 (cont.): NSSC subspace for the nonsharp separation sequences for a four-component mixture. (r)–(v) are corresponding to the DSSs of (r)–(v) in Table 2.2. (s′) is from (s) by mechanism 1 instead of mechanism 2.

vided there are no thermal couplings introduced in the NSSC subspace, the configurations of the SSC subspace and the NSSC subspace are both called traditional distillation configurations (TDCs).

2.9 Subspace of the original thermally coupled configurations

This subspace is for synthesis of distillation configurations from the SSC subspace and the NSSC subspace. The distillation configurations generated in this subspace are called the original thermally coupled configurations (OTC). An OTC is defined as having the same columns as the original SSC or NSSC, but a reduced number of heat exchangers [35]. An OTC is distinguished from other possible thermodynamically equivalent structures (TES) by the column arrangements [35].

The NSSC subspace has shown that PS generates many more alternatives from the distinct nonsharp sequences than from the SSC subspace. The fundamental principle behind the NSSC subspace is that the nonsharp splits can reduce thermodynamic irreversibility and thus improve separation efficiency and reduce energy consumption. Meanwhile, mechanisms 1 and 2 are based on process integration principles and can reduce the number of simple columns and heat exchangers so that the NSSCs have the same number of $N - 1$ columns and $2(N - 1)$ heat exchangers as the SSCs.

Basically, for a zeotropic mixture, one can search the SSC and NSSC subspaces to find an optimal configuration based on the given evaluation indicator, such as minimum total vapor flow rate, total energy consumption, or total annual cost.

However, like introducing nonsharp splits instead of only sharp splits, from a PS point of view, one would naturally look for the fundamental thermodynamic principle that is universally applicable to any multicomponent distillations and can generate new distillation alternatives to improve thermodynamic efficiency. One such universal principle is the thermal coupling mechanism for multicomponent distillation. For any configuration in the SSC and NSSC subspaces, there are always condensers and/or reboilers associated with submixtures. For example, for the configuration of Figure 2.4(e) in the SSC subspace, the condenser of the first column is associated with the submixture ABC (condenser ABC), whereas the reboiler of the second column is associated with the submixture BC (reboiler BC). For the NSSC shown in Figure 2.9(o) in the NSSC subspace, one condenser and two reboilers are associated with submixtures: condenser ABC, reboiler BCD, and reboiler CD. Such condensers and reboilers are associated with those column sections whose functions are not for producing the final products but for generating the corresponding submixtures to the condensers and reboilers and then transporting to the subsequent columns. Even if these column sections do not produce final products, they still serve as either rectifying sections or stripping sections to perform prefractionations of the corresponding submixtures. However, such prefractionations are destroyed in these condensers and reboilers. Therefore, the condensers and reboilers associated with submixtures inevitably incur remixing, which increases thermodynamic irreversibility and reduces separation efficiency. To avoid such remixing, these condensers and reboilers can be eliminated.

The mechanism to eliminate a submixture condenser is such that the submixture vapor stream is directly transported to the subsequent column, while a liquid stream is taken from the subsequent column and transported to the prior column. These paired vapor and liquid streams in two-way transport between the two columns to replace a submixture condenser are called thermal coupling streams (or a thermal coupling). Similarly, the mechanism to eliminate a submixture reboiler is such that the submixture liquid stream is directly transported to the subsequent column, while a vapor stream is taken from the subsequent column and transported to the prior column. Again, the paired liquid and vapor streams in two-way transport between the two columns to replace a submixture reboiler are called thermal coupling streams (or a thermal coupling).

It is clear that thermal coupling provides a universal mechanism to coordinate mass and heat transport between the different columns in a multicomponent distillation configuration. The fundamental principle of thermal coupling is to reduce remixing and thus reduce the thermodynamic irreversibility of the separation system. Therefore, theoretically, thermally coupled distillation systems can improve the separation efficiency and reduce energy consumption.

Mechanism 3: Thermal coupling is a mechanism to eliminate the condensers and reboilers associated with submixtures and coordinate mass and heat transport between different columns with two-way thermal coupling streams, which will generate thermally coupled configurations.

For multicomponent distillation synthesis, the universal mechanism of thermal coupling was first used to generate the subspace of the 14 thermally coupled configurations from the subspace of 14 sharp SCCs for five-component distillations [32–34]. We also developed the systematic synthesis procedure to generate the complete subspace of the thermally coupled distillation configurations. Subsequently, a methodology for the synthesis of thermally coupled configurations for both sharp and nonsharp separation sequences was developed [35]. For a given traditional distillation configuration, a formula for predicting the number of thermal coupling configurations was developed [33]. This equation is valid for any configurations of SSC subspace and NSSC subspace. If we use N_{cr} to represent the total number of condensers and reboilers associated with submixtures, then the Equation (2.2) is derived for calculation of the number of thermally coupled distillation schemes for an N-component traditional configuration.

$$C_N = \sum_{j=1}^{N_{cr}-1} \frac{N_{cr}!}{j!(N_{cr}-j)!} + 1 = 2^{N_{cr}} - 1 \tag{2.2}$$

Where, C_N is the total number of thermally coupled configurations generated for an SSC or an NSSC. For $N_{cr} = 1$, $C_N = 1$. The term $\sum_{j=1}^{N_{cr}-1} \frac{N_{cr}!}{j!(N_{cr}-j)!}$ represents the number of thermally coupled schemes with a lower number of thermal couplings than the one in which all thermal couplings are introduced for the N_{cr} condensers and reboilers.

Applying mechanism 3, the generated thermally coupled configuration is defined as the original thermally coupled configuration (OTC), which has the same columns as the original SSC or NSSC. This distinguishs it from the other possible thermodynamically equivalent structures in which the column arrangements are different from the SSC or NSSC.

For example, the three thermally coupled configurations generated from the configuration of Figure 2.4(e) are shown in Figure 2.10.

The seven thermally coupled configurations generated from the configuration of Figure 2.9(o) are shown in Figure 2.11.

Notice that, so far, the thermal couplings are introduced to those condensers whose submixtures contain the most volatile component, and to those reboilers whose submixtures contain the least volatile component. Therefore, the thermal couplings are located at the two ends of the columns.

Now, pay attention to those NSSCs from nonsharp sequences where there is a submixture BC transferring between the intermediate locations of two columns. Figure 2.9(f, o, p, r) shows such NSSC configurations. There, the BC submixture is transported in one direction, which is due to the OHI mechanism to combine simple columns and which is done the same way as for the single middle component products (B or C). It is known that the heat integration is between the reboiler BC of the upper column and the condenser BC of the lower column. With the thermal coupling mechanism, instead of one-way transport, submixture BC can be transported in two-way thermal coupling streams, which can produce different configurations from those of the original NSSCs, as shown in Figure 2.12.

Obviously, the thermal coupling BC increases the total number of OTCs. For example, for the NSSC in Figure 2.9(o), Equation (2.2) gives the number of OTCs as seven, as shown in Figure 2.11. If the one-way transport submixture BC is replaced with two-way thermal coupling, then the total number of OTCs is 15. Therefore, the number of such

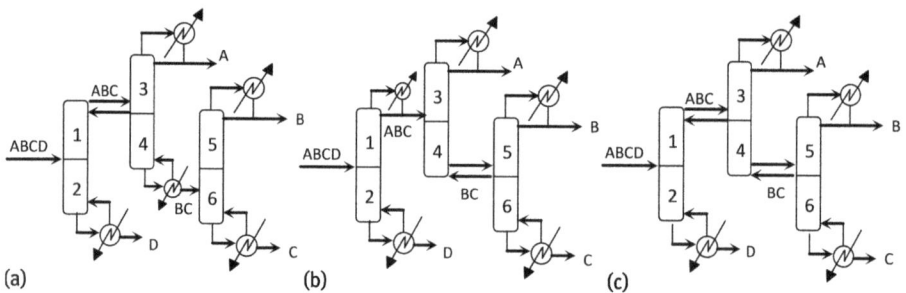

Fig. 2.10: Original thermally coupled configurations for the SSC of Figure 2.4(e). (a) eliminating condenser ABC, (b) eliminating reboiler BC, (c) eliminating both condenser ABC and reboiler BC.

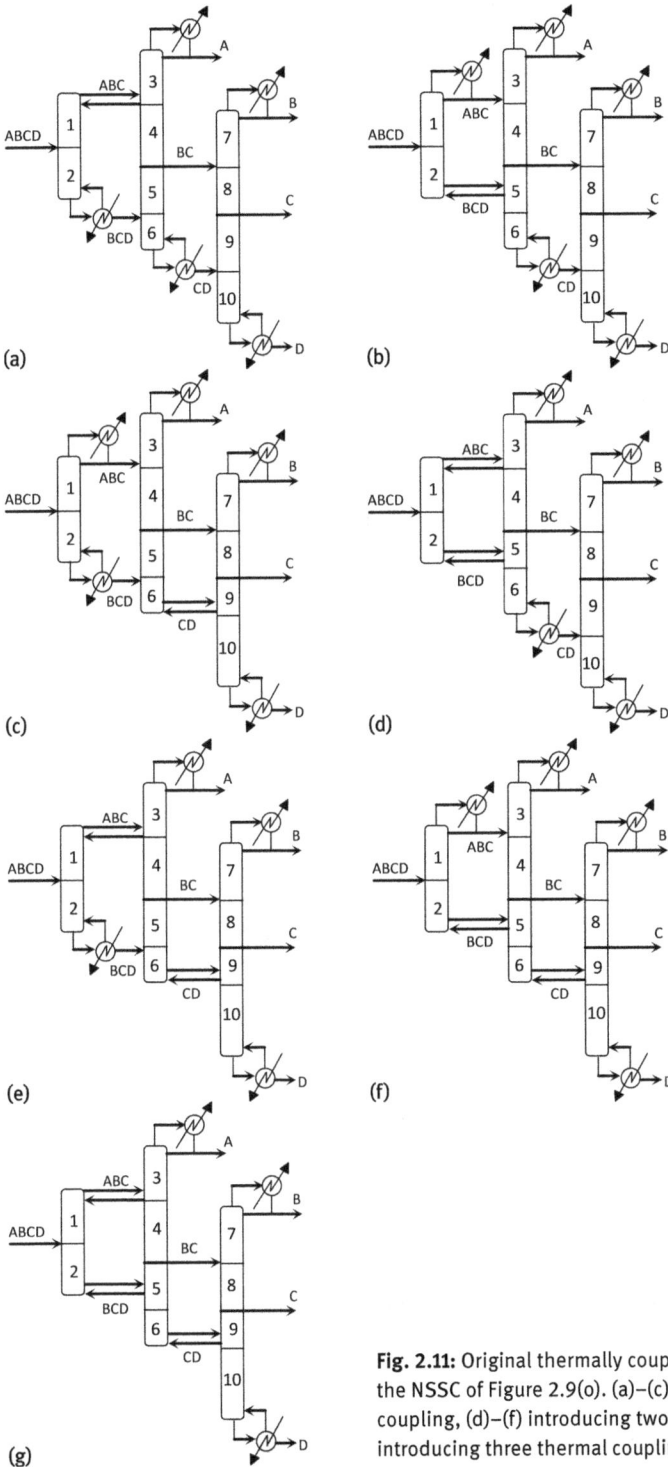

(a)

(b)

(c)

(d)

(e)

(f)

(g)

Fig. 2.11: Original thermally coupled configurations for the NSSC of Figure 2.9(o). (a)–(c) introducing one thermal coupling, (d)–(f) introducing two thermal couplings, (g) introducing three thermal couplings.

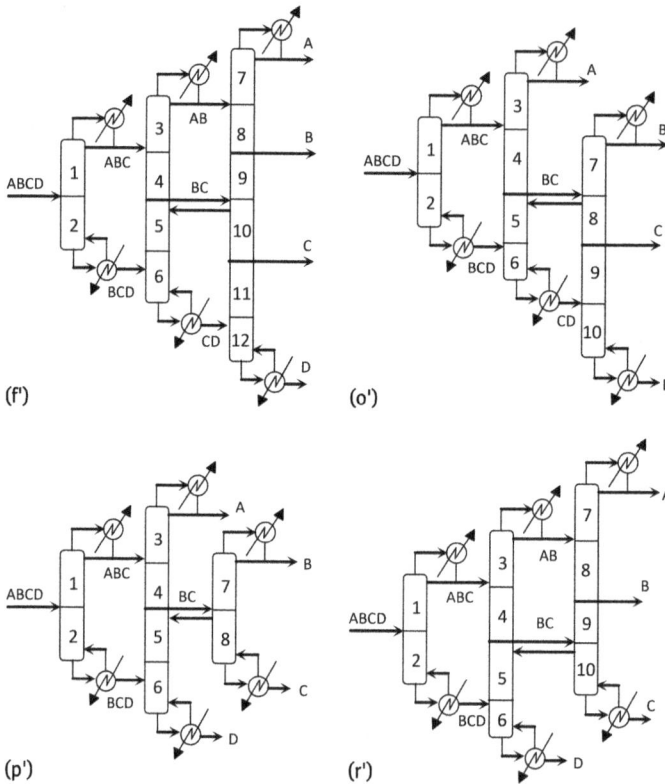

Fig. 2.12: NSSCs with two-way transporting of the intermediate submixture BC. (f'), (o'), (p') and (r') are corresponding to (f), (o), (p) and (r) in Figure 2.9.

intermediate transport submixtures can be added to generate the OTCs. If we use N_{im} to represent the total number of such intermediate submixtures in an NSSC, together with N_{cr}, the total number of OTCs can be calculated with the following equation:

$$OTC_N = \sum_{j=1}^{N_{cr}+N_{im}-1} \frac{(N_{cr} + N_{im})!}{j!(N_{cr} + N_{im} - j)!} + 1 = 2^{N_{cr}+N_{im}} - 1 \qquad (2.3)$$

When mechanism 3 is applied to all the configurations in the SSC and NSSC subspaces, a new OTC subspace can be generated. Therefore, with mechanism 3, a systematic procedure is developed for the synthesis of the subspace of the thermally coupled configurations from the SSC and NSSC subspaces. The procedure is presented in Figure 2.13.

The distinct features of the OTCs can be summarized as follows:
– An OTC has less heat exchangers than its SSC or an NSSC counterpart.
– An OTC has exactly the same columns as its SSC or NSSC counterpart.

A short summary for thermal coupling principle is as follows:

Step 0. Given a zeotropic mixture with N components.

↓

Step 1. Set the SSC subspace and the NSSC subspace of all the alternatives.

↓

Step 2. Select an SSC configuration or an NSSC configuration, identifying and counting the total number of condensers and reboilers associated with the submixtures as Ncr, and identify and counting the intermediate transport submixtures as Nim.

↓

Step 3. Apply equation 3 to calculate the total number of original thermally coupled configurations (OTC$_N$).

↓

Step 4. Replacing the identified Ncr condensers and reboilers and the identified Nim one-way intermediate transport submixtures with thermal couplings in a combinatorial way, and draw all the thermally coupled configruations.

↓

Step 5. Do you want to generate OTC for another SSC or NSSC? If yes go to step 2, if no go to step 6. Repeat this cycle until all the configurations in the SSC and the NSSC subspaces have been examined.

↓

Step 6. Store all the generated OTC configurations in the OTC subspace and stop.

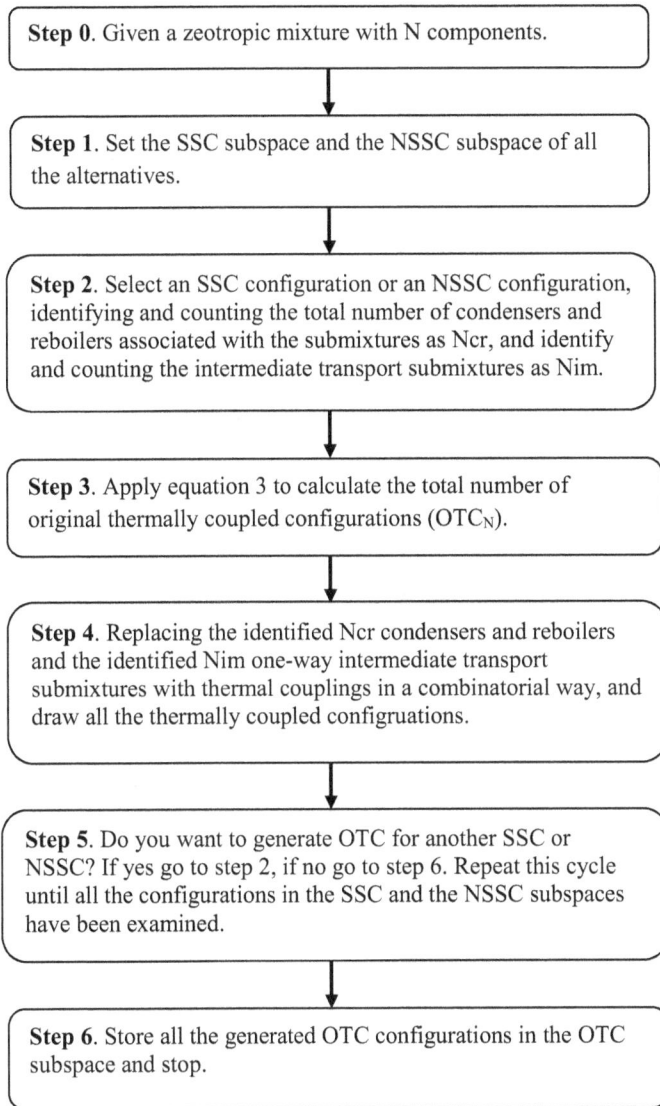

Fig. 2.13: Systematic procedure for synthesis of the OTC subspace from the SSC and NSSC subspaces.

- Thermal coupling is a mechanism to coordinate mass and heat transports between different columns.
- The fundamental of thermal coupling is to reduce remixing or thermodynamic irreversibility of the separation system.
- Following the systematic procedure of Figure 2.13, the complete subspace of the OTCs can be generated for all the configurations in the SSC and NSSC subspaces.

As discussed above, the introduced thermal couplings can theoretically improve the thermodynamic efficiency, thus OTCs have the potential to reduce the energy consumption than traditional configurations. Moreover, the reduced number of heat exchangers is also expected to reduce the capital costs. Most significantly, such thermal couplings can give the structural degrees of freedom to rearrange the column sections, which will generate thermodynamically equivalent structures and dramatically change the column equipment designs.

2.10 Subspace of the thermodynamically equivalent structures

This subspace is for synthesis of the distillation configurations from the OTC subspace. The distillation configurations generated in this subspace are called thermodynamically equivalent structures (TESs). A TES is defined as having different structural arrangement of column sections among the columns than the OTC [35, 47].

The design of a multicomponent distillation system should consider individual columns and their combinations as well as coordination between all the column sections. After all, it is the column sections that perform the corresponding functions in a multicomponent distillation system. In the NSSC subspace, by mechanisms 1 and 2, a column in an NSSC can have more than two column sections due to combinations of simple columns. In the OTC subspace, by mechanism 3, the column units in an OTC are the same as those of its corresponding NSSC. However, the interconnections of the column units in an OTC are changed by the introduced thermal coupling streams. Such thermal coupling streams perform simultaneous mass and heat transports between the thermally linked columns.

In an earlier work, we observed that there are intrinsic uneven distributions of vapor and liquid flows between columns in OTCs [47]. A distinct feature of an OTC is that it has structural degrees of freedom to rearrange its column sections. The number of structural degrees of freedom in an OTC is equal to the number of thermal couplings introduced into its NSSC. The structural degrees of freedom in a thermally coupled configuration are related to those movable column sections designated by the introduced thermal couplings. The rearrangements of the column sections can produce TESs. The TESs provide the opportunity to redistribute the vapor and liquid flows between the columns compared to the OTCs. This not only improves column equipment design, but also improves the hydraulic performance and operability of a thermally coupled distillation system [47].

To produce all the possible TES structures, we developed the following two rules to predict the number of structural degrees of freedom and the number of movable column sections:

Rule 1: the number of structural degrees of freedom rule

The number of structural degrees of freedom in a thermally coupled distillation configuration is equal to the number of thermal couplings introduced into its traditional distillation configuration. Each of the structural degrees of freedom corresponds to one of the thermal couplings.

Rule 2: the number of movable column sections rule

Each of the thermal couplings in a thermally coupled distillation configuration designates a movable column section (or a counterpart of movable column sections). The number of movable column sections (including the counterparts of movable column sections) in a thermally coupled configuration is equal to the number of the structural degrees of freedom. In other words, the number of movable column sections (including the counterparts of movable column sections) in a thermally coupled distillation configuration is equal to the number of thermal couplings introduced into its NSSC.

A general mechanism was discovered and applied for generating thermodynamically equivalent structures. The mechanism is based on the functional equivalence of the corresponding column sections between the distillation configurations [31–35, 47].

Mechanism 4: Mechanism 4 is for the rearrangement of column sections among the columns in the thermally coupled configuration. The introduced thermal couplings create the structural degrees of freedom that designate the movable column sections. The movable column sections create the opportunities to recombine and rearrange the column sections among the columns. The recombination and rearrangement of column sections is a universal mechanism to produce thermodynamically equivalent structures for a thermally coupled configuration.

A formula has also been developed to calculate the number of TES structures for an OTC. For any OTC of an N-component distillation, as long as the number of thermal couplings (N_{TC}) is determined, the number of all of its possible TES structures (TTC$_N$) is readily calculated using the following equation:

$$\text{TTC}_N = \sum_{i=0}^{N_{TC}} C^i_{N_{TC}} = 2^{N_{TC}} \tag{2.4}$$

Notice that the number of TESs calculated from Equation (2.4) contains the OTC. Notice also that the OTC is thermodynamically equivalent to all other TESs in terms of thermal couplings. However, the columns in the OTC are the same as in its corresponding traditional distillation configuration (TDC; either SSC or NSSC). There is no reaarangement of the column sections in the OTC. All the other TESs have different columns from those in the OTC and its TDC.

With mechanism 4 and the above two rules, a systematic procedure to generate the complete TES subspace is developed, as shown in Figure 2.14.

With the systematic procedure shown in Figure 2.14, we can generate TES systems for any OTC in the OTC subspace. Taking the OTC in Figure 2.10(c) as an example, this OTC was generated from the SSC in Figure 2.4(e). In this OTC, two thermal couplings are introduced through eliminating condenser ABC and reboiler BC. Therefore, two structural degrees of freedom are created, which make column sections 3 and 6

Step 0. Given a zeotropic mixture with N components.

Step 1. Set the OTC subspace of all the alternatives.

Step 2. Select an OTC configuration, identifying and counting the total number of thermal couplings (N_{TC}).

Step 3. Apply equation 4 to calculate the total number of thermodynamically equivalent structures (TTC_N).

Step 4. Identify the corresponding movable column sections for all the thermal couplings in the OTC, and determine the distinct combinations of the structural degrees of freedom.

Step 5. Determine the specific movable sections for each of the distinct combinations and define the structure of each TES. Draw the TES configurations from the defined movable column sections.

Step 6. Do you want to generate TES for another OTC ? If yes go to step 2, if no go to step 7. Repeat this cycle until all the configurations in the OTC subspace have been examined.

Step 7. Store all the generated TES configurations in the TES subspace and stop.

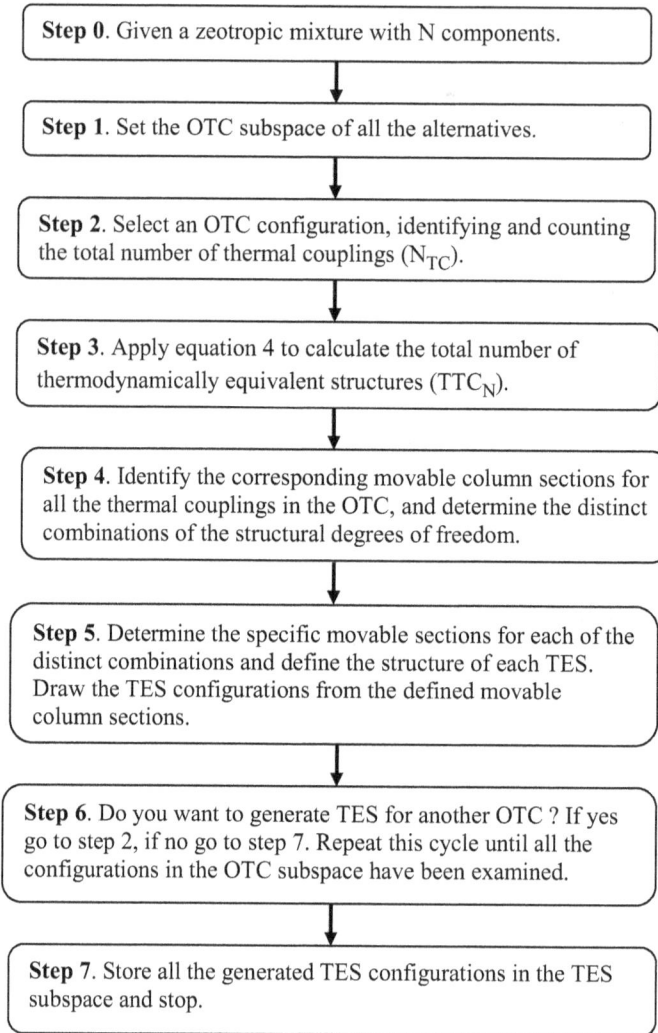

Fig. 2.14: Systematic procedure for synthesis of the TES subspace from the OTC subspace.

movable. We can move the movable sections 3 and 6 separately or together, which generates three TES structures, as shown in Figure 2.15.

Considering the OTC in Figure 2.16(a) as another example, this OTC was generated from the NSSC in Figure 2.9(o). Three heat exchangers are eliminated and replaced with three thermal couplings. Eliminating condenser ABC makes column section 3 movable, eliminating reboiler BCD makes column section 6 movable, and eliminating reboiler CD makes section 10 movable. The one-way intermediate transport of BC sub-mixture is replaced with two-way thermal coupling streams, which makes a pair of column sections (7, 8 + 9) movable. Therefore, in total four structural degrees of free-

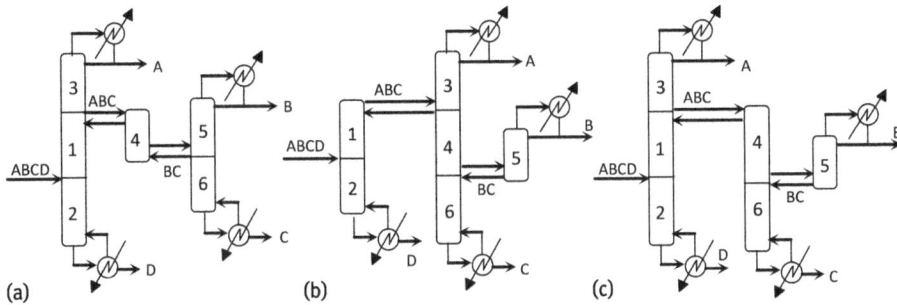

Fig. 2.15: TESs for the OTC of Figure 2.10(c). (a) move section 3, (b) move section 6, (c) move both sections 3 and 6.

dom are created. By Equation (2.4), it is calculated that there are 16 TES structures in total. These 16 TESs are presented in Figure 2.16.

Of the 16 TESs, Figure 2.16(a) is the OTC without section movement. Figure 2.16(b–e) is obtained by once using one structural degree of freedom. Figure 2.16(f–k) is obtained by once using two structural degrees of freedom. Figure 2.16(l–o) is obtained by once using three structural degrees of freedom. Finally, Figure 2.16(p) is obtained by simultaneously using all of the four structural degrees of freedom. Like Figure 2.15(c), Figure 2.16(p) is the thermodynamically equivalent side-column structure of the OTC in Figure 2.16(a), which was obtained by simultaneous movement of all the movable column sections. For any OTC, there is one unique thermodynamically equivalent side-column structure. A special subspace of such thermodynamically equivalent side-column arrangements of thermally coupled configurations has been formulated for four-component mixtures [35].

The distinct features of the TES structures are as follows:
- They have the same number of columns as the OTC.
- They have different arrangements of column sections compared with columns from the OTC.
- A column in the TES can have a very different number of column sections compared with a column in the OTC.
- Products can be produced from different columns than in the OTC.

The generation of the TES structures tells us that, by introducing thermal couplings to coordinate the mass and heat transports between the columns, the column sections do not belong to the column units as in the NSSC. The functional column sections can be flexibly recombined to formulate different column units. This gives an approach for optimal design of a multicomponent distillation system, which can be done by determination of the optimal arrangement of column sections in the TESs. This will not only ensure savings in energy and capital costs, but also improve the hydraulic performance and operability of the thermally coupled distillation system [47].

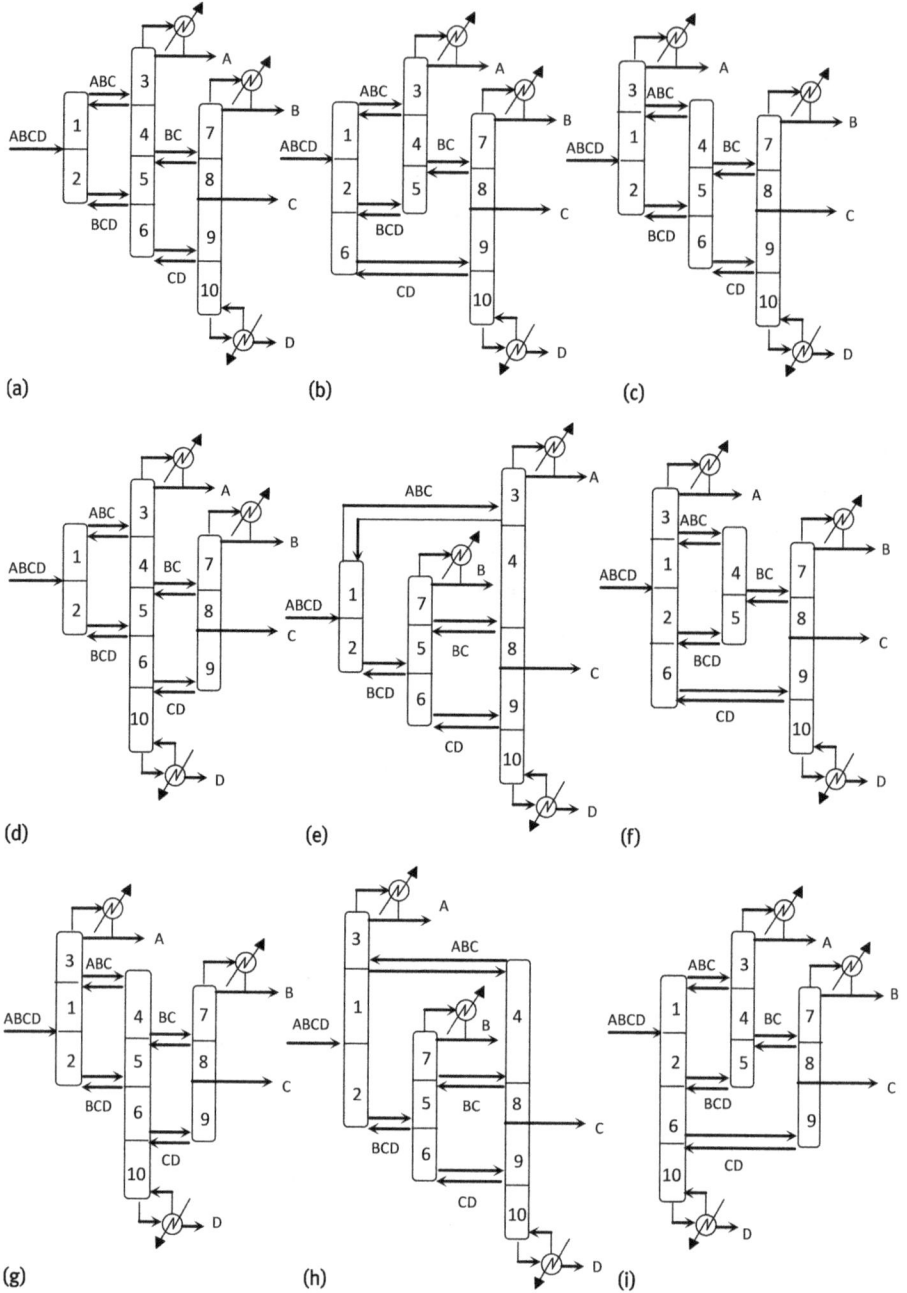

Fig. 2.16: The 16 TESs for the OTC of Figure 2.16(a).

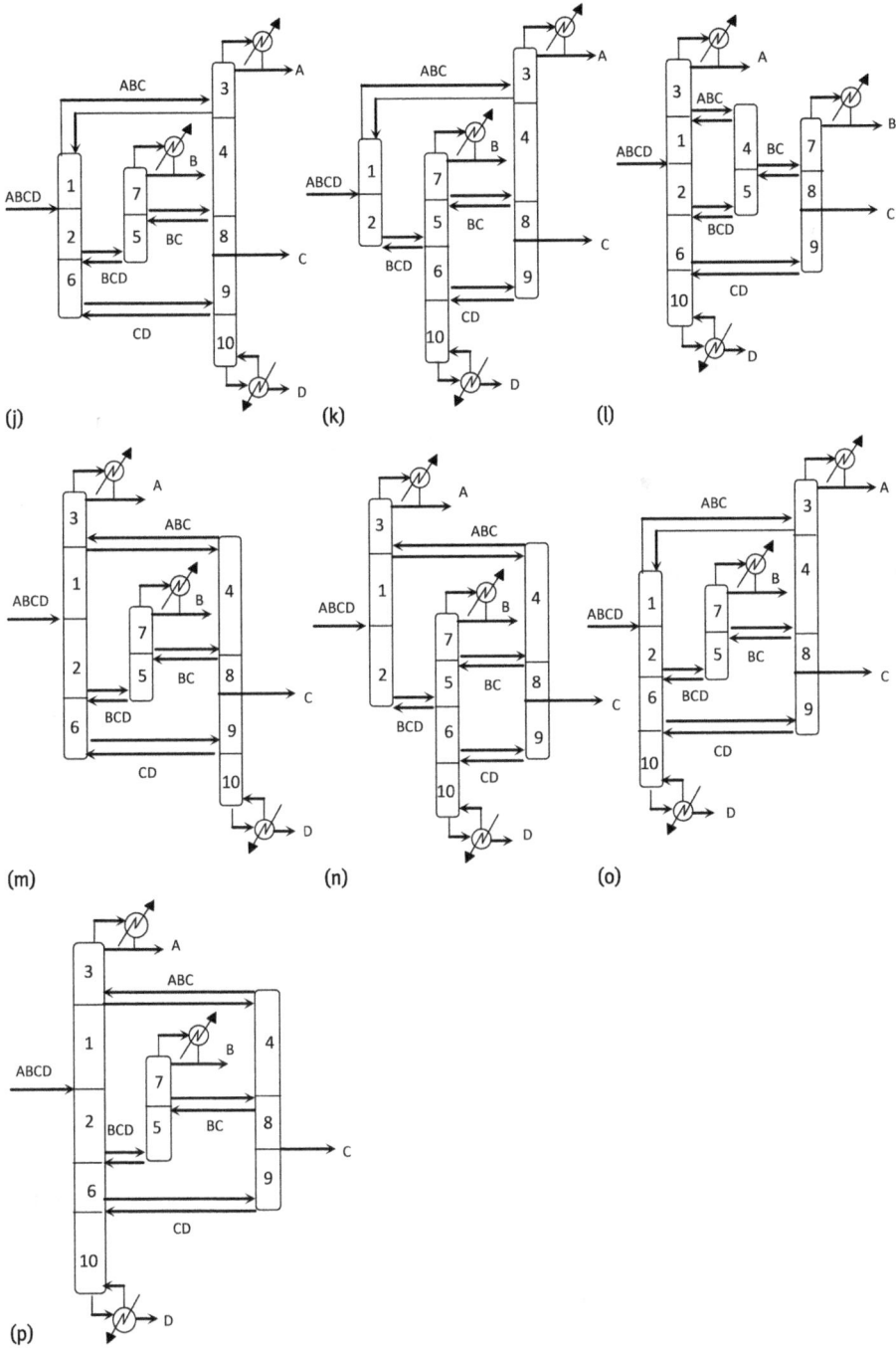

Fig. 2.16 (cont.): The 16 TESs for the OTC of Figure 2.16(a).

Thermal coupling is not just used to eliminate condensers and reboilers, it is essential for coordinating the mass and heat transports simultaneously among the functional column sections in the distillation system. Thus, it is a mechanism to simultaneously change the software and hardware assembly mechanism in a distillation system. Such simultaneous changes in software and hardware has resulted in dramatic changes in the recombination of functional sections, which has manifested in TESs.

However, a TES does not change the number of columns in a distillation system. As a consequence, a TES has the same number of columns as its OTC and its NSSC.

2.11 Intensified distillation systems with fewer columns

This subspace is for production of intensified distillation systems (IDSs) with fewer columns.

In terms of PI, the desire is always to use less to do the same or more. For multicomponent distillation, given the mixture and required products of the separation task, it is intriguing to pursue PI with the aim of using less equipment units to accomplish the task.

Distillation columns are some of the most expensive items of equipment in process industries; therefore, it is an attractive achievement for PI to result in a distillation system that uses fewer columns to accomplish the same multicomponent distillation task. This section focuses on the subspace of IDSs with fewer columns. The benchmark number of columns in a multicomponent distillation configuration is the SSC. For an N-component mixture, the SSCs use $N-1$ columns to produce N products, which is the minimum number of simple columns. Therefore, it serves as the benchmark for synthesis of IDSs. An IDS is defined as a system that can produce at least N products with less than $N-1$ columns for an N-component distillation. Because energy efficiency and capital investment are the dominant criteria for evaluation, a distillation configuration with less than $N-1$ columns for an N-component separation is considered an intensified system with the potential to reduce both energy and capital costs.

We consider two types of distinct intensified systems with fewer columns, which formulate two subspaces: the intensified simple column configurations (ISC subspace), and the intensified distillation systems (IDS subspace).

These two subspaces show that, for multicomponent distillation, it is proper to use column sections as the individual functional units for constructing the whole system. Both a rectifying section and a stripping section can serve as functional units. The functionality of a column section depends on the situation; it could serve as a transport section, transient function, or appendix for a specific separation case with specified feed mixture and products.

2.11.1 Subspace of the intensified simple column configurations

Taking into account the simplicity of the distillation configurations in the SSC and NSSC subspaces, as shown in Figures 2.4 and 2.9, this section presents what we call the intensified simple column configuration (ISC). The ISC is defined as having similar structural simplicity to the SSC and NSSC in that each column produces an overhead product with a condenser and a bottom product with a reboiler, but it uses less columns than the SSCs and NSSCs (less than $N - 1$) [48]. The synthesized ISC has a similar structure to the TDC but uses less than $N - 1$ columns.

The mechanism to produce ISCs is to generate TES structures in which there are single-section side columns that serve for transport of the submixtures. In any multicomponent distillation configuration, there are a certain number of column sections that give the final products and also a certain number of column sections that give submixtures instead of final products. We designate the former as *product column sections* and the latter as *nonproduct column sections*. In TESs, both the product sections and nonproduct sections can be recombined and rearranged so that individual columns can have a different number of column section(s). There are some special TESs in which at least one single-section side column consists of a nonproduct section. We can generate such special TESs from the two SSCs in Figure 2.4(b, e). Their OTCs, which can generate the special TESs, are generated as shown in Figures 2.17 and 2.18, respectively.

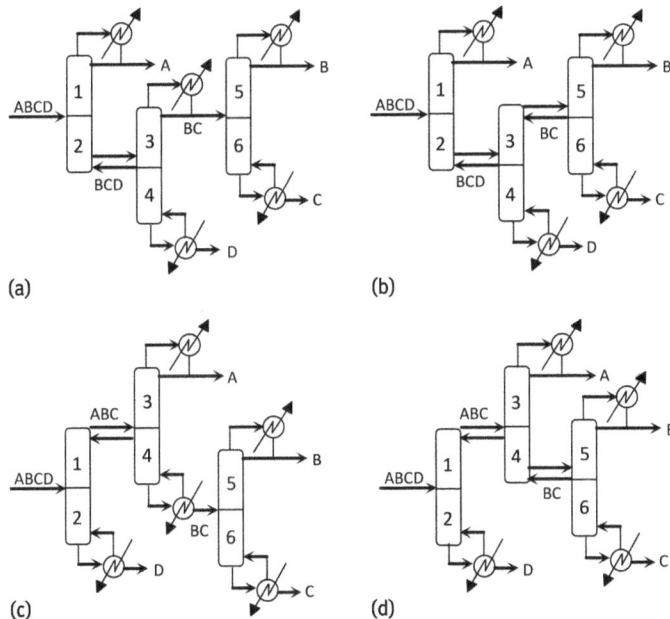

Fig. 2.17: OTCs for generation of the special TESs in Figure 2.18.

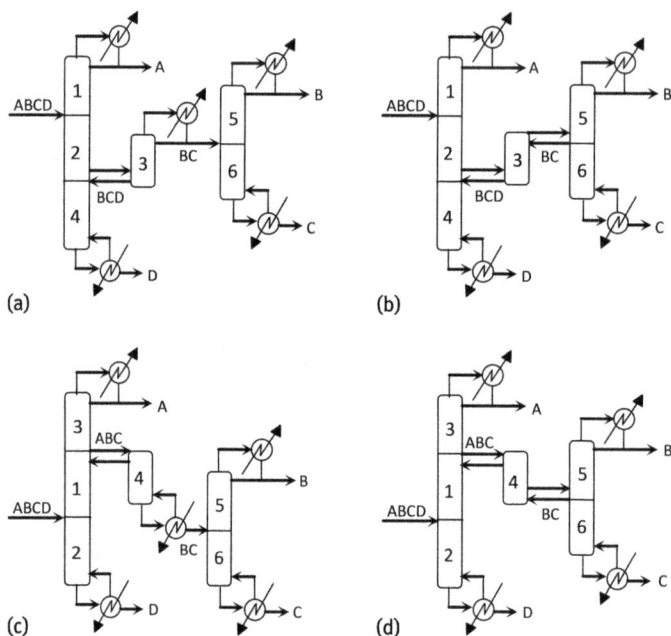

Fig. 2.18: The special TESs generated from the OTCs in Figure 2.17.

For the TESs in Figure 2.18, the single-section side columns do not produce any products, but mainly transport intermediate submixtures between the columns. In Figure 2.18(a), the side-rectifier transports the submixture BC in one direction after the condenser; we define it as a *one-way transport side-rectifier* (TSR1). In Figure 2.18(b), the side-rectifier transports the submixture BC in two directions through thermal coupling streams; we define it as a *two-way transport side-rectifier* (TSR2). Similarly, in Figure 2.18(c), the side-stripper transports the submixture BC in one direcction after the reboiler; we define it as a *one-way transport-side-stripper* (TSS1). In Figure 2.18(d), the side-stripper transports the submixture BC in two directions through thermal coupling streams; we define it as a *two-way transport side-stripper* (TSS2).

Functionally, the side-rectifiers in TSR1 and TSR2 in Figure 2.18(a, b) are for further rectifying the heaviest component D in the mixture BCD, so that a mixture with only intermediate components B and C is transported to the next column. However, depending on the relative volatilities of the feed components and on product purity requirements, when the relative volatility between C and D is large and separation between C and D is easy, a mixture with only B and C can be withdrawn and the side column can be eliminated. Similarly, the side-strippers in TSS1 and TSS2 in Figure 2.18(c, d) further strip the lightest component A in the mixture ABC, so that a mixture with only intermediate components B and C is transported to the next column. Also, depending on the relative volatilities of the feed components and on product purity requirements, when the relative volatility between A and B is large and separation between A and B is easy, a mixture with only B and C can be withdrawn and the side column can be eliminated. Therefore, in all cases in Figure 2.18, the side columns can be eliminated to allow direct submixture transport between columns. Transport of the submixture is one-way when an TSR1 or TSS1 is eliminated, as shown in Figure 2.19(a, c). Correspondingly, when an TSR2 or TSS2 is eliminated, transport of the submixture is two-way, as shown in Figure 2.19(b, d).

For the derived configurations shown in Figure 2.19, each column produces two products, one distillate from the top condenser and another from the bottom reboiler. Therefore, they have similar structural simplicity to the SCCs. However, both the number of columns and the number of heat exchangers are less than that of their counterpart SSCs. Therefore, they fulfill the PI principle that the systems can accomplish the same separation task with a reduced number of equipment units. We define the configurations derived in Figure 2.19 as ISCs.

From the derivation of ISCs for quaternary mixtures, as shown in Figure 2.19, it is clear that the ISCs are obtained from the corresponding TESs shown in Figure 2.18 by eliminating the single-section transport side columns. There are the following two criteria for TES structures that can lead to ISCs:

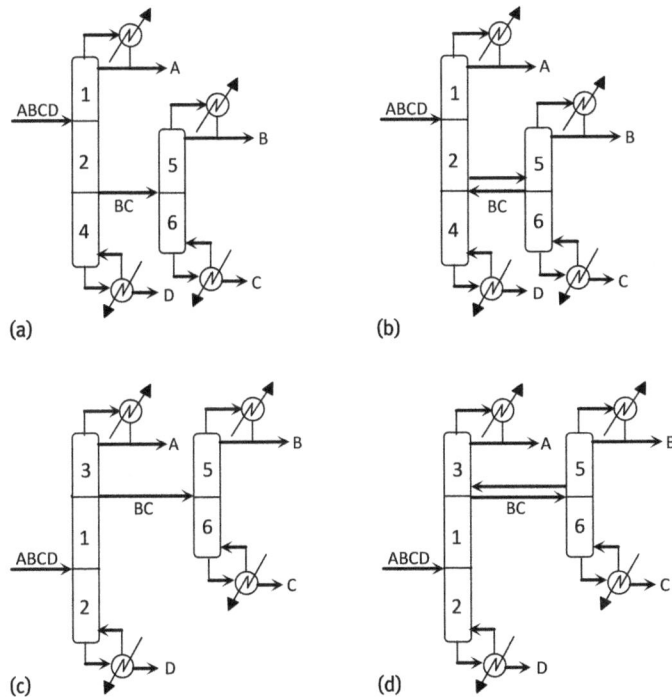

Fig. 2.19: ISCs derived from the TES configurations shown in Figure 2.18.

Criterion 1: The TES must contain single-section transport side columns.

Criterion 2: Except for the transport side columns, each of the other columns in the TES must produce an overhead product with a condenser and a bottom product with a reboiler (i.e., have the features of a simple column).

Figure 2.18 also shows for TESs that fulfill these two criteria, there are the following four cases of single-section transport side columns that can produce the ISCs:

Side column case 1: TES structure with one-way transport side-rectifier (TSR1)

Side column case 2: TES structure with two-way transport side-rectifier (TSR2)

Side column case 3: TES structure with one-way transport side-stripper (TSS1)

Side column case 4: TES structure with two-way transport side-stripper (TSS2)

Therefore, to achieve the possible ISCs for an N-component mixture, we need to systematically examine the TESs and identify those that fulfill the two criteria. This gives mechanism 5 for PI to synthesize ISCs.

Mechanism 5: ISCs are generated from TESs that include single-section side columns. When the single-section side columns only serve to transport intermediate submixtures, they can be eliminated to generate the ISC systems.

For any SSC or NSSC, we can generate all of its OTCs and TESs by the systematic precedures presented in Figures 2.13 and 2.14, respectively. Therefore, a systematic procedure can be formulated for step-by-step synthesis of ISCs, which is presented in Figure 2.20. Without further clarification, the systematic procedure aims to identify those TES structures that fulfill the two criteria and contain at least one of the four cases of the transport side columns.

A distinct feature of the NSSCs in Figure 2.9 is the combination of simple columns. To generate all the possible ISCs, we should not combine simple columns where there is a paired condenser and reboiler involving the same intermediate submixture. Depending on the number of such intermediate submixtures, there is flexibility to generate different NSSCs that contain $N - 1$ or more than $N - 1$ columns, both of which can generate the ISC systems.

The procedure for synthesis of ISCs for the nonsharp DSS sequence (p) in Table 2.2 is outlined as follows for a four-component mixture:

Steps 1–3: The NSSC with four columns (more than $N - 1$) is generated, as shown in Figure 2.21; the two columns involving submixture BC are not combined.

Step 4: Generation of the OTCs is straightforward, achieved by replacing the heat exchangers associated with submixtures with thermal coupling streams. Figure 2.22(a) shows the OTC obtained by eliminating condenser ABC and reboiler BCD. The reboiler BC and condenser BC in Figure 2.21 can be replaced with thermal coupling streams, individually or simultaneously, to generate the OTCs shown in Figure 2.22(b–d).

Step 5: TESs are generated for the OTCs in Figure 2.22, which are shown in Figure 2.23.

Steps 6 and 7: The generated ISCs are shown in Figure 2.24.

Step 0. Given a zeotropic mixture with N components.

Step 1. Formulate a distinct separation sequence (DSS) by introducing a set of intended individual splits, assign a simple column for each of the intended individual splits (both sharp and nonsharp splits).

Step 2. Determine the traditional distillation configurations by sequencing or combining the simple columns for the introduced individual splits in step 1.

Step 3. Select a NSSC from the generated NSSCs in step 2 for the formulated distinct separation sequence.

Step 4. Generate the original thermally coupled configurations (OTCs) for the selected NSSC in step 3.

Step 5. Generate the thermodynamically equivalent structures (TESs) for the OTCs generated in step 4.

Step 6. Identify the TES structures among the TESs generated in step 5 that fulfill the two criteria.

Step 7. Generate the intensified simple configurations (ISCs) by eliminating the transport side columns in the identified TES structures in step 6. Store the generated ISCs in step 10.

Step 8. Select a remaining NSSC and repeat steps from 4 to 7 until all the generated NSSCs in step 2 for the formulated distinct separation sequence have been selected.

Step 9. Do you want to examine another distinct separation sequence (DSS)? If yes go to step 1, if no go to step 10. Repeat this cycle until all the feasible separation sequences have been examined.

Step 10. Summarize all the generated ISC configurations and stop.

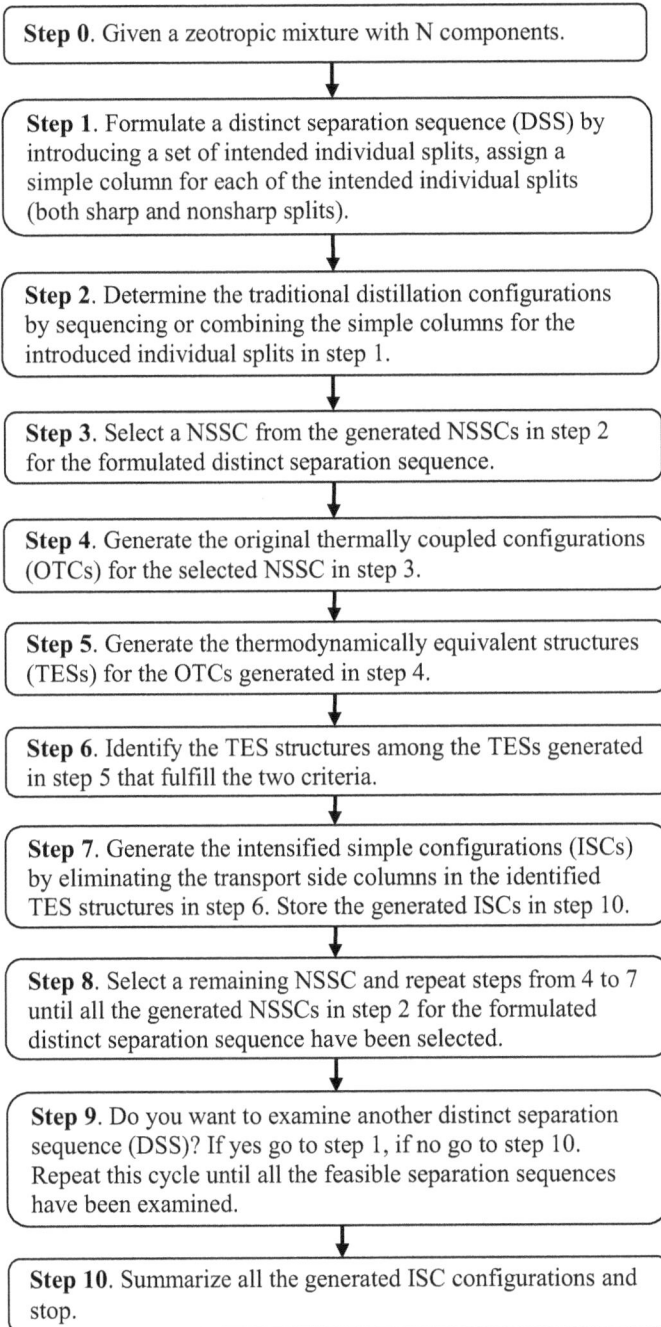

Fig. 2.20: Generalized procedure for systematic synthesis of the ISCs.

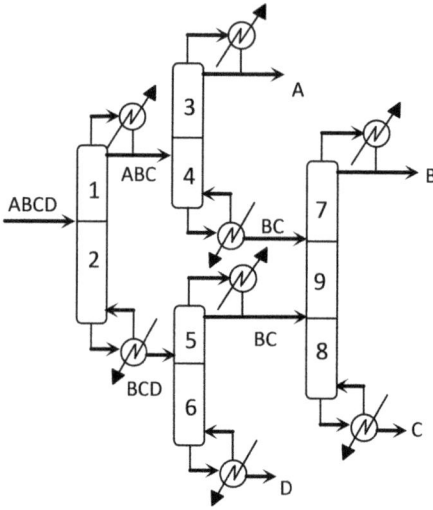

Fig. 2.21: NSSC with four columns for the non-sharp DSS sequençe (p) in Table 2.2.

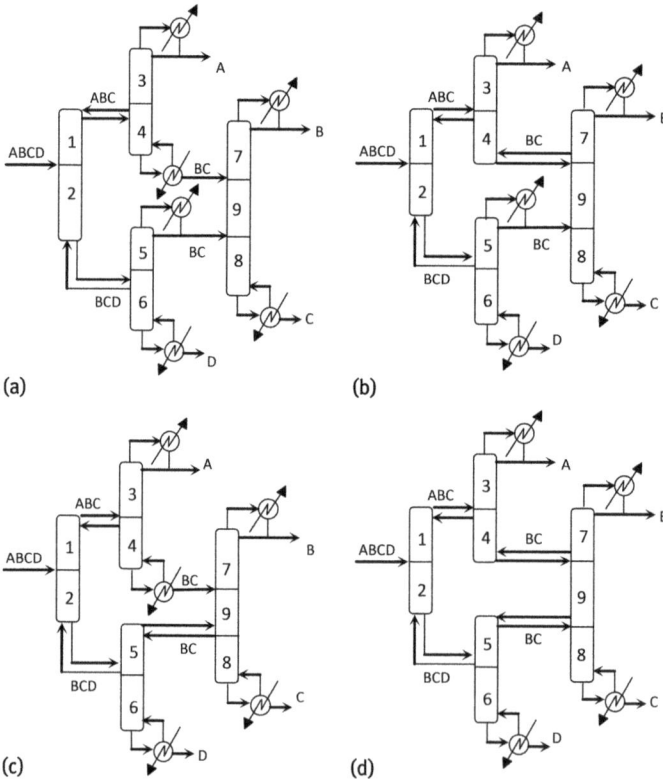

(a)

(b)

(c)

(d)

Fig. 2.22: OTCs for the NSSC in Figure 2.21.

Steps 8–10: For this example of the selected DSS, this is the only NSSC that can generate ISCs, so the procedure stops here.

The following distinct structural features can be observed for the ISCs:
- An ISC uses fewer columns (less than $N - 1$) for an N-component distillation than an SSC or NSSC.
- Each column has an overhead condenser and a bottom reboiler.
- There are submixtures transferring between intermediate locations of columns. An intermediate submixture can transport between two columns in one or two directions.
- A column in an ISC can have more than two sections.

Note that the ISCs with two-way transport of an intermediate submixture have no structural degrees of freedom to rearrange the column sections. Unlike the thermal coupling two-way communications in OTCs, the two-way transport of a submixture in

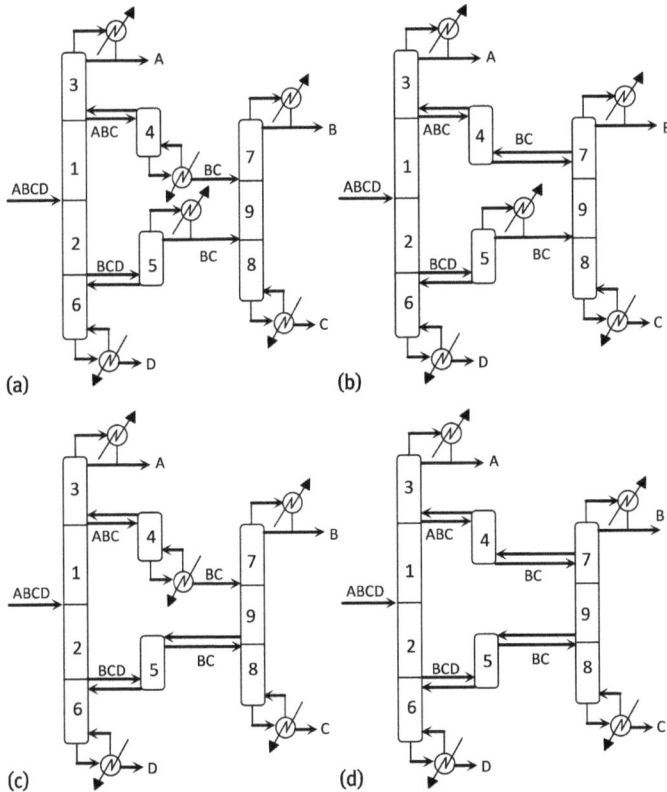

Fig. 2.23: TESs for the OTCs in Figure 2.22.

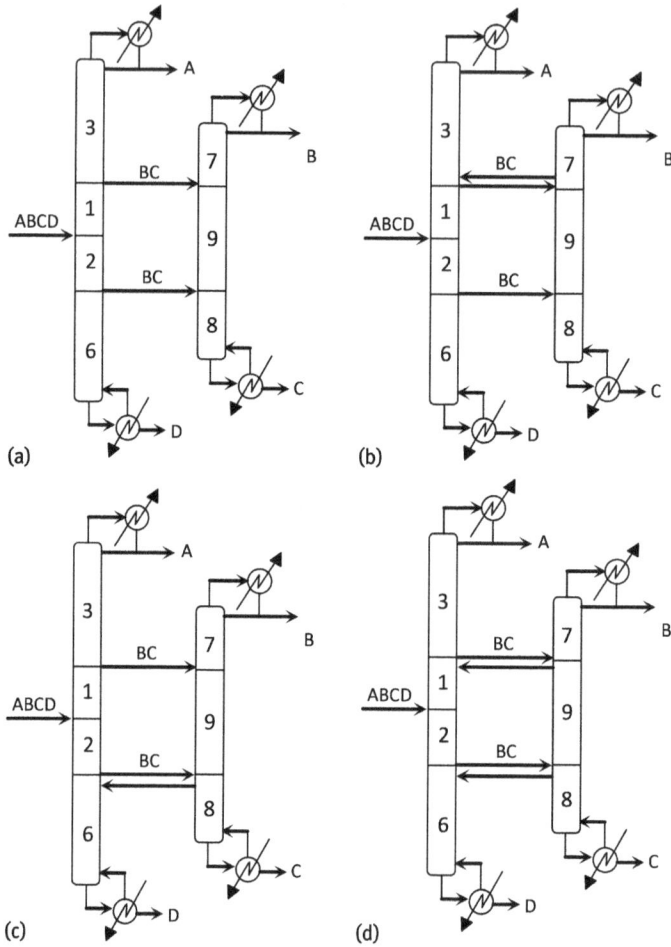

Fig. 2.24: ISCs from the TESs in Figure 2.23.

an ISC is created through eliminating the side columns. Therefore, ISCs with two-way transport of submixtures cannot produce TESs.

2.11.2 Subspace of intensified distillation systems

In the described procedure for synthesis of ISC intensified distillation systems, we have focused on the single-section-side columns that serve to transport submixtures between columns. The single-section side columns are not product columns and can be eliminated to produce the ISCs with less than $N − 1$ columns.

Distillation configurations with less than $N − 1$ columns as the intensified systems have the potential to reduce both energy and capital costs. It is important to further develop the systematic procedure so that all possible IDSs with less than $N-1$ columns can be synthesized. This section presents such a general systematic procedure. An IDS is defined as using less than $N-1$ columns to produce at least N products for an N-component distillation [22, 36]. The IDS subspace contains many more alternatives than the ISC subspace. In the TESs, there are the following two kinds of single-section side columns:

- column for transporting a submixture of middle components between two columns
- column for purifying a final product of a middle component

As illustrated earlier, there are four types of transport single-section side columns (TSS1, TSS2, TSR1, TSR2). There are also two types of product single-section side columns, as follows:

Side column case 5: single-section side-rectifier product column (SRPC)
Side column case 6: single-section side-stripper product column (SSPC)

The synthesis of IDSs must deal with all six types of single-section side columns. Here, we can formulate a new mechanism for PI to synthesize IDSs. The mechanism is based on considering all the single-section side columns among the TES structures.

Mechanism 6: IDSs are generated from the TESs where there exist single-section side columns, which serve either to transport an intermediate submixture or to purify a final product of a middle component. Both kinds of single-section side columns can be eliminated to produce the IDSs.

To generate all the possible IDSs with less than $N-1$ columns, it is a precondition that all possible TES structures with single-section side columns are produced. This is guaranteed by keeping the maximum structural flexibility to rearrange the column sections. It has been demonstrated that an SCC representation for a DSS can keep the maximum structural flexibility [36].

Starting from the SCC, there are two major steps to obtain all IDSs, i.e., generation of OTCs, and generation of TESs. Mechanism 3 is a universal mechanism for generating OTCs, and mechanism 4 is a universal mechanism for generating TESs. Therefore, a generalized systematic procedure can be developed that produces IDS configurations starting from the SCC for any DSS (both sharp and nonsharp). The generalized procedure is presented in Figure 2.25.

Note that when applying the systematic procedure to generate the IDS subspace for all the DSSs, all the six different types of single-section side columns can appear among the TES structures. Because each of the single-section side columns has a unique section number, we name the single-section side columns by their column section numbers.

We present an example of the generation of IDSs with the above generalized systematic procedure. We consider the distinct nonsharp sequence with five intended individual splits of $\underline{ABCD} \rightarrow A/BC \rightarrow B/CD \rightarrow B/C \rightarrow C/D$. Its SCC representation is shown in Figure 2.5. This means we have implemented steps 1 and 2 of the systematic proce-

Step 0. Given a zeotropic mixture with N components.

↓

Step 1. Formulate a distinct separation sequence (DSS) by introducing a set of intended individual splits for the given mixture with both sharp and nonsharp splits.

↓

Step 2. Represent the DSS with a simple column configuration (SCC) by using one simple column for each of the individual splits in the DSS.

↓

Step 3. Examine if heat-integration to combine the simple columns between condensers and reboilers associated with intermediate components is implemented. If yes then mechanism 1 or mechanism 2 or both are employed; If no then go to step 4.

↓

Step 4. Identify the condensers and/or reboilers associated with submixtures to be eliminated with mechanism 3 and generate the thermally coupled configurations.

↓

Step 5. Generate the thermodynamically equivalent structures with single-section-side columns for the thermally coupled configurations with mechanism 4.

↓

Step 6. Generate the intensified distillation systems by eliminating the single-section side columns with mechanism 6.

↓

Step 7. Repeat steps 3 to 6 until all the intensified distillation systems are generated for the SCC of the formulated DSS.

↓

Step 8. Do you want to examine another distinct nonsharp separation sequence (DSS)? If yes go to step 1, if no go to step 9. Repeat this cycle until all the feasible DSS sequences have been examined.

↓

Step 9. Store all the generated IDS configurations in the IDS subspace and stop.

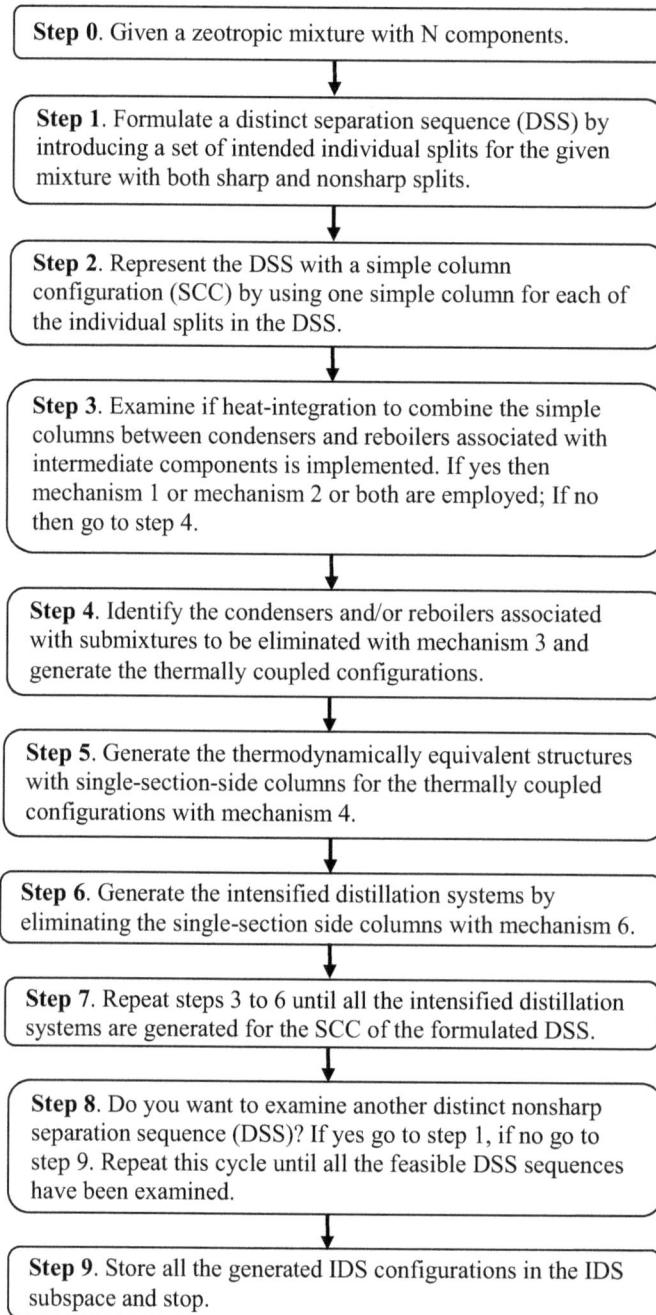

Fig. 2.25: Generalized systematic procedure for synthesis of the intensified distillation systems.

dure outlined in Figure 2.25. There are five cases which focus on implementing the four steps (steps 3–6 in Figure 2.25) for generating the possible IDS cases for the selected DSS.

Case 1: Starting from the SCC in Figure 2.5 we can use the following four steps to generate the first IDS:

Step 1: Combine the two columns coproducing the middle component C through open-heat integration of mechanism 1, as shown in Figure 2.26(a).

Step 2: Eliminate condenser ABC and reboiler BCD through the thermal couplings of mechanism 3, as shown in Figure 2.26(b).

Step 3: Remove the movable sections 3 and 6 to rearrange the column sections by mechanism 4, as shown in Figure 2.26(c).

Step 4: Eliminate the single-section side columns 4 and 5 through mechanism 6, as shown in Figure 2.26(d).

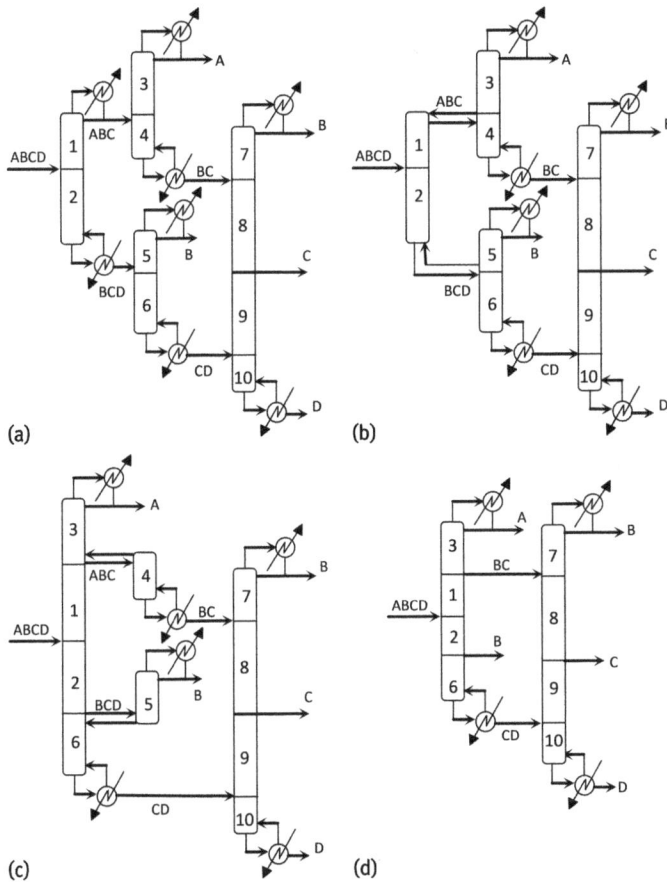

Fig. 2.26: Generation of the first intensified distillation system for the SCC in Figure 2.5.

Case 2: Starting from the SCC in Figure 2.5 we can use the following four steps to generate the second IDS:

Step 1: Combine the two columns coproducing the middle component C through open-heat integration of mechanism 1, as shown in Figure 2.27(a);

Step 2: Eliminate reboiler BCD and reboiler BC through the thermal couplings of mechanism 3, as shown in Figure 2.27(b).

Step 3: Remove the movable sections 6 and 8+9+10 together to rearrange the column sections by mechanism 4, as shown in Figure 2.27(c).

Step 4: Eliminate the single-section side columns 5 and 7 through mechanism 6, as shown in Figure 2.27(d).

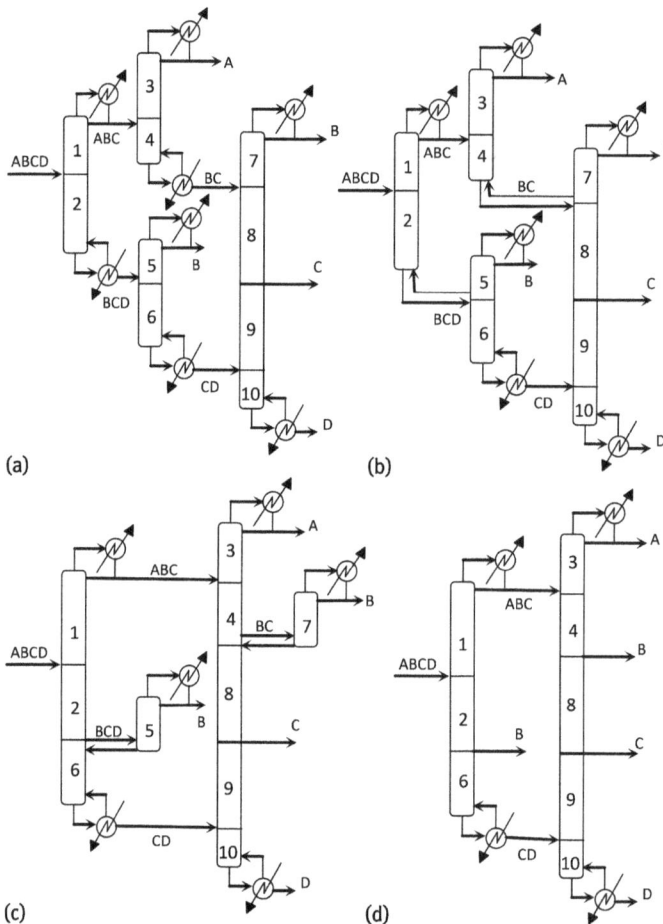

Fig. 2.27: Generation of the second intensified distillation system for the SCC in Figure 2.5.

Case 3: Starting from the SCC in Figure 2.5 we can use the following four steps to generate the third IDS:

Step 1: Combine the two columns between the reboiler with component C and the condenser with component B through closed-heat integration of mechanism 2, as shown in Figure 2.28(a).

Step 2: Eliminate condenser ABC and reboiler CD through the thermal couplings of mechanism 3, as shown in Figure 2.28(b).

Step 3: Remove the movable sections 3 and 10 to rearrange the column sections by mechanism 4, as shown in Figure 2.28(c).

Step 4: Eliminate the single-section side columns 4 and 9 through mechanism 6, as shown in Figure 2.28(d).

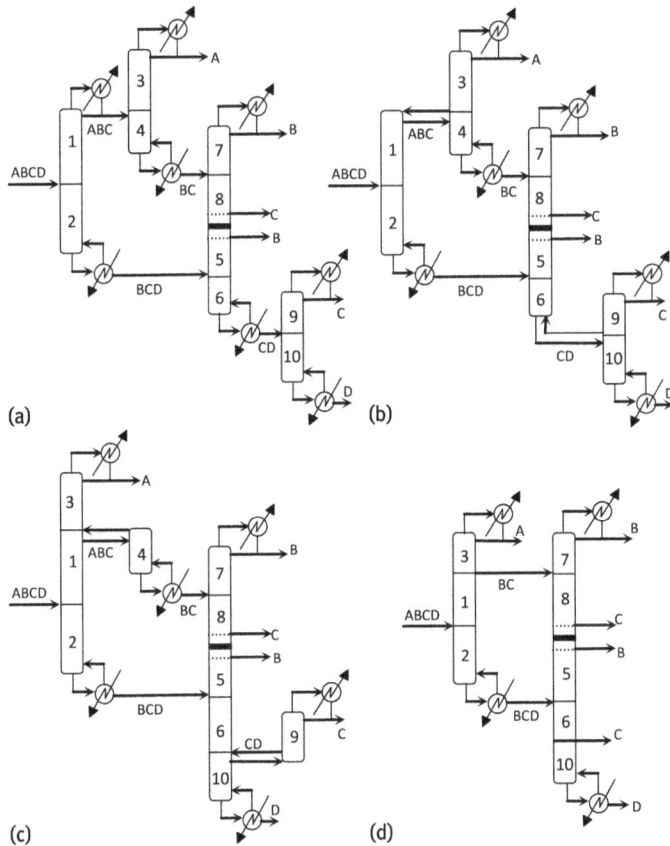

Fig. 2.28: Generation of the third intensified distillation system for the SCC in Figure 2.5.

Case 4: Starting from the SCC in Figure 2.5 we can use the following three steps to generate the fourth IDS:

Step 1: Eliminate condenser ABC, reboiler BCD, and reboiler CD through the thermal couplings of mechanism 3, as shown in Figure 2.29(a).

Step 2: Remove the movable sections 3, 6, and 10 to rearrange the column sections by mechanism 4, as shown in Figure 2.29(b).

Step 3: Eliminate the single-section side columns 4, 5, and 9 through mechanism 6, as shown in Figure 2.29(c).

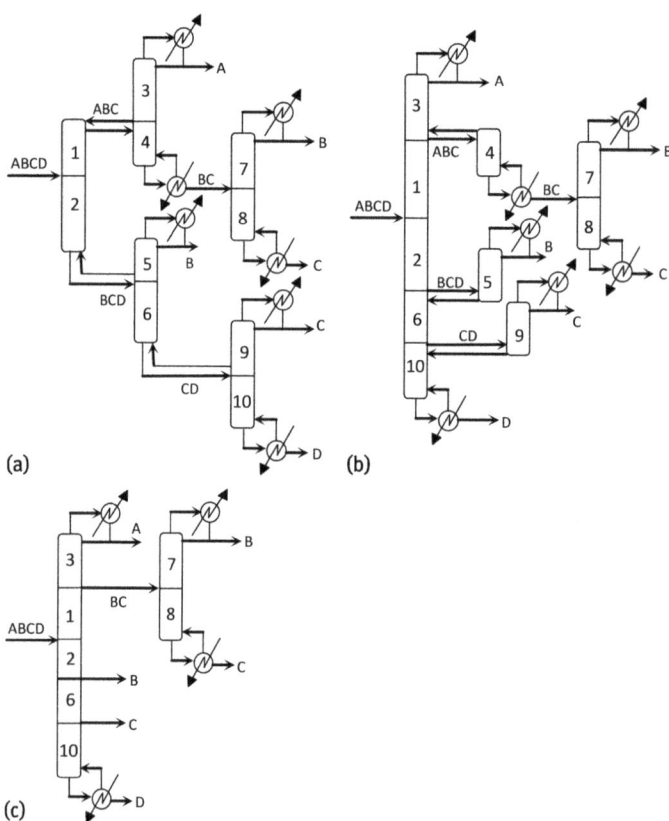

Fig. 2.29: Generation of the fourth intensified distillation system for the SCC in Figure 2.5.

Case 5: Starting from the SCC in Figure 2.5 we can use the following three steps to generate the fifth IDS.

Step 1: Eliminate reboiler BCD, reboiler CD, and reboiler BC through the thermal couplings of mechanism 3, as shown in Figure 2.30(a).

Step 2: Remove the movable sections 6, 10, and 8 to rearrange the column sections by mechanism 4, as shown in Figure 2.30(b).

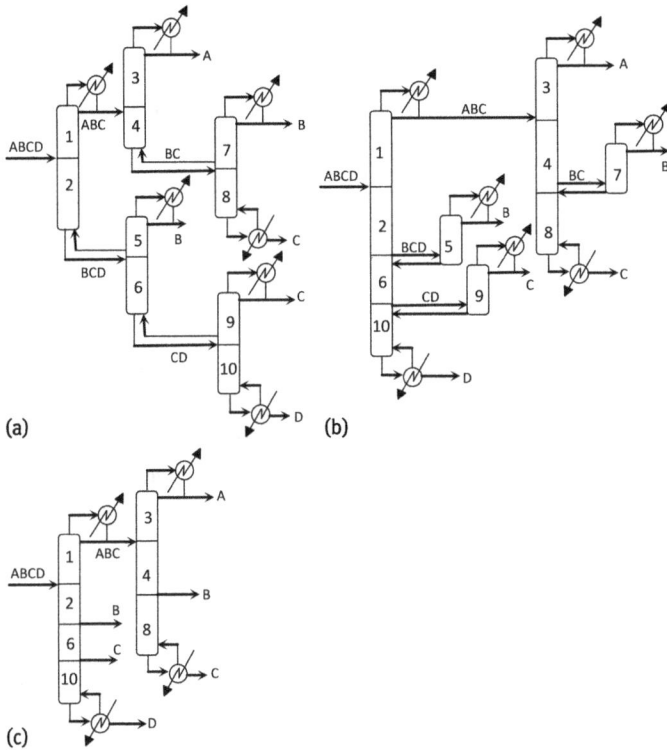

Fig. 2.30: Generation of the fifth intensified distillation system for the SCC in Figure 2.5.

Step 3: Eliminate the single-section side columns 5, 9, and 7 through mechanism 6, as shown in Figure 2.30(c).

The IDSs have the following features:
1. The IDS systems use less than $N-1$ columns for an N-component distillation. This is true for both sharp and nonsharp sequences.
2. Each column in an IDS system has a condenser and a reboiler. When necessary, those condensers and reboilers associated with submixtures can be replaced with thermal couplings, which can then generate the TESs. However, in all cases, the condenser for the lightest component product and the reboiler for the heaviest component product are always needed.
3. For nonsharp sequences, some middle components are produced more than once in different locations of the IDSs. This is different from the IDSs for sharp sequences, where each middle component is produced only once. When products with different purities or fractions with different composition spectra are needed from the separation processes, nonsharp sequences should be considered.

4. After eliminating single-section side columns that transport submixtures of middle components, the submixture of middle components can transport in one or two directions between the columns.
5. In the steps to approach an IDS system, all configurations obtained in the intermediate steps with structural changes from the SCC are feasible and useful configurations. If the product purity of the final IDS does not satisfy the separation specification, the earlier configurations derived in the intermediate steps may be considered. For such cases, the side columns can be recovered to ensure purity of the products.

From the ISC and IDS subspaces, it is emphasized that thermal couplings enable column sections in a distillation configuration to be rearranged in the TESs, and also allow the functionality of the column sections to be re-examined on the basis of the separation tasks and product specifications. In certain cases, some column sections can be eliminated and the multicomponent distillation system can be intensified to use fewer columns.

2.12 Subspace of dividing-wall columns

This subspace is for intensifying distillation configurations to produce dividing-wall columns (DWC). The DWC is considered to be intensified distillation equipment because, in certain cases, it has the potential to save energy and capital costs significantly. The DWC was presented in a patent by Wright [49]. The ternary Petlyuk column [43] and its equivalent concentric column was also an early DWC. Kaibel [50] discussed the DWC for quaternary distillation. The DWC implementation for the quaternary Petlyuk column was investigated by Christiansen et al. [51]. Today, DWCs are used in industry as a result of intensive steady-state and dynamic performance studies. However, several of the above-mentioned DWCs were developed through some chaotic inventive activities. Nevertheless, we can summarize the following points concerning DWCs:

– DWC is intensified equipment for multicomponent distillation.
– DWC is now widely applied in industry and the DWC equipment design is mostly owned by the companies.
– Several known DWCs were developed by chaotic inventive activities.
– DWC is one of the most attractive intensified alternatives for multicomponent distillation.

We have investigated the synthesis of DWCs for multicomponent distillation using systematic synthesis methods [37, 52]. The synthesis method has shown that a DWC is not obtained with a single-step inventive activity; an evolutionary approach, following certain steps, is needed to determine the physical structure of the column sections.

This section provides a generalized systematic procedure to conceptually design a DWC for a mixture with four or more components. The conceptual design of DWCs involves the physical definition and arrangement of the functional sections of the DWC equipment.

In the literature, there is no consistent definition of a DWC. There are different terms, such as divided-wall column, distillation column with vertical partitions [50], partitioned distillation column [53], and partitioned Petlyuk arrangement [51], which generally refer to a column with an internal dividing wall or a vertical partition wall. There are also arrangements such as the concentric column [43] and triangular wall structure [51].

We present the systematic synthesis procedure for generation of DWCs from PS and PI points of view. We show that a DWC is produced either from a thermally coupled configuration or from its TES. Therefore, a DWC is defined as an intensified distillation equipment of either a thermally coupled configuration or its TES. A DWC has the same functional sections as the thermally coupled configuration and its TES, but a reduced number of column units. Strict correspondence between the column sections is required for the systematic procedure to generate DWCs from the thermally coupled configurations.

We will develop systematic procedures for the synthesis of DWCs, first from the SSC subspace and then extended to the NSSC subspace.

2.12.1 DWCs from thermodynamically equivalent structures with side columns

Note that thermodynamically equivalent side column structures are from both the SSC and NSSC subspaces. As defined above, a DWC is an intensified thermally coupled configuration with a reduced number of columns. On the other hand, the DWC has the same functional sections as its thermally coupled configuration. Therefore, we need a mechanism to keep the functional sections of the thermally coupled configuration and, at the same time, to reduce the number of the columns. To do so, we first examine the TESs of the two thermally coupled configurations shown in Figure 2.31.

Note that the thermally coupled configurations in Figure 2.31(a, b) with two-section simple columns cannot produce DWCs. This is because each column is a simple column with one rectifying section and one stripping section and cannot accommodate a dividing wall.

The TESs shown in Figure 2.31(c, d) have the same number of columns (three). However, in TES (Figure 2.31c), there is a side-rectifier column with one section 3. Functionally, this side-rectifier must rectify its feed submixture BCD so that the pure component B is obtained through the condenser. Similarly, in TES (Figure 2.31d), there is a side-stripper column with one section 6. Functionally, this side-stripper must strip its feed submixture BC so that the pure component C is obtained through the reboiler.

In earlier procedures to generate ISC and IDS subspaces, such side-rectifier and side-stripper columns were simply eliminated to generate intensified systems with fewer columns. In many cases, this results in impure side-stream products. In this procedure, such a side column can be incorporated into its thermally linked column through a dividing wall to produce the DWC configuration. For instance, the two DWCs generated from the two TESs of Figure 2.31(c, d) are presented in Figure 2.32 (a, b), respectively.

In the example, the TES structure must simultaneously create the main column and the side column. The main column accommodates the dividing wall, and the side column serves as the dividing wall. The main column has more than two sections, and a side column has one or more sections. Therefore, the following criteria can be identified for generation of DWCs:

Criterion 1: In the thermally coupled configuration, there must be a column with more than two sections to accommodate the dividing wall(s).

Criterion 2: In the thermally coupled configuration, there must be a column that can serve as the dividing wall to be incorporated into its thermally linked column.

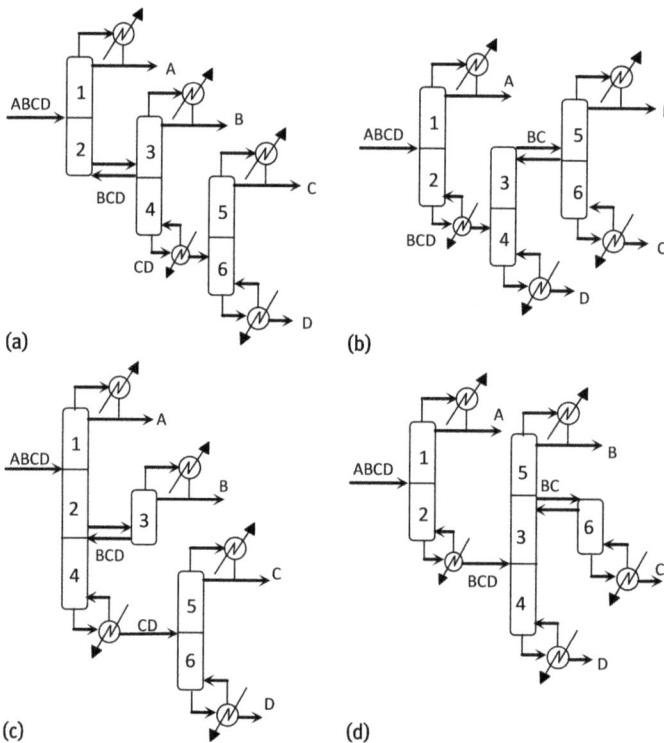

Fig. 2.31: Thermally coupled configurations (a) and (b) (OTC) and their corresponding thermodynamically equivalent structures (c) and (d) (TES).

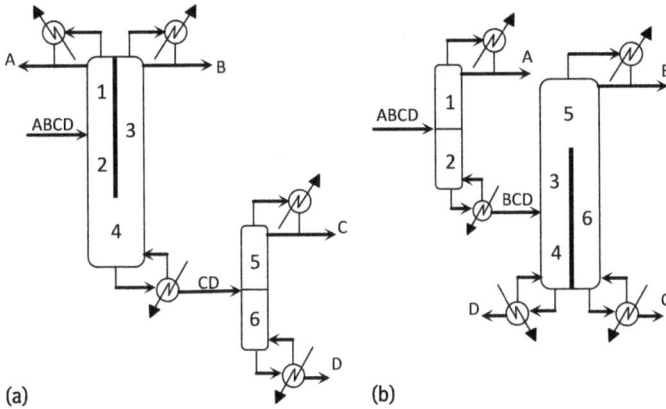

Fig. 2.32: DWCs generated from the TESs of Figure 2.31.

For any thermodynamically equivalent structure, such a side column, either a side-rectifier or a side-stripper can be incorporated into its thermally linked column to produce the DWC. Therefore, incorporating a side column provides a mechanism to produce DWCs.

Mechanism 7: This mechanism involves incorporating a side column. A side column, either a side-rectifier or a side-stripper in a TES, can be incorporated into its thermally linked column to produce a DWC.

Note that for those TESs with more than one side column, it is possible to incorporate the side columns into the thermally linked column one at a time or more than one at the same time. Thus, such a TES can generate more than one DWC. Note also that we have demonstrated that the thermodynamically equivalent side column structures can be generated from both the SSC subspace and the NSSC subspace. Therefore, mechanism 7 is valid for both sharp and nonsharp sequences.

On the basis of mechanism 7, a systematic procedure can be formulated to generate DWCs, as presented in Figure 2.33.

We present an example of DWC generation using the systematic procedure in Figure 2.33. First, we select the SSC in Figure 2.4(b). Implementing steps 1–5 can straightforwardly generate the TES structures, as shown in Figure 2.34.

In step 6, DWCs are generated from the TESs in Figure 2.34 and are presented in Figure 2.35. In step 7, no more DWCs can be generated for this selected sharp sequence and the procedure stops.

Step 0. Given a zeotropic mixture with N components.

↓

Step 1. Formulate a distinct separation sequence (DSS) by introducing a set of intended individual splits for the given mixture with both sharp and nonsharp splits.

↓

Step 2. Represent the DSS with a simple column configuration (SCC) with one simple column for each of the individual splits in the DSS.

↓

Step 3. Examine if heat-integration to combine the simple columns between condensers and reboilers associated with intermediate components is implemented. If yes then mechanism 1 or mechanism 2 or both are employed; If no then go to step 4.

↓

Step 4. Identify the condensers and/or reboilers associated with submixtures to be eliminated with mechanism 3 and generate the thermally coupled configurations (OTCs).

↓

Step 5. Generate the thermodynamically equivalent structures (TESs) with side columns for the thermally coupled configurations with mechanism 4.

↓

Step 6. Generate the dividing wall columns by incorporating the side columns into their thermally linked columns with mechanism 7.

↓

Step 7. Repeat steps 3 to 6 until all the DWC columns are generated for the SCC of the formulated DSS.

↓

Step 8. Do you want to examine another distinct separation sequence (DSS)? If yes go to step 1, if no go to step 9. Repeat this cycle until all the feasible DSS sequences have been examined.

↓

Step 9. Store all the generated DWC columns in the subspace and stop.

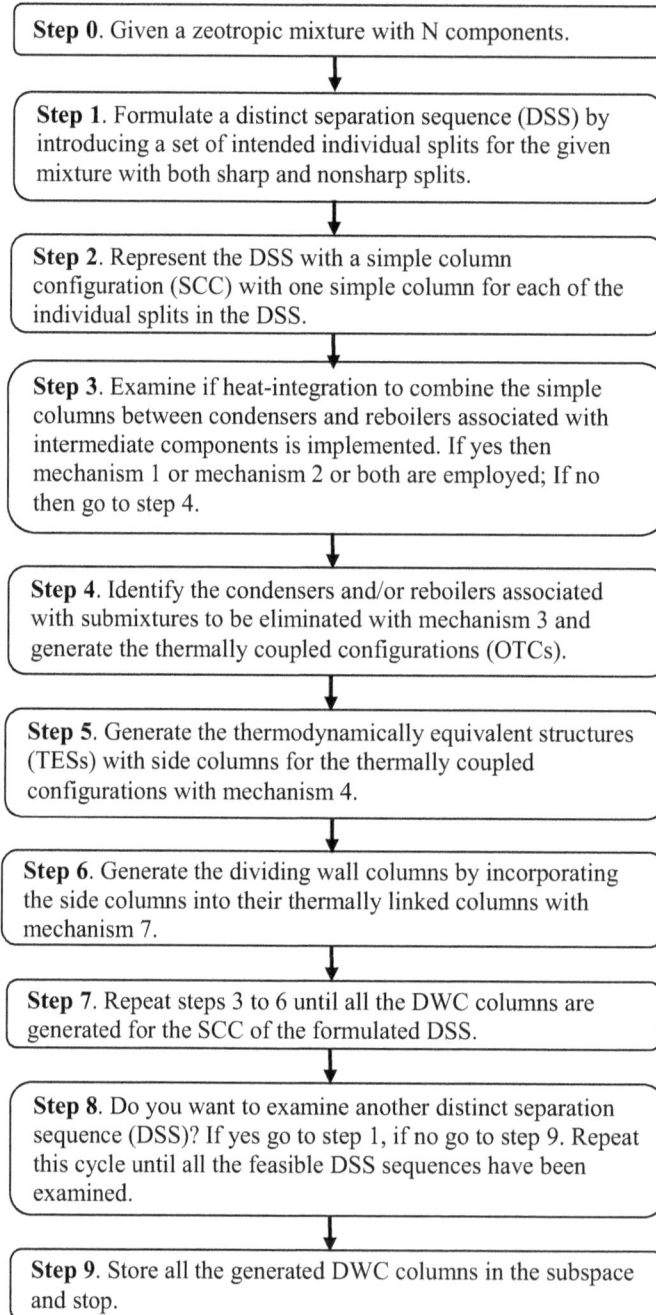

Fig. 2.33: Systematic procedure for synthesis of DWCs by incorporating side columns.

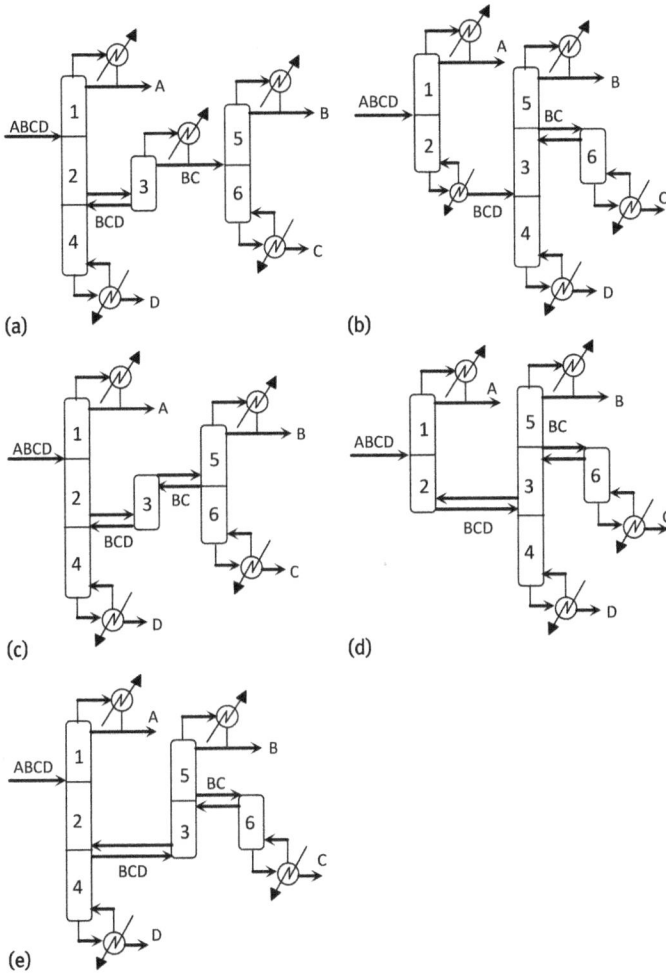

Fig. 2.34: TESs for the selected SSC of Figure 2.4(b).

The distinct features of the DWCs are as follows:

- A DWC uses less than $N - 1$ columns for an N-component distillation.
- The number of dividing walls in a DWC is equal to the number of incorporated side columns.
- The total number of functional zones and sections in a DWC is equal to the total number of column sections in its TES.
- The condensers and reboilers in the DWCs are unchanged and remain as in the TES structures.

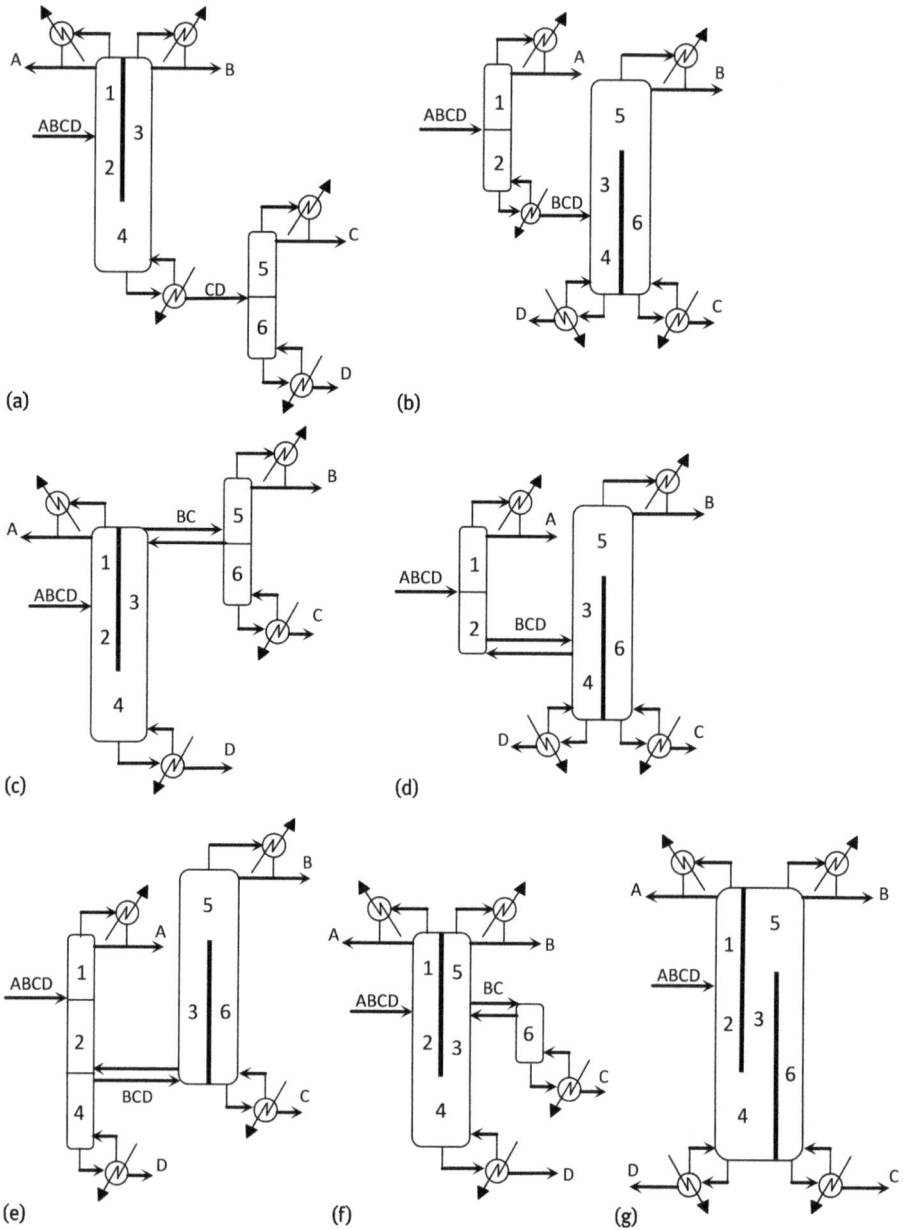

Fig. 2.35: DWCs generated from the TESs in Figure 2.34. DWCs of (a)–(d) are from the TESs of (a)–(d) in Figure 2.34, correspondingly; DWCs of (e)–(g) are from the TES of Figure 2.34(e).

2.12.2 DWCs from prefractionation columns

In the above procedure for the synthesis of DWCs, we have focused on the side columns that are thermally linked as either side-strippers or side-rectifiers. These side columns are incorporated into their thermally linked columns to produce the DWCs.

Now we consider thermally coupled configurations from nonsharp sequences. There are two distinct features: First, there is the combined column obtained through mechanisms 1 and 2, which has more than two sections. Second, a nonsharp split is always used for the prefractionation of a submixture, which introduces a column without pure products. Such a column serves to prefractionate a submixture and is usually called a prefractionator. A prefractionator is defined as having both its overhead condenser and its bottom reboiler associated with a submixture. When replacing the condenser and reboiler with thermal couplings in a prefractionator, such a prefractionation column can link its two ends with other column(s) by thermal couplings. Moreover, the two ends of a prefractionator can be connected in the following two ways:

Case 1: Both ends are connected to the same column, which is a combined column with more than two sections.

Case 2: The two ends are connected to two different columns.

For case 2, we will show that the TES can make the prefractionator link to the same column.

Therefore, a thermally coupled configuration from a nonsharp sequence can directly fulfill the two criteria for generation of DWCs; that is, the prefractionator can be directly incorporated into its thermally linked column. This creates a new mechanism for generating DWCs.

Mechanism 8: This mechanism involves incorporating a prefractionator. A prefractionator, either in an original thermally coupled configuration or in a TES, can be incorporated into its thermally linked column to produce the DWC.

The opportunity to apply mechanism 8 can exist in both the OTC and TES. For example, for the OTCs of Figure 2.36(a, b), there is a prefractionator for the submixture ABC and for the submixture BCD. The prefractionator can be incorporated into its thermally linked subsequent column to generate the DWCs, as shown in Figure 2.36(c, d).

With mechanisms 7 and 8, a generalized systematic procedure for synthesis of DWCs is formulated in Figure 2.37.

We present an example of DWC generation from nonsharp sequences. Steps 1 and 2 select the DSS of A$\underline{B}$CD → BC/D → A/B → B/C and represent the nonsharp sequence with the SCC, as shown in Figure 2.38. Then, steps 3–7 generate the specific DWCs for this SCC.

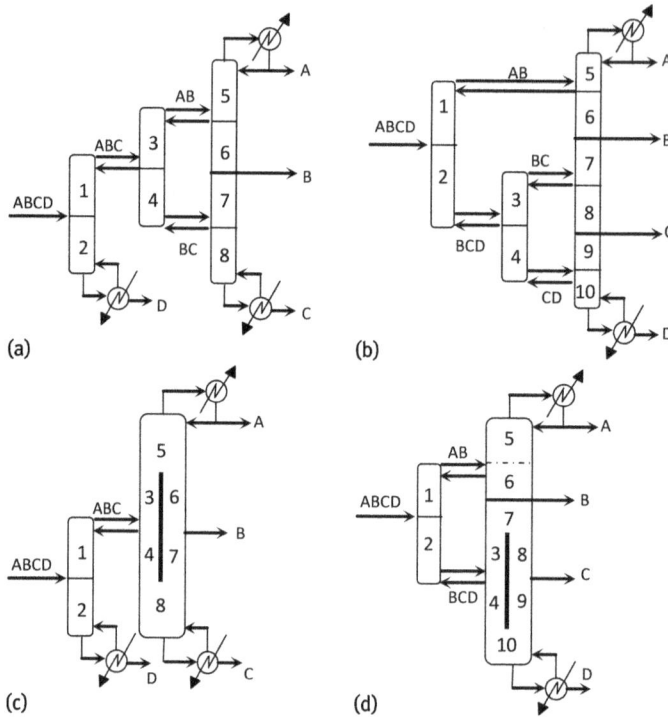

Fig. 2.36: DWCs generated from the OTCs with prefractionators. (a) and (b): OTC with a prefractionator; (c) and (d): DWC generated from (a) and (b) correspondingly.

Case 1: Starting from the SCC in Figure 2.38, we can use the following four steps to generate the first DWC:

Step 1: Combine the two columns coproducing B through mechanism 1.

Step 2: Replace condenser BC with thermal coupling through mechanism 3.

Step 3: Remove sections 5, 6, and 7 together to produce a TES with side column 8, as in Figure 2.39(a).

Step 4: Incorporate side column 8 to produce the DWC, as in Figure 2.39(b).

Case 2: Starting from the SCC in Figure 2.38, we can use the following four steps to generate the second DWC:

Step 1: Combine the two columns coproducing B through mechanism 1.

Step 2: Replace reboiler BCD with thermal coupling through mechanism 3.

Step 3: Remove section 4 to produce a TES with side column 3, as in Figure 2.40(a).

Step 4: Incorporate side column 3 to produce the DWC, as in Figure 2.40(b).

Cases 3 and 4: Starting from the SCC in Figure 2.38, we can use the following four steps to generate the third and fourth DWCs:

Step 1: Combine the two columns coproducing B through mechanism 1.

Step 0. Given a zeotropic mixture with N components.

Step 1. Formulate a distinct separation sequence (DSS) by introducing a set of intended individual splits for the given mixture with both sharp and nonsharp splits.

Step 2. Represent the DSS with a simple column configuration (SCC) with one simple column for each of the individual splits in the DSS.

Step 3. Examine if heat-integration to combine the simple columns between condensers and reboilers associated with intermediate components is implemented. If yes then mechanism 1 or mechanism 2 or both are employed; If no then go to step 4.

Step 4. Identify the condensers and/or reboilers associated with submixtures to be eliminated with mechanism 3 and generate the thermally coupled configurations (OTCs).

Step 5a. Identify the prefractionation columns, which can be directly incorporated into their thermally linked columns.

Step 5b. Generate the thermodynamically equivalent structures (TESs) with side columns or with prefractionation columns for the thermally coupled configurations with mechanism 4.

Step 6. Generate the dividing wall columns by incorporating the side columns or the prefractionation columns into their thermally linked columns with mechanisms 7 and 8.

Step 7. Repeat steps 3 to 6 until all the DWC columns are generated for the SCC of the formulated DSS.

Step 8. Do you want to examine another distinct separation sequence (DSS)? If yes go to step 1, if no go to step 9. Repeat this cycle until all the feasible DSS sequences have been examined.

Step 9. Store all the generated DWC columns in the subspace and stop.

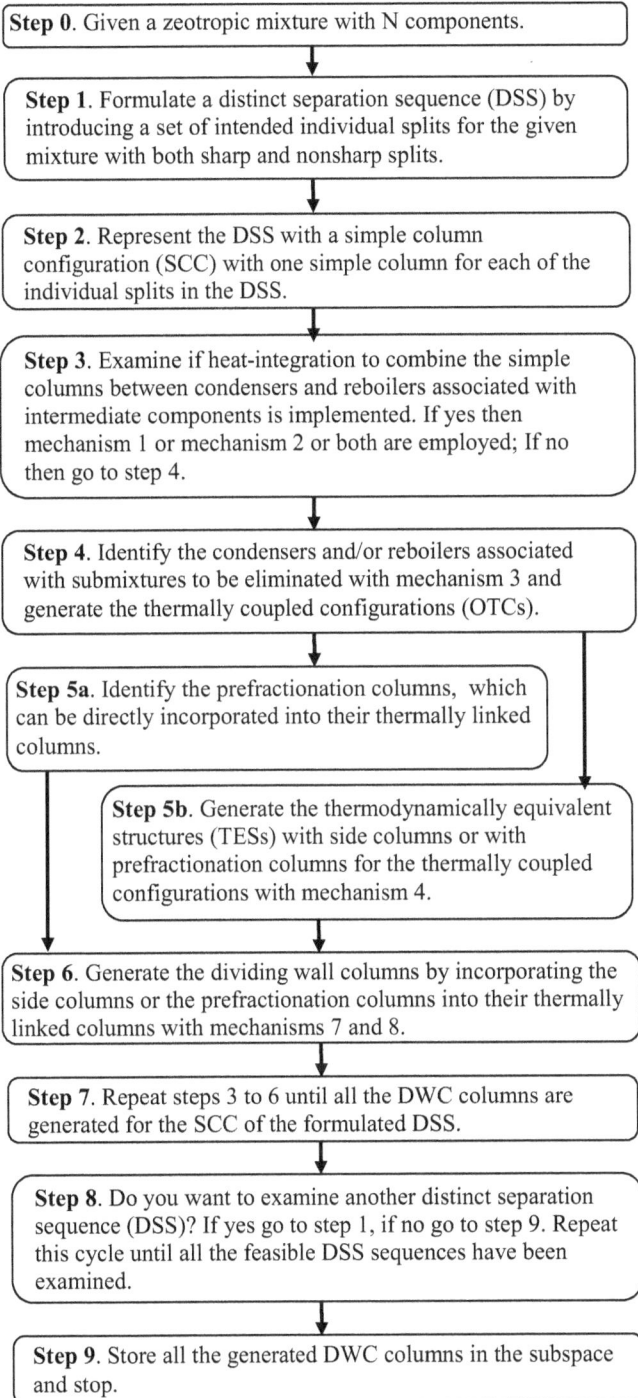

Fig. 2.37: Generalized systematic procedure for synthesis of DWCs.

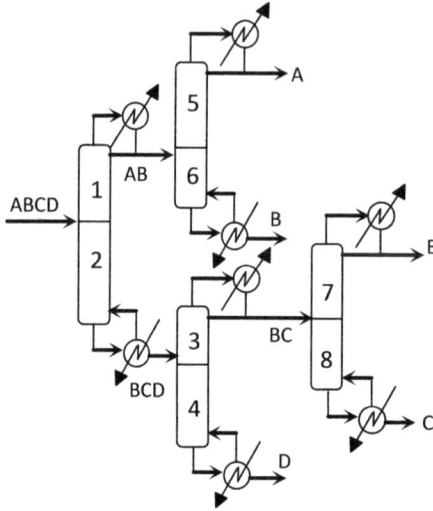

Fig. 2.38: The SCC for the selected nonsharp sequence.

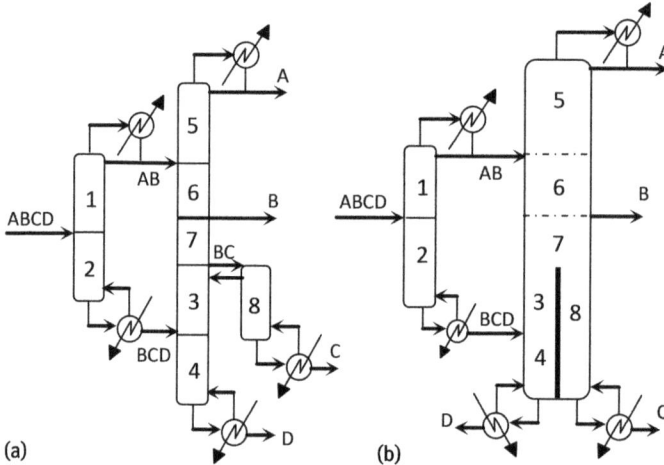

(a) (b)

Fig. 2.39: Generation of the first DWC from the SCC in Figure 2.38.

Step 2: Replace condenser AB, condenser BC, and reboiler BCD with thermal couplings through mechanism 3.

Step 3: Remove sections 5, 6, and 7 together to produce a TES with side column 8, as in Figure 2.41(a).

Step 4: Incorporate the prefractionator to produce the DWC, as in Figure 2.41(b). Or

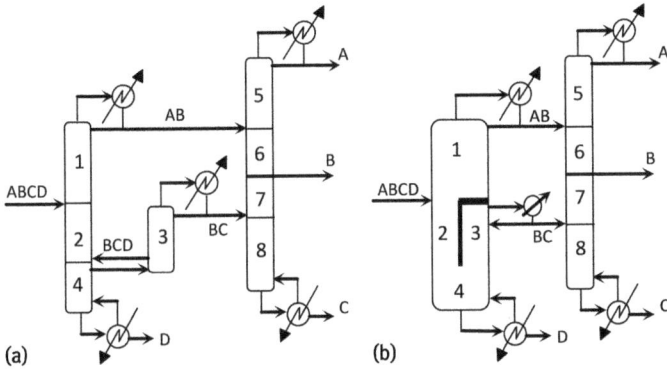

Fig. 2.40: Generation of the second DWC from the SCC in Figure 2.38.

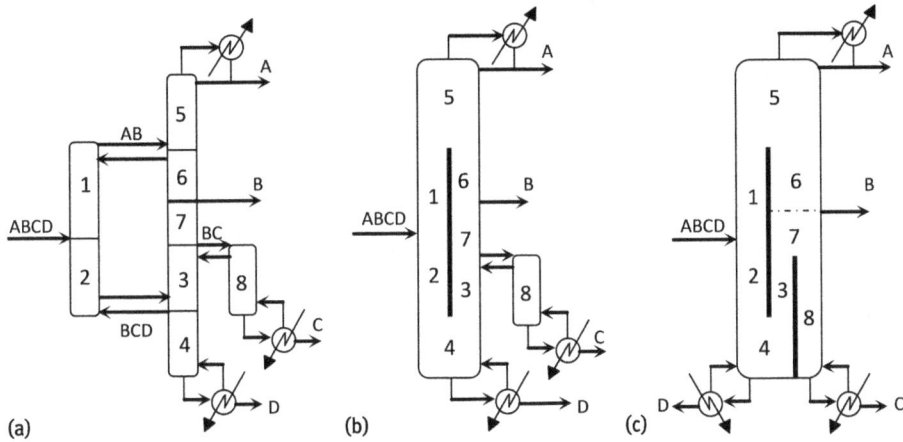

Fig. 2.41: Generation of the third and fourth DWC from the SCC in Figure 2.38.

Step 4a: Incorporate the prefractionator and side column 8 simultaneously to produce the DWC, as in Figure 2.41(c).

Case 5: Starting from the SCC in Figure 2.38, we can use the following three steps to generate the fifth DWC:

Step 1: Combine the two columns between reboiler B and condenser BC through mechanism 2.

Step 2: Replace condenser AB and reboiler BCD with thermal couplings through mechanism 3, as shown in Figure 2.42(a).

Step 3: Incorporate the prefractionator to produce the DWC, as in Figure 2.42(b).

Note that for the sharp sequences, there may also be prefractionation columns. For example, for the sharp sequence AB/CD → A/B → C/D, its SCC is shown in Figure 2.4(c); it is obvious that the first column is a prefractionator. For such sharp sequences, the

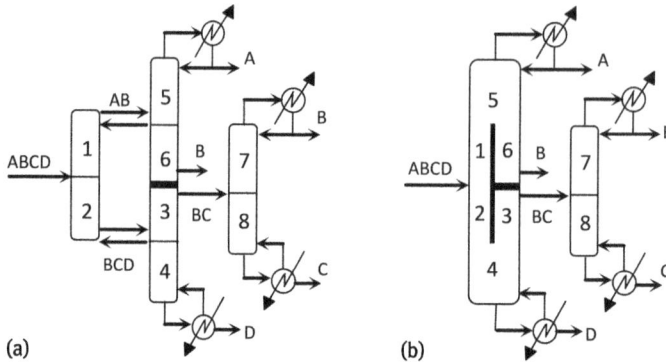

Fig. 2.42: Generation of the fifth DWC from the SCC in Figure 2.38.

procedure in Figure 2.37 can also be applied to generate DWCs. For example, we can use the following three steps to generate the DWC shown in Figure 2.43(b):

Step 1: Combine the two columns between reboiler B and condenser C through mechanism 2.

Step 2: Replace condenser AB and reboiler CD with thermal couplings through mechanism 3, as shown in Figure 2.43(a).

Step 3: Incorporate the prefractionator to produce the DWC, as in Figure 2.43(b).

The features of DWCs are summarized as follows:

- The DWCs from both sharp and nonsharp sequences use less than $N - 1$ columns for an N-component mixture.
- The DWCs have functional sections and zones that are equivalent to those of either the OTC or its TES. Therefore, the total number of functional zones and sections in a DWC is equal to the total number of column sections in the OTC or its TES.
- The number of dividing walls in a DWC is equal to the number of incorporated side columns and prefractionators.

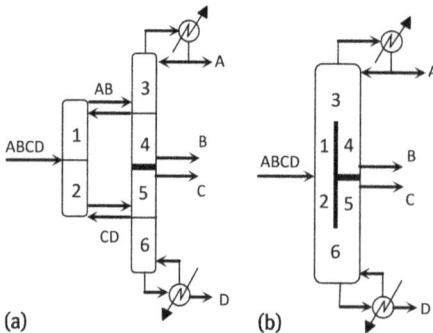

Fig. 2.43: Generation of a DWC from the sharp sequence for a four-component mixture.

- The condensers and reboilers in the DWCs remain the same as in the OTC or the TES.
- The external thermal coupling in the OTC or TES is incorporated into the DWC as an internal thermal coupling. In some cases, this results in loss of degrees of freedom in terms of control and operability.
- Compared with ISC and IDS systems, where equipment intensification is achieved through reduction in the number of functional column sections, the DWC retains all of the functional zones/sections in the intensified system. Therefore, the inherent difference between a DWC and an ISC or IDS is that the DWC simultaneously retains the thermal couplings between the original column sections and merges thermally linked sections in the same equipment. This gives the advantages of both thermal couplings and equipment intensification.

2.13 Summary and remarks

2.13.1 Primary results

In this chapter, we have presented PS and PI for multicomponent distillation using systematic methodology. The primary results are presented as the following two categories:

Category I: distinct subspaces of multicomponent distillation configurations
Category II: systematic procedures for generation of distinct subspaces

The following are major points in obtaining the primary results in the above two categories:

- We emphasize the conceptual design for both PS and PI.
- We rely on the different mechanisms to define distinct structures for multicomponent distillation systems.
- The systematic procedures are the methods for both PS and PI to obtain the distinct subspaces. However, all the systematic procedures are fundamentally synthesis-dominated methods (in terms of synthesis and analysis as basic methods of scientific inquiry).
- The mechanisms are universal mechanisms and the systematic procedures are generalized methods for synthesis and intensification of multicomponent distillation systems.

The following subsections include summaries and remarks on various aspects of the synthesis and intensification of multicomponent distillation systems.

2.13.2 Distinct subspaces for multicomponent distillation

In this work, we have obtained seven multicomponent distillation subspaces, each consisting of distinct distillation configurations. They are named as follows:
1. sharp sequence configurations subspace (SSC subspace)
2. nonsharp sequence configurations subspace (NSSC subspace)
3. original thermally coupled configurations subspace (OTC subspace)
4. thermodynamically equivalent structures subspace (TES subspace)
5. intensified simple column configurations subspace (ISC subspace)
6. intensified distillation systems subspace (IDS subspace)
7. intensified dividing-wall columns subspace (DWC subspace)

Except the SSC subspace directly from the well-known sharp separation sequences, the other six subspaces are obtained by the developed systematic synthesis procedures with specific mechanisms.

2.13.3 About the subspaces and their relationships

It is clear that the alternatives in one subspace are structurally different from those in other subspaces. However, before a specific separation problem is defined by the specific mixture and separation requirements, there is no guarantee that the optimal system in a particular subspace will be found. All the distinct configurations in the distinct subspaces might be applied, depending on the mixture and separation requirements. However, for a conceptual design to determine the structures of the systems, the relationships between the subspaces are given, as shown in Figure 2.44.

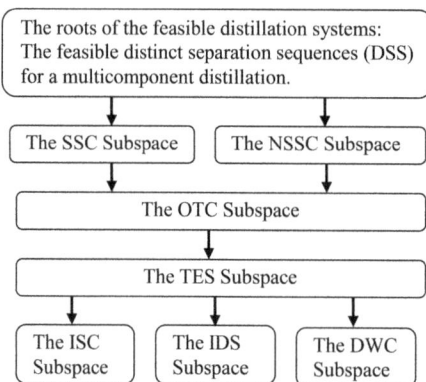

Fig. 2.44: Relationships between the subspaces of multicomponent distillation systems.

2.13.4 About the method scopes of the subspaces for process synthesis and process intensification

In this chapter, PS focuses on the systematic synthesis procedure for generation of feasible process alternatives, whereas PI focuses on intensifying multicomponent distillation systems by use of fewer columns.

Considering the methods used to generate the various subspaces, we have shown that each of the subspaces can be generated by a systematic synthesis procedure. The systematic procedure can explicitly define the physical structures of the alternatives. Consequently, all generated alternatives are feasible systems for the subspace. On the other hand, considering the configurations obtained, there are dramatic differences in terms of the number of columns. In the subspaces of ISC, IDS, and DWC, the systems use fewer columns.

The subspaces of NSSC, OTC, and TES are included in the scope of PS in terms of both the alternatives and the systematic procedures. The subspaces of ISC, IDS, DWC are included in the scope of PI because of the intensified alternatives with fewer columns. However, the methods are still systematic synthesis procedures for obtaining intensified alternatives. There is a strong interplay between PS and PI for the ISC, IDS, and DWC subspaces. PS calls for a systematic synthesis procedure to generate all feasible alternatives, whereas PI calls for specific mechanisms to reduce the number of columns.

2.13.5 About the mechanisms and systematic procedures

We have proposed eight mechanisms as key strategies for changing the structures of the distillation configurations. The eight mechanisms are universal mechanisms and can be used for generation of distillation configurations for a mixture with any number of components. Similarly, all the systematic procedures are generalized procedures for synthesis of distinct subspaces for any N-component mixture. Each systematic procedure incorporates distinct mechanisms to define the distinct distillation systems. Referring to synthesis and analysis as the basic methods of scientific inquiry, all the systematic procedures developed in this chapter for the synthesis and intensification of distillation systems are fundamentally synthesis-dominated methods.

2.13.6 Common elements of the methods for distillation subspaces

The methodology for both PS and PI in this chapter is the systematic synthesis procedure for conceptual design of the systems' structures. For a specific PS work (like subspaces of NSSC, OTC, and TES) or a specific PI work (like subspaces of ISC, IDS, and DWC), the methodology can involve strong interplay between PS and PI.

However, in either a specific PS work or a specific PI work, the methodology must account for the synthesis and intensification process from the concrete concept and mechanism to the concrete technical systems. Considering the methodology for generation of specific subspaces and their systematic procedures for multicomponent distillation, we have identified the following general elements of the methodology:
-   specific mechanisms
-   systematic synthesis procedure
-   generated subspace (distinct systems)
-   distinct features of the generated systems

The addressed functionality and utility of the technical system discussed here is for multicomponent distillation, and the performance target is to reduce energy consumption and capital investment.

If we wish to generalize these basic elements from distillation systems to any technical system (reaction systems, other separation systems, heat transfer systems, etc.), for any concrete PS or PI work, we can identify the following general elements of the methodology:
-   process or equipment with targeted functionality and utility (problem scope)
-   concept, mechanism, and principle (should emerge or be given)
-   evaluation criteria of the technical systems (performance, goals, indicators)
-   changes in the software aspects (should emerge or be given)
-   changes in the hardware aspects (should emerge or be given)
-   procedure and method (should be formulated to assemble the software and hardware systems)
-   technical systems obtained (should present or show their states)
-   distinct features of the obtained systems (should analyze and highlight)
-   evaluation of the specified evaluation indicators (qualitative and quantitative)
-   other specific analyses to check the feasibility, performance, and efficiency of the obtained system(s), depending on the systems and methods used

In summary, the concept of specific PS and PI works means that, on the one hand, there are basic elements to be addressed when doing a specific PS or PI work; on the other hand, there are basic elements that can be examined in both kinds of works.

2.13.7 About further studies and applications

The discussed distinct subspaces and their systematic synthesis procedures focus on the structural designs of multicomponent distillation systems. Because the mechanisms are universal mechanisms for any N-component mixture and the synthesis procedures are generalized procedures for obtaining the corresponding subspaces, they can be applied to study both the distinct alternatives in a subspace and separation

problems in which the feed mixture and product specifications are defined. These studies need to include the following:
–   methods for systems parameter design
–   optimization methods to search among the subspaces
–   optimization methods for parameter design of the systems
–   heuristics for selecting the optimal alternatives
–   dynamics and control of the selected alternatives
–   application in industrial separation problems and processes

2.14 Conclusions

Pursuing novel systems for multicomponent distillation is both intriguing and challenging. It is intriguing because multicomponent distillation is widely used in both fossil-based manufacturing and renewable biomass-based manufacturing and, therefore, an optimal system can save significant energy and capital for a specific application. The challenge is that it is difficult to find a global optimal system for a specific application before a complete alternative space is generated. In this chapter, we have made the attempt to generate the complete alternative space for a multicomponent distillation.

The complete space consists of seven distinct subspaces of distillation configurations, ranging from the conventional distillation subspace to the DWC subspace. In total, seven systematic procedures are developed and each can generate a subspace consisting of distinct systems. The systematic procedures can do both PS and PI works in the generation of the distinct distillation systems.

PS and PI are highly relevant in terms of methodology for muticomponent distillation systems. Their combination is an efficient way to synthesize distillation alternatives and to intensify distillation equipment. Therefore, a systematic synthesis procedure that integrates PS and PI is an efficient methodology for intensifying process systems.

2.15 Nomenclature

CHI: closed heat integration
DSS: distinct separation sequence
DWC: dividing-wall column
ESA: energy separating agent
ETC: external thermal coupling
IDS: intensified distillation system
ISC: intensified simple column configuration
ITC: internal thermal coupling

MSA: mass separating agent

N_{cr}: number of condensers and reboilers with mixtures

N_{im}: number of intermediate transporting mixtures

NSSC: nonsharp sequence configuration

N_{TC}: number of thermal couplings

OHI: open heat integration

OTC: original thermally coupled configuration

OTC_N: number of original thermally coupled configurations

PI: process intensification

PS: process synthesis

SCC: simple column configuration

SRPC: side-rectifier product column

SSC: sharp sequence configuration

SSC_N: number of sharp sequence configurations

SSPC: side-stripper product column

TDC: traditional distillation configuration

TES: thermodynamically equivalent structure

TSR1: one-way transport-side-rectifier

TSR2: two-way transport-side-rectifier

TSS1: one-way transport-side-stripper

TSS2: two-way transport-side-stripper

TTC_N: number of thermodynamically equivalent structures

2.16 Bibliography

[1]   Hendry JE, Rudd DF, Seader JD. Synthesis in the design of chemical processes. AIChE J, 1973, 19, 1–15.

[2]   Nishida N, Stephanopoulos G, Westerberg AW. A review of process synthesis. AIChE J, 1981, 27, 321–351.

[3]   Tedder DW, Rudd DF. Parametric studies in industrial distillation: Part 1. design comparisons. Part 2. Heuristic optimisation. Part 3. Design methods and their evaluation. AIChE J, 1978, 24, 303.

[4]   Fidkowski ZT, Krolikowski L. Minimum energy requirement of thermally coupled distillation dystems. AIChE J, 1987, 33, 654.

[5]   Vu LD, Gadkari PB, Govind R. Analysis of ternary distillation column sequences. Sep Sci Technol, 1987, 22, 1659.

[6]   Glinos KN, Malone MF. Optimality regions for complex column alternatives in distillation systems. Chem Eng Res Des, 1988, 66, 229.

[7]   Carlberg NA, Westerberg AW. Temperature-heat diagrams for complex columns. 2. Underwood's Method for Side Strippers and Enrichers. Ind Eng Chem Res, 1989, 28, 1379.

[8]   Carlberg NA, Westerberg AW. Temperature-heat diagrams for complex columns. 3. Underwood's method for the Petlyuk configuration. Ind Eng Chem Res, 1989, 28, 1386.

[9] Agrawal R, Fidkowski ZT. Are thermally coupled distillation columns always thermodynamically more efficient for ternary distillations? Ind Eng Chem Res, 1998, 37, 3444.

[10] Re'v E, Emtir M, Szitkai Z, Mizsey P, Fonyo' Z. Energy savings of integrated and coupled distillation systems. Comput Chem Eng, 2001, 25, 119.

[11] Fidkowski ZT. Distillation configurations and their energy requirements. AIChE J, 2006, 52, 2098–2106.

[12] Heaven DL. Optimum sequencing of distillation columns in multicomponent fraction. M.S. Thesis, University of California, Berkeley, 1969.

[13] Thompson RW, King CJ. Systematic synthesis of separation schemes. AIChE J, 1972, 18, 941–948.

[14] Rathore RNS, VanWormer KA, Powers GJ. Synthesis strategies for multicomponent separation systems with energy integration. AIChE J, 1974, 20, 491–502.

[15] Seader JD, Westerberg AW. A combined heuristic and evolutionary strategy for the synthesis of simple separation sequences. AIChE J, 1977, 23, 951–954.

[16] King CJ. Separation processes. Second edition, McGraw-Hill Book Company, 1980.

[17] Henley EJ, Seader JD. Equilibrium-stage separation operations in chemical engineering. New York: John Wiley & Sons, Inc. 1981.

[18] Nath R, Motard RL. Evolutionary synthesis of separation processes. AIChE J, 1981, 27, 578–587.

[19] Nadgir VM, Liu YA. Studies in chemical process design and synthesis: Part V: A simple heuristic method for systematic synthesis of initial sequences for multicomponent separations. AIChE J, 1983, 29, 926–934.

[20] Rong BG, Kraslawski A. Optimal design of distillation flowsheets with a lower number of thermal couplings for multicomponent separations. Ind Eng Chem Res, 2002(b), 41, 5716–5726.

[21] Kim JK, Wankat PC. Quaternary distillation systems with less than $N-1$ columns. Ind Eng Chem Res, 2004, 43, 3838–3846.

[22] Errico M, Rong BG, Tola G, Turunen I. A method for systematic synthesis of multicomponent distillation systems with less than $N-1$ columns. Chem Eng Process, 2009, 48, 907–920.

[23] Kaibel G, Schoenmakers H. Process synthesis and design in industrial practice. Computer Aided Process Engineering-12, Grievink J, Schijndel JV (eds). Elsevier Science, Amsterdam, 2002.

[24] Agrawal R. Synthesis of multicomponent distillation column configurations. AIChE J, 2003, 49, 379–401.

[25] Shah VH, Agrawal R. A matrix method for multicomponent distillation sequences. AIChE J, 2010, 56, 1759–1775.

[26] Shenvi AA, Shah VH, Zeller JA, Agrawal R. A synthesis method for multicomponent distillation sequences with fewer columns. AIChE J, 2012, 58, 2479–2494.

[27] Caballero JA, Grossmann IE. Generalized disjunctive programming model for the optimal synthesis of thermally linked distillation columns. Ind Eng Chem Res, 2001, 40, 2260–2274.

[28] Caballero JA, Grossmann IE. Thermodynamically equivalent configurations for thermally coupled distillation. AIChE J, 2003, 49, 2864–2884.

[29] Caballero JA, Grossmann IE. Design of distillation sequences: from conventional to fully thermally coupled distillation systems. Comput Chem Eng, 2004, 28, 2307–2329.

[30] Caballero JA, Grossmann IE. Structural considerations and modeling in the synthesis of heat-integrated thermally coupled distillation sequences. Ind Eng Chem Res, 2006, 45, 8454–8474.

[31] Rong BG, Kraslawski A, Nyström L. The Synthesis of thermally coupled distillation flowsheets for separations of five-component mixtures. Comput Chem Eng, 2000, 24, 247–252.

[32] Rong BG, Kraslawski A, Nyström L. Design and synthesis of multicomponent thermally coupled distillation flowsheets. Comput Chem Eng, 2001, 25, 807–820.

[33] Rong BG, Kraslawski A. Optimal design of distillation flowsheets with a lower number of thermal couplings for multicomponent separations. Ind Eng Chem Res, 2002, 41, 5716–5726.

[34] Rong BG, Kraslawski A. Partially thermally coupled distillation systems for multicomponent separations. AIChE J, 2003, 49(5), 1340–1347.

[35] Rong BG, Kraslawski A, Turunen I. Synthesis of functionally distinct thermally coupled configurations for quaternary distillations. Ind Eng Chem Res, 2003, 42, 1204–1214.

[36] Rong BG. A systematic procedure for synthesis of intensified nonsharp distillation systems with fewer columns. Chem Eng Res Design, 2014, 92, 1955–1968.

[37] Rong BG. Synthesis of dividing wall columns (DWC) for multicomponent distillations – A systematic approach. Chem Eng Res Des 2011, 89(8), 1281–1294.

[38] Doherty MF, Malone MF. Concentual design of distillation systems. The McGraw-Hill Companies, Inc. NY, 2001.

[39] Rong BG, Turunen I. A New method for synthesis of thermodynamically equivalent structures for Petlyuk arrangements. Chem Eng Res Design, 2006, 84, 1095–1116.

[40] Rong BG, Turunen I. New heat-integrated distillation configurations for Petlyuk arrangements. Chem Eng Res Design, 2006, 84, 1117–1133.

[41] Rong BG, Kraslawski A. Synthesis of thermodynamically efficient distillation schemes for multicomponent separations. Computer-Aided Chem Eng, 2002, 10, 319–324.

[42] Rong BG, Kraslawski A, Turunen I. Systematic synthesis of functionally distinct new distillation systems for five-component separations. Computer-Aided Chem Eng, 2005, 20, 823–828.

[43] Petlyuk FB, Platonov VM, Slavinsk DM, Thermodynamically optimal method for separating multicomponent mixtures. Int Chem Eng, 1965, 5, 555–561.

[44] Fidkowski ZT, Królikowski L. Minimum energy requirements of thermally coupled distillation systems. AIChE J, 1987, 33(4), 643–653.

[45] Rong BG, Kraslawski A, Turunen I. Synthesis of heat-integrated thermally coupled distillation systems for multicomponent separations. Ind Eng Chem Res, 2003, 42, 4329–4339.

[46] Rong BG, Turunen I. Synthesis of new distillation systems by simultaneous thermal coupling and heat integration. Ind Eng Chem Res, 2006, 45, 3830–3842.

[47] Rong BG, Kraslawski A, Turunen I. Synthesis and optimal design of thermodynamically equivalent thermally coupled distillation systems. Ind Eng Chem Res, 2004, 43, 5904–5915.

[48] Rong BG, Errico M. Synthesis of intensified simple column configurations for multicomponent distillations. Chem Eng Proc: Process Intensification 2012, 62, 1–17.

[49] Wright RO. Fractionation Apparatus. U.S. Patent 2,471,134, 1949.

[50] Kaibel G. Distillation columns with vertical partitions. Chem Eng Technol, 1987, 10, 92.

[51] Christiansen AC, Skogestad S, Lien K. Complex distillation arrangements: extending the Petlyuk ideas. Comput Chem Eng, 1997, 21, S237.

[52] Rong BG, Turunen I. Process intensification for systematic synthesis of new distillation systems with less than N − 1 columns. Computer-Aided Chem Eng, 2006, 21, 1009–1014.

[53] Becker H, Godorr S, Kreis H, Vaughan J. Partitioned distillation columns – Why, When & How. Chem Eng, 2001, 108, 68.

Adriana Freites Aguilera, Pasi Tolvanen, Victor Sifontes Herrera,
Jean-Noël Tourvielle, Sébastien Leveneur, and Tapio Salmi

3 Reaction intensification by microwave and ultrasound techniques in chemical multiphase systems

Abstract: The chemical industry of today is rather conservative concerning reactor technology and the forms of energy used in chemical synthesis. However, ultrasound and microwave technologies are attractive methods for the intensification of chemical processes, which is a subarea of process intensification known as reaction intensification. The aim is to create processes that are more rapid, energy saving, and cleaner through the use of microwaves and ultrasound, which should lead to more compact factories in the future. This chapter briefly describes the history and mechanisms underlying these two types of intensification methods. The effects of microwave and ultrasound technologies on selected green processes are highlighted by some illustrative case studies, including fatty acid epoxidation, enantioselective hydrogenation, sugar alcohol production, and starch oxidation.

Keywords: Reaction intensification, microwave irradiation, ultrasonification, sonochemistry, vegetable oil epoxidation, starch oxidation, sugar hydrogenation

3.1 Microwave irradiation

3.1.1 Background

Microwave irradiation is a form of electromagnetic energy and, like every kind of electromagnetic energy, is composed of both electric and magnetic elements. Microwave frequencies are located in the electromagnetic spectrum between 300 MHz and 30 GHz and the wavelengths vary from 1 m to 1 cm. Microwave frequencies are situated between infrared radiation and radio frequencies, as illustrated in Figure 3.1.

The prefix "micro-" in the term "microwave" denotes that the wavelengths are "small" compared with waves used in radio broadcasting. Domestic microwave heaters use the specific frequency of 2.45 GHz and a wavelength of 12.2 cm to avoid interfering with radar and telecommunication operations.

An American physicist, Percy LeBaron Spencer, accidentally discovered the heating effect of microwaves while working with radar applications. Spencer was working with an active radar device when he noticed that a chocolate bar in his pocket had melted. He decided to investigate the performance of microwaves in food heating and constructed the first microwave oven. In subsequent years, a company named Raytheon patented the invention. In 1952, the first commercial microwave oven was

https://doi.org/10.1515/9783110465068-003

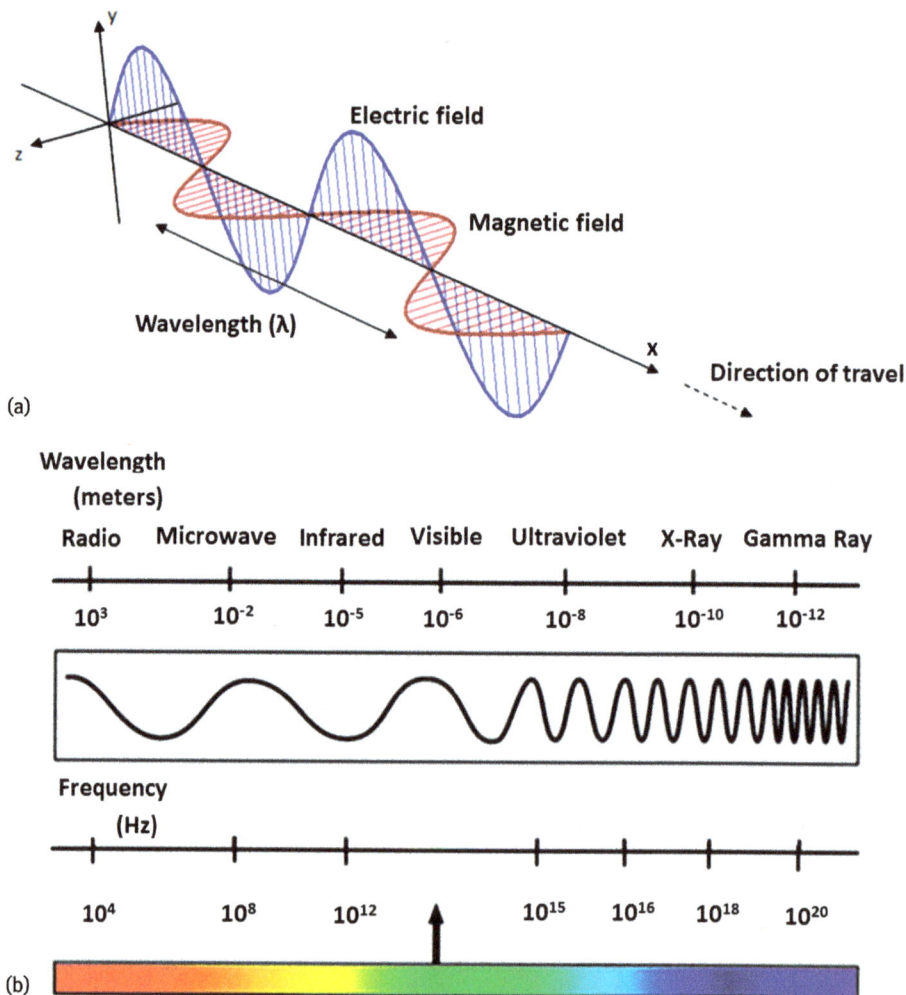

(a)

(b)

Fig. 3.1: Electromagnetic irradiation: (a) electric and magnetic components and (b) electromagnetic spectrum.

put on sale, although the high cost and large dimensions did not make it accessible to the general public. By the late 1960s, commercial microwave ovens were of a reasonable size (counter top) and an affordable price, which has since made them one of the most common household items. A massive increase in the use of microwave ovens occurred between 1970 and 1980 [1, 2].

In addition to domestic applications, the use of microwaves has extended to industrial applications and research purposes. A huge increase in microwave-assisted studies in the scientific field has occurred since the late 1990s, as demonstrated by Figure 3.2.

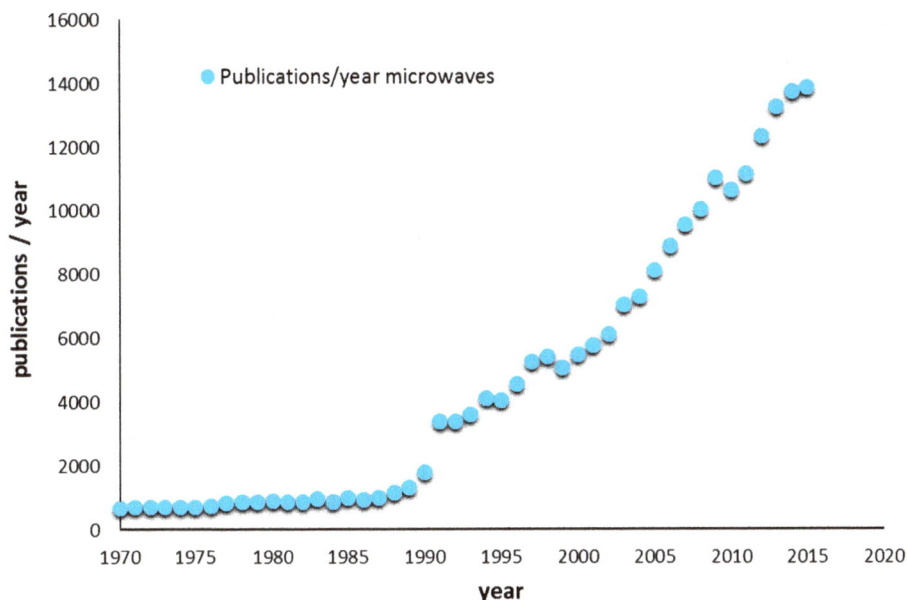

Fig. 3.2: Scientific publications on microwave applications, by year [3].

Unlike conventional heating methods (convection or conduction), microwave irradiation produces heating generated in the bulk, which offers more uniform heating patterns. There is no direct contact between the energy source and the reaction mixture, hence no hot spots appear in the reaction system (e.g., vessels, pipes, catalysts, and any other surface in contact with the reaction mixture) that could contribute to generation of side products or loss of energy. Therefore, microwave heating assures optimal energy consumption [4].

3.1.2 Heating mechanisms

Microwave heating is caused by interaction of an alternating electromagnetic field with a material. Microwaves can generate an increase in temperature in a sample by several mechanisms, which are briefly summarized next.

3.1.2.1 Dipolar polarization

When a molecule has a dipole moment, it is sensitive to external electric fields and attempts to rotate in order to align itself with the field. Depending on the properties of the substance and the frequency of the waves, this rotation generates heat. For waves in the microwave spectrum, the frequency can create a rotation response in the dipoles; however, such frequencies are too high for the particles to follow the field precisely.

This creates a phase difference between the frequency and the dipoles that are constantly trying to align themselves with the field. The constant movement of the particles increases collisions and friction between particles, which generates a temperature increase in the substance.

This description does not apply to gases because gas molecules are too far apart to interact with each other. Alignment of the dipoles with the field happens practically instantaneously and, thus, no heating is produced [1, 5].

3.1.2.2 Ionic conduction

For a sample containing ions in solution, an applied electric field causes the ions to travel from one side to another through the solution. This phenomenon causes an increase in the kinetic energy of the particles, which translates into a rise in temperature. The amount of ions present in the solution is proportional to the increasing temperature effect generated by the ionic conductive loss [4, 6].

3.1.2.3 Interfacial polarization

Interfacial polarization is a mechanism of limited importance in microwave heating. Interfacial polarization plays a part in dielectric heating in the case of a material that comprises a nonconductive phase with conductive inclusions (e.g., suspensions). At microwave frequencies, dipolar polarization has a minor effect on temperature increase [7].

3.1.3 Dielectric loss and permittivity

Depending on its properties, every material has the ability to convert electromagnetic energy into potential and kinetic energy, which results in an increase in temperature (thermal energy). Dielectric loss (δ) quantifies this dissipation of energy and can be calculated through the measurable constants ε' and ε'' as follows:

$$\delta = \tan^{-1} \frac{\varepsilon''}{\varepsilon'} \tag{3.1}$$

where ε' represents the dielectric constant (also called relative permittivity), which measures the ability of a molecule to be polarized by an electric field. The constant ε'' represents the loss factor, characterizing how efficiently the energy is converted into heat. The permittivity (ε) merges both dielectric constant and dielectric loss and expresses the ability of a material to transmit or permit an electric field. It is defined by the following expression:

$$\varepsilon(\omega) = \varepsilon'(\omega) + \varepsilon''(\omega)i \tag{3.2}$$

where, ω is the angular frequency. These properties depend on the frequency, composition, and temperature [5].

3.1.4 Selective heating

Each kind of material has a different response to microwave irradiation. Depending on the interactions, materials can be roughly classified into three groups as follows (Figure 3.3) [1, 9]:

Transparent materials: These are also called insulators because they do not heat up in response to microwaves; the radiation is completely transmitted through the material. Some examples are glass, alumina, silica, and plastics.

Absorbent materials: The electromagnetic energy is received by the material, causing an increase in temperature. They are also called dielectric materials and are composed of polar molecules, which easily interact with microwaves. Water is a microwave absorbent par excellence. Other microwave absorbents are ethylene glycol, ethanol, methanol, 2-propanol, and formic acid [9]. Some of these components are commonly used as solvents for microwave syntheses [10–13].

Reflective materials: These are mostly metals and they possess properties that cause deviation of the microwaves when they hit the material surface; therefore, they are not heated. They are also called conductors because they transfer the electromagnetic energy received.

Materials in nature can (and usually do) possess the properties of more than one group in this classification; in fact, there are no perfectly transparent materials in nature. Table 3.1 shows the most prominent interactions for substances present in the reactions studied in this chapter.

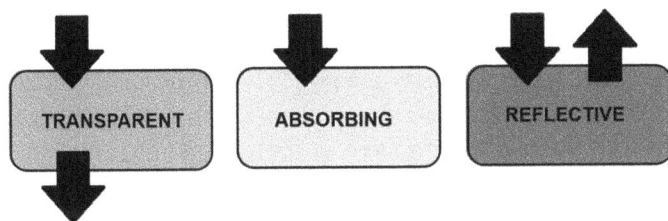

Fig. 3.3: Types of material interactions with microwaves.

Tab. 3.1: Examples of interactions of materials with microwaves.

Substance	Interaction
Water	Absorber
Acetic acid	Absorber
Peroxyacetic acid	Absorber
Hydrogen peroxide	Moderate absorber
Oleic acid	Transparent
Epoxyoleic acid	Transparent

The different interactions of microwaves with materials enable another peculiar kind of thermal effect, which is often called selective heating. Selective heating is produced as a result of the properties and complexity of the materials, which sometimes create favorable conditions for reaction under microwave irradiation [14]. This can be especially useful for heterogeneously catalyzed systems. For example, if microwaves are applied to a catalyzed system, the catalyst surface (in this case metal) is heated instead of the adsorbed substance on the metal, which results in an increase in reaction rate.

3.1.4.1 Nonthermal effect

Microwaves are principally used to increase temperature through irradiation. However, some studies claim the existence of a nonthermal effect (also called a non-Arrhenius effect) that produces a rate enhancement. It has being stated that perturbation of the dipole moments influences interfacial reactions, generating higher reaction rates and yields [14, 15]. This phenomenon is often argued to be nonexistent by many researchers [16–21], but others claim that the non-Arrhenius effect enhances the collision efficiency by joint orientation of polar molecules involved in the reaction, thereby decreasing the activation energy [22–25]. Several studies describe the different conditions and requirements to be fulfilled for this effect to be valid [1, 22–24, 26].

3.1.4.2 Superheating

In microwave heating, it is sometimes possible to heat a liquid above its normal boiling point. This phenomenon is commonly called superheating. It occurs because the rapid heating in the solution prevents adequate dissipation of energy through the liquid, limiting the formation of boiling nuclei. Nucleation of the vapor bubbles occurs in points where surface tension is low and internal energy is at a minimum (Figure 3.4). Generally, this happens in the scratches or pits in the surface of the vessel (heterogeneous nucleation). Under conventional heating, because the temperature of the walls increases, nucleation occurs in the specific places where vaporization is initiated more easily. In the case of microwave heating, energy dissipation occurs in the bulk of the reaction mixture and the temperature of the walls of the container is lower. For this reason, the vaporization process begins when the internal energy overcomes the intermolecular attractive forces (homogeneous nucleation), which often requires a temperature higher than the normal boiling point [1, 27].

Several studies claim that microwaves can increase the boiling point by up to 26 °C. It is assumed that superheating is responsible for the enhancement of chemical reactions [6].

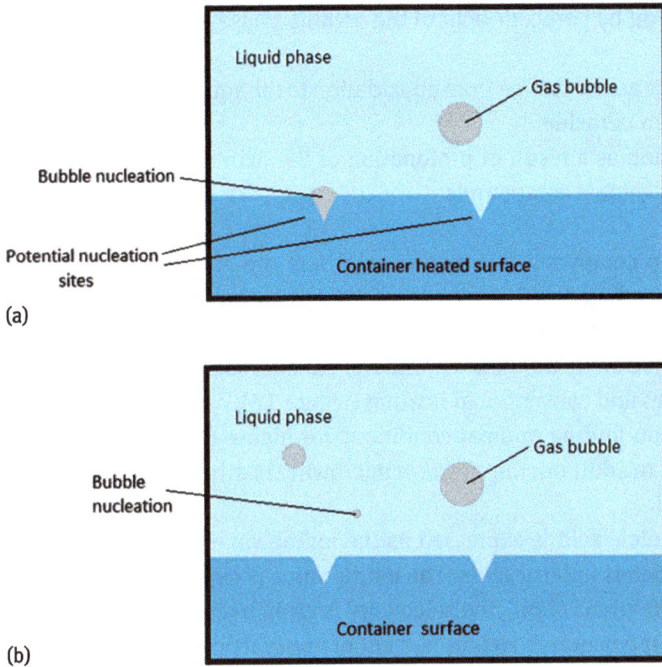

(a)

(b)

Fig. 3.4: (a) Heterogeneous nucleation and (b) homogeneous nucleation.

3.1.5 Case 1: conventional heating versus microwaves for epoxidation of vegetable oils

Epoxidized vegetable oils are used for producing plasticware and intermediates for the synthesis of polyols, glycols, olefinic compounds, and stabilizers for polymers. Moreover, they are promising products for the development of biolubricants.

Epoxidation of oleic acid, as model compound for vegetable oils, was studied at the University of Åbo Akademi in Finland [28]. The reaction was performed in a semi-batch reactor. The oxidative reagent was peroxyacetic acid formed in situ. The system comprised an aqueous phase containing hydrogen peroxide, acetic acid, and water and an organic phase containing oleic acid and the generated epoxidized oleic acid. Moreover, some side products were obtained when the oxirane ring opened.

Epoxides are very promising products for the chemical industry because the epoxy group can be easily functionalized. Due to the reactivity of the highly strained ring, they can be used for the synthesis of a variety of chemicals. However, this characteristic reactivity also triggers a series of side reactions that lead to ring opening.

The scheme of the reaction involves the following steps (Figure 3.5) [28]:
- formation of peracetic acid by acetic acid and hydrogen peroxide in the aqueous phase
- transfer of peracetic acid into the organic phase

- epoxidation of the oil by peracetic acid in the organic phase to regenerate acetic acid
- transfer of the acetic acid resulting from epoxidation to the aqueous phase to react again with hydrogen peroxide
- ring-opening reaction as a result of protonation of the oxirane, producing poly-alcohols, esters, or ketonic compounds

The experimental setup comprised a loop reactor, where the mixture was pumped through a cavity and irradiated with microwaves. A heat exchanger was integrated into the system to replace microwave heating when necessary. Experiments were con-ducted in duplicate with exactly the same conditions but alternating the type of heat-ing between microwaves and conventional heating (Figure 3.6).

The work focused on finding optimal conditions for higher reactant conversion and product selectivity. In addition, the effects of microwave heating and conventional heating were compared.

The conversion of oleic acid is expressed as the iodine value, which represents the amount of double bonds in fatty acids. The iodine value is expressed as grams of iodine consumed by 100 grams of oil. The selectivity is given by the oxirane number, which expresses the composition in weight percent of epoxyoleic acid in the organic phase.

Experimental results showed that a clear enhancement in epoxidation kinetics was accomplished with microwave heating as compared with conventional heating (Figure 3.7). This phenomenon was attributed to selective heating of the aqueous phase by microwaves, which enabled a higher interfacial mass transfer. A tempera-ture gradient appeared between the phases because the components in the aqueous

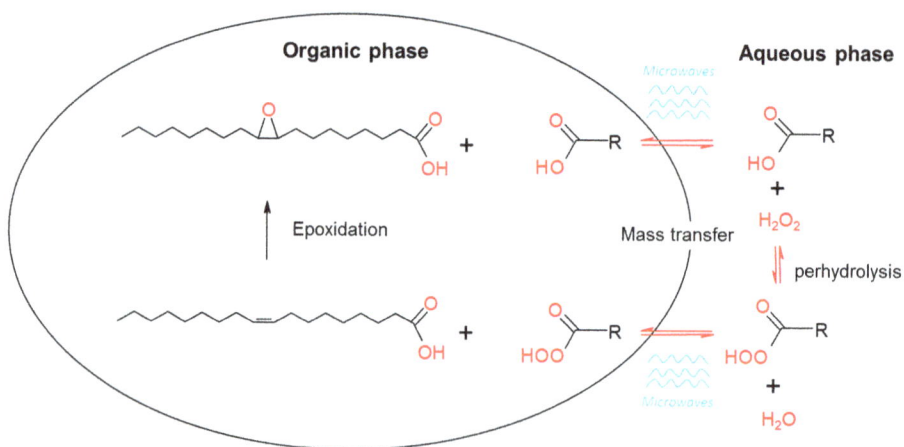

Fig. 3.5: Epoxidation in situ of oleic acid with acetic acid, assisted by hydrogen peroxide [28].

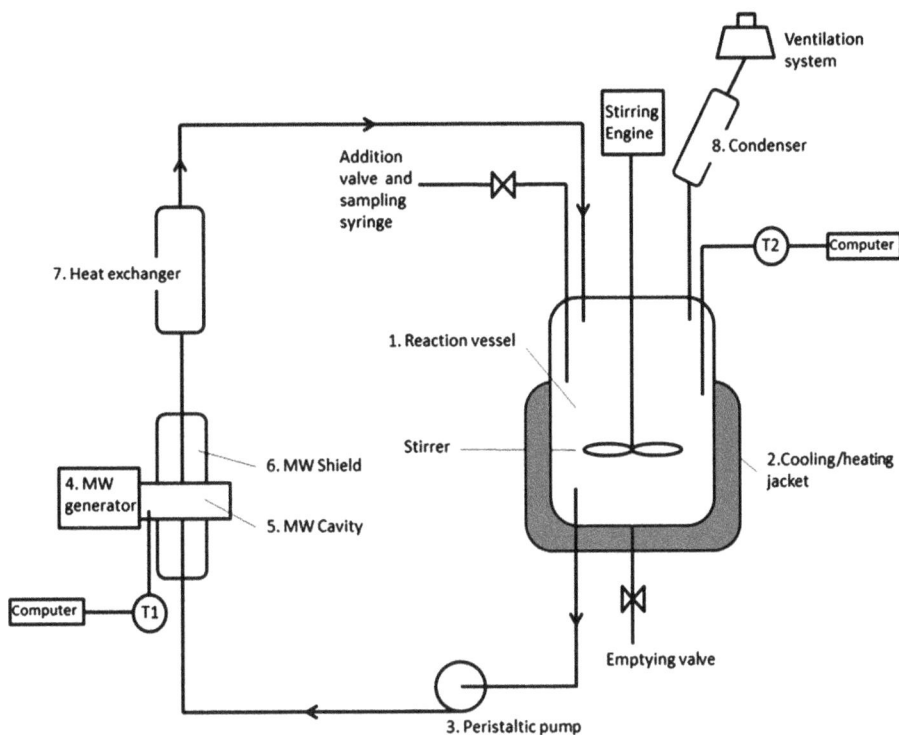

Fig. 3.6: Configuration of the reactor set-up [28].

phase are absorbers or moderate absorbers of microwaves, whereas components in the organic phase are transparent (Table 3.1).

It is important to note that, at higher temperatures, the selectivity decreased as a result of side reactions, which led to decomposition of the oxirane ring. However, the conversion at high temperatures was still remarkably higher than that obtained with conventional heating.

3.1.6 Case 2: effect of resonant microwave fields on temperature distribution in time and space

The mechanisms and physics of microwave heating differ profoundly from those in conventional heating. Heating occurs rapidly and its patterns are difficult to predict. A joint study by the University of Delft and the University of Leuven revealed the characteristics of heating within time and space using a resonant microwave field [30].

Many studies [10, 25, 26, 31–33] report unexpectedly high conversions over short periods using microwaves. The higher reaction rates are often explained by nonuniform heating. Even though nonuniform microwave heating is sufficient in some appli-

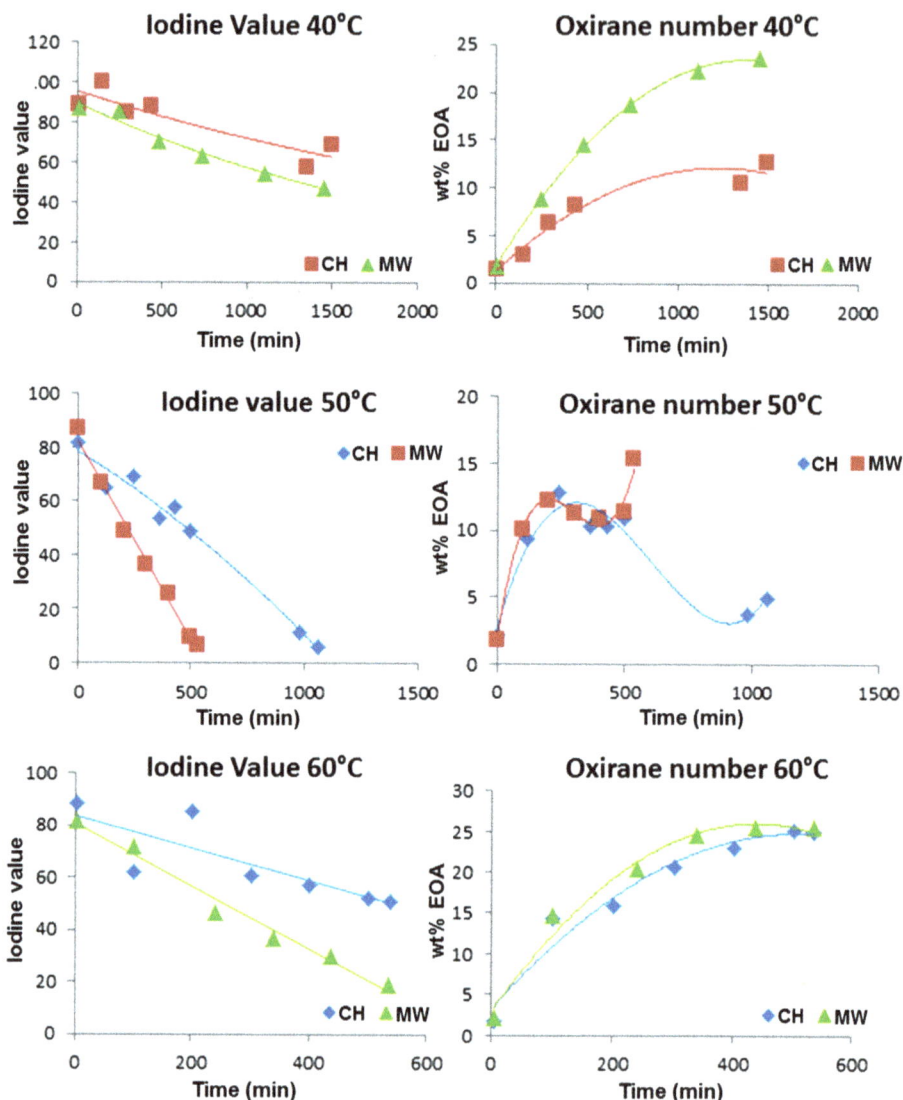

Fig. 3.7: Iodine values for 40, 50, and 60 °C with conventional heating and microwaves (left). Epoxy content for 40, 50, and 60 °C with conventional heating and microwaves (right) [28].

cations (e.g., domestic microwave ovens), processes such as packed beds, continuous flow (milli)reactors, or polymerization systems that are too viscous to be stirred effectively require knowledge and control of the spatial distribution of the microwave field. The same is required for reliable fundamental studies on microwave chemistry interactions.

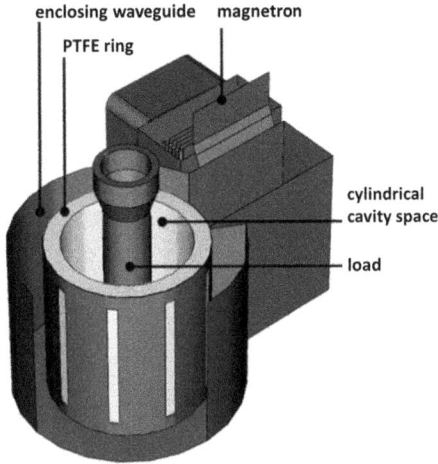

enclosing waveguide   magnetron

PTFE ring

cylindrical
cavity space

load

Fig. 3.8: Microwave circuit in Discover: cutaway view with the main parts of the microwave circuit indicated [30].

The microwave setup consisted of a cylindrical cavity space in which the load was placed [30], as seen in Figure 3.8. The loads under study were three round-bottomed insulated vials containing water. The pairs of inner and outer diameters of the vials were 10 and 13 mm, 12 and 16 mm, and 20 and 24 mm.

The cylindrical cavity comprised a PTFE ring placed inside its circumference. The cylindrical space was enclosed on the outside by a waveguide that was closed at one end and connected to the microwave source (i.e., the magnetron) at the other end. The frequency of the field generated by the magnetron was 2.45 GHz. The metal wall between the cylindrical space and the waveguide was slotted. Electromagnetic energy was exchanged between the waveguide and cylindrical cavity through these slots. Four temperature sensors were used for analysis, as well as thermosensitive paper strips fitted into the vials.

From the results obtained with both analyzing techniques (fiber optic sensors and thermal paper), it was concluded that the samples were heated in a nonuniform way. Moreover, it was shown that the heating rate not only varies in space, but also in time. It is important to note that the temperature rise would not have been registered by a probe placed on the bottom of the vial; this highlights the risk of missing the dominant temperature trends in microwave-heated systems by unfortunate sensor placement.

The heating patterns presented in Figures 3.9 and 3.10 fluctuate considerably with the diameter of the vials. Heating is distributed nonuniformly in space and is highly sensitive to variations in geometric and operating parameters.

This study revealed that the field behaves in a highly complex manner. The behavior was attributed to the resonant nature of the microwave field. The standing wave patterns associated with resonance were shown to be strongly dependent on the geometry of the vials and the properties of the medium in which the microwave field is present.

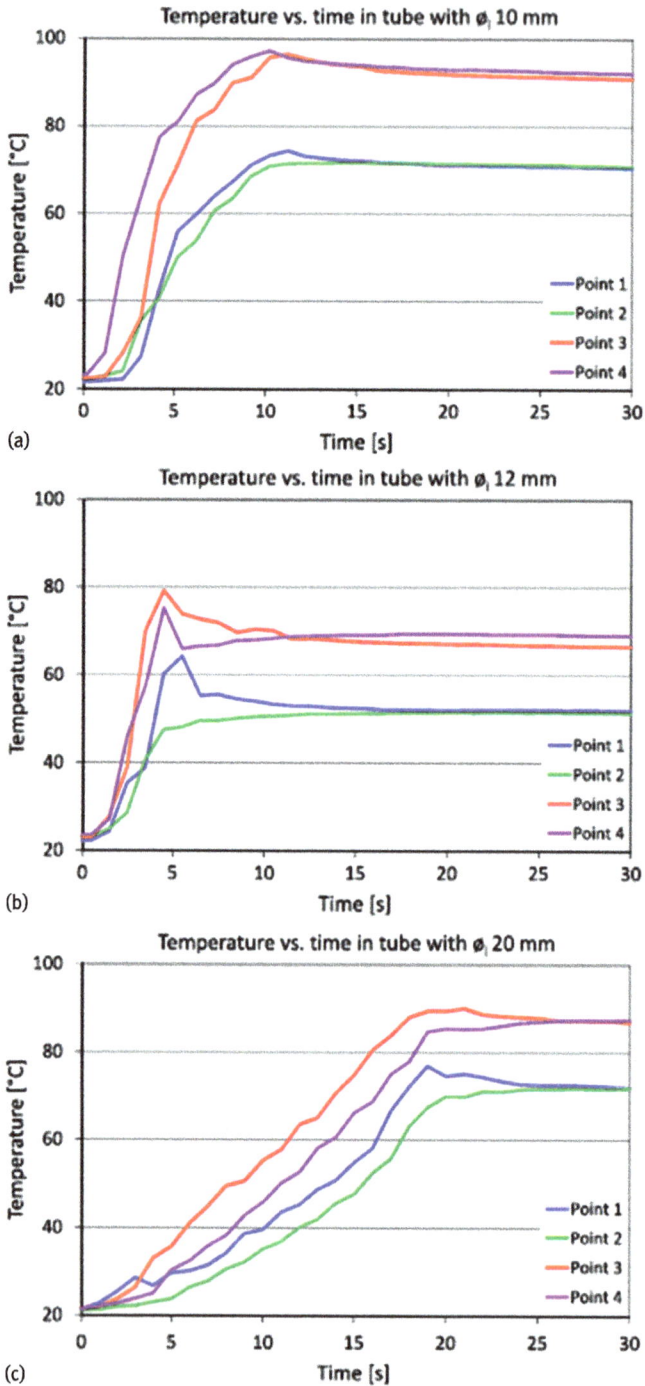

(a)

(b)

(c)

Fig. 3.9: Temperature transients for vials of different diameters, as measured by fiber optic sensors in the vials [30]. Tube diameters (a) 10 mm, (b) 12 mm, and (c) 20 mm.

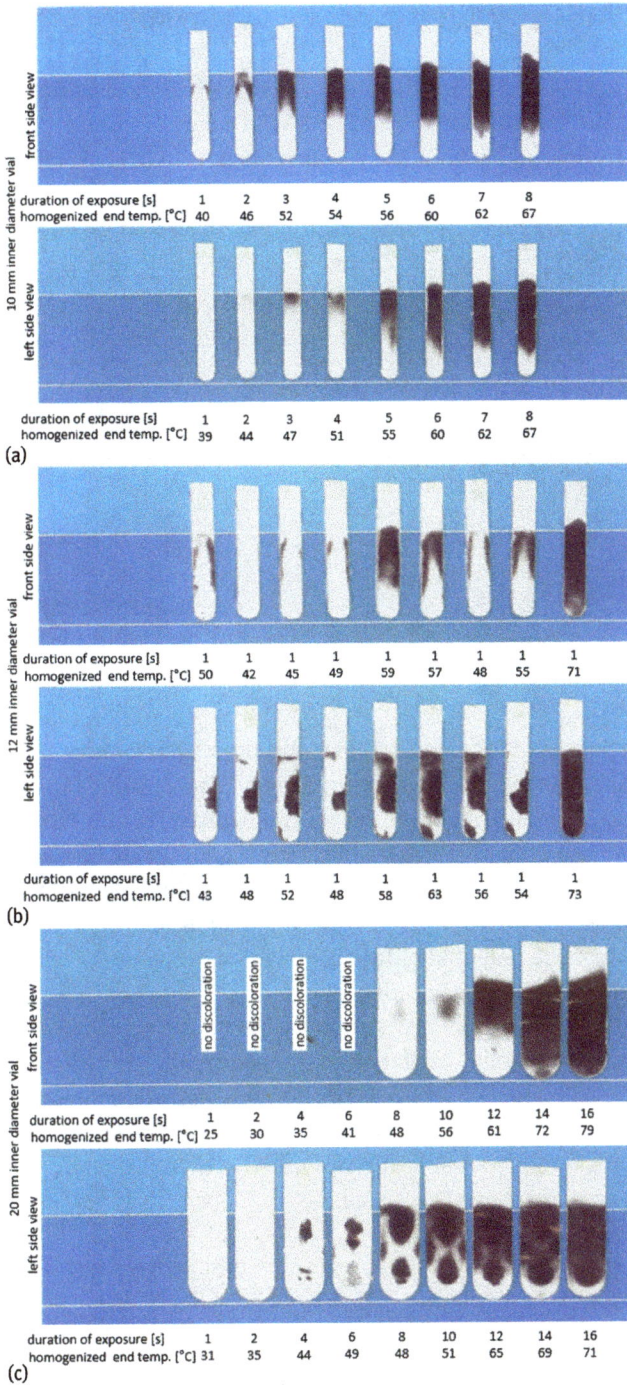

Fig. 3.10: Discoloration of strips of thermal fax paper placed in vials of different diameters and containing water to a level of 40 mm [30]. Vial diameter (a) 10 mm, (b) 12 mm and (c) 20 mm.

3.2 Process intensification by ultrasound: sonochemistry

Chemical processes that are of heterogeneous character tend to suffer from mass transfer limitations. Moreover, if the reaction is performed on a solid catalytic surface, then catalyst deactivation can also occur. Such undesired phenomena can be avoided by using some kind of process intensification. Ultrasound irradiation is a promising tool for this purpose. The enhancement obtained using ultrasound is based on the cavitational effects introduced in the liquid, as shown in Figure 3.11 [34].

Fig. 3.11: Imploding cavity in a liquid irradiated with ultrasound [34].

Ultrasound is utilized by humankind for many different purposes. Ultrasonic devices are used to detect objects and measure distances in the form of echo or sonar (sound navigation and ranging) and to produce real-time images (sonography) of human tendons, muscles, and joints. In obstetric ultrasonography, ultrasound can create visual images of unborn fetuses. Industrially, ultrasound is used for cleaning, mixing, and accelerating chemical processes. A broad range of application fields are listed in Table 3.2. After discovery of the huge potential of ultrasonification in the chemical industry, the new term "sonochemistry" was introduced.

The history of interest in ultrasound and cavitational effects dates back over 100 years. The first description of ultrasonic cavitation was given by the British shipbuilders Thornycroft and Barnaby in 1895, who described the erosion of a submarine propeller. The first time ultrasound was utilized to enhance a chemical reaction dates back to 1927 [37]. Since then, several excellent reviews have been published on the effect of ultrasound on chemical reactions [35, 37, 38].

3.2.1 What is ultrasound?

Ultrasound is simply sound pitched above the frequency border of human hearing [39]. It is the part of the sonic spectrum ranging from 20 kHz to 10 MHz, corresponding to wavelengths of 10 cm to 10^{-3} cm. Sonochemistry is the general term for

Tab. 3.2: Applications of ultrasound [36].

Chemical and allied industries	Other applications
Air scrubbing	Abrasion
Atomization	Cleaning
Cell disruption	Coal–oil mixtures
Crystal growth	Cutting
Crystallization	Degradation of powders
Defoaming	Dental descaling
Degassing	Drilling
Depolymerization	Echo-ranging
Dispersion of solids	Erosion
Dissolution	Fatigue testing
Drying	Flaw detection
Emulsification	Flow enhancement
Extraction	Imaging
Filtration	Medical inhalers
Flotation	Metal–grain refinement
Homogenization	Metal tube drawing
Sonochemistry	Nondestructive testing of metals
Stimulus for chemical reactions	Physiotherapy
Treatment of slurries	Plastic welding
	Powder production
	Soldering
	Sterilization
	Welding

the use of ultrasound to enhance chemical reactions and processes. Unlike the use of microwaves, ultrasound is not radiation; it does not travel at the speed of light but at the speed of sound, and therefore needs a physical medium in which to travel. Microwaves can excite both outer and inner electrons of a reactant, leading to the desired chemical reaction. The utilization of ultrasound has a different effect; the ultrasound waves produce local cavitations and "hot spots," which, in turn, can create huge thermal and shear stress for the substance (Figure 3.12). Cavitation can be described as the generation, subsequent growth, and collapse of cavities, leading to a release of large amounts of energy over a very small area [39]. This produces very high energy densities, of the order of $10^{18}\,\mathrm{kW\,m^{-3}}$. Conditions such as a few thousand atmospheres of pressure and several thousand degrees Kelvin are believed to exist for a very short time [40]. Moreover, implosion of the formed cavitation bubble results in liquid jets, which can have velocities of up to $280\,\mathrm{m/s}$. Therefore, ultrasound and microwave irradiation can be said to be the simplest, cheapest, and most efficient nonconventional heating methods available.

Although the formation, growth, and collapse of ultrasonic cavities is a complex system, detailed mathematical expressions have been formulated to describe acoustic

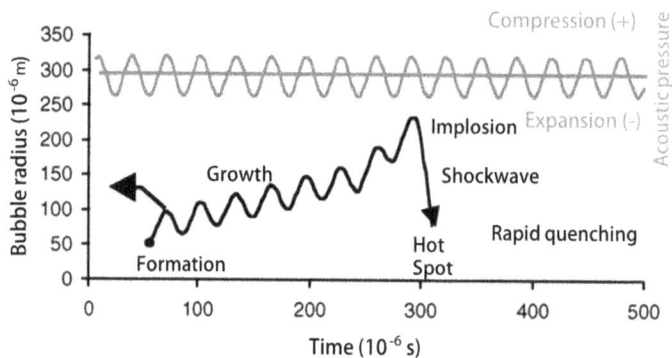

Fig. 3.12: Cavitation process [36]: formation, growth, and violent collapse of bubbles, generating very high pressures and temperatures.

bubble dynamics and have been reviewed [36]. Here, we discuss the following factors that affect the intensity of the cavitation: (1) ultrasonic frequency, (2) applied power, (3) the solvent used, (4) gas type and content, and (5) external temperature and pressure. Additional reviews that provide a general overview of sonochemistry can be found in the literature [35, 41–43].

The interest in sonification is still growing. A renaissance in sonochemistry started in the 1980s. Since then, the annual number of publications related to sonochemistry has constantly increased, as illustrated by Figure 3.13.

The emergence of new sonification technology has attracted interest in a broad range of fields. The subjects of recent reviews include organic chemistry synthesis [45], the influence of ultrasound on clay minerals [46], destructive sonification of pharmaceuticals in wastewater [47, 48], and ultrasound-assisted extraction [49]. The effect of sonification on the growth rate of nanoparticles (e.g., gold nanoparticles from chloroauric acid, $HAuCl_4$) has also been reported [50, 51]. Ultrasound-enhanced enzymatic oxidation of a variety of compounds is also reviewed [52]. In the light of these reviews, it can be concluded that ultrasound technology is being intensively researched in a broad range of scientific fields, demonstrating a lucrative opportunity for attempts aimed at process intensification.

3.2.2 Ultrasonification techniques

Several different types of equipment can be utilized for achieving ultrasonification. The most common is the cleaning ultrasonic bath, which is widely used to homogenize small amounts of substance or to break the crystal structure of solid particles. This equipment is classified as a low intensity (1–2 W/m^2) system, and typical applications are in liquid–solid systems. An ultrasonic bath is standard equipment in most laboratories.

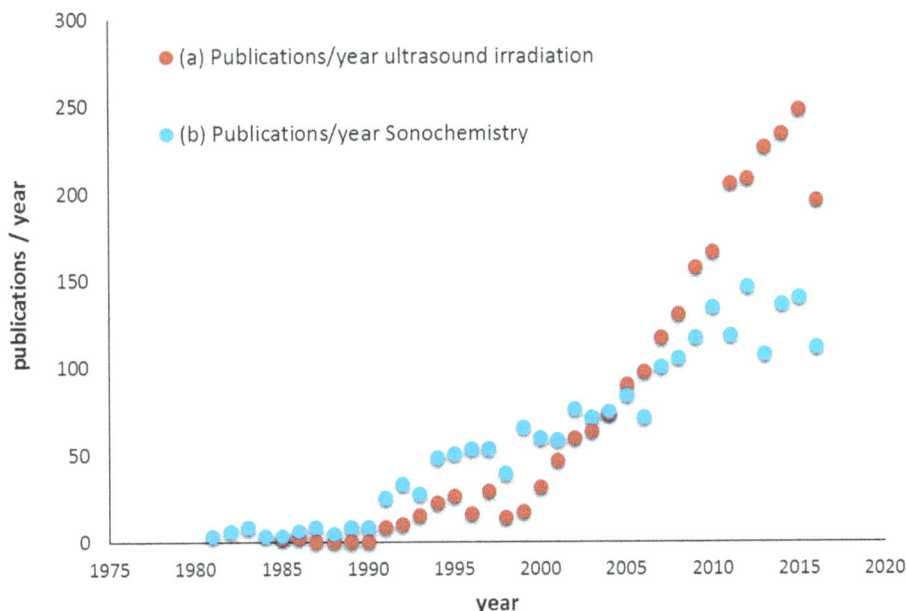

Fig. 3.13: Publications per year with topic keyword (a) sonochemistry and (b) ultrasound irradiation (field restriction: chemistry), fetched from Web of Science [44].

For reactor purposes, several different constructions exist. The ultrasonic device can be applied either inside the chemical reactor (i.e., probe or horn system; see Section 3.2.3.2) or outside the reactor. In this case, the transmitters are pointed at the reactor wall (see Section 3.2.4). Typically, the horn type is several times more efficient at transmitting energy into the reactor solution, but scale-up is problematic.

In the last 15 years, several different novel ultrasonic reactor designs have been developed. The different types, such as multiple transducers with or without multiple frequencies and use of direct or indirect sonication, are nicely reviewed [35]. For a broad range of ultrasonic devices, the reader is referred to the book by Papadakis [53]. Nevertheless, despite the wide range of applications of ultrasound and sonochemistry, this type of reactor is mostly used at the laboratory scale. There are still difficulties for practical use of sonochemical reactors at the industrial scale. The greatest challenge for scale-up is in obtaining a uniform distribution of cavitational activity in large reactors [54].

3.2.3 Case 1: catalyst activation by ultrasonification

Heterogeneous catalysts tend to lose their activity during use. This so-called deactivation is a problem for experiments at both laboratory and industrial scales. The declining activity is a result of complex kinetic and thermodynamic phenomena, result-

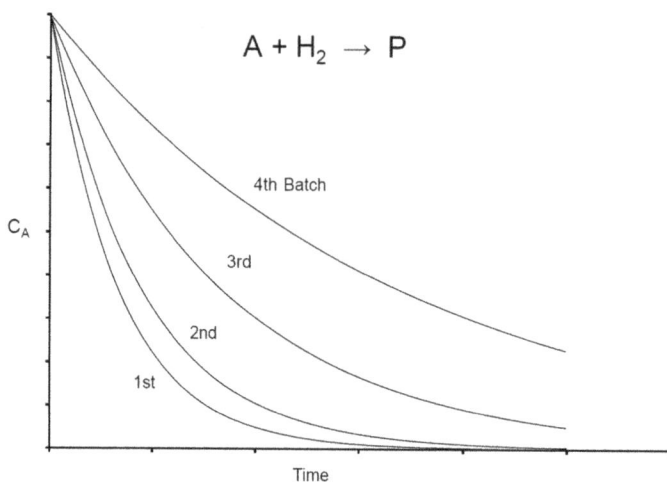

$$A + H_2 \rightarrow P$$

Fig. 3.14: Deactivation of the catalyst after successive experiments performed in batch mode.

ing in a decrease in product capacity and quality in the synthesis of fine chemicals. When an active catalyst is used for the first time, the desired reaction is carried out rapidly and reactant is efficiently converted into the desired product. However, when the same (already spent) catalyst is put to work for a second time, the reaction velocity is significantly slower and the needed reaction time is longer (Figure 3.14). With each successive experiment, the activity becomes monotonically lower until the catalyst is useless. Usually, the only choice for the plant engineer is to keep adding new, fresh catalyst to maintain the reaction time at a reasonable level. This eventually leads to larger solid catalytic material, which in turn leads to problematic mass transfer issues.

To take a closer look at the reason for deactivation, one must look at exactly what happens on the surface of the catalytic solid particles. The chemical reason for catalyst deactivation is most often strong adsorption of organic components or inorganic impurities (e.g., sulfur and phosphor compounds) onto the solid catalyst particle. Free vacant sites on the surface become occupied by so-called coke. Hence, the desired surface reactions become impossible. Furthermore, the deposited coke is extremely difficult to remove. Catalyst deactivation (e.g., coke formation, poisoning, sintering) can be experimentally determined by performing tests in a continuous packed bed reactor, into which fresh reactant is constantly fed. If the catalyst is stable (i.e., no deactivation occurs), a stable steady state is reached and the final concentrations at the reactor outlet no longer change (Figure 3.15).

A remedy to the problem of deactivation is to clean the catalyst particle surface during the course of the reaction. Ultrasonification is a possible method for achieving this. Ultrasound waves produce strong local impulses of energy (heat, pressure) in the reactant solution, leading to the formation of small gas bubbles (cavitation bub-

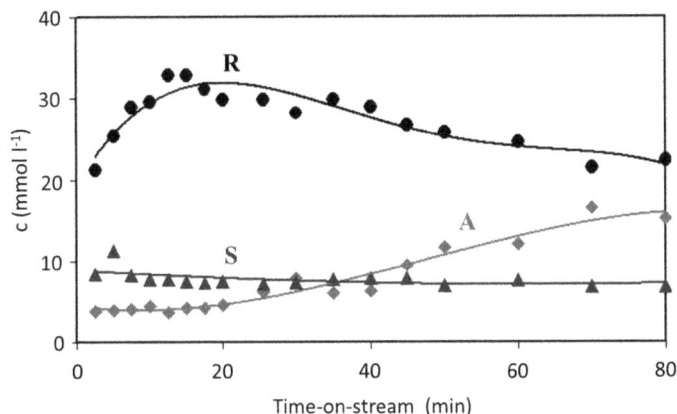

Fig. 3.15: Illustration of catalyst deactivation testing in a continuous packed bed reactor for a parallel reaction system: A → R and A → S [55].

bles), which collide and eventually collapse on the catalyst surface. Through this, the surface is mechanically cleaned and the catalyst is regenerated.

Several studies have been performed on the possibility of regenerating the catalyst surface using ultrasound. An excellent example is the case of hydrogenation of 1-phenyl-1,2-propanedione, an intermediate in pharmaceutical production. The reaction scheme is illustrated in Figure 3.16. The hydrogenation reaction takes place on the surface of a Pt catalyst particle (i.e., fiber). One of the products (B) is of significance because it is the component used for the synthesis of ephedrine; the products C–G are undesired byproducts. It is possible to shift the product distribution towards the R-isomer (B) by introducing a catalyst modifier (cinchonidine), which steers the adsorption of reactant A in such a way that an excess of the desired isomer (B) is formed [56, 57]. Without this modifier, a 50/50 distribution of isomers B and C is obtained on the Pt catalyst.

Application of ultrasound (horn-type sonification in an autoclave) produced a clear enhancement of reactant conversion; 75% was converted in only 53 min in the presence of ultrasound, whereas the same conversion without ultrasound required 87 min (Figure 3.17). This corresponds to an intensification factor of 1.64, meaning that the reactor volume can be reduced by a factor of 1.64.

3.2.3.1 Mathematical modeling of ultrasound intensification

Ultrasound intensification can be mathematically modeled by introducing a constant k_{US} for the removal of impurities on the catalyst surface and a constant k_d for catalyst deactivation. The ratio k_{US}/k_d is an indicator of catalyst deactivation and reaction intensification. As shown in Figure 3.18, the complex reaction system can be modeled

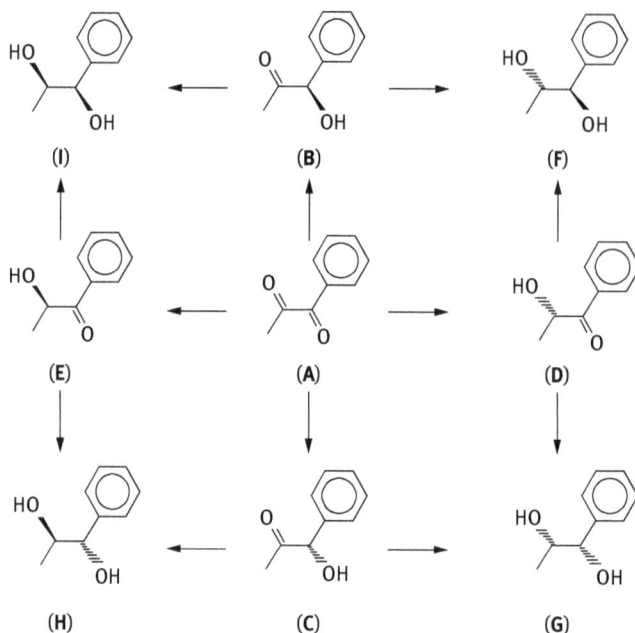

Fig. 3.16: Catalytic hydrogenation of 1 phenyl-1,2-propanedione on a modified Pt catalyst [57]. (**A**) 1-Phenyl-1,2-propanedione, (**B**) (*R*)-1-hydroxy-1-phenylpropanone, (**C**) (*S*)-1-hydroxy-1-phenylpropanone, (**D**) (*S*)-2-hydroxy-1-phenylpropanone, (**E**) (*R*)-2-hydroxy-1-phenylpropanone, (**F**) (1*R*,2*S*)-1-phenyl-1,2-propanediol, (**G**) (1*S*,2*S*)-1-phenyl-1,2-propanediol, (**H**) (1*S*,2*R*)-1-phenyl-1,2-propanediol, (**I**) (1*R*,2*R*)-1-phenyl-1,2-propanediol.

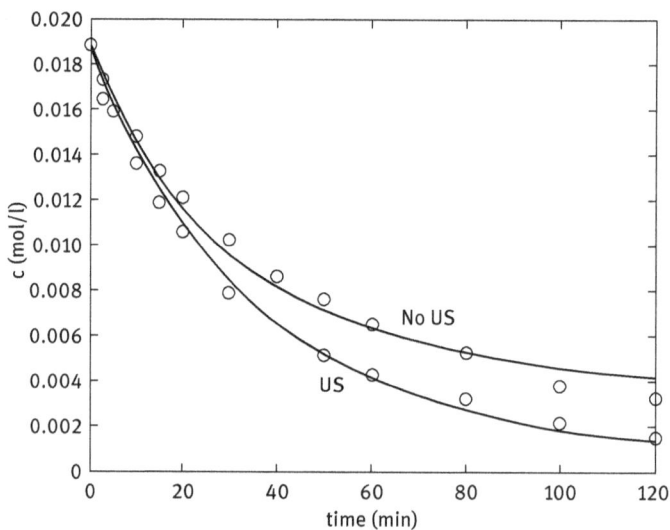

Fig. 3.17: Effect of ultrasound (US) on catalyst activity [57].

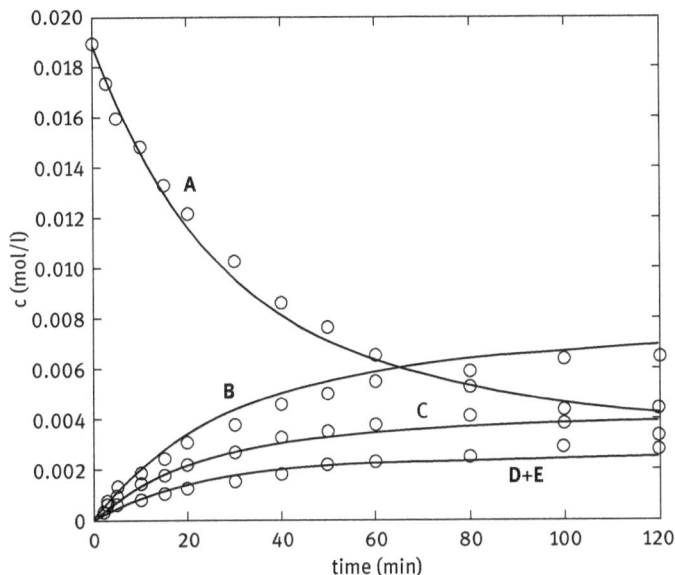

Fig. 3.18: Modeling of the effect of ultrasound on catalyst activity [56]. (**A**) 1-Phenyl-1,2-propanedione, (**B**) (*R*)-1-hydroxy-1-phenylpropanone, (**C**) (*S*)-1-hydroxy-1-phenylpropanone, (**D**) (*S*)-2-hydroxy-1-phenylpropanone, (**E**) (*R*)-2-hydroxy-1-phenylpropanone.

by the proposed approach [56]. The rate constants were determined by nonlinear regression analysis, and the model coincided very well with experimental data.

3.2.3.2 Effect of ultrasound on hydrogenation of sugars

Carbohydrates constitute an important fraction (75%) of the available renewable biomass and are thus becoming a key target in the field of green chemistry [58]. Naturally occurring polysaccharides, such as hemicelluloses, are the starting molecules for derivatization to simpler mono- and disaccharides. These, in turn, act as substrates for the production of value-added molecules.

Sugar alcohols are a particularly interesting group of molecules derived from carbohydrates. The reduction of sugar molecules to sugar alcohols has become a topic of evolving importance, as demonstrated by the increase in global production of sorbitol from 700,000 tons per year in 2007 [59] to over 1,000,000 tons per year in 2009 [60].

These compounds are obtained through the reduction of the carbonyl group present in the sugar molecule by means of either chemical agents (such as sodium borohydride) or molecular hydrogen in the presence of a homogeneous or heterogeneous catalyst. The catalytic route is preferred for industrial hydrogenation of sugars into sugar alcohols, where heterogeneous catalysts based on Ni, Pd, Pt, or Ru are used [61, 62]. Heterogeneous catalysts are preferred because the formation of stoichiometric coproducts is avoided. Thus, the selective catalytic hydrogenation of naturally occurring sugar molecules to their corresponding sugar alcohols is an environmentally friendly route for production of alternative sweeteners. Sugar hydrogenation in the last 40 years has focused mainly on glucose, fructose, xylose, and lactose [62–66]. Typically, selectivity for the sugar alcohol is high on both sponge Ni and supported Ru catalysts, exceeding 95% under optimal conditions. Similar data have been reported for the hydrogenation of arabinose, galactose, glucose, lactose, maltose, rhamnose, and xylose [66–69].

A research group from Finland determined the impact of ultrasound on the hydrogenation of L-arabinose and D-galactose to the corresponding sugar alcohols in terms of rate enhancement and catalyst rejuvenation. The experiments were conducted in the same way as in an earlier study [71], but with the additional application of ultrasound irradiation. Hydrogenation experiments were carried out with a Ru/active carbon catalyst at 105 °C and 50 bar of hydrogen. Three methods of acoustic irradiation were applied (direct irradiation, indirect online irradiation, and catalyst pretreatment with ultrasound). The power calibration was studied in the case of indirect irradiation. The effect of pressure and concentration were also investigated. Characterization of fresh and spent catalyst was carried out using nitrogen physisorption, transmission electron microscopy (TEM), and scanning electron microscopy (SEM).

Indirect irradiation was carried out with the device illustrated in Figure 3.23 (starch section), consisting of six transducers mounted on the center of each face of a hexagonal hull. Each transducer delivered 100 W of power at a frequency of 20 kHz. The stainless steel reactor autoclave was partially submerged in the bath, which had been previously filled with silicone oil Rhodorsil® 47V 350 and heated to 170 °C on a heating plate.

The sonification system with the internal ultrasound probe is displayed in Figure 3.19. The purpose of installing an internal probe was to enable direct irradiation of the medium and control exactly how much power was delivered to the system. The internal probe was factory calibrated and permitted the selection of different output powers. A high stirring rate (1800 rpm) during the reaction guaranteed thorough contact of the reaction medium with the ultrasonic probe.

To measure the ultrasonic power, the formation of triiodide ions from a KI solution was monitored by UV spectrometry at 354 nm (Weissler reaction [70]). For calibrating the spectrometer, an I_2 solution (0.05 M) was prepared by dissolving the corresponding amount of resublimed iodine in KI solution. In aqueous iodide solution, the insoluble iodine (I_2) molecules form soluble triiodide (I_3^-) ions, which were titrated against a 0.01 M standardized solution of sodium thiosulfate. The solution was then diluted to obtain three different standards (0.1, 0.075, and 0.05 mM), which were used to calibrate the spectrometer detector (Shimadzu 2550).

After calibration of the spectrometer, 100 mL of a KI solution (25% w/v) was poured into the standard reactor and the thermostat set to 23 °C. Ultrasonic irradiation using the external source was carried out for 30 min at different output settings. The amount of produced triiodide was then quantified using the UV spectrometer.

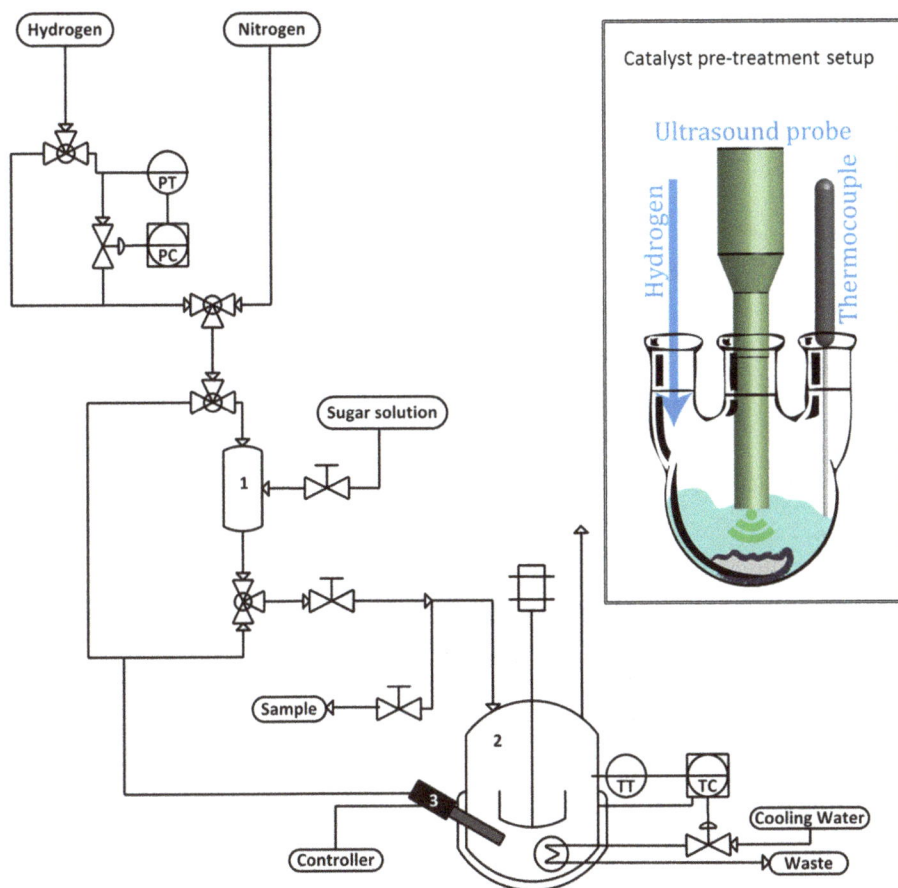

Fig. 3.19: Modified hydrogenation reactor system with an internal ultrasound source: (1) bubbling chamber, (2) reactor, and (3) internal probe.

The results were compared with those obtained under the same conditions using the internal ultrasonic source (factory calibrated) in a glass reactor (Table 3.3). The comparison provided an approximate evaluation of acoustic power (Figure 3.20).

Examination of the catalyst cluster size by TEM and SEM showed growth of the catalyst particles and agglomeration of metal nanoparticles when ultrasound was applied, but not in the silent experiments.

Tab. 3.3: Comparison of results for internal and external ultrasound devices.

Device	Amplitude [%]	Power [W]	Absorbance	I_3^- conc. [mmol/L]
Internal	10	12	0.445	0.01443
External	100	9	0.349	0.01133

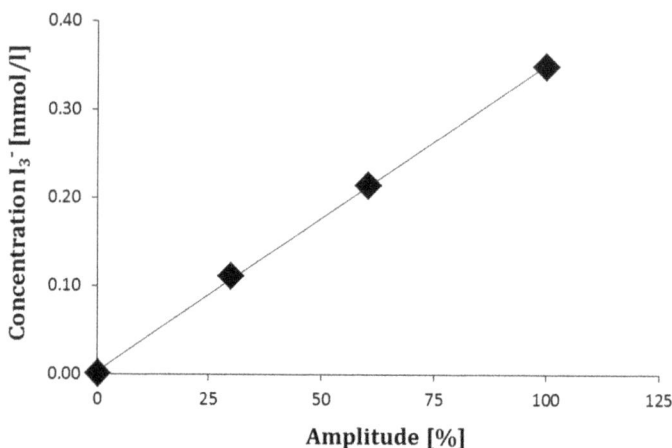

Fig. 3.20: Evolution of complex at a frequency of 20 kHz.

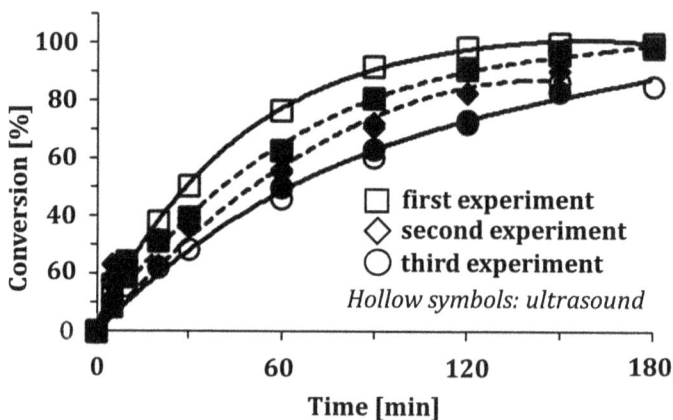

Fig. 3.21: Impact of ultrasound (external source) on catalyst deactivation with arabinose (at 105 °C and 50 bar).

The hydrogenation experiments revealed that ultrasound does not always enhance the hydrogenation kinetics of L-arabinose and D-galactose on a Ru/C catalyst. For the range of power delivered through the indirect and direct methods (10–30 W), no favorable impact was observed. In contrast to what was expected, ultrasound seemed to intensify catalyst deactivation (Figure 3.21).

Characterization of fresh and spent catalysts indicated that particle agglomeration occurred together with suspected poisoning as a result of iron leaching. Whether iron adsorbs on the catalyst metal sites or on active carbon remains an open question. Moreover, sintering of the active sites was observed. Nonetheless, this phenomenon is less understood than agglomeration of the support, which in turn leads to a decrease in the specific surface area. The effect of pressure, otherwise absent in typical silent conditions, was clearly observable under ultrasonic irradiation, the reaction rates being faster with increasing pressure. Higher sugar concentrations did not impact the effect that ultrasound had on the reaction rate. To complete the study, catalyst deactivation kinetics were modeled under silent and sonicated conditions. A pseudo first-order model gave an acceptable fit for sugar hydrogenation.

3.2.4 Ultrasonification of starch

Starch is commonly used in the paper industry as a coater and binder. However, prior to use, it must be chemically modified, usually by partial oxidation. Conventional methods are inexpensive and effective, but they are not environmentally friendly. Stoichiometric oxidants, such as chlorites, and heavy metal catalysts such as iron sulfates are used. It is therefore necessary to invent more sustainable methods for starch oxidation. A special iron complex (FePcS) has recently been shown to perform this kind of reaction with peroxides [71–73] and provides a possible way to improve the starch oxidation process. A study of starch oxidation revealed that starch can be oxidized in an environmentally friendly manner by FePcS catalyst and H_2O_2, but the yield was moderate. The reaction probably took place mainly on the outer surface of the starch granule, but also inside the granule after a prolonged time [74]. In this research, when the hydroxyl groups of potato starch were partially oxidized to the corresponding carboxyl and carbonyl groups, there was an increase in the degree of oxidation when the solid starch was first pretreated by ultrasound [75]. The reaction mechanism is illustrated in Figure 3.22.

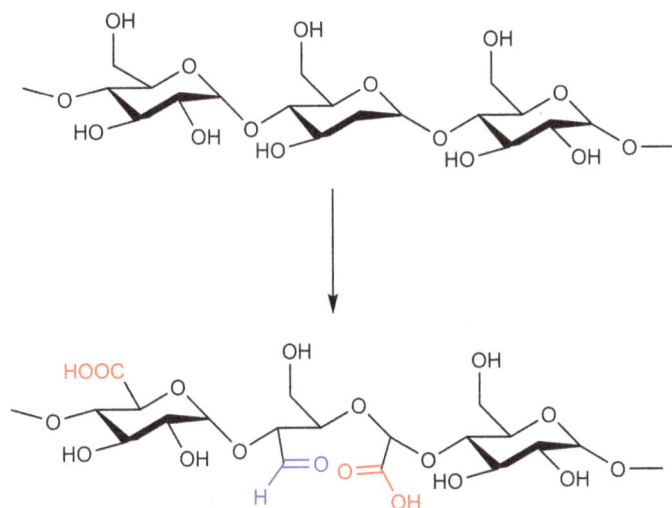

Fig. 3.22: Reaction mechanism for partial oxidation of starch (substitution of carbonyl or carboxyl groups).

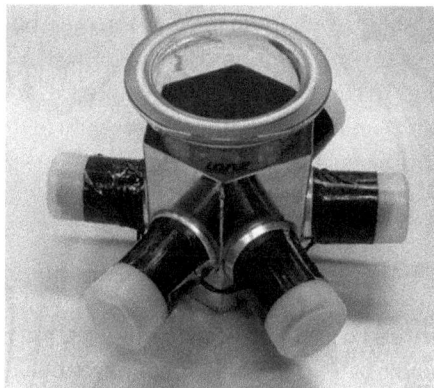

Fig. 3.23: Transducer ultrasonic reactor vessel (100 W at 20 kHz frequency; six transducers) [77].

An ultrasonification device was constructed in-house to provide indirect sonication to the reaction vessel, the design being similar to that of an ultrasonic bath (Figure 3.23). The outer hull of the reactor was hexagonal and fitted with transducers mounted on the center of each face of the hull. Each transducer delivered nominally 100 W of power at a frequency of 20 kHz, which was aimed toward the center of the bath. The reactor used in this particular study was a glass reactor vessel. It was partially submerged in the bath, which had been previously filled with water. Ultrasound treatment was initiated at room temperature and stopped after 24 h. At this point, the reactor temperature had increased to 36 °C due to the absorption of ultrasound. However, the moderate increase in temperature was not relevant because potato starch gelatinization, which should be avoided, occurs at a much higher temperature of 57–58 °C.

Analysis of the amount of carboxylic acid groups attached to starch pretreated with ultrasound demonstrated a significant increase (Figure 3.25) compared with native, untreated starch, (60% higher degree of substitution). The reason was partial destruction of the structure of starch granules, leading to pores and ruptures. These evidently increased the reactive surface area, as confirmed by nitrogen adsorption measurements (Figure 3.24). The Brunauer–Emmett–Teller (BET) surface area increased by

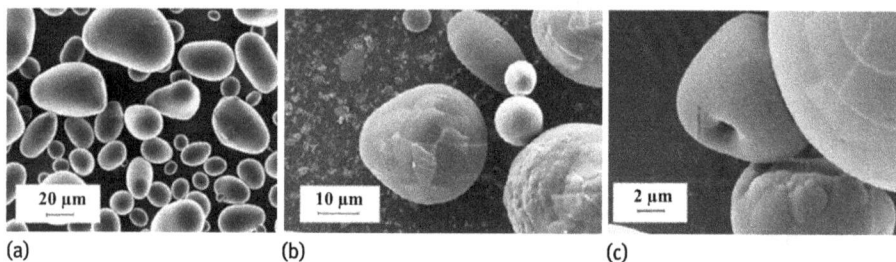

(a) (b) (c)

Fig. 3.24: (a) Native potato starch (BET surface area 10 m^2/g; zoom 500×). (b) Native potato starch treated with ultrasound for 24 h at 38 °C (BET surface area 18 m^2/g; zoom 1500×). (c) Ultrasound-treated starch after reaction at pH 8.4 (zoom 5000×) [77].

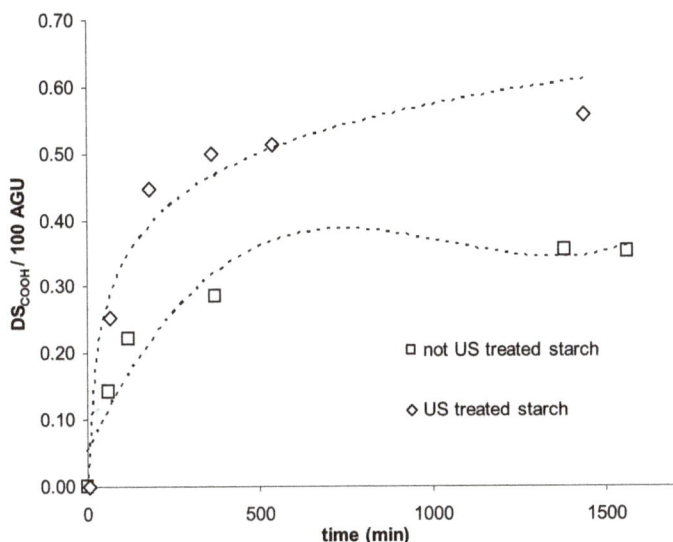

Fig. 3.25: The degree of substituted carboxyl groups per 100 AGU units (DS_{COOH}) when potato starch was pretreated with ultrasound. The results are compared with a standard normal batch experiment without pretreating the starch, using the same pH of 8.4.

a similar order of magnitude (from 10 m^2/g to 18 m^2/g). Moreover, this study also investigated the effect of the origin of the starch (potato or corn starch), solid-to-liquid ratio, and pH. A mathematical model was developed that accurately described the complex product formation on both the outer and inner parts of the starch particles [76, 77]. However, these studies unfortunately did not incorporate ultrasonification.

A possible reason for the increase in surface area could be degradation of one of the starch polymers, amylose. Potato starch typically contains around 70% highly branched amylopectin and 30% long linear amylose. It has recently been reported that ultrasound treatment preferentially degrades the amorphous regions and attacks the linear amylose of potato starch more easily than branched amylopectin [78, 79]. A 20-kHz Branson digital Sonifier S-450D (Germany) coupled with a horn was used in this study. The ultrasonic effect is shown in Figure 3.26. Potato starch granules with defects appear to have rough surfaces, presenting dents, cuts, or cracks after being exposed to ultrasound.

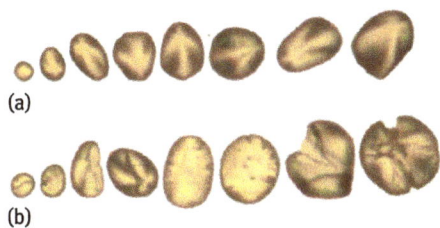

Fig. 3.26: Images of starch granules (a) without defects and (b) with defects [78].

3.3 Conclusions

Ultrasound and microwave technologies are attractive methods for the intensification of chemical processes, which is a subarea of process intensification known as reaction intensification. This chapter describes the effect of microwave and ultrasound technologies on selected green processes, such as fatty acid epoxidation, enantioselective hydrogenation, sugar alcohol production, and starch oxidation. The effect of ultrasound is in most cases based on the retardation of catalyst deactivation by surface cleaning. The effect of microwaves can be explained by the improved interfacial mass transfer in multiphase systems. A bottleneck in the scale-up of ultrasound and microwave technologies is the result of a lack of sufficiently advanced mathematical models for these processes. Therefore, future development should focus on modeling approaches to predict quantitatively the effects of microwaves and ultrasound on chemical systems.

Acknowledgment: The work is a part of the activities of Johan Gadolin Process Chemistry Centre (PCC), a center of excellence in scientific research financed by Åbo Akademi. Financial support from the Academy of Finland (Tapio Salmi), Raisio Research Foundation (Pasi Tolvanen), and Fortum Foundation (Adriana Freites) is gratefully acknowledged.

3.4 Bibliography

[1] Loupy A. Microwaves in Organic Synthesis. Wiley-VCH 2002.
[2] Zhu Y, Chen F. Microwave-assisted preparation of Inorganic Nanostructures in Liquid phase, Chem Rev American Chemical Society, 2014, 114, 6462–6470.
[3] Web of science. (Accessed on September 9th, 2016 at: http://apps.webofknowledge.com/ RAMore.do?product=WOS&search_mode=GeneralSearch&SID=P1652XW5mi4MDCbc76Q& qid=1&ra_mode=more&ra_name=PublicationYear&colName=WOS&viewType=raMore).
[4] Toukoniitty B. Utilization of electromagnetic and acoustic irradiation in enhancing heterogeneous catalytic reactions. Doctoral Thesis, Åbo Akademi, Finland, 2007.
[5] Leveneur S, Ledoux A, Estel L, Taouk B, Salmi T. Epoxidation of vegetable oils under microwave irradiation. Chemical Engineering Research and Design, 2014, 92, 1495–1502.
[6] Lidström P, Tierney J, Wathey B, Westman J. Microwave assisted organic synthesis – a review. Tetrahedron, 2001, 57, 9225–9283.
[7] Robson H. Verified Syntheses Of Zeolitic Materials, 2nd ed(s). LA, USA. 2001, 39.
[8] Ertl G, Knözinger H, Weitkamp J. Handbook of Heterogeneous Catalysis Volume 3. Wiley-VCH.
[9] Kappe CO. Review. Controlled Microwave Heating in Modern Organic Synthesis. Angew Chem Int, 2004, 43, 6250–6284.
[10] Majetich G, Hicks R. Applications of microwave-accelerated organic synthesis. Radial Phys Chem, 1995, 45, 567–579.
[11] Chen D, Tang K, Shen G, et al. Microwave-assisted synthesis of metal sulfides in ethylene glycol. Mater Chem Phys, 2003, 82, 206–209.

[12] Ding T, Zhu J. Microwave heating synthesis of HgS and PbS nanocrystals in ethanol solvent. Mat Sci Eng B, 2003, 100, 307–313.

[13] Horikoshia S, Matsuzakia S, Sakamotoa S, Serpone N. Efficient degassing of dissolved oxygen in aqueous media bymicrowave irradiation and the effect of microwaves on a reaction catalyzed by Wilkinson's catalyst. Radiat Phys Chem, 2014, 97, 48–55.

[14] Golea VL, Gogate PR. Intensification of glycerolysis reaction of higher free fatty acid containing sustainable feedstock using microwave irradiation. Fuel Process Technol, 2014, 118, 110–116.

[15] Kingston HM, Haswell SJ. Microwave Enhanced Chemistry: Fundamentals, Sample Preparation, and Applications. 1st ed(s). NW, USA, American Chemical Society, 1997, 415–418.

[16] Kappe CO, Pieber B, Dallinger D. Microwave effects in organic synthesis: Myth or reality. Angew Chem Int, 2013, 52, 1088–1094.

[17] Hosseini M, Stiasni N, Barbieri V, Kappe CO. Microwave-Assisted Asymmetric Organocatalysis. A Probe for Nonthermal Microwave Effects and the Concept of Simultaneous Cooling. J Org Chem, 2007, 72, 1417–1424.

[18] Herrero MA, Kremsner JM, Kappe CO. Nonthermal Microwave Effects Revisited: On the Importance of Internal Temperature Monitoring and Agitation in Microwave Chemistry. J Org Chem, 2008, 73, 36–47.

[19] Razzaq T, Kremsner JM, Kappe CO. Investigating the Existence of Non-thermal/Specific Microwave Effects Using Silicon Carbide Heating Elements as Power Modulators. J Org Chem, 2008, 73, 6321–6329.

[20] Dallinger D, Irfan M, Suljanovic A, Kappe CO. An Investigation of Wall Effects in Microwave-Assisted Ring-Closing Metathesis and Cyclotrimerization Reactions. J Org Chem, 2010, 75, 5278–5288.

[21] Bacsa B, Horváti K, Bõsze S, Andreae F, Kappe CO. Solid-Phase Synthesis of Difficult Peptide Sequences at Elevated Temperatures: A Critical Comparison of Microwave and Conventional Heating Technologies. J Org Chem, 2008, 73, 7532–7542.

[22] De la Hoz A, Díaz-Ortiz Á., Moreno A. Microwaves in organic synthesis. Thermal and non-thermal microwave effects. Chem Soc Rev, 2005, 34, 164–178.

[23] Maoz R, Cohen H, Sagiv J. Specific Nonthermal Chemical Structural Transformation Induced by Microwaves in a Single Amphiphilic Bilayer Self-Assembled on Silicon. Langmuir, 1998, 14, 5988–5993.

[24] Pagnotta, M., Pooley, C.L.F., Gurland, B. and Choi, M., Microwave activation of the mutarotation alpha-d-glucose – an example of an intrinsic microwave effect, J. Phys. Org. Chem, 1993, 6, 407.

[25] Perreux L, Loupy A. A tentative rationalization of microwave effects in organic synthesis according to the reaction medium and mechanistic considerations. Tetrahedron, 2001, 57, 9199–9228.

[26] Kappe CO, Dallinger D. Controlled microwave heating in modern organic synthesis: highlights from the 2004–2008 literature. Mol Divers, 2009, 12, 71–193.

[27] Berlan J. Microwaves In Chemistry: Another Way Of Heating Reaction Mixtures. Radiat Phys Chem, 1995, 45, 581–589.

[28] Freites A, Tolvanen P, Eränen K, Leveneur S, Salmi T. Epoxidation of oleic acid under conventional heating and microwave radiation. Chem Eng Process, 2016, 102, 70–87.

[29] Saurabh T, Patnaik M, Bhagt SL, Renge V. Epoxidation Of Vegetable Oils: A Review. International Journal of Advanced Engineering Technology, 4, 2011, 491–501.

[30] Sturm G, Verweij M, van Gerven T, Stankiewicz A, Stefanidis G. On the effect of resonant microwave fields on temperature distribution in time and space. International Journal of Heat and Mass Transfer, 2012, 55, 3800–3811.

[31] Bogdal D. Microwave-Assisted Organic Synthesis: One Hundred Reaction Procedures, Elsevier, Amsterdam 2005.

[32] Leadbeater NE, Barnard TM, Stencel LM. Batch and Continuous-Flow Preparation of Biodiesel Derived from Butanol and Facilitated by Microwave Heating. Energy Fuels, 2008, 22, 2005–2008

[33] Arvela RK, Leadbeater NE, Mack TL, Kormos CM. Microwave-promoted Suzuki coupling reactions with organotrifluoroborates in water using ultra-low catalyst loadings. Tertrahedron, 2006, 47, 217–220

[34] Suslick KS. The Chemical Effects of Ultrasound. Sci Am, 1989, 80–86.

[35] Sancheti SV, Gogate PR. A review of engineering aspects of intensification of chemical synthesis using ultrasound. Ultrasonics Sonochemistry 2016.

[36] Thompson LH, Doraiswamy LK. Sonochemistry: Science and Engineering. Ind Eng Chem Res, 1999, 38, 1215–1249

[37] Richards WT, Loomis AL. The Chemical Effects of High Frequency Sound Waves. I. A preliminary Study. J Am Chem Soc, 1927, 49, 3086–3100.

[38] Toukoniitty B, Mikkola JP, Murzin DYu, Salmi T. Utilization of electromagnetic and acoustic irradiation in enhancing heterogeneous catalytic reactions. Applied Catalysis A, 2005, 279, 1–22.

[39] Gogate P, Mujumdar S, Pandit A. Large-scale sono-chemical reactors for process intensification: design and experimental validation. J Chem Technol Biotechnol, 78, 685–693.

[40] Suslick KS. Sonochemistry. In Kirk–Othmer encyclopedia of chemical technology, 4th edn, 26, 517–541. New York: John Wiley 1998.

[41] Mason TJ, Lorimer JP. An introduction to sonochemistry. Endeavour, 1989, 13, 123–128.

[42] Mason TJ. Sonochemistry and sonoprocessing: the link, the trends and (probably) the future. Ultrasonics Sonochemistry, 2003, 10, 175–179.

[43] Suslick KS. Sounding out new chemistry. New Sci, 1990, 125, 50–53.

[44] Web of knowledge. (Accessed on October 3rd, 2016 at: https://wcs.webofknowledge.com/).

[45] Cravotto G, Borretto E, Oliverio M, Procopio A, Penoni A. Organic reactions in water or biphasic aqueous systems under sonochemical conditions. A review on catalytic effects. Catalysis Communications, 2015, 63, 2–9.

[46] Chatel G, Novikova L, Petit S. How efficiently combine sonochemistry and clay science? Applied Clay Science, 2016, 119, 193–201.

[47] Tran N, Drogui P, Brar SK. Sonochemical techniques to degrade pharmaceutical organic pollutants. Environ Chem Lett, 2015, 13, 251–268.

[48] Pétrier C, The use of power ultrasound for water treatment. Power Ultrasonics, 2015, 939–972.

[49] Chemat F, Rombaut N, Sicaire AG, Meullemiestre A, Fabiano-Tixier AS, Abert-Vian M. Ultrasound assisted extraction of food and natural products. Mechanisms, techniques, combinations, protocols and applications. A review. Ultrasonics Sonochemistry, 2017, 34, 540–560.

[50] Pokhrel N, Vabbina PK, Pala N. Sonochemistry: Science and Engineering. Ultrason Sonochem, 2016, 29, 104–128

[51] Okitsu K, Ashokkumar M, Grieser F, Sonochemical Synthesis of Gold Nanoparticles: Effects of Ultrasound Frequency. J Phys Chem B, 2005, 109, 20673–20677.

[52] Gonçalves I, Silva C, Cavaco-Paulo A. Ultrasound enhanced laccase applications. Green Chem, 2015, 17, 1362–1374.

[53] Papadakis EP. Ultrasonic Instruments and Devices, 1 edn. Academic Press 2000.

[54] Asgharzadehahmadi S, Raman AAA, Parthasarathy R, Sajjadi B. Sonochemical reactors: Review on features, advantages and limitations. Renew Sust Energ Rev, 2016, 63, 302–314.

[55] Toukoniitty B, Toukoniitty E, Mäki-Arvela P, Salmi T, Murzin D, Kooyman PJ. Effect of ultrasound in enantioselective hydrogenation of 1-phenyl-1,2-propanedione: comparison of catalyst activation, solvents and supports. Ultrason Sonochem, 2006, 13, 68–75.

[56] Toukoniitty B, Mäki-Arvela P, Kalantar Neyestanaki A, et al. Enantioselective hydrogenation of 1-phenyl-1,2-propanedione. J Catal, 2001, 204, 281–291.

[57] Toukoniitty B, Kangas M, Wärnå J, et al. Suppression of catalyst deactivation by acoustic irradiation – Kinetic approach. Chimica oggi, 2006, 24, 62–65.

[58] Lichtenthaler FW, Peters S. Carbohydrates as green raw materials for the chemical industry. C R Chimie, 2004, 7, 65–90.

[59] Corma A, Iborra S, Velty A. Chemical Routes for the Transformation of Biomass into Chemicals. Chem Rev, 2007, 107, 2411–2502.

[60] Eisenbeis C, Guettel R, Kunz U, Turek T. Monolith loop reactor for hydrogenation of glucose. Catal Today, 2009, S147, S342–S346.

[61] Déchamp N, Gamez A, Perrard A, Gallezot P. Kinetics of glucose hydrogenation in a trickle-bed reactor. Catal Today, 1995, 24, 29–34.

[62] Wisniak J, Hershkowitz M, Leibowitz R, Stein S. Hydrogenation of Xylose to Xylitol. Ind Eng Chem, 1974, 13, 75–79.

[63] Makkee M, Kieboom APG, van Bekkum H. Combined action of an enzyme and a metal catalyst on the conversion of d-glucose/d-fructose mixtures into d-mannitol. Carbohydr Res, 1985, 138, 225–236.

[64] Kuusisto J, Mikkola JP, Sparv M et al. Hydrogenation of lactose over sponge nickel catalysis-kinetics and modelling. Ind Eng Chem, 2006, 45, 5900–5910.

[65] Mikkola JP, Sjöholm R, Salmi T, Mäki-Arvela P. Xylose hydrogenation: kinetic and NMR studies of the reaction mechanisms. Catal Today, 1999, 48, 73–81.

[66] Crezee E, Hoffer BW, Berger RJ, Makkee M, Kapteijn F, Moulijn JA. Three-phase hydrogenation of d-glucose over a carbon supported ruthenium catalyst – mass transfer and kinetics. Appl Catal A-Gen, 2003, 251, 1–17.

[67] Mikkola JP, Salmi T, Sjöholm R. Effects of solvent polarity on the hydrogenation of xylose. J Chem Technol Biotechnol, 2001, 76, 90–100.

[68] Kuusisto J, Mikkola JP, Sparv M, Wärnå J, Karhu H, Salmi T. Kinetics of the catalytic hydrogenation of D-lactose on a carbon supported ruthenium catalyst. Chem Eng J, 2008, 139, 69–77.

[69] Sifontes V, Oladele O, Kordás K, et al. Sugar hydrogenation over a Ru/C catalyst. J Chem Technol Biotechnol, 2011, 86, 658–668.

[70] Weissler A, Cooper HWHW, Snyder S. J Am Chem Soc, 1950, 72, 1769–1775.

[71] Sorokin A, De Suzzoni-Dezard S, Poullain D, Noël JP, Meunier B. CO_2 as the Ultimate Degradation Product in the H_2O_2 Oxidation of 2,4,6-Trichlorophenol Catalyzed by Iron Tetrasulfophthalocyanine. J Am Chem Soc, 1996, 118, 7410–7411.

[72] Zalomaeva OV, Kholdeeva OA, Sorokin A B. Preparation of 2-methyl-1,4-naphthoquinone (vitamin K3) by catalytic oxidation of 2-methyl-1-naphthol in the presence of iron phthalocyanine supported catalyst. C R Chimie, 2007, 10, 598–603.

[73] Kachkarova-Sorokina SL, Gallezot P, Sorokin AB. A novel clean catalytic method for waste-free modification of polysaccharides by oxidation. Chem Commun, 2004, 2844–2845.

[74] Tolvanen P, Mäki-Arvela P, Sorokin AB, Salmi T, Murzin DY. Kinetics of starch oxidation using hydrogen peroxide as an environmentally friendly oxidant and an iron complex as a catalyst. Chem Eng J, 2009, 154, 52–59.

[75] Tolvanen P, Mäki-Arvela P, Sorokin AB, Leveneur S, Salmi T, Murzin DY. Batch and semi-batch partial oxidation of starch by hydrogen peroxide in the presence of iron tetrasulfophthalocyanine catalyst: the effect of ultrasound and catalyst addition policy. Ind Eng Chem Res, 2011, 50, 749–757.

[76] Salmi T, Tolvanen P, Wärnå J, Mäki-Arvela P, Murzin D, Sorokin A. Mathematical modeling of starch oxidation by hydrogen peroxide in the presence of an iron catalyst complex. Chem Eng Sci, 2016, 146, 19–25.

[77] Tolvanen P, Mäki-Arvela, P, Sorokin AB, Salmi T, Murzin DY. Oxidation of starch by H_2O_2 in the presence of iron tetrasulfophthalocyanine catalyst: the Effect of catalyst concentration, pH, solid liquid ratio, and origin of starch. Ind Eng Chem Res, 2013, 52, 9351–9358.

[78] Zuo YYJ, Hébraud P, Hemar Y. Ashokkumar M. Quantification of high-power ultrasound induced damage on potato starch granules using light microscopy. Ultrason Sonochem, 2012, 19, 421–426.

[79] Chan HT, Bhat R, Karim AA. Effect of sodium dodecyl sulphate and sonication treatment on physicochemical properties of starch. Food Chem, 2010, 120, 703–709.

Anton A. Kiss

4 Process intensification by reactive distillation

Abstract: Reactive distillation (RD) is a widespread process intensification method that combines reaction and separation into a single unit, which allows the simultaneous production and removal of products. This improves productivity and selectivity, reduces energy use, eliminates the need for solvents, and leads to intensified and highly efficient systems with green engineering attributes. Some of these key benefits are realized by using the reaction to improve separation, whereas others are obtained by using the separation to improve reaction performance. This chapter addresses the working principles, modeling, design, control, and applications of RD. Several industrial case studies are provided to illustrate the key features of RD processes.

Keywords: Design and control, industrial applications, process modeling and simulation, reactive distillation, specific equipment, working principle

4.1 Introduction

In the chemical industry, most processes involve reaction and separation operations that are carried out typically in different sections of the plant and use different equipment types operated under a wide variety of conditions [1]. Recycle streams between these operating units are often used to improve conversion and selectivity, minimize production of undesired byproducts, reduce energy requirements, and enhance process controllability. Recent economic and environmental considerations have encouraged industry to focus on novel technologies based on process intensification [2–4].

Reactive distillation (RD) integrates the reaction and separation steps into a single unit, allowing the simultaneous production and removal of products. This results in an improvement in productivity and selectivity, reduction in energy use, and elimination of the need for solvents, which leads to intensified, high-efficiency systems with green engineering attributes [5, 6]. Some of these benefits are realized by using reactions to improve separation (e.g., overcoming azeotropes, removing contaminants, avoiding or eliminating difficult separations) whereas others are realized by using separation to improve reactions (e.g., enhancing overall rates, overcoming reaction equilibrium limitations, improving selectivity, removing catalyst poisons). The best synergistic effect is achieved when both aspects are important [5, 7, 8]. However, application of RD is limited by constraints such as the need for a common operation range for distillation and reaction (similar temperature and pressure), proper boiling point sequence (product should be the lightest or heaviest component, whereas side or byproducts boil at intermediate temperatures), and proper residence time characteristics [9, 10].

https://doi.org/10.1515/9783110465068-004

Although a number of industrial RD applications have been around for many decades [1], the RD crown is still carried by the Eastman process that reportedly replaced a methyl acetate production plant with a single RD column using 80% less energy at only 20% of the investment costs [11, 12]. Nowadays, the application with the largest number of installations is production of methyl tertiary butyl ether (MTBE), which is used in gasoline blending. Other esters such as ethyl tertiary butyl ether (ETBE), tertiary amyl methyl ether (TAME), and fatty acid methyl esters (FAME) are also produced by RD [7, 8].

Considering the large extent of the RD field, this chapter aims to give a brief overview of what process intensification means for RD. For more detailed information, there are several books and reviews that have been published in the past decade and cover a large range of subjects related to RD, including process synthesis, conceptual design, control, optimization, operation, and industrial application [1, 5, 7, 8, 13–21].

4.2 Principles of reactive distillation

RD is particularly attractive in systems where specific chemical and phase equilibrium conditions coexist. Reaction and distillation take place in the same zone of a distillation column, and the reactants are converted with simultaneous separation of products and recycle of unused reactants. Because products must be separated from reactants by distillation, products should be lighter and/or heavier than the reactants. The ideal case is when one product is the lightest component and the other product is the heaviest, with the reactants being intermediate boiling components [1].

Moreover, as both operations occur simultaneously in the same unit, there must be a good match between the temperatures and pressures required for reaction and separation [10, 22], as clearly illustrated by Figure 4.1 [2]. The operating window is typically limited by the thermodynamic properties (e.g., boiling point) of the components involved. Furthermore, the windows in which the (catalytic) reaction delivers acceptable yields and selectivity usually have a limited overlap with the separation window. This overlap may be reduced further by the feasibility window concerning equipment design. This usually leads to a very restricted area in which reactive separation is technically and economically feasible [19]. The combination of reaction and distillation is clearly not possible if there is no significant overlap of the operating conditions of reaction and separation (e.g., a high pressure reaction cannot be combined with vacuum distillation). One must also consider that working in the limited overlapping window of operating conditions is not always the optimal solution, but merely a trade-off [22]. By contrast, in a conventional multiunit flowsheet, the reactors can be operated at parameters that are most favorable for the chemical kinetics, while the distillation columns can be operated at pressures and temperatures that are most favorable for the vapor–liquid equilibrium properties [1].

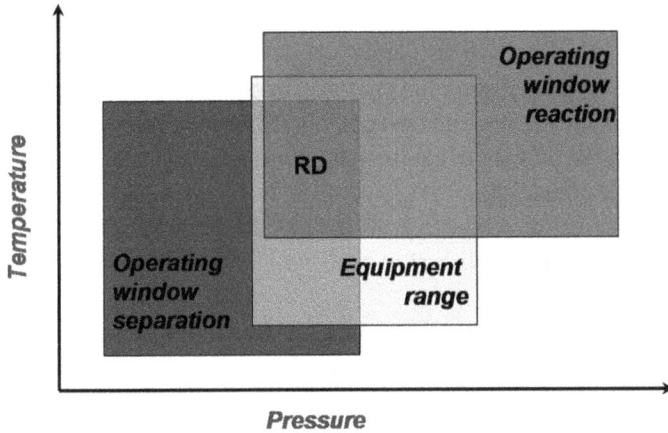

Fig. 4.1: Overlapping of operating windows for reaction, separation, and equipment.

The pressure and temperature effects are much more pronounced in RD than in conventional distillation because these parameters affect both the phase equilibrium and chemical kinetics [1, 23–25]. A low temperature that gives high relative volatilities may provide low reaction rates that require large amounts of catalyst or liquid holdups to achieve the required conversion. In contrast, a high temperature may promote undesirable side reactions or give a low equilibrium constant that makes it difficult to drive the reaction to completion [1, 22].

RD is typically applied to equilibrium reactions such as esterification, etherification, hydrolysis, and alkylation. Luyben and Yu [1] reported that over 1100 articles and 800 US patents on RD have been published during the past 40 years, covering in total over 235 reaction systems. Most of the reactions types belong to one of the following stoichiometries:

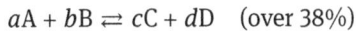

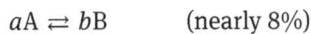

$$aA + bB \rightleftharpoons cC + dD \quad \text{(over 38\%)}$$
$$aA + bB \rightleftharpoons cC \quad \text{(over 25\%)}$$
$$aA \rightleftharpoons bB + cC \quad \text{(about 9\%)}$$
$$aA \rightleftharpoons bB \quad \text{(nearly 8\%)}$$

The rest fall into other categories of two- or three-stage reactions [1].

RD has attracted considerable attention, especially for chemical equilibrium-limited, liquid-phase reactions, which conventionally require a large excess of one of the reactants. To aid the consideration of RD as an option for a new process, Shah et al. [26] proposed a systematic framework for the feasibility and technical evaluation of RD processes, based on fundamental explanations as well as industrial applications of RD reported in the scientific literature. To perform a feasibility evaluation, some basic information on the chemical process is required, such as vapor–liquid equilibrium, stoichiometry of reactions, kinetics, and enthalpy of reactions [2]. However, in the process of deciding whether RD is the right choice, one must keep in mind the following benefits (+) and limitations (–) of reactive distillation [19]:

+ *Process simplification*: Complex conventional processes can be reduced to a RD column, which is typically much easier and cheaper to operate and maintain.

+ *Capital savings*: Reducing the number of equipment units used leads to a considerable reduction in capital expenditure.

+ *Increased conversion*: Removal of products from the reactive liquid phase causes chemical equilibrium to be re-established at a higher conversion rate, and full conversion is attainable.

+ *Increased selectivity*: Undesired side reactions can be suppressed by removing the target product from the reactive section of the RD column. This also means less waste and fewer byproducts.

+ *Reduced energy usage*: The heat of an exothermic reaction can be used in situ for vaporization of the liquid, thus reducing the reboiler duty and, consequently, the heat transfer area of the reboiler.

+ *Reduced degradation of chemicals*: Due to the lower residence time (compared with a classic reactor), the chemicals are exposed to high temperatures for shorter periods.

+ *Intrinsic safety*: By nature, RD is a boiling system in which hot-spots and runaways are avoided. Also, the relatively low liquid holdup means that RD can be effectively used for hazardous chemicals.

+ *Overcoming of azeotropes*: Under carefully selected conditions, azeotropes can be consumed by the reaction occurring in the liquid phase.

+ *Separation of close-boiling components*: A reactive entrainer reacts with one component to form another product (with a larger difference in boiling point) that can be easily separated. The reverse reaction to recover the original component is then performed in a subsequent RD column.

– *Reduced flexibility*: This is a result of the higher degree of integration compared with classic units.

– *Volatility constraints*: An appropriate order of volatilities is required to ensure high concentrations of reactants and low concentrations of products in the reactive zone. RD should be applied only if the difference in boiling points between the reactants and products is larger than 20 K.

– *Operating-window constraints*: Both reaction and separation processes take place under the same conditions (pressure and temperature); hence, reasonable conversion levels must be attainable at operating conditions that are suitable for distillation (Figure 4.1).

– *Occurrence of reactive azeotropes*: A reactive azeotrope is formed when the change in concentration caused by distillation is fully compensated by the reaction. Reactive azeotropes can sometimes create distillation boundaries that make separation difficult or even infeasible.

– *Occurrence of multiple steady states*: The complex interplay between reaction, separation, and other phenomena leads to strongly nonlinear process behavior with
possible multiple steady states. For this reason, the same RD column configuration operated under the same conditions can exhibit different steady-state column
profiles and, thus, different conversions.

4.3 Modeling reactive distillation

RD is a multicomponent process; therefore, it is qualitatively more complex than similar binary processes. Thermodynamic and diffusional coupling in the phases and at
the interface are accompanied by complex chemical reactions [10]. Consequently, RD
models must take into consideration the column hydrodynamics, mass transfer resis-
tances, and reaction kinetics. The complexity of an RD model depends on its application and the problem to be solved (i.e., plant design, model-based control, or real-time
optimization). A process model consists of submodels for mass transfer, reaction, and
hydrodynamics of various complexities. The mass transfer between the vapor and liquid phase can be described by a rigorous rate-based approach (Maxwell–Stefan diffusion equations) or accounted for by the simple equilibrium (EQ) stage model, assuming thermodynamic equilibrium between the vapor and liquid phases [10].

The detailed modeling of RD has been covered by several studies [10, 27–37]. Figure 4.2 illustrates both the EQ and nonequilibrium (NEQ) stage models [2, 29]. The EQ
model assumes that each vapor stream leaving a tray or packing segment is in thermodynamic equilibrium with the correspondent liquid stream leaving the same tray
or segment. In the case of RD, the chemical reaction is additionally considered via
reaction equilibrium equations or via rate expressions integrated into the mass and
energy balances [10, 29]. Use of the Hatta number, representing the reaction rate in
reference to that of mass transfer, helps to discriminate between very fast, fast, aver-

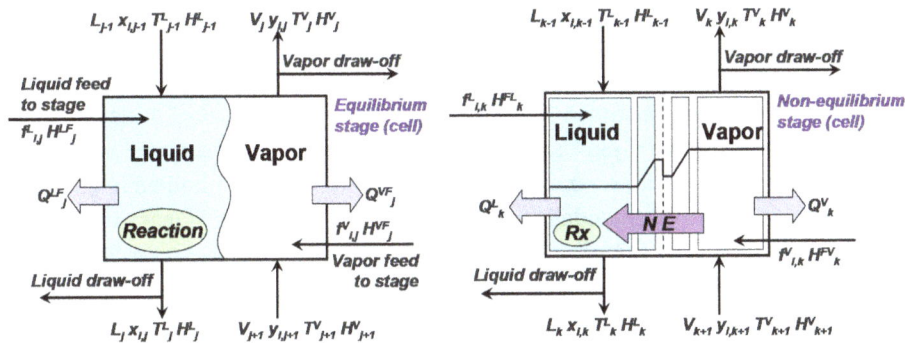

Fig. 4.2: Modeling reactive distillation: equilibrium (left) and nonequilibrium (right) stages.

age, and slow chemical reactions. If a (very) fast reaction system is considered, then RD can be satisfactorily described by assuming reaction equilibrium.

A more physically consistent way to describe a column stage is the rate-based approach, which directly takes into account the actual rates of multicomponent mass and heat transfer and chemical reactions [10, 28, 29, 34, 36, 37]. Both approaches can be conveniently simulated in commercial process simulators such as Aspen Plus (e.g., RADFRAC distillation unit with RateSep model), Aspen Custom Modeler (AspenTech, 2009 [38]), ChemCAD, Pro II, and gPROMS.

It is also important to note that the introduction of an in-situ separation function within the reaction zone leads to complex interactions between vapor–liquid equilibrium, vapor–liquid mass transfer, catalyst diffusion, and chemical kinetics. Such interactions can lead to the phenomenon of multiple steady states and complex dynamics, which have been verified in experimental laboratory and pilot plant units [29, 39, 40].

Summarizing, the simulation of RD processes can consider two types of fundamental models, EQ and NEQ stage models. The benefits and drawbacks of each approach, as well as a direct comparison, are discussed in the paper of Taylor and Krishna [29]. EQ modeling can be formulated at two levels: (1) simultaneous phase and chemical equilibrium and (2) phase equilibrium with chemical kinetics.

The full equilibrium model requires only thermodynamic knowledge. Residue curve maps (RCMs) can help to highlight the range of feasible design in term of pressure, temperature, and separation of products. The simulation is relatively easy, but care should be paid to the accuracy of thermodynamic properties, phase equilibrium, and chemical equilibrium. Such a model based on phase and chemical equilibrium allows rapid assessment of the feasibility of an RD process [10]. The simulation becomes more realistic when knowledge of the chemical kinetics is added. The progression of reaction on each stage can be followed, and thus the number of theoretical stages for achieving a target conversion can be obtained. A key parameter in the kinetic approach is the reaction holdup. Accordingly, both the selection of internals and hydraulic predesign are necessary. An accurate knowledge of the reaction rate expression is also necessary and can be extrapolated over the interval of composition and temperature. This is a crucial point in RD and a major source of failure. The reaction rate must be expressed adequately, either on a pseudo-homogeneous basis (volume) or per mass of (solid) catalyst. Using concentration instead of activity can also introduce large errors, especially when highly nonideal mixtures are handled (e.g., when containing water). In NEQ modeling, the intensity of the interfacial mass transfer in liquid and vapor phases can be accounted for by using the Maxwell–Stefan equations. Specific correlations for calculating the mass transfer coefficients are needed, which in turn depend on the selected internals. The potential accuracy of this approach is paid for by the much more elaborate procedure that needs customized programming. Comparison with experimental results showed that NEQ modeling gives good results if accurate model parameters are employed [41].

4.3.1 Residue curve map

With respect to the design of RD systems, residue curve maps (RCMs) are an invaluable tool for initial screening and flowsheet development [41]. RD is characterized by the simultaneous occurrence of chemical and phase equilibrium. This should be the starting point of a feasibility analysis. Useful insights into chemical and phase equilibrium can be found by graphical representations such as RCMs. Only the general frame is presented here; for more details, the reader is referred to the books by Stichlmair and Fair [23], Doherty and Malone [24], and Dimian and Bildea [42].

Let us consider the general equilibrium reaction

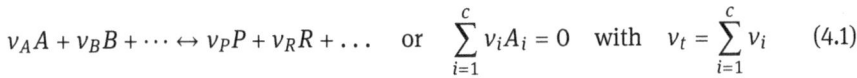

$$\nu_A A + \nu_B B + \cdots \leftrightarrow \nu_P P + \nu_R R + \ldots \quad \text{or} \quad \sum_{i=1}^{c} \nu_i A_i = 0 \quad \text{with} \quad \nu_t = \sum_{i=1}^{c} \nu_i \quad (4.1)$$

The chemical equilibrium constant formulated by means of activities is given by the following expression:

$$K_{eq} = \frac{a_P^{\nu_P} a_R^{\nu_R} \cdots}{a_A^{\nu_A} a_B^{\nu_B} \cdots} = \frac{x_P^{\nu_P} x_R^{\nu_R} \cdots}{x_A^{\nu_A} x_B^{\nu_B} \cdots} \frac{\gamma_P^{\nu_P} \gamma_R^{\nu_R} \cdots}{\gamma_A^{\nu_A} \gamma_B^{\nu_B} \cdots} = \prod_i (x_i \gamma_i)^{\nu_i} = K_x K_\gamma \quad (4.2)$$

The composition can be expressed with respect to a reference species k, as follows:

$$x_i = \frac{x_{i0}(\nu_k - \nu_t x_k) + \nu_i(x_k - x_{k0})}{\nu_k - \nu_t x_{k0}} \quad (4.3)$$

This relation describes the so-called *stoichiometric lines* [23], which help the graphical representation, and the introduction of transformed variables. These lines converge into a pole π, whose location is given by the following expression:

$$x_i = \frac{\nu_i}{\sum_i \nu_i} = \frac{\nu_i}{\nu_t} \quad (4.4)$$

Note that when the total number of moles does not change by reaction ($\nu_t = 0$), the stoichiometric lines are parallel.

A *residue curve* characterizes the evolution of the liquid composition in a vessel during a batchwise reactive distillation experiment. The RCM is obtained by considering different initial mixture compositions. For nonreactive mixtures, the RCM is obtained by solving the following differential equation [41]:

$$\frac{dx_i}{d\xi} = x_i - y_i \quad (4.5)$$

where $\xi = H/V$ is a "warped-time" defined as the ratio of molar liquid holdup H by the molar vapor rate V, and x_i and y_i are vapor and liquid compositions, respectively. A similar representation based on *distillation lines* describes the composition on successive trays of a distillation column with infinite number of stages at infinite reflux

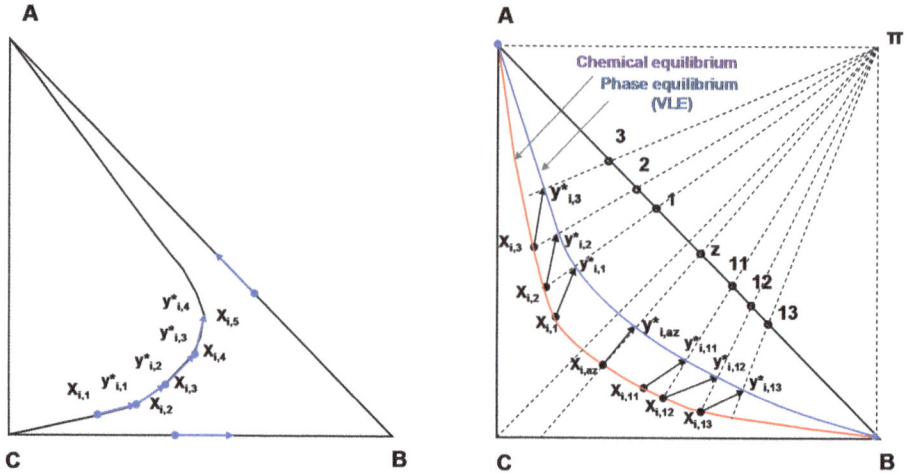

Fig. 4.3: Construction of the distillation lines for nonreactive (left) and reactive mixtures (right).

(∞/∞ analysis). In contrast to the relation describing stoichiometric lines, the distillation lines may be obtained by algebraic computations involving series of bubble and dew points, as follows:

$$x_{i,1} \rightarrow y^*_{i,1} = x_{i,2} \rightarrow y^*_{i,2} = x_{i,3} \rightarrow y_{i,3} \cdots \qquad (4.6)$$

Figure 4.3 (left) shows the construction of a distillation line for an ideal ternary mixture in which A and C are the light (stable node) and heavy (unstable node) boilers, respectively, and B is an intermediate boiler (saddle) [41, 42]. The initial point $x_{i,1}$ produces the vapor $y^*_{i,1}$, which, by condensation, gives a liquid with the same composition such that the next point is $x_{i,2} = y^*_{i,1}$, etc. Accordingly, the distillation line describes the evolution of composition on the stages of a distillation column at equilibrium and total reflux from the bottom to the top. The slope of a distillation line is a measure of the relative volatility of components. Analysis by RCM or distillation curve map leads to similar results.

When a chemical reaction takes place, the residue curves can be found using the following equation:

$$\frac{dx_i}{d\xi} = x_i - y_i + Da(v_i - v_t x_i)R \qquad (4.7)$$

where Da is the Damköhler number, given by the ratio of the characteristic process time H/V to the characteristic reaction time $1/r_0$. The reaction rate r_0 is the reference value at the system pressure and an arbitrary reference temperature, usually the lowest or highest boiling point. For catalytic reactions, r_0 includes a reference value of the catalyst amount. R is the dimensionless reaction rate, equal to r/r_0. The kinetics of a liquid-phase reaction is described as a function of activities, as follows:

$$r = k\left[\left(\prod a_i^{v_i}\right)_{pr} - \left(\prod a_j^{v_j}\right)_{eq} / K_{eq}\right] \qquad (4.8)$$

Thus, parameter Da is a measure of the reaction rate, but its absolute value cannot be taken as the basis for comparing different systems.

Analogous with this procedure, *reactive distillation lines* can be computed by a series of dew and bubble points incorporating the chemical equilibrium, as follows:

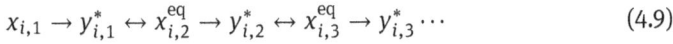

$$x_{i,1} \rightarrow y_{i,1}^* \leftrightarrow x_{i,2}^{eq} \rightarrow y_{i,2}^* \leftrightarrow x_{i,3}^{eq} \rightarrow y_{i,3}^* \cdots \tag{4.9}$$

Graphical construction of reactive distillation lines at equilibrium is shown in Figure 4.3 (right) for the reversible reaction $A + B \rightleftharpoons C$ [41, 42]. The initial point $x_{i,1}$, at chemical equilibrium produces a vapor $y_{i,1}^*$, which by condensation and equilibrium reaction gives a liquid with composition $x_{i,2}$. This is found by crossing the stoichiometric line passing through $y_{i,1}^*$ with the chemical equilibrium curve. Then, the liquid $x_{i,2}$ produces the vapor $y_{i,2}^*$, and so on. Similarly, points 11, 12, 13, . . . N show the situation in which the mixture becomes richer in B and poorer in A and C. Figure 4.3 (right) emphasizes the particular position where phase equilibrium and stoichiometric lines are colinear. The liquid composition remains unchanged because the resulting vapor, after condensation, is converted into the original composition. This point is a *potential reactive azeotrope*, but it becomes a true *reactive azeotrope* when the composition also satisfies the chemical equilibrium.

Figure 4.4 illustrates the construction of an RMC for the reversible reaction $A + B \rightleftharpoons C$, for which the relative volatilities are in the order $3/2/1$ and the equi-

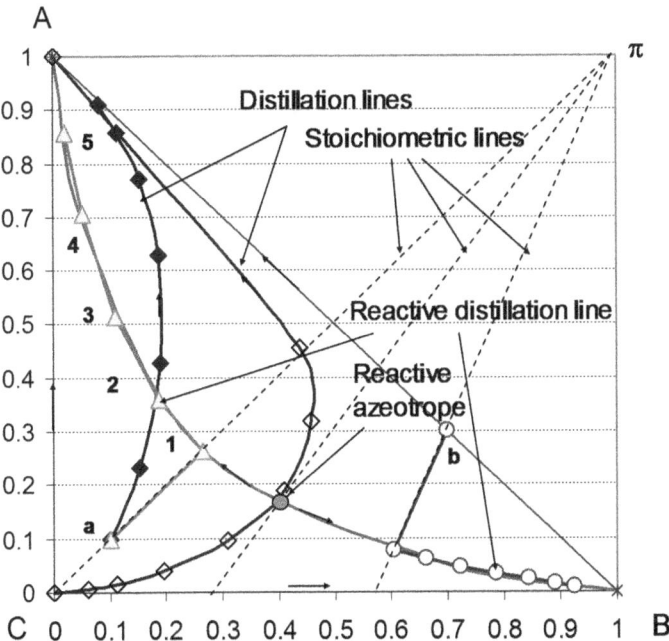

Fig. 4.4: Residue curves for a ternary mixture involving the equilibrium reaction $A + B \rightleftharpoons C$.

librium constant K_x is 6.75 [41, 42]. The physical and reactive distillation lines can be obtained simply by computation in Microsoft Excel using the above equations. In this case, the starting point (point a in Figure 4.4) is a liquid with composition (0.1, 0.1, 0.8) that is not at chemical equilibrium. The coordinates of the triangle are in normal mole fractions. After a short straight path, the reactive distillation line superposes the chemical equilibrium curve. The same trend is also observed when starting from other points. Figure 4.4 also graphically illustrates the formation of a reactive azeotrope as the point where a particular stoichiometric line becomes tangential to the nonreactive residue curve and simultaneously intersects the chemical equilibrium curve. For more than three species, a change of variables appears useful in reducing the dimensionality of a graphical representation. More generally, the composition of a reacting system characterized by c molar fractions can be reduced to $(c - 1)$ new composition variables by the following transformation:

$$X_i = \frac{v_k x_i - v_i x_k}{v_k - v_t x_k} \tag{4.10}$$

Preferably, the reference k component should be a product. The kinetics of a reaction rate has a substantial influence on RCMs. Venimadhavan et al. [43] provide more information on the effect of kinetics on RCMs for RD. Distillation boundaries and physical azeotropes can vanish, while other singular points might appear as a result of kinetic effects.

The RCMs derived from the chemical and phase equilibria for reactive and nonreactive sections of the RD column represent variations of the stationary points (corresponding to pure components and azeotropes) between equilibrium boundaries for a given temperature and pressure. The variations in these singular points can be represented by the Da number [43, 44]. Hence, the influence of kinetics on RCMs can be studied by integrating the equation for finite Da numbers. In addition, the singular points satisfy the following relation:

$$DaR = \frac{x_i - y_i}{x_i - v_t x_i} \tag{4.11}$$

Note that the influence of the kinetics of a chemical reaction on the vapor–liquid equilibrium is very complex. Consequently, evaluating the kinetic effects on RCMs is of great importance for conceptual design of RD systems. However, in practice, the reaction rate is sufficiently fast that chemical equilibrium is reached quickly. The RCM simplifies considerably, but even in this case the analysis may be complicated by the occurrence of reactive azeotropes [41].

4.4 Design and control

Compared with conventional distillation, RD sets specifications on both product compositions and reaction conversion. Accordingly, the degrees of freedom in an RD column must be adjusted to accomplish these specifications while optimizing an objective function such as the total annual cost. The following specifications are typically required [1, 2, 16, 41, 45]:

- Column pressure and pressure drop on column or stage. Setting the pressure is constrained by the temperatures of the top and bottom, more specifically by the availability of on-site cold and hot utilities. If the bottom temperature is excessive, the solution may be the dilution with reactant, which will be recovered separately. In general, working at the highest acceptable temperature is recommended because of its accelerating effect on the reaction rate.
- Number of stages, feed locations of reactants, and exit points of the product streams. The RD configuration is set up in term of rectification, reaction, and stripping sections.
- Top distillate (liquid, vapor, mixed) or bottom product; absolute or ratio values.
- Condenser and reboiler types.
- Reflux ratio or boilup ratio.
- Holdup distribution on the reactive stages.

Typical design specifications in conventional distillation are the concentration of heavy key component in the distillate and the concentration of light key component in the bottom product. The holdup has no effect whatsoever on the steady-state design of a distillation column, only on the dynamic behavior. The column diameter is easily determined from the maximum vapor-loading correlations, after calculating the vapor rates required to achieve the desired separation. However, the holdup is very important in RD as the reaction rates depend directly on the liquid holdup and the amount of catalyst on each tray. Accordingly, RD requires an iterative design procedure because the liquid holdup must be known before the column can be designed [41]. This means that a tray holdup is assumed and the column is designed to achieve the desired conversion and product purities. Then, the column diameter is calculated as well as the required height of liquid on the reactive trays corresponding to the assumed holdup. Liquid heights of over 10–15 cm are not recommended because of hydraulic pressure-drop limitations. If the calculated liquid height is too large, then a smaller holdup is assumed and the calculations are repeated [1].

The following shortcut method for the hydraulic design of an RD column can be used [41]:

1. Estimate a mean volumetric liquid flow rate for operation.
2. Assume an initial value for the superficial liquid velocity at the "load point" (U_{LP0}); recommended value is 10 m^3/m^2 h.
3. Assume an initial value for the number of stages per meter (NSTM).
4. Determine the column diameter. Knowing the packing specifications, estimate the volume of packing and the catalyst holdup per reaction stage.
5. Introduce the above values in simulation, in which the reaction rate is expressed in units compatible with the holdup (mass, molarity, or volume).

6. Determine the total number of reactive stages needed to achieve the target conversion. Pay attention to the profiles of temperature, concentration, and reaction rate. Extract liquid and gas flows, as well as fluid properties.
7. Recalculate the load-point velocity and the liquid holdup from the above information by using specific correlations and diagrams. Check the hydraulic design by selecting packaging with similar characteristics.
8. Verify that the gas load and pressure drop are within optimal regions.

Afterwards, check all values and repeat points 4–8 until acceptable values are achieved.

A number of other design and feasibility check methods for RD systems are also available and can be classified into three main groups: (1) graphical and topological considerations, (2) optimization techniques, and (3) heuristic and evolutionary approaches. Table 4.1 conveniently summarizes these methods, including their principles, assumptions, and a brief pro/con analysis.

   For a long time it was assumed that reaction and distillation can be favorably combined in a column enhanced with special internals or additional exterior volume. However, distillation columns are an appropriate solution only for reactions that are sufficiently fast, so that high conversions can be reached within the residence time range of the columns. Schoenmakers and Bessling [14] showed that two operating conditions could be distinguished: (1) distillation-controlled range, where conversion is influenced by the concentration of the components to be separated, and (2) kinetics-controlled range, in which the conversion is influenced mainly by the residence time and the reaction constant. The industrial design of RD should aim at operating conditions within these two ranges, that is, just sufficient residence time and only the necessary expenditure. Figure 4.5 presents the resulting principles for the choice of equipment for homo- and heterogeneous catalysis [2, 14]. In the case of heterogeneous catalysis, additional separate reaction volumes are needed to retain the catalyst inside the column. These volumes can be arranged either within the equipment or in a side reactor unit coupled by recycle streams.

   For the most frequently encountered type of reaction (A + B ⇌ C + D), there are several groups of possible RD configurations, as illustrated in Figure 4.6. For this, Luyben and Yu [1] provided a theoretical case study using the following assumptions: equimolar reactant feed, equilibrium constant K_{eq} = 2 (at 366 K), kinetic constants k_F = 8 mol/s and k_B = 4 mol/s, liquid holdup defined from tray sizing, maximum kinetic holdup of 20 times the tray holdup, and relative volatility of LLK/LK/HK/HHK fixed at a value of 2. The results based on the total annual cost ($I_p < III_p < II_p < III_r < II_r < I_r$) indicated that group I_p is the most favorable and I_r is the most unfavorable.

Tab. 4.1: Main design and feasibility check methods for reactive distillation.

	Description and assumptions	Pro/con (+/-) analysis	References
Residue curve map (RCM)	Phase equilibrium representations for a ternary or quaternary system. RCMs can be constructed analytically based on physical properties of the system, and for cases with reaction this is expressed by the Damköhler number (Da). The residue curve will always end at a stable node. For $Da = 0$, the RC will end at a pure component or nonreactive azeotrope. For high Da numbers, the RC will end at a pure component, chemical equilibrium point, or reactive azeotrope. In this way the feasibility of RD can be analyzed	+ Couples the reaction and vapor–liquid equilibrium + Reliable tool for feasibility analysis in RD + Requires little data (feed composition, phase equilibrium model parameters, chemical equilibrium and reaction stoichiometry) − Cannot analyze the feasibility of hybrid processes − Limited to four components due to its graphical nature − Accurate thermodynamic data is required to correctly describe the reactive distillation process	[22, 79–82]
Static analysis	Considers the composition on each stage to be constant, with the total column being a succession of reaction and distillation operations. The assumptions allow one to estimate the liquid composition on a stage and the vapor composition below. The composition profiles are estimated by distillation lines. The number of theoretical stages is determined from the intersection of the distillation line with the chemical equilibrium manifold (CEM). This CEM line represents the boundary of the forward and backward reactions. − Vapor–liquid flow rates in RD are infinitely large − Reactive zone is large enough to achieve a set conversion − RD column is assumed to be in steady state and theoretical stages are chosen − Uses only one reversible reaction	+ Straightforward solution on whether RD is feasible + Allows a first estimation of the flowsheet + Requires little data (feed composition, phase equilibrium model parameters, reaction stoichiometry) + An effective tool for studying nonideal mixtures with multiple chemical reactions and components + Allows selection of appropriate steady states from their complete set + Simplifies troublesome calculations, reducing computational time + Does not require fixed variables such as column structure, number of stages, extent of reaction, and product composition + The RCM technique allows one to determine the stability of products and product compositions and the column structure for the entire feed region − Assumes infinite separation efficiency − Matching the operating lines with assumed product composition can be troublesome	[22, 82]

	Description and assumptions	Pro/con (+/-) analysis	References
Attainable regions	The processes of reaction, mixing, and separation are taken into account. For a given system, the attainable region is the region/portion of concentration space that can be achieved from a given composition by any combination of reaction and stripping. This technique aims to identify the attainable regions by describing them with feasible reactor networks of model reactors (CISTR and PFR). For the assumptions, c1 and c2 are two attainable product compositions. – Vector c1c2 belongs to attainable region – On the attainable region, the reaction vector points inwards, is a tangent, or is zero – No reaction vector intersects the attainable region	+ Takes into account the mixing effect + Can be used to point out promising flow sheets + Includes the theory derived for model reactors – Complicated graphical analysis – Economics not taken into account	[22, 82–84]
Thermo-based approach	Based on the existence of RD lines and potential reactive azeotropes. First, the potential products are determined from the knowledge of distillation lines. Then, the column mass balance is checked and the possible separation borders are identified. In this way, the feasibility of RD is checked	+ Can be applied to fast and slow equilibrium reactions + Feasible products can be determined within RD – Finding reactive azeotropes might be difficult – Detailed knowledge of phase equilibrium, reaction kinetics, and residence time within column is needed	[82]
Fixed point algorithm	Based upon changes in location of fixed points that occur at pure component and azeotrope compositions. They move to new positions as the reflux ratio, boil-up ratio, or Damköhler number change. These fixed points represent chemical and physical equilibrium conditions, and their location is dependent on process parameters. By analyzing the change of fixed points, the feasibility can be checked. – Ideal vapor–liquid equilibrium – Negligible heat of mixing – Equal latent heats – Constant liquid holdup on reactive stages	Highly flexible in generating alternative designs at various design parameters + Mass and energy balance can be decoupled, allowing determination of column profile by means of the mass balance and equilibrium equations – Limited implicitly by its graphical nature	[82, 85]

	Description and assumptions	Pro/con (+/-) analysis	References
Graphical techniques	Based on the conventional McCabe–Thiele or Ponchon–Savarit methods. Difference from traditional methods is the reaction on stages. For the reaction, the conversion is regulated by adjusting the liquid holdup on each stage. – Binary system with a single reaction – Constant molar overflow – Vapor–liquid equilibrium is reached	+ Gives a visualization of what happens on each stage – Exclusively applied to binary systems, which are usually isomerization reactions – Limited by their graphical nature	[82, 86]
Reactive cascades	Each section of the reactive distillation column is represented by a cascade of flashes either in cocurrent or counter-current. In this way, the product compositions can be estimated. – Each flash stage has the same residence time – The same fraction of feed is vaporized in each stage	+ Global feasibility analysis can be performed as a function of production rate, catalyst concentration, and liquid holdup + The method does not have any restriction in the number of components or reactions + The method is easily implemented – Damköhler number is assumed to be constant – Effect of different operating conditions or designs is not taken into account – Energy balances are not taken into account	[82, 84]
Phenomena-based approach	All three phenomena (mixing, separation, and reaction) are written in vectors. When there is no change in composition, the sum of these vectors is 0 and a kinetic fixed point occurs. Fixed points are not desired in RD because they represent flat concentration profiles. Therefore, several methods were developed to move away from this fixed point	+ Only physical and chemical data are required + Can be used independently of methodology that incorporates composition change – Feasibility is not fully proven	[82, 87]
Scalar/vectorial difference points	An approach that combines the phenomena-based approach with the reactive cascade methodology. The assumptions include: – Known top and bottom compositions – Known reflux ratio – Constant molar overflow – Chemical reaction of type: $2A \rightleftarrows C + D$	Combination of pro/cons of phenomena-based approach and reactive cascades	[82]

	Description and assumptions	Pro/con (+/-) analysis	References
Memetic Algorithms	This algorithm combines a problem-specific evolutionary algorithm with a mathematical programming method. The advantage of this system is the ability to compute and identify multiple optima	+ Can solve complex RD designs + Possibility to compute and identify local optimum	[88, 89]
Mixed-integer nonlinear programming (MINLP)	A design approach based on rigorous calculations. The objective function tries to minimize the equipment costs and operating costs. However, the function is constrained by the MESH equations on each stage, material balances, kinetic and thermodynamic relationships, and other relationships between process variables and the number of stages. The function consists of master and subproblem. A master program selects the integer variables, and the optimal column design is obtained in the primal problem. – Vapor and liquid phases are in equilibrium – No reaction takes place in the vapor phase – Liquid phase is homogeneous – Enthalpy of liquid streams is negligible – Heat of evaporation is constant – Temperature dependence of reaction rates can be expressed in power-law form – Cost of separating products downstream is given by an analytical function	+ Wide application field (e.g., multiple reactions, reactive equilibrium, or thermal neutrality cannot be assured) + Allows calculations based on economics and controllability – Need for complicated numerical tools – Difficult initialization – Computational time – Multiple (local) optimum points	[82, 88–91]

	Description and assumptions	Pro/con (+/-) analysis	References
Orthogonal collocation on finite elements (OCFE)	MINLP sometimes has difficulties solving the equations for a high number of stages and for simultaneous design of more than one distillation unit. The OCFE approach transforms the number of stages from a discrete number in the MINLP approach to a continuous variable and treats the temperature as a function of position. − Complete mixing of each phase − Constant liquid holdup and no vapor holdup on each stage − No liquid entrainment on stages − Adiabatic stages − Thermal equilibrium between liquid and vapor	+ Converts complex problems to simpler formulations − Not always applicable	[50, 82, 92]
Mixed-Integer dynamic optimization (MIDO)	A method that combines design and control tasks. This simultaneous approach takes advantage of the interaction between design and control	+ Simultaneous design and control − No general procedure that guarantees convergence to a global solution	[82, 93]

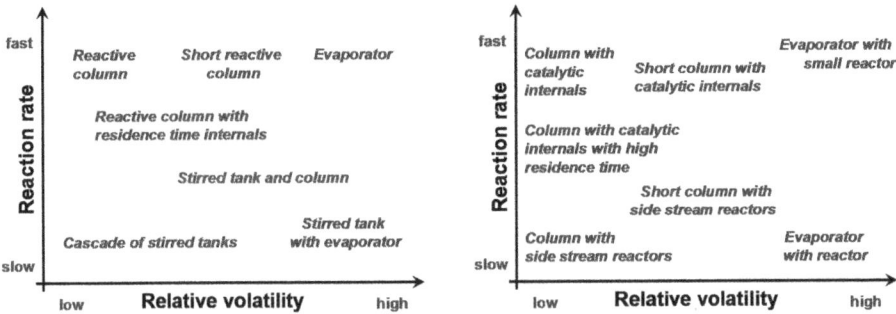

Fig. 4.5: Equipment selection for homogeneous (left) and heterogeneous (right) catalysis.

Additionally, Figure 4.7 [2] shows the most common RD configuration alternatives, ranging from conventional RD and heterogeneous azeotropic RD columns [22, 46–50] to reactive dividing-wall columns [37, 51, 52], and an RD column combined with a pre-reactor and/or side reactors [53–55].

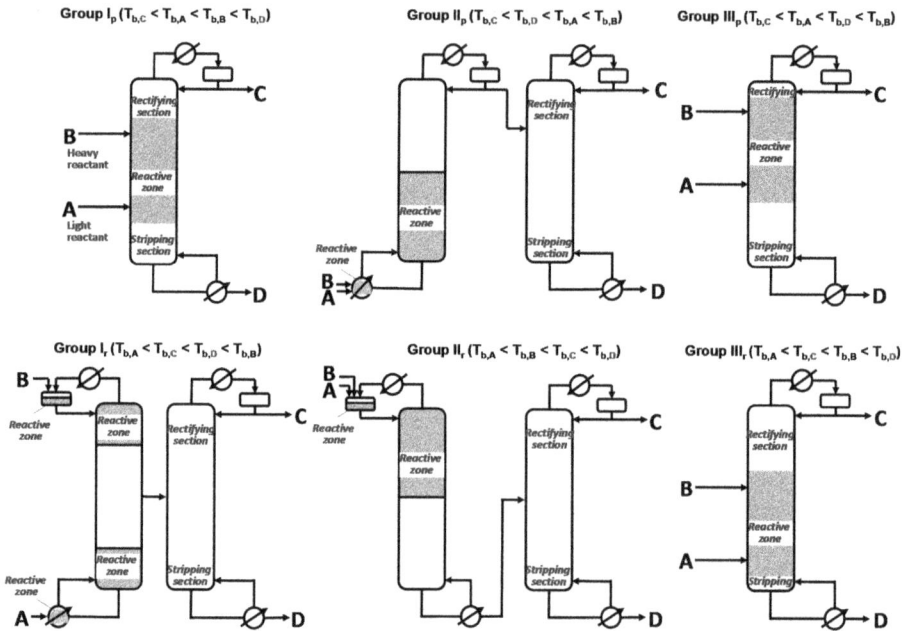

Fig. 4.6: Reactive distillation groups for quaternary reversible reactions (A + B ⇌ C + D), based on various orders of relative volatility between components.

RD provides an excellent example of the ever-present interaction between design and control. Both steady-state and dynamic aspects of a chemical process must be considered at all stages of the development and commercialization of a chemical process (laboratory, pilot plant, and production plant) [1, 56, 57]. The controllability of an RD column is usually improved by adding additional reactive trays, while considering the conflict between the steady-state design and dynamic controllability [1]. In neat operation mode, the reactants are fed according to the stoichiometric ratio. Consequently, the control system must be able to detect any imbalance that would inevitably result in gradual accumulation of one of the reactants, loss of conversion, and reduction in product purity [58]. An alternative is operation with an excess of one reactant. This makes control of the RD column easier, but requires the recovery and recycle of the reactant that is in excess [1].

Control of an RD distillation column is a challenging task due to process nonlinearity and complex interactions between the vapor–liquid equilibrium and chemical reactions. Different types of control methodologies can be used for RD processes, ranging from simple proportional-integral-derivative (PID) controllers to advanced model predictive controllers (MPC), such as dynamic matrix control (DMC), quadratic dynamic matrix control (QDMC), robust multivariable predictive control technology (RMPCT), generalized predictive control (GPC), and other advanced controllers [56].

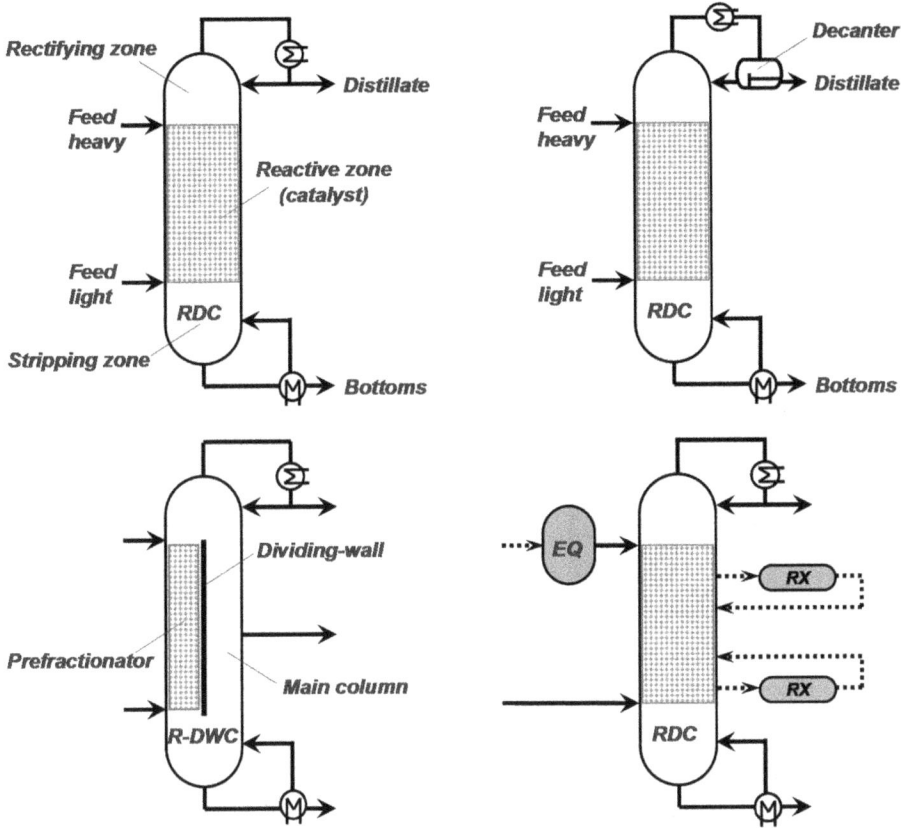

Fig. 4.7: Conventional reactive distillation (top left), heterogeneous azeotropic RD (top right), reactive DWC (bottom left), and RD column with prereactor and/or side-reactors (bottom right).

4.5 Reactive distillation equipment

RD processes can be divided into homogeneous (autocatalyzed or homogeneously catalyzed) and heterogeneous (catalyzed by a solid catalyst; catalytic distillation) processes. Note that the rate of autocatalytic reactions can only be influenced by the temperature and pressure of the RD equipment. Homogeneous catalysis also allows the reaction rate to be influenced by changing the concentration of catalyst; thus, the reaction rate can be adapted over a wide range to the needs of the RD equipment. Although homogeneous catalysis is more flexible, it requires expensive separation steps for catalyst recovery. Solid catalysts require a special construction (e.g., catalyst packed in "tea bags" on trays, or sandwiched in structured packing such as Sulzer Katapak) to fix the catalytic particles in the reactive zone, thus limiting the catalyst concentration that can be achieved. The reaction rate can be enhanced only to the limit set by the attainable concentration range. Heterogeneous catalysis is simpler in principle, but it

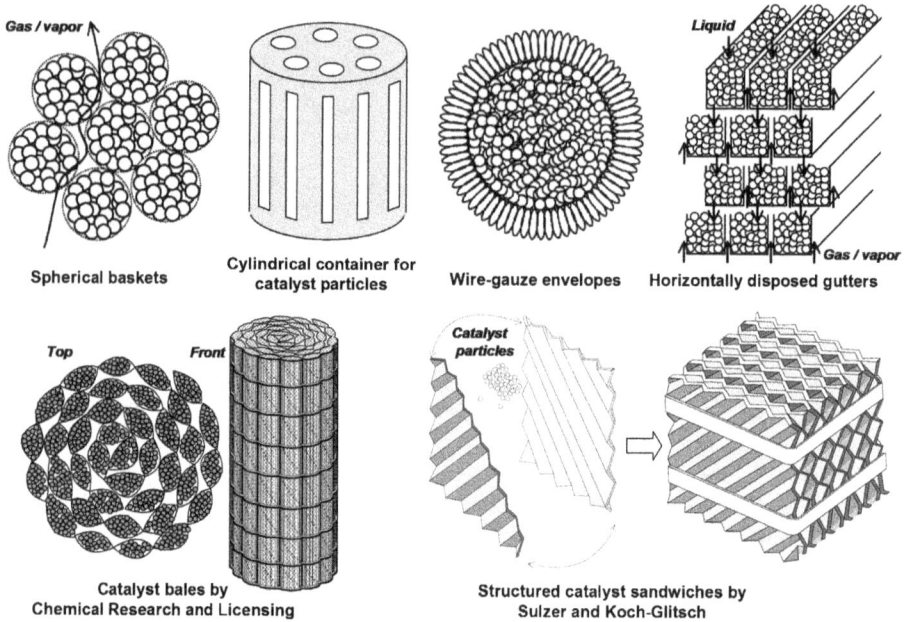

Fig. 4.8: Examples of internals for catalytic distillation: various "tea-bag" configurations for trays, gutters (top), and structured catalyst sandwiches for packed columns (bottom).

needs a larger equipment volume and higher operating temperature and suffers from limited catalyst life time [9, 14].

Unlike conventional distillation, the choice of internals for reactive distillation is much more limited [9]. Figure 4.8 presents some examples of catalytic packing [2, 9, 29], the most important being the following:

- catalytic Raschig rings (random packing) with surface-coated catalyst
- catalyst bales, formed by wrapped wire sheets filled with catalyst
- catalytic structured packing

In the latter, the elements have the shape of sandwiches manufactured from corrugated wire gauze sheets hosting catalyst bags, assembled as cylinders or rectangular boxes. The packing structure consists of alternating catalyst bags and open channel spaces. To ensure higher efficiency of combined reaction and diffusion, the catalyst particles should have a diameter of about 0.8–1 mm. The advantages of structured packing are uniform flow conditions with minimum back-mixing and maldistribution; good radial dispersion, an order of magnitude better than in conventional packed beds, ensuring a longer residence time; and efficient maintenance and replacement of catalyst. Commercial products include Katapak-S®, manufactured by Sulzer ChemTech [59], MultiPak® supplied by Julius Montz [60], and Katamax® from Koch-Glitsch.

4.6 Applications of reactive distillation

The reasons for the many benefits of RD are the synergistic effects of the simultaneous chemical reaction and distillation. However, these synergies make RD extraordinarily complex [19]. Conceptually, RD belongs to the category of multifunctional reactors and is a very complex system because of simultaneous interactions between vapor and liquid phases. In cases where solid catalysts are used, there are also interactions with the solid phase. Owing to its complexity, there are several constraints and difficulties (discussed earlier) that limit the widespread application of RD. Nonetheless, it is worth noting that RD has made already a significant impact at the industrial scale, with many large-scale applications.

Table 4.2 [2, 61] lists the most important applications of RD, which include (trans-) esterification, etherification, hydrolysis, (de-)hydration, alkylation, isomerization, (de-)hydrogenation, amination, condensation, nitration, and chlorination [10, 16]. Currently, the application with the largest number of installations is production of methyl tertiary butyl ether (MTBE), used as a gasoline additive. CDTECH (a partnership between ABB Lummus Global and Chemical Research & Licensing Company) has licensed over 200 commercial processes operating worldwide at capacities of 100–3000 kt/year for the production of ethers (MTBE, TAME, ETBE), hydrogenation of aromatics, hydro-desulfurization (HDS), and production of ethyl benzene, and isobutylene [7]. Sulzer ChemTech also reports several industrial-scale applications such as synthesis of ethyl, butyl, and methyl acetates; hydrolysis of methyl acetate; synthesis of methylal; methanol removal from formaldehyde; and production of fatty acid esters [7].

4.7 Case study: biodiesel production by heat-integrated reactive distillation

Biodiesel is an established renewable and biodegradable fuel that is produced mainly from green sources (e.g., vegetable oils, animal fat, waste cooking oils from the food industry). The current trend is to use cheap raw materials such as waste cooking oils, but these can contain a substantial amount (up to 100%) of free fatty acids (e.g., palm-oil fatty acid distillate from hydrolysis of triglycerides). Accordingly, development of an efficient continuous process for biodiesel production is required. In such a process, the use of a solid catalyst is especially necessary to suppress the costly processing steps and waste treatment. The common problem of all conventional biodiesel processes is the use of liquid catalysts that require neutralization and an expensive multistep separation that generates salt waste streams [62].

To solve these problems, solid acids can be applied in an esterification process based on RD. RD is an integrated process that is able to shift the chemical equilibrium

Tab. 4.2: Main industrial applications of reactive distillation.

Reaction type	Catalyst/internals
Alkylation	
Alkyl benzene from ethylene/propylene and benzene	Zeolite β, molecular sieves
Amination	
Amines from ammonia and alcohols	H_2 and hydrogenation catalyst
Carbonylation	
Acetic acid from CO and methanol/dimethyl ether	Homogeneous
Condensation	
Diacetone alcohol from acetone	Heterogeneous
Bisphenol-A from phenol and acetone	N/A
Trioxane from formaldehyde	Strong acid catalyst, zeolite ZSM-5
Esterification	
Methyl acetate from methanol and acetic acid	H_2SO_4, Dowex 50, Amberlyst-15
Ethyl acetate from ethanol and acetic acid	N/A
2-Methyl propyl acetate from 2-methyl propanol and acid	Katapak-S
Butyl acetate from butanol and acetic acid	Cation-exchange resin
Fatty acid methyl esters from fatty acids and methanol	H_2SO_4, Amberlyst-15, metal oxides
Fatty acid alkyl esters from fatty acids and alkyl alcohols	H_2SO_4, Amberlyst-15, metal oxides
Cyclohexyl carboxylate from cyclohexene and acids	Ion-exchange resin bags
Etherification	
MTBE from isobutene and methanol	Amberlyst-15
ETBE from isobutene and ethanol	Amberlyst-15/pellets, structured
TAME from isoamylene and methanol	Ion-exchange resin
DIPE from isopropanol and propylene	ZSM 12, Amberlyst-36, zeolite
Hydration/dehydration	
Mono ethylene glycol from ethylene oxide and water	Homogeneous
Hydrogenation/dehydrogenation	
Cyclohexane from benzene	Alumina-supported Ni catalyst
Methyl isobutyl ketone from benzene	Cation-exchange resin with Pd/Ni
Hydrolysis	
Acetic acid and methanol from methyl acetate and water	Ion-exchange resin bags
Acrylamide from acrylonitrile	Cation exchanger, copper oxide
Isomerization	
Iso-paraffins from *n*-paraffins	Chlorinated alumina and H_2
Nitration	
4-Nitrochlorobenzene from chlorobenzene and nitric acid	Azeotropic removal of water
Transesterification	
Ethyl acetate from ethanol and butyl acetate	Homogeneous
Diethyl carbonate from ethanol and dimethyl carbonate	Heterogeneous
Vinyl acetate from vinyl stearate and acetic acid	N/A
Unclassified reactions	
Monosilane from trichlorsilane	Heterogeneous
Methanol from syngas	$Cu/Zn/Al_2O_3$ and inert solvent
Diethanolamine from monoethanolamine and ethylene oxide	N/A
Polyesterification	Autocatalytic

to completion and preserve the activity of solid catalyst by continuously removing the products. Thus, RD also leads to lower investment and operating costs [58, 62–65].

The integrated RD process presented here was designed according to previously reported process synthesis methods for reactive separations [10, 66]. Rigorous simulations embedding experimental results were performed using Aspen Plus simulator (AspenTech, 2009 [38]). The RD column was simulated using the RADFRAC unit with RateSep (rate-based) model enabled, and considering three-phase balances.

The physical properties required for the simulation and the binary interaction parameters for methanol–water and acid–ester pairs were available in the Aspen Plus database of pure components. The other interaction parameters were estimated using the UNIFAC-DMD group contribution method. The fatty components were conveniently lumped into a single fatty acid and its fatty acid methyl ester (FAME), according to the reaction $R–COOH + CH_3OH \rightleftharpoons R–COO–CH_3 + H_2O$. Dodecanoic (lauric) acid/ester was selected as lumped component because of the availability of experimental results, kinetics, and vapor–liquid–liquid equilibrium parameters for this system [67, 68]. The lumping of components is reasonable because fatty acids and their corresponding esters have similar properties. This approach has already been successfully used to simulate production processes for other fatty esters [64, 68, 69].

In this case study, sulfated zirconia is considered as solid acid catalyst because of the availability of kinetic data [64, 68]. The esterification reaction is a second-order reversible reaction; hence, the reaction rate accounts for both direct and reverse reactions.

$$r = (k_1 W_{cat}) C_{Acid} C_{Alcohol} - (k_2 W_{cat}) C_{Ester} C_{Water} \qquad (4.12)$$

where k_1 and k_2 are the kinetic constants for the direct (esterification) and reverse (hydrolysis) reactions, W_{cat} is the amount of catalyst, and $C_{Component}$ is the molar concentration of the components present in the system. Because water is continuously removed from the system, the reverse hydrolysis reaction is extremely slow; hence, the second term of the reaction rate can be practically neglected. Note that the presence of other compounds and impurities may influence the kinetics, so this has to be checked and quantified experimentally for various types of feedstock.

The conceptual design of the process is based on an RD column that integrates the reaction and separation steps into a single operating unit. The system is able to shift the reaction equilibrium toward product formation by continuous removal of reaction products, instead of using excess reactant. An additional flash and a decanter are used to guarantee the high purity of the products. The RD column consists of a core reactive zone completed by rectifying and stripping separation sections, whose extent depends on the separation behavior of the reaction mixture. Because methanol and water are much more volatile than the fatty ester and acid, they separate easily in the top. Figure 4.9 (top) presents the flowsheet of this biodiesel process based on conventional RD, as reported by Kiss et al. [64] and Kiss and Bildea [70].

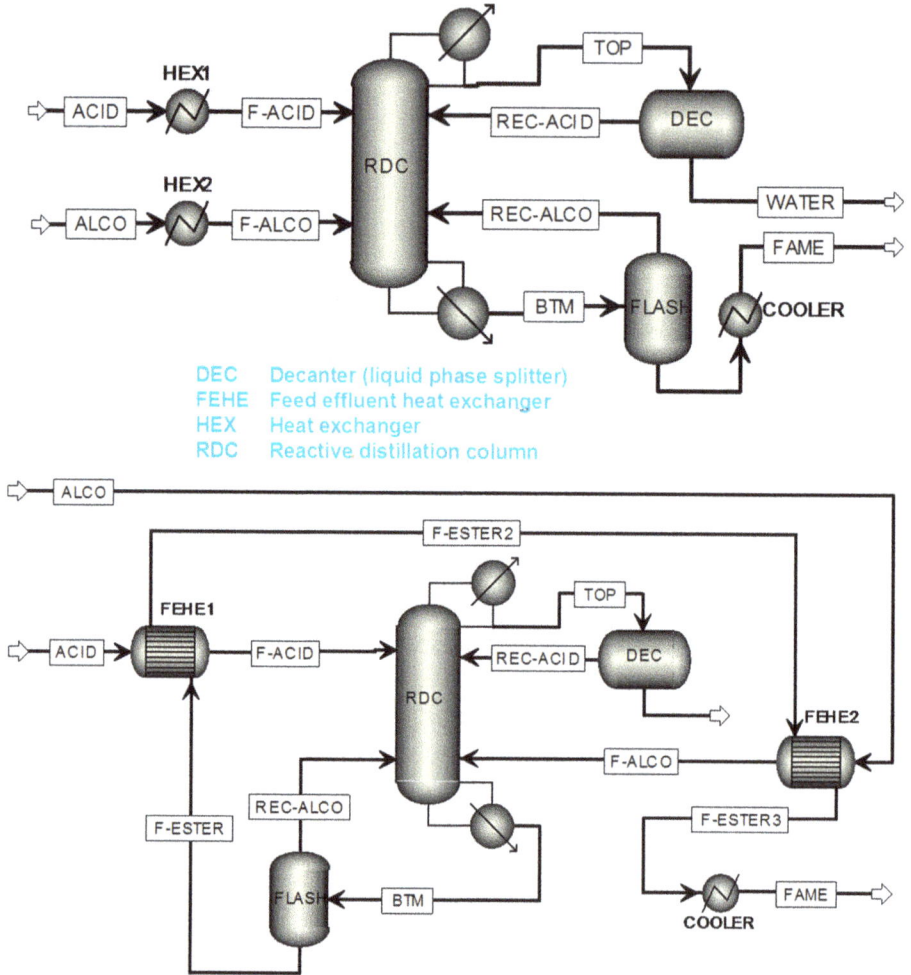

DEC   Decanter (liquid phase splitter)
FEHE Feed effluent heat exchanger
HEX   Heat exchanger
RDC   Reactive distillation column

Fig. 4.9: Synthesis of fatty esters by reactive distillation: base case flowsheet (top) and heat-integrated RD flowsheet (bottom).

The reference flowsheet presented in Figure 4.9 (top) is relatively simple, with just a few operating units, two cold streams that need to be preheated (fatty acid and alcohol), and two hot streams that have to be cooled (top water and bottom fatty esters). Heat integration was performed by applying previously reported heuristic rules [42, 71]. Consequently, a feed-effluent heat exchanger (FEHE) should replace each of the two heat exchangers HEX1 and HEX2. Figure 4.9 (bottom) illustrates the improved process design that includes heat integration around the RD column [58]. The hot bottom product of the column (FAME) is used to preheat both reactants (i.e., fatty acid and

Tab. 4.3: Design parameters for simulating the heat-integrated reactive distillation column.

Parameter	Value	Units/remarks
Total number of theoretical stages	15	Reactive from 3 to 12
Column diameter	0.4	m
HETP	0.5	m
Valid phases	—	Vapor, liquid, liquid
Volume liquid holdup per stage	18	L
Mass catalyst per stage	6.1	kg
Catalyst bulk density	1050	kg/m^3
Fatty acid conversion	>99.99	%
Fatty acid feed, on stage 3 (liquid, at 145 °C)	1167	kg/h
Methanol feed, on stage 10 (liquid, at 65.4 °C)	188	kg/h
Reboiler duty	136	kW
Condenser duty	−72	kW
Reflux ratio (mass ratio R/D)	0.10	kg/kg
Boil-up ratio (mass ratio V/B)	0.12	kg/kg
Production of biodiesel (FAME)	1250	kg/h
RD column productivity	20.4	kg FAME/kg catalyst/h
Specific energy requirements	108.8	kWh/t ester

alcohol feed streams). Notably, there is no longer need for an external hot utility to preheat the reactant feed streams and no additional heat exchanger is required by this heat-integrated setup [2].

The main design parameters (e.g., column size, catalyst loading, and feed condition) are listed in Table 4.3 [58]. High conversion of the reactants is achieved, with the productivity of the RD unit exceeding 20 kg fatty ester/kg catalyst/h. Purity specifications over 99.9%wt were achieved for the final biodiesel product (FAME stream). Figure 4.10 (top) shows the liquid and vapor composition, as well as the reaction rate and temperature profiles along the reactive distillation column [58]. The RD column is operated in the temperature range of 70–210 °C, at ambient pressure. Because the reaction takes place mainly in the reactive zone, the reaction rate exhibits a maximum in the middle of the column. The concentration of water increases from the bottom to the top of the column, whereas the concentration of fatty esters increases from the top to bottom. Therefore, in the top of the reactive separation column there is mainly water with negligible amounts of fatty acids, whereas in the bottom there is liquid fatty ester product (biodiesel) with a very limited amount of methanol.

Heat-integrated RD offers major advantages such as reduced capital investment and operating costs, in addition to no catalyst-related waste streams and no soap formation. However, the controllability of the process is just as important as the savings in capital and operating costs. In processes based on reactive separations, feeding the reactants according to their stoichiometric ratio is essential in achieving high product purity [68, 72]. This constraint must be fulfilled not only during normal operation,

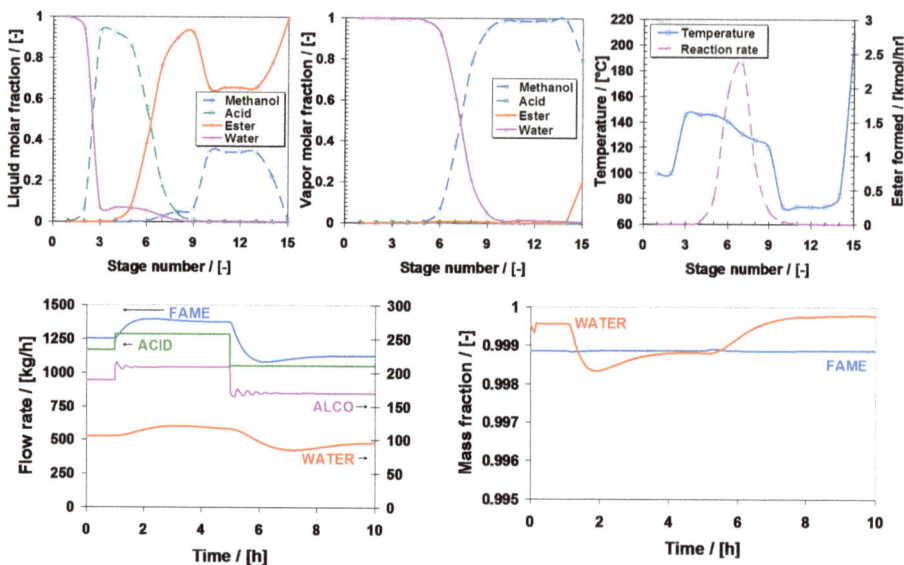

Fig. 4.10: Steady state profiles (top) show liquid/vapor molar composition, temperature, and reaction rate profiles along the reactive column. Dynamic simulation results (bottom) show an acid flow rate disturbance of +10% at 1 h and -10% at 5 h.

but also during the transitory regimes arising due to planned production rate changes or unexpected disturbances. In spite of the high degree of integration, the heat-integrated RD process is very controllable. A key result is an efficient control structure that can ensure the ratio of reactants required for total conversion of fatty acids and for the prevention of difficult separations (for details see Kiss [58]).

Figure 4.10 (bottom) depicts the dynamic simulation results [58]. The simulation starts from the steady state. At time $t = 1$ h, the acid flow rate is increased by 10%, from 1168 kg/h to 1284.4 kg/h. Then, at time $t = 5$ h, the acid flow rate is decreased to 1051.2 kg/h, representing a 10% decrease with respect to the nominal value. The new production rate is achieved in about 2 h. The purity of FAME remains practically constant throughout the dynamic regime, the main impurity being methanol. Notably, the acid concentration stays below the 2000 ppm requirement of the ASTM D6751-08 standard (i.e., acid number < 0.50 mg KOH/g biodiesel).

It is worth noting that both designs (base case and heat-integrated flowsheet) presented here are suitable for a large range of fatty acids and alcohol feedstocks. These RD-based processes have no additional separation steps and produce no waste salt streams because water is the only byproduct.

4.8 Case study: fatty esters synthesis by dual reactive distillation

Fatty esters are key products of the chemical process industry, being incorporated in a wide variety of products with high added value, from cosmetics to plasticizers and biodetergents. A key problem in the synthesis of fatty esters by RD is effective water removal. This is necessary to protect the solid catalyst and avoid costly recovery of the alcohol excess. This case study presents a novel approach based on dual esterification of lauric acid with light and heavy alcohols, namely methanol and 2-ethylhexanol. These two complementary reactants have an equivalent reactive function, but synergistic thermodynamic features. The setup behaves similarly to a reactive absorption combined with reactive azeotropic distillation, with the heavy alcohol as co-reactant and water-separation agent. Superacid solid catalyst based on sulfated zirconia,

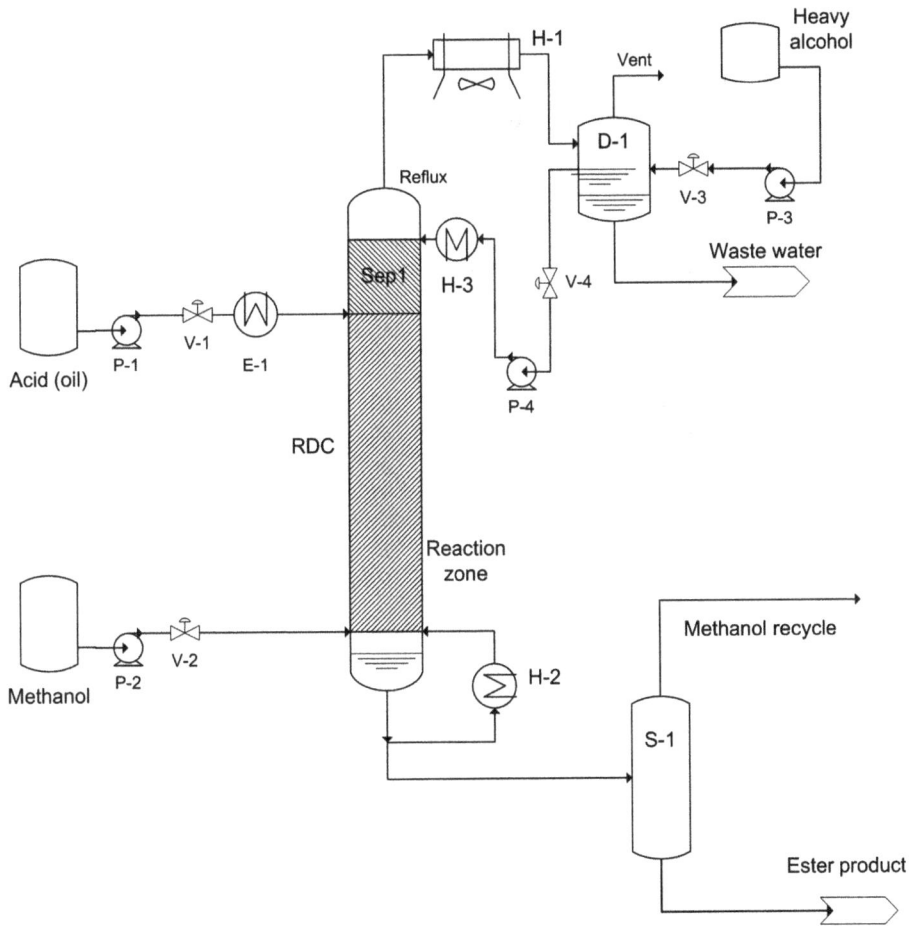

Fig. 4.11: Flowsheet configuration and control strategy of the dual reactive distillation setup.

whose activity is comparable for the two alcohols, can be used at temperatures of 130–200 °C and moderated pressure. Control of the inventory of alcohols is realized by fixing the reflux of heavy alcohol and the inflow of the light alcohol column. This strategy achieves both stoichiometric reactant feed rate and large flexibility in ester production. The distillation column for recovering light alcohol from water is no longer necessary. The result is a compact, efficient, and easy-to-control multiproduct reactive setup, as clearly illustrated in Figure 4.11 [68]. Other important design details of the RD column are conveniently provided in Table 4.4 [68].

In this control structure, the reactants are fed into the process in a ratio that satisfies the overall mass balance imposed by the reaction stoichiometry and the phase equilibrium at the top and the bottom of the RD column. In contrast, control structures fixing the feed rates of all reactants (acid, light alcohol, and heavy alcohol) will not work in the presence of small control implementation errors, the failure manifesting in accumulation or depletion of one reactant [73].

Figure 4.12 (top) compares the temperature profiles for the base case and for a 10% increase in lauric acid flow rate, with and without temperature control [68]. Accurate control of lauric acid concentration in the bottom stream is achieved using a concentration controller that prescribes the setpoint of the temperature controller in a cascade structure. For this control configuration, a change in lauric acid feed flow

Tab. 4.4: Design parameters for simulating the dual reactive distillation column.

Parameter	Value	Units/remarks
Number of theoretical stages	25	Reactive from 5 to 24
Lauric acid feed (on stage 5, at 3.5 bar, 150 °C)	100	kmol/h
Methanol feed (directly in reboiler, at 4 bar, 100 °C)	130	kmol/h
2-Ethylhexanol (fed in decanter, at 3.5 bar, 130 °C)	13.48	kmol/h
Catalyst bulk density	1050	kg/m^3
Volume holdup per stage	0.050	m^3
Mass catalyst per stage	55	kg
Reflux flow rate	2500	kg/h
Column diameter	1.2	m
HETP	0.5	m
Fatty acid conversion	>99.99	%
Reboiler duty	1750	kW
Condenser duty	−1492	kW
Production rate	22,333	kg ester/h
Productivity of RD column	20.3	kg ester/kg catalyst/h
Bottom product composition (mass fraction)	650 ppb acid	kg/kg
	11 ppm water	kg/kg
	0.058 methanol	kg/kg
	0.788 methyl ester	kg/kg
	0.174 2 EH ester	kg/kg
Specific energy requirements	166.8	kWh/t ester

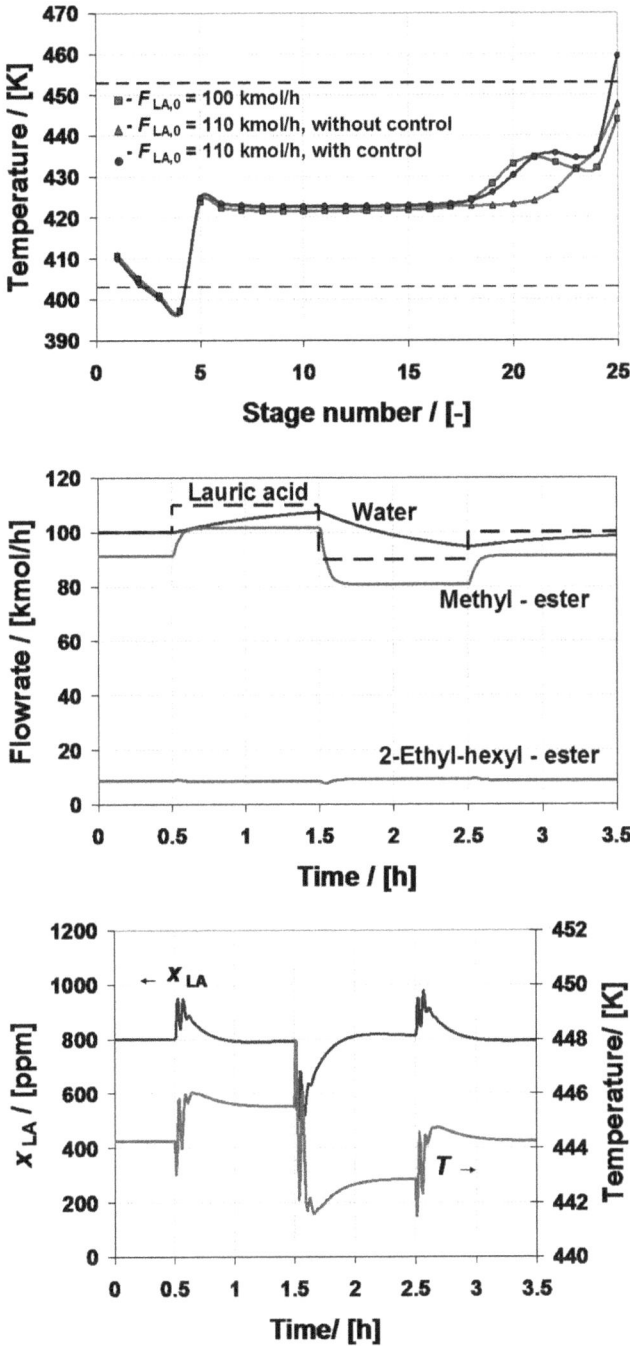

Fig. 4.12: Temperature profiles along the reactive distillation column (top). Dynamic simulation results show the increase and decrease in production rate (middle). Temperature and the concentration of lauric acid in the bottom stream during production rate changes (bottom).

rate leads to a change in methyl ester production rate. In contrast, when both flow rate ratios (i.e., lauric acid feed/methanol entering the column and lauric acid feed/heavy alcohol reflux) are constant, the change in lauric acid feed flow rate leads to changes in both the methyl ester and ethylhexyl ester production rates.

Figure 4.12 (middle) presents the performance of the control system for the following scenario: the simulation starts from the steady state (feed rate of lauric acid, 100 kmol/h), which is maintained for 0.5 h [68]. Then, the feed rate of lauric acid is increased from 100 to 110 kmol/h and, after 1 h, is decreased to 90 kmol/h. Finally, the initial flow rate of 100 kmol/h is restored. The change in acid feed flow rate leads to a change in light ester production rate, with the same magnitude of change. The heavy ester production rate remains constant. The dynamics is fast, only 20 min being necessary to achieve the new production rate. The amount of water obtained at the top of the column reflects the amount of ester formed. During the entire transient period, the concentration of water on the reactive trays remains below the 2 wt% limit.

Figure 4.12 (bottom) presents the temperature and concentration of lauric acid in the bottom stream for the same scenario. Both variables remain very close to the nominal values. Notably, tuning of the controllers is not crucial for the performance of the control system. In this case study, the parameters of the controllers were set as follows: The range of the controlled variable was set to the nominal value of ±10 °C for the temperature control loops and the nominal value of ±50% for the level control loops. For all loops, the range for the manipulated variable was set to twice the nominal value. The gain of the feedback controllers was set to 1%/%. An integration time of 20 min was selected for the temperature controllers. In conclusion, the control structure presented here achieves stable operation and is able to modify the throughput while keeping the characteristics of the products at their design values [68].

At optimal operation, the highest yield and purity can be achieved using stoichiometric feeds in the desired ratio of fatty esters. At this point, the amount of methanol lost in the top product is practically negligible. The heavy ester plays the role of a solvent and prevents escaping methanol in the top product. On the top stages, the heavy alcohol enhances water concentration in the vapor phase, from which water is separated by condensation and decanting, while heavy ester is produced in an amount proportional to the reflux flow rate. Optimal operation is based on controlling the inventory of reactants by using the principle of fixed recycle flows of co-reactants, in this case the reflux of the organic phase and methanol inflow to the RD column. This strategy allows large changes in the production rate. The control strategy is generic and can be employed for esterifications involving formation of azeotropes, as for ethanol and (iso-)propanol. The overall result in integrating design and process control is a compact, efficient, and easy-to-control multiproduct RD setup.

4.9 Case study: industrial reactive distillation process for methyl acetate production

Methyl acetate is a high volume commodity chemical with applications as an intermediate in the manufacture of a variety of polyesters, such as photographic film base, cellulose acetate, Tenite cellulosic plastics, and Estron acetate [74]. The industrial process for methyl acetate (MeOAc) synthesis is based on the equilibrium-limited esterification reaction of methanol (MeOH) with acetic acid (AcOH). The reaction takes place in the liquid phase in the presence of an acid catalyst such as sulfuric acid or a sulfonic acid ion-exchange resin [75], as follows:

$$MeOH + AcOH \rightleftharpoons MeOAc + H_2O \tag{4.13}$$

The expression of reaction rate in the form of activities is strongly preferred because water and methanol have higher polarity than methyl acetate, leading to strongly non-ideal solution behavior. An activity-based rate model for the reaction chemistry is given by the following relation [76]:

$$r = k_f \left(a_{AcOH} a_{MeOH} - \frac{a_{MeOAc} a_{H_2O}}{K_{eq}} \right) \tag{4.14}$$

where the reaction equilibrium constant (K_{eq}) and the forward rate constant (k_f) are given by the following expressions:

$$K_{eq} = 2.32 \times \exp(782.98/T); \qquad \text{with } T \text{ expressed in K} \tag{4.15}$$
$$k_f = 9.732 \times 10^8 \times \exp(-6287.7/T); \qquad h^{-1} \tag{4.16}$$

Due to the commercial success of the RD process in replacing the conventional process (both shown in Figure 4.13) [61] and the potential of ion-exchange resins, a rate expression for an ion-exchange resin-catalyzed reaction was proposed [74]. The following expression for the reaction rate (r) is based on kinetic data generated over a range of molar feed ratios more typical of RD conditions:

$$r = \frac{k \left(a_{AcOH} a_{MeOH} \left(-a_{MeOAc} a_{H_2O}/K_{eq} \right) \right)}{\left(1 + K_{AcOH} a_{AcOH} + K_{MeOH} a_{MeOH} + K_{MeOAc} a_{MeOAc} + K_{H_2O} a_{H_2O} \right)} \tag{4.17}$$

where k is the rate constant and K_{eq} the equilibrium constant, which is equal to 5.2 according to Agreda et al., [11]. The K_i values (i being the component AcOH, MeOH, MeOAc, and H_2O) are the adsorption coefficients of the Langmuir–Hinshelwood–Hougen–Watson (LHHW) model. The expression has been successfully used to verify the experimentally observed RCMs of this system. The heat of reaction is low ($\Delta H = -3.0165\,kJ/mol$), indicating a slightly exothermic reaction that is typical for acetate esterifications. The liquid-phase activity coefficients are well represented by the Wilson model [76]. The reaction can be carried out at 310–393 K and atmospheric

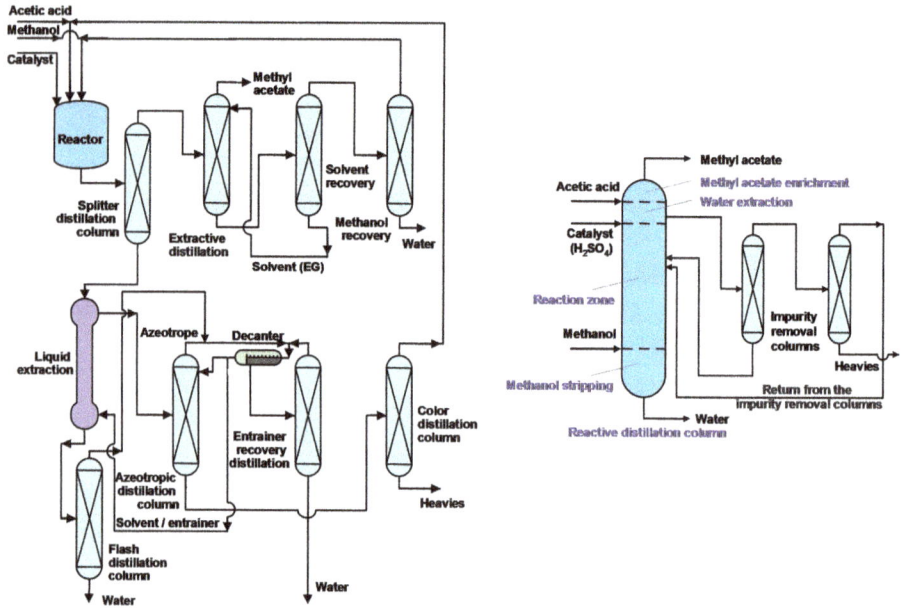

Fig. 4.13: Methyl acetate production: conventional process (left) compared with reactive distillation (right).

pressure. The only main side reaction is the formation of dimethyl ether (DME) by the etherification of methanol, which is predominant at high temperatures [74].

Conventional processes used reactors with a large excess of a reactant to achieve high conversion, followed by an energy-intensive downstream separation due to the formation of MeOAc/MeOH and MeOAc/H$_2$O azeotropes. A typical process employs one or two reactors, eight distillation columns, and one liquid extraction column, making it rather complex and capital intensive. Eastman Kodak has developed a RD process that delivers high purity product using a near-stoichiometric ratio of MeOH to AcOH [11]. The RD column (shown in Figure 4.13) has the following sections:

Methyl acetate enrichment section: This is where acetic acid and methyl acetate are separated above the acetic acid feed, allowing pure methyl acetate to be recovered as overhead product.

Water extraction section: This is where acetic acid acts as a mass separating agent (extractant) and extracts water (thus breaking the methyl acetate/water azeotrope) and some methanol.

Reaction section: This is where the reaction occurs in a series of counter-current flashing stages with sulfuric acid as homogeneous catalyst.

Methanol stripping section: This is where methanol is stripped from water as the bottom product.

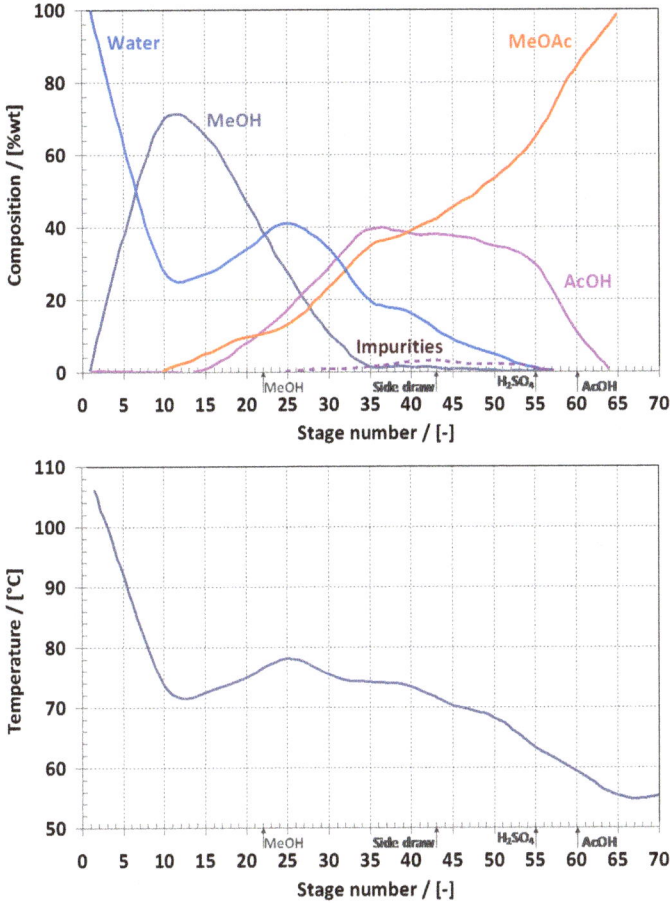

Fig. 4.14: Composition and temperature profiles in the RD column for methyl acetate production.

Figure 4.14 shows the composition and temperature profiles along the RD column [11, 75]. High purity methyl acetate is obtained as distillate and water byproduct is removed as bottom product. Some mid-boiling components are formed due to impurities present in the feed. For this reason, a small side stream is withdrawn above the catalyst feed point and treated separately in an impurity-removal system. Impurities are stripped and concentrated while a methanol and methyl acetate stream is recycled to the reaction zone of the RD column. Notably, the RD column is operated at a near-stoichiometric molar ratio of acetic acid and methanol, but is able to yield high purity methyl acetate as product. The whole process is practically integrated in a single column, thus eliminating the need for a complex distillation column system and recycle of the methanol/methyl acetate azeotrope. Remarkably, a single RD column built at Eastman Kodak's Tennessee plant produces 180 kt/year of high purity methyl acetate [74].

4.10 Concluding remarks

Reactive distillation is a process intensification method that has attracted considerable attention over the past decades as a promising alternative to conventional reaction–separation sequences. RD technology has already been successfully applied in the process industry, especially to overcome the conversion of chemical equilibrium-limited reactions [19]. However, the potential of RD technology has not been fully tapped and there is ongoing research to improve RD processes by means such as ultrasound-assisted RD [77], use of high-gravity fields [78], or coupling RD with other operations such as membrane separations [19].

The main drivers for RD applications are economic (over 20% reduction in variable cost, capital expenditure, and energy use), environmental (lower CO_2 emissions, no or reduced salt waste), and social (improved safety and health due to reduced carbon footprint, lower reactive content, and lower less thermal sensitivity and runaway). It is thus clear that RD contributes to all three pillars of sustainability in the chemical industry. The industrial implementation of RD by collaboration of various partners in research, scale-up, design, and operation can be considered as a model for the rapid implementation of other process intensification techniques in the chemical process industry [7]. Taking into account the remarkable progress in hardware development, modeling for design and simulation, control strategies, and real-time optimization, and also considering the pace at which new applications are being explored, RD remains one of the most important tools for process intensification using green chemistry and engineering [2, 7, 8, 16].

Despite the more complex design, control, and equipment needs, RD is a process intensification technology that fulfills the principles of green engineering (e.g., prevention instead of treatment, design for separation, maximization of efficiency, output-pulled versus input-pushed, "meet the need and minimize excess," integration of local materials and energy flows, design for a commercial afterlife, renewable versus depleting). RD offers key unique features such as a reduced number of processing units, enhanced overall rates, improved selectivity, reduced energy requirements, little or no solvent use, and the ability to overcome unfavorable equilibrium and avoid difficult separations [6, 16].

RD is currently an established unit operation in chemical process technology and is also a front-runner in the field of process intensification [7, 8]. Nowadays, there are a variety of models for screening, analysis, design, and optimization of RD columns. Moreover, RCMs are an invaluable tool for initial process screening and flowsheet development. Equilibrium models are useful for preliminary design, whereas nonequilibrium models are used for the final RD design, development of control strategies, and commercial RD plant design and simulation [7, 14, 29, 70].

Future research and development should focus on quick evaluation of RD applicability, easy methodology for choosing the most suitable equipment, and scale-up of equipment because this is where the great deficiencies still lie. Moreover, considering the green opportunities created by the use of reactive separations, future prospects aim to spread the use of RD technologies in the production of bulk or specialty chemicals such as ethers, esters, polyesters, and biofuels.

4.11 Bibliography

[1]   Luyben WL, Yu CC. Reactive distillation design and control, Wiley-AIChE, Hoboken, New-Jersey, US, 2008.
[2]   Kiss AA. Advanced distillation technologies – Design, control and applications. Wiley, Chichester, UK, 2013.
[3]   Reay D, Ramshaw C, Harvey A. Process intensification – Engineering for efficiency, sustainability and flexibility. Second Edition, Butterworth-Heinemann, UK, 2013.
[4]   Patrut C, Bildea CS, Kiss AA. Catalytic cyclic distillation – A novel process intensification approach in reactive separations. Chemical Engineering and Processing: Process Intensification, 2014, 81, 1–12.
[5]   Malone MF, Doherty MF. Reactive distillation. Industrial & Engineering Chemistry Research, 2000, 39, 3953–3957.
[6]   Malone MF, Huss RS, Doherty MF. Green chemical engineering aspects of reactive distillation. Environmental Science & Technology, 2003, 37, 5325–5329.
[7]   Harmsen GJ. Reactive distillation: The front-runner of industrial process intensification: A full review of commercial applications, research, scale-up, design and operation. Chemical Engineering and Processing, 2007, 46, 774–780.
[8]   Harmsen GJ. Process intensification in the petrochemicals industry: Drivers and hurdles for commercial implementation. Chemical Engineering and Processing, 2010, 49, 70–73.
[9]   Krishna R. Reactive separations: more ways to skin a cat. Chemical Engineering Science, 2002, 57, 1491–1504.
[10] Noeres C, Kenig EY, Gorak A. Modelling of reactive separation processes: reactive absorption and reactive distillation. Chemical Engineering & Processing, 2003, 42, 157–178.
[11] Agreda VH, Partin LR, Heise WH. High-purity methyl acetate via reactive distillation. Chemical Engineering and Processing, 1990, 86, 40–46.
[12] Siirola JJ. Industrial applications of chemical process synthesis. Advances in Chemical Engineering, 1996, 23, 1–62.
[13] Tuchlenski A, Beckmann A, Reusch D, Dussel R, Weidlich U, Janowsky R. Reactive distillation – industrial applications, process design & scale-up. Chemical Engineering Science, 2001, 56, 387–394.
[14] Schoenmakers HG, Bessling B. Reactive and catalytic distillation from an industrial perspective. Chemical Engineering and Processing, 2003, 42, 145–155.
[15] Stankiewicz A. Reactive separations for process intensification: an industrial perspective. Chemical Engineering and Processing, 2003, 42, 137–144.
[16] Sundmacher K, Kienle A (eds). Reactive distillation: Status and future directions, Wiley-VCH, Weinheim, 2003.
[17] Hiwale RS, Bhate NV, Mahajan YS, Mahajani SM. Industrial applications of reactive distillation: Recent trends. International Journal of Chemical Reactor Engineering, 2004, 2, R1.
[18] Sundmacher K, Kienle A, Seidel-Morgenstern A (eds). Integrated chemical processes: Synthesis, operation, analysis, and control, Wiley-VCH, Weinheim, 2005.
[19] Keller T. Reactive distillation, in Gorak A, Olujic Z (eds), Distillation: Equipment and processes, Elsevier, 2014.

[20] Segovia-Hernández JG, Hernández S, Bonilla Petriciolet A. Reactive distillation: A review of optimal design using deterministic and stochastic techniques. Chemical Engineering and Processing: Process Intensification, 2015, 97, 134–143.

[21] Mansouri SS, Sales-Cruz M, Huusom JK, Woodley JM, Gani R. Integrated process design and control of reactive distillation processes, 9th IFAC Symposium on Advanced Control of Chemical Processes ADCHEM 2015, Whistler, Canada, 7–10 June 2015, IFAC-PapersOnLine, 2015, 48, 1120–1125.

[22] Thery R, Meyer XM, Joulia X. Analysis of the feasibility, synthesis and conception of processes of reactive distillation: State of the art and critical analysis. Canadian Journal of Chemical Engineering, 2005, 83, 242–266.

[23] Stichlmair JG, Fair JR. Distillation, Principles and Practice, Wiley-VCH, New York, 1998.

[24] Doherty MF, Malone MF. Conceptual design of distillation systems, McGraw-Hill, New York, 2001.

[25] Schmidt-Traub H, Gorak A. Integrated reaction and separation operations, Springer, New York, 2006.

[26] Shah M, Kiss AA, Zondervan E, de Haan AB. A systematic framework for the feasibility and technical evaluation of reactive distillation processes. Chemical Engineering and Processing, 2012, 60, 55–64.

[27] Grosser JH, Doherty MF, Malone MF. Modeling of reactive distillation systems. Industrial & Engineering Chemistry Research, 1987, 26, 983–989.

[28] Taylor R, Krishna R. Multicomponent mass transfer, Wiley, New York, 1993.

[29] Taylor R, Krishna R. Modelling reactive distillation. Chemical Engineering Science, 2000, 55, 5183–5229.

[30] Lee J-H, Dudukovic MP. A comparison of the equilibrium and nonequilibrium models for a multicomponent reactive distillation column. Computers & Chemical Engineering, 1998, 23, 159–172.

[31] Kreul LU, Górak A, Barton PI. Modeling of homogeneous reactive separation processes in packed columns. Chemical Engineering Science, 1999, 54, 19–34.

[32] Baur R, Higler AP, Taylor R, Krishna R. Comparison of equilibrium stage and nonequilibrium stage models for reactive distillation. Chemical Engineering Journal, 2000, 76, 33–47.

[33] Klöker M, Kenig EY, Hoffmann A, Kreis P, Górak A. Rate-based modelling and simulation of reactive separations in gas/vapour–liquid systems. Chemical Engineering and Processing, 2005, 44, 617–629.

[34] Kenig EY, Gorak A, Pyhalahti A, Jakobsson K, Aittamaa J, Sundmacher K. Advanced rate-based simulation tool for reactive distillation. AIChE Journal, 2004, 50, 322–342.

[35] Egorov Y, Menter F, Klöker M, Kenig EY. On the combination of CFD and rate-based modelling in the simulation of reactive separation processes. Chemical Engineering and Processing, 2005, 44, 631–644.

[36] Mueller I, Pech C, Bhatia D, Kenig EY. Rate-based analysis of reactive distillation sequences with different degrees of integration. Chemical Engineering Science, 2007, 62, 7327–7335.

[37] Mueller I, Kenig EY. Reactive distillation in a Dividing Wall Column – rate-based modeling and simulation. Industrial & Engineering Chemistry Research, 2007, 46, 3709–3719.

[38] Aspen Technology, Aspen Plus: User guide – Volume 1 & 2; Aspen physical property system – Physical property models, 2009.

[39] Güttinger TE. Morari M. Predicting multiple steady states in distillation: Singularity analysis and reactive systems. Computers & Chemical Engineering, 1997, 21, S995–S1000.

[40] Venimadhavan G, Malone MF, Doherty MF. Bifurcation study of kinetic effects in reactive distillation. AIChE Journal, 1999, 45, 546–556.

[41] Dimian AC, Bildea CS, Kiss AA. Integrated design and simulation of chemical processes, 2nd Edition, Elsevier, 2014.

[42] Dimian AC, Bildea CS. Chemical process design – Computer-aided case studies, Wiley-VCH, Weinheim, 2008.

[43] Venimadhavan G, Buzad G, Doherty MF, Malone MF. Effect of kinetics on residue curve maps for reactive distillation. AIChE Journal, 1994, 40, 1814–1824.

[44] Okasinski MJ, Doherty MF. Design method for kinetically controlled, staged reactive distillation columns. Industrial & Engineering Chemistry Research, 1998, 37, 2821–2834.

[45] Nagy ZK, Klein R, Kiss AA, Findeisen R. Advanced control of a reactive distillation column. Computer Aided Chemical Engineering, 2007, 24, 805–810.

[46] Serafimov LA, Pisarenko Yu A, Kulov NN. Coupling chemical reaction with distillation: Thermodynamic analysis and practical applications. Chemical Engineering Science, 1999, 54, 1383–1388.

[47] Subawalla H, Fair JR. Design guidelines for solid-catalyzed reactive distillation systems. Industrial & Engineering Chemistry Research, 1999, 38, 3696–3709.

[48] Tuchlenski A, Beckmann A, Reusch D, Dussel R, Weidlich U, Janowsky R. Reactive distillation – industrial applications, process design & scale-up. Chemical Engineering Science, 2001, 56, 387–394.

[49] Daza OS, Perez-Cisneros ES, Bek-Pedersen E, Gani R. Graphical and stage-to-stage methods for reactive distillation column design. AIChE Journal, 2003, 49, 2822–2841.

[50] Damartzis T, Seferlis P. Optimal design of staged three-phase reactive distillation columns using non-equilibrium and orthogonal collocation models. Industrial & Engineering Chemistry Research, 2010, 49, 3275–3285.

[51] Kiss AA, Pragt H, van Strien C. Reactive dividing-wall columns – How to get more with less resources? Chemical Engineering Communications, 2009, 196, 1366–1374.

[52] Hernandez S, Sandoval-Vergara R, Barroso-Munoz FO, Murrieta-Duenasa R, Hernandez-Escoto H, Segovia-Hernandez JG, Rico-Ramirez V. Reactive dividing wall distillation columns: Simulation and implementation in a pilot plant. Chemical Engineering & Processing, 2009, 48, 250–258.

[53] Klöker M, Kenig EY, Schmitt M, Althaus K, Schoenmakers H, Markusse AP, Kwant G. Influence of operating conditions and column configuration on the performance of reactive distillation columns with liquid-liquid separators. Canadian Journal of Chemical Engineering, 2003, 81, 725–732.

[54] Baur R, Krishna R. Distillation column with reactive pump arounds: an alternative to reactive distillation. Chemical Engineering and Processing, 2004, 43, 435–445.

[55] Kaymak DB, Luyben WL. Effect of the chemical equilibrium constant on the design of reactive distillation columns. Industrial & Engineering Chemistry Research, 2004, 43, 3666–3671.

[56] Sharma N, Singh K. Control of reactive distillation column – A review. International Journal of Chemical Reactor Engineering, 2010, 8, R5.

[57] Luyben WL. Distillation design and control using Aspen simulation, 2nd Edition, Wiley-AIChE, Hoboken, New-Jersey, US, 2013.

[58] Kiss AA. Heat-integrated reactive distillation process for synthesis of fatty esters. Fuel Processing Technology, 92 (2011), 1288–1296.

[59] Goetze L, Bailer O, Moritz C, von Scala C. Reactive distillation with Katapak. Catalysis Today, 2001, 69, 201–208.

[60] Hoffman A, Noeres C, Gorak A. Scale-up of reactive distillation columns with catalytic packings. Chemical Engineering and Processing, 2004, 43, 383–395.

[61] Kiss AA. Applying reactive distillation. NPT Procestechnologie, 2012, 19(1), 22–24.

[62] Kiss AA. Process intensification technologies for biodiesel production – Reactive separation processes. Springer Briefs in Applied Sciences and Technology series, 2014.

[63] Kiss AA, Dimian AC, Rothenberg G. Solid acid catalysts for biodiesel production – towards sustainable energy. Advanced Synthesis & Catalysis, 2006, 348, 75–81.

[64] Kiss AA, Dimian AC, Rothenberg G. Biodiesel by reactive distillation powered by metal oxides. Energy & Fuels, 2008, 22, 598–604.

[65] Kiss AA. Separative reactors for integrated production of bioethanol and biodiesel. Computers & Chemical Engineering, 2010, 34, 812–820.

[66] Schembecker G, Tlatlik S. Process synthesis for reactive separations. Chemical Engineering and Processing, 2003, 42, 179–189.

[67] Kiss AA, Rothenberg G, Dimian AC, Omota F. The heterogeneous advantage: biodiesel by catalytic reactive distillation. Topics in Catalysis, 2006, 40, 141–150.

[68] Dimian AC, Bildea CS, Omota F, Kiss AA. Innovative process for fatty acid esters by dual reactive distillation. Computers & Chemical Engineering, 2009, 33, 743–750.

[69] Kiss AA. Novel process for biodiesel by reactive absorption. Separation & Purification Technology, 2009, 69, 280–287.

[70] Kiss AA, Bildea CS. A review on biodiesel production by integrated reactive separation technologies. Journal of Chemical Technology and Biotechnology, 2012, 87, 861–879.

[71] Chen YH, Yu CC. Design and control of heat-integrated reactors. Industrial & Engineering Chemistry Research, 2003, 42, 2791–2808.

[72] Bildea CS, Kiss AA. Dynamics and control of a biodiesel process by reactive absorption. Chemical Engineering Research and Design, 2011, 89, 187–196.

[73] Kiss AA, Bildea CS, Dimian AC. Design and control of recycle systems by non-linear analysis. Computers & Chemical Engineering, 2007, 31, 601–611.

[74] Mahajani SM, Chopade SP. Reactive distillation: Processes of commercial importance, in Wilson ID, Edlard TR, Poole CA, Cooke M (eds), Encyclopedia of Separation Science. Academic Press, London, UK, pp 4075–4082, 2000.

[75] Kiss AA. Process intensification: Industrial applications, in Segovia-Hernandez JG, Bonilla-Petriciolet A (eds), Process intensification in chemical engineering: Design, optimization and control. Springer International Publishing, 221–260, 2016.

[76] Huss RS, Chen F, Malone MF, Doherty MF. Reactive distillation for methyl acetate production. Computers & Chemical Engineering, 2003, 27, 1855–1866.

[77] Wierschem M, Skiborowski M, Gorak A, Schmuhl R, Kiss AA. Techno-economic evaluation of an ultrasound-assisted enzymatic reactive distillation process. Computers & Chemical Engineering, (2017), Article in press.

[78] Krishna G, Min TH, Rangaiah GP. Modeling and analysis of novel reactive HiGee distillation. Computer Aided Chemical Engineering, 2012, 31, 1201–1205.

[79] Fien G-JAF, Liu YA. Heuristic synthesis and shortcut design of separation processes using residue curve maps: A review. Industrial & Engineering Chemistry Research, 1994, 33, 2505–2522.

[80] Ung S, Doherty MF. Synthesis of reactive distillation systems with multiple equilibrium chemical reactions. Industrial & Engineering Chemistry Research, 1995, 34, 2555–2565.

[81] Espinosa J, Aguirre P, Pérez G. Some aspects in the design of multicomponent reactive distillation columns with a reacting core: Mixtures containing inerts. Industrial & Engineering Chemistry Research, 1996, 35, 4537–4549.

[82] Almeida-Rivera CP, Swinkels PLJ, Grievink J. Designing reactive distillation processes: Present and future. Computers & Chemical Engineering, 2004, 28, 1997–2020.

[83] Glasser D, Crowe C, Hildebrandt D. A geometric approach to steady flow reactors: The attainable region and optimization in concentration space. Industrial & Engineering Chemistry Research, 1987, 26, 1803–1810.

[84] Gadewar SB, Chadda N, Malone MF, Doherty MF. Feasibility and process alternatives for reactive distillation, in Sundmacher K, Kienle A (eds), Reactive Distillation: Status and Future Directions. Wiley, 2003.

[85] Frey T, Stichlmair J. Thermodynamic fundamentals of reactive distillation. Chemical Engineering & Technology, 1999, 22, 11–18.

[86] Lee JW, Hauan S, Westerberg AW. Graphical methods for reaction distribution in a reactive distillation column. AIChE Journal, 2000, 46, 1218–1233.

[87] Hauan S, Lien KM. A phenomena based design approach to reactive distillation. Chemical Engineering Research and Design, 1998, 76, 396–407.

[88] Urselmann M, Barkmann S, Sand G, Engell S. Optimization-based design of reactive distillation columns using a memetic algorithm. Computers & Chemical Engineering, 2011, 35, 787–805.

[89] Urselmann M, Engell S. Design of memetic algorithms for the efficient optimization of chemical process synthesis problems with structural restrictions. Computers & Chemical Engineering, 2015, 72, 87–108.

[90] Zondervan E, Shah M, de Haan AB, Optimal design of a reactive distillation column. Chemical Engineering Transactions, 2011, 24, 295–300.

[91] Amte V, Nistala SH, Mahajani SM, Malik RK. Optimization based conceptual design of reactive distillation for selectivity engineering. Computers & Chemical Engineering, 2013, 48, 209–217.

[92] Seferlis P, Damartzis T, Dalaouti N. Efficient reduced order dynamic modeling of complex reactive and multiphase separation processes using orthogonal collocation on finite elements, in Georgiadis MC, Banga JR, Pistikopoulos EN (eds), Process Systems Engineering: Dynamic Process Modeling, vol. 7. Wiley, 2011.

[93] Paramasivan G, Kienle A. Inferential control of reactive distillation columns – An algorithmic approach. Chemical Engineering & Technology, 2011, 34, 1235–1244.

Massimiliano Errico

5 Process synthesis and intensification of hybrid separations

Abstract: Hybrid flowsheets are defined, in the context of process intensification, as alternatives suitable for replacing energy-intensive separation methods through the combination of more than one unit operation. Distillation is one of the first options considered for achieving a required separation; therefore, this chapter examines distillation-based hybrid alternatives. Due to the extent of the topic, this analysis is limited to the separation of bioalcohols utilizing pervaporation and liquid–liquid extraction as assisted distillation methods. For each case, different hybrid flowsheets are reported and commented on. The corresponding distillation-based processes are considered for comparison. Synthesis of the possible hybrid flowsheets appears to be important, especially when multicomponent mixtures are considered. This aspect is discussed for the combination of liquid–liquid extraction and distillation as applied to the separation of biobutanol from its fermentation broth. The synthesis of alternative hybrid flowsheets is reported, showing that one configuration can realize a 43% reduction in the total annual cost.
Bioalcohol production by fermentation perfectly represents the case where distillation alone is penalized by the thermodynamics of the mixture, but its applicability can be extended using valid alternative hybrid flowsheets.

Keywords: Hybrid flowsheet, process intensification, distillation, bioethanol, biobutanol

5.1 Introduction

There are many possible definitions of crisis, but Coyne [1] stated, "A crisis is an unexpected event that creates uncertainty and poses a direct or perceived threat to the goals and norms of an organization or society." Offe [2] pointed out that crisis are "processes in which the structure of a system is called into question." Being specific, by coupling the words "energy" and "crisis," the latter acquires the metaphoric meaning of "some turning point in energy resources" [3]. Events like the 1973 Arab oil embargo, the 1976/1977 shortage of natural gas, and the 1977 New York blackout (even if they were not the result of a resource crisis) impacted all levels of society, from governments to citizens. Is our energy availability unlimited? Even if insulation is cheaper than heat, is it sustainable to keep wasting energy? Are we spoiled energy users? Can we change into wise energy users without compromising our habits?

These questions, together with increases in the cost of energy, swiftly reached the industrial sector, where a change in the energy price can determine production prof-

https://doi.org/10.1515/9783110465068-005

itability. Therefore, great research efforts were focused on rational energy usage and on developing alternative production methods. One of the first results of this approach was introduction of pinch technology [4, 5]. This method is based on the second law of thermodynamics and aims to determine the best heat exchanger network for reducing utility consumption. Since 1983, pinch technology has achieved energy savings in the range of 10–70%, reductions in the capital cost of new plants of up to 25%, increased process capacity by the removal of bottlenecks, and greater flexibility and operability. In particular, cost-savings associated with crude oil units were quantified at 1.75 million US$, with a payback time of 1.6 years [6]. Pinch analysis still represents a valid design and retrofit methodology and has achieved outstanding results in industries such as biofuel production [7, 8] and food processing [9].

Since the time of the energy crisis, society has faced different issues associated with energy price, energy availability, environmental quality, and control or process risk due to dangerous or toxic substances. All these drivers have merged into a new approach for process development, called process intensification. A commonly accepted definition of process intensification is difficult, or maybe impossible, to report, but does not limit the importance of process intensification within the chemical engineering community [10]. Stankiewicz and Moulijn [11] defined process intensification as "the development of novel apparatus and techniques that, compared to those commonly used today, are expected to bring dramatic improvements in manufacturing and processing, substantially decreasing equipment-size/production-capacity ratio, energy consumption, or waste production, and ultimately resulting in cheaper, sustainable technology." This approach defines two main dimensions: (1) the possibility to reduce equipment size (process-intensifying equipment), microreactors and divided wall columns being excellent examples [12, 13]; and (2) the possibility to develop multifunction types of equipment (process-intensifying methods), as in the case of reactive distillation [14].

Beyond the definition of process intensification, it is interesting to define how this concept or philosophy matches or differs from the fundamental areas related to process system engineering.

Process system engineering concerns "the improvement of decision-making processes for the creation and operation of the chemical supply chain. It deals with the discovery, design, manufacture, and distribution of chemical products in the context of many conflicting goals." This means that the action and focus of process system engineering takes place along all the product creation chain, from the molecular scale to particles, compartments, process units, process plants, and enterprise [15]. Process intensification aims to increase the efficiency of single steps in the chain by proposing new mechanisms, materials, and structural building blocks for process synthesis [16]. Process intensification opens up new opportunities for process systems engineering in terms of model development and inclusion of innovative types of equipment [17].

Process system engineering methods and tools such as process integration, process optimization, process synthesis, and design have been somehow combined with process intensification. Process integration, as reported by Gundersen [18], is defined as "systematic and general methods for designing integrated production systems, ranging from individual processes to total sites, with special emphasis on the efficient use of energy and reducing environmental effects." Reductions in heat, mass, and power are the most common applications of process integration and, because the final objective is to obtain a more efficient process, it is easy to imagine that the borders between process intensification and process integration are very blurred. Process integration principles are used in process intensification and, as Babi et al. [19] reported, process integration and intensification are considered concurrently. However, in contrast to process intensification, process integration techniques do not include any phenomena addition or enhancement. The term "process integration" is also used to describe the interconnection of equipment by means of recycle streams in an industrial plant. Baldea [20] proved that intensification represents a limit case in tight integration through significant material recycling.

Process optimization is defined as "the use of specific methods to determine the most cost-effective and efficient solution to a problem or design for a process" [21] and it is a core part of process system engineering. Process intensification does not include explicit methods for process optimization, but optimization tools have been successfully applied to the design of complex intensified systems, for example, reactive distillation as reported by Taylor and Krishna [22].

The interaction between process intensification and process synthesis and design is an open challenge where much work remains to be done. A systematic methodology for the development of intensified processes could help the designer to generate all possible alternatives. In the specific case of intensified distillation systems, Rong developed a systematic procedure for generation of divided-wall columns [23], whereas Almeida-Rivera et al. [24] focused their studies on the reactive distillation option.

Nishida et al. [25] defined process system synthesis as "an act of determining the optimal interconnection of processing units as well the optimal type and design of the units within a process system." Babi et al. [26] specified that the objective of process synthesis is to define "the best process route, from among numerous alternatives, to convert given raw material to specific (desired) products, subject to predefined performance criteria." Including process intensification into process synthesis requires considering not only the unit operation scale, but also the task and phenomena scales in order to include new intensified alternatives. A possible multiscale approach was discussed by Babi et al. [27] and Lutze et al. [28], who considered a unit operation-based methodology and later introduced a phenomena-based method [29].

Regardless of the definition used to describe process intensification, the ultimate aim is the innovation and improvement of a process through considering contrasting objectives. Process intensification actions can be divided into three main groups or levels. The phase level is the most detailed and considers the molecules that build up a thermodynamic phase. At the process unit level, all phases are embedded in equipment. At the plant level, the interconnection between equipment units is considered [30].

This chapter explores different process intensification options at the plant level. In particular, it is possible to distinguish between hybrid unit operations following the definition of Babi et al. [19]: "A hybrid/intensified unit operation, is an operation that enhances the function of one or more unit operations performing a task or a set of tasks through a new design of the unit operation or the combination of more than one unit operations." In these hybrid systems, mass and/or energy exchanges are integrated in the same unit or hybrid flowsheet so that different unit operations are combined to obtain a more efficient process. The latter option is explored in this chapter. However, considering the number of available unit operations and their possible combinations, the analysis is limited to distillation-assisted hybrid flowsheets. In particular, the following hybrid systems are considered:

– pervaporation-assisted distillation
– liquid–liquid extraction-assisted distillation

The possible applications of hybrid systems can also be very different. For this reason, to give a homogeneous view of the different processes, applications based on bioalcohol production have been chosen.

For each process considered, a limited number of references are given. These references were selected as the starting point for further literature research.

5.2 Pervaporation-assisted distillation

The term "pervaporation" was introduced by Kober to describe a combination of permeation and evaporation [31]. In this process, a hot fluid contacts one side of a semipermeable membrane, and a vacuum is applied on the other side where permeate vapor is collected. The driving force of the process is the difference in the vapor pressure between the hot fluid and the permeate vapor. Depending on the phase of the hot fluid, it is possible to distinguish between pervaporation and vapor permeation. This difference is depicted in Figure 5.1.

Figure 5.1(a) represents the pervaporation process. In this case, the hot liquid feed is in contact with the membrane that separates the low-pressure permeate vapor. The vapor enriched in one or more components is cooled and condensed. Figure 5.1(b) shows the vapor permeation process. In this case, the membrane is in contact with the vapor stream in equilibrium with the hot liquid feed, while the low-pressure permeate vapor is cooled and condensed. Vapor permeation is thermodynamically equivalent to pervaporation because both processes are subjected to the same driving force [32]. If not specified otherwise, only pervaporation is considered here because most biofuels are produced in the liquid phase as a result of fermentation processes.

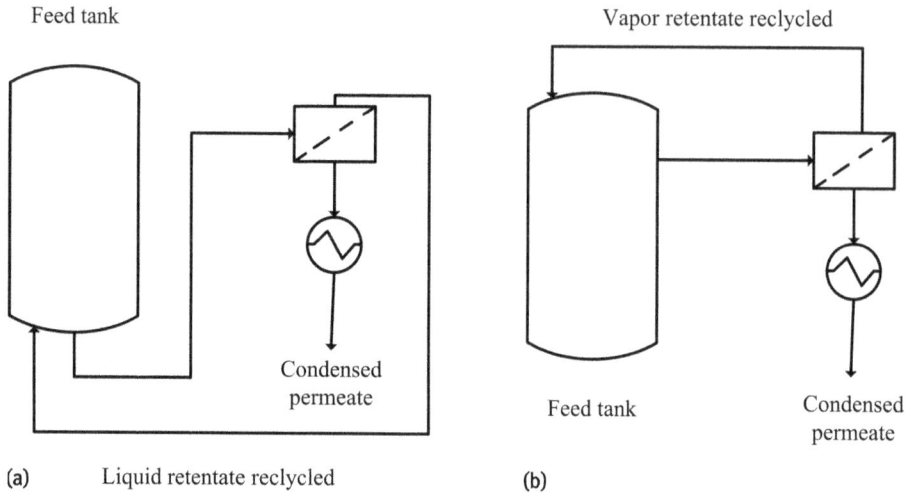

Fig. 5.1: Systems for (a) pervaporation and (b) vapor permeation.

The degree of separation achieved by pervaporation mainly depends on three factors: vapor–liquid equilibria, membrane selectivity, and feed-to-permeate partial vapor pressure ratio [33]. The first parameter is related to the feed composition, and a large amount of data is available from distillation studies. Membrane selectivity is an intrinsic property of the membrane material and is related to the solubility and diffusivity. Solubility selectivity is related to the interactions between feed components and the membrane. For example, hydrophobic membrane materials preferentially sorb hydrophobic compounds. The diffusivity efficiency depends on the size and shapes of permeating components. Because the membrane is the core of the pervaporation process, an essential step for industrial application is development of sorption- or diffusion-controlling materials [34].

The feed-to-permeate partial vapor pressure ratio is related to the separation achievable by pervaporation and can be considered an operational parameter. Usually, the permeate component is very dilute in the feed and its vapor pressure can be increased by increasing the feed temperature. The permeate-side vapor pressure can be reduced by means of a vacuum pump, using a condenser, or with a sweeping fluid. The three alternatives are reported in Figure 5.2.

In Figure 5.2(a), the low pressure on the permeate side is assured by a vacuum pump. This solution is applied for low permeate volumes and low capacities, mainly at laboratory scale. In Figure 5.2(b), the vacuum is spontaneously generated by permeate condensation. This is the most cost-effective solution. Lastly, in Figure 5.2(c), the membrane permeate side is swept with an inert carrier gas that is usually recycled and conditioned.

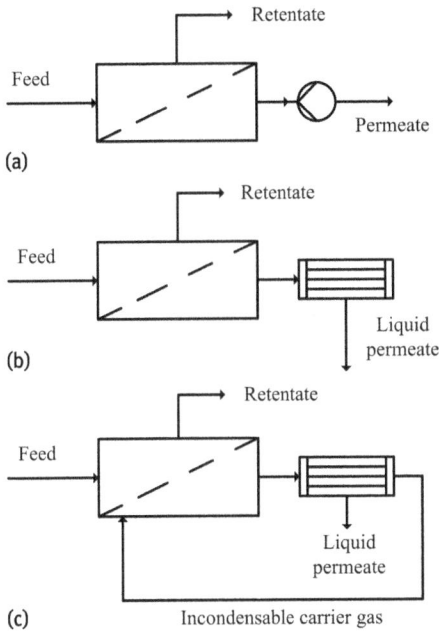

Fig. 5.2: Different arrangements to achieve a vapor pressure gradient: (a) vacuum pump, (b) condenser, and (c) carrier gas.

The required feed-to-permeate partial pressure ratio should be higher than 1, but to assure sufficient efficacy in industrial applications, a value between 7 and 10 is recommended [35].

Pervaporation is mainly applied for the separation of volatile organic components, organic/organic separations, organic/water separations, or food-related applications.

In the first category, typical examples are the separation of compounds such as chloroform, benzene, and toluene from aqueous dilute solutions [36]. For organic/organic separations, many different systems have been studied, including methanol/benzene, methanol/toluene, and benzene/cyclohexane, as reviewed by Smitha et al. [37]. The majority of pervaporation systems installed worldwide are for organic/water separations. Most of the mixtures studied are azeotropic solutions of alcohols, due to the need to reduce energy consumption associated with typical separation methods. A review of different aqueous mixtures separated by pervaporation was published by Chapman et al. [38]. Recently, pervaporation has been also applied in food technology, mostly for the separation of aroma compounds [39].

The main reasons to explore the combination of pervaporation and distillation are the widespread use of distillation as a separation method and its efficiency in large-scale production processes. However, a shift of part of the separation work to less energy-intensive methods could bring hybrid solutions that are more efficient. Different possibilities have been proposed for pervaporation-assisted distillation flowsheets. Four alternatives are reported in Figure 5.3.

The configuration in Figure 5.3(a) is usually referred to as a predistillation scheme whereby the column feed is passed through the membrane module. The permeate and retentate are re-introduced in the column in such a way that the operation of the column is not disturbed. For this reason, it is necessary to optimize the permeate temperature and pressure. The configuration of Figure 5.3(b) is called a parallel scheme. When the side stream has the same flow rate and feed withdrawal point, it is mathematically equivalent to the predistillation scheme. The configurations in Figure 5.3(c, d) are called post-distillation schemes, whereby the distillate or the bottom stream are fed to the membrane to reach the desired purity. The energy consumption of different configurations were compared by Alshehri and Lai [40], considering separation of propylene and propane as a case study.

For the same separation case, David et al. [41] proposed a hybrid pilot plant and quantified 20–50% savings in capital costs and up to 50% in operating expenses, compared with the distillation-based processes. Gottschlich and Roberts [42] also examined the separation of propene and propylene, proving the convenience of hybrid pervaporation/distillation systems when high purities are required.

Moreover, different design procedures have been proposed for the design of pervaporation-assisted distillation systems, such as the minimum area method, which is based on the well-known McCabe–Thiele diagram [43, 44].

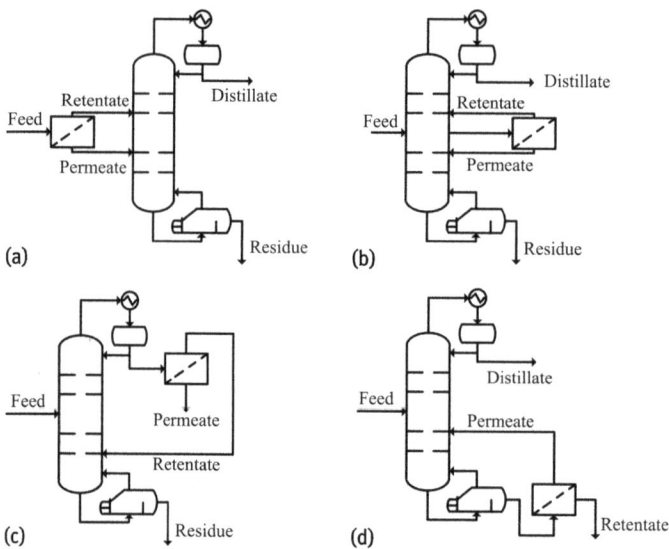

Fig. 5.3: Hybrid pervaporation-assisted distillation configurations: (a) predistillation, (b) parallel scheme, (c) distillate post-distillation, and (d) residue post-distillation.

5.2.1 Hybrid distillation/pervaporation processes for bioethanol purification

Bioethanol is produced by fermentation of sugars contained in agriculture residue or energy crops. It has been recognized as a sustainable transport fuel because it has the potential to act partially as a substitute for oil-derived gasoline. It is produced as a dilute aqueous solution, resulting in an energy-demanding concentration step. The presence of the ethanol–water azeotrope makes the separation even more challenging.

For the separation of homogeneous azeotropic mixtures, pervaporation has the advantage of eliminating the need for an external mass-separating agent. Moreover, the energy request is limited only to the latent heat of the permeate components. By contrast, in the case of distillation it is necessary to supply heat to the whole feed. If extractive distillation is considered, the entrainer cost should also be considered [34]. Different works have addressed the synthesis and optimization of distillation alternatives to reduce the energy consumption of separation in order to increase the profitability of bioethanol production [45, 46]. Standalone pervaporation is probably not economically feasible, but it can improve the efficiency of the overall process when combined with other unit operations, such as distillation [47].

The bioethanol production process can be broken down into the five main steps of pretreatment, cellulose hydrolysis, concentration and detoxification, fermentation, and product separation [48]. Pervaporation can be integrated into different parts of the process sequence. One alternative is the integration of pervaporation and the fermentation step. The benefit of this kind of coupling is to reduce inhibition and poisoning of the microorganism by the bioalcohol, with a consequent increase in bioconversion. Usually, a filtration system is placed between the fermentor and the pervaporation unit to remove suspended solids in the broth. The retentate is recycled back to the fermentor, and the permeate is a stream enriched in alcohol. Because distillation hybrid flowsheets are the main focus of this chapter, details on this case can be found in specific references [49–52].

Different integrations between pervaporation and distillation are possible in the product separation section of the general bioethanol production line. The aim is to obtain fuel-grade quality bioethanol, with a purity grade equal to or higher than 99.5 wt%. With ordinary distillation, it is impossible to reach this purity because of the presence of the ethanol–water azeotrope. Different alternatives are available for overcoming this limitation (pressure swing distillation, extractive distillation, etc.), resulting in different integration possibilities.

A hybrid distillation/pervaporation system was patented by Tusel and Ballweg in 1983 [53]. It consists of a distillation column followed by two pervaporation units, as schematically represented in Figure 5.4. The distillation column is fed by a 8.8 wt% solution of ethanol and water preheated to its boiling point. The distillation column separates a distillate stream with a purity of 80 wt% ethanol, which is condensed and brought to a pressure of 3 bar to be fed to the first pervaporation module. The vapor

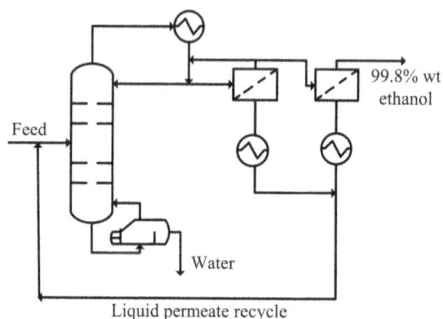

Fig. 5.4: Post-distillation scheme for the production of fuel-grade ethanol.

side of the module is kept at a pressure of 70 mbar. This module is constituted of a lower selective membrane so the water separates quickly. The permeate from the first module contains 10 wt% ethanol and is recycled back to the distillation column. The retentate has an ethanol content of 95 wt% and, using a heat exchanger circuit, is sent to the upper part of the distillation column to recover part of the column's vapor condensation heat, which is necessary for the second pervaporation module. Part of the retentate is recycled to the first pervaporation module and the remaining part is fed to the second module. This module is equipped with a higher selectivity membrane to give a final ethanol concentration of 99.8 wt%. As for the first module, after condensation, the permeate is recycled to the distillation column feed.

The scheme proposed has the advantage of increasing the ethanol concentration in the feed of the first pervaporation module from 80 to 92.5 wt%. Moreover, recovery of the condensation heat by the retentate allows reduction of the reflux ratio required by the distillation column. The authors quantified a reduction in steam consumption of between 5 and 1.6 kg/L alcohol produced.

Other post-distillation arrangements have been reported by several authors. Examples include the Lurgi pervaporator, where a plate-type membrane module and the permeate condenser are combined into a compact unit [54], and two heat-integrated distillation columns coupled with a pervaporation module [55]. The control properties of pervaporation modules integrated into the distillate side of an ordinary distillation column were also examined, proving their applicability on an industrial scale [56].

A different configuration was also proposed, whereby the pervaporation module was included between two distillation columns [57, 58]. This system is reported in Figure 5.5 and fits well with the separation of minimum temperature azeotropic mixtures such as ethanol–water. In the first column of the hybrid flowsheet, pure water is recovered as the bottom stream, while the distillate approaches the azeotropic composition. The distillate of the first column is fed to the pervaporation module where, by means of hydrophilic membranes, a water-rich permeate is separated and recycled back to the first column after being condensed. The retentate phase, rich in ethanol, is fed to the second distillation column. In this column, ethanol is the heaviest component and is obtained pure as the bottom stream. The distillate stream, with a composition

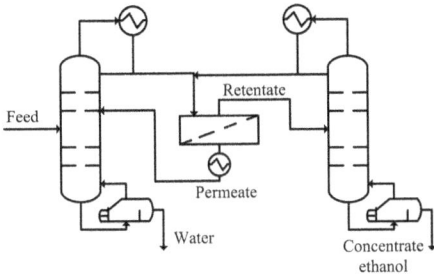

Fig. 5.5: Hybrid configuration with the pervaporation module included between two distillation columns.

close to the azeotrope, is recycled to the feed for the pervaporation module. Gooding and Bahouth [57] reported a case study based on this configuration, considering a 5 mol% ethanol feed, distillate compositions of 81 and 95 mol% for the first and second columns, respectively, a retentate stream with 97 mol% ethanol, and a final product of 99.5 mol% purity. Unfortunately, no economic data were reported for comparison with azeotropic or extractive distillation.

Brüschke and Tusel [59] considered the same system for concentration of a mass flow rate of 2000 kg/h containing 94 wt% ethanol to produce a stream of 1867 kg/h with a purity of 99.85 wt% ethanol. The authors claimed a 28% saving in the capital cost and a 40% reduction in the operative cost compared with separation by distillation using an entrainer.

Note that integrated systems using vapor permeation are also possible. For example, Huang et al. [60] discussed mechanical vapor-recompression distillation combined with membrane vapor permeation. In mechanical vapor compression, the overhead vapor is compressed and used as auxiliary fluid in the column reboiler where it exchanges its latent heat of condensation. Part of the condensed vapor is used as liquid reflux and the other part is the final product. This technology alone is not able to reach fuel-grade purity, so other separation units are normally required. In this case, coupling of distillation with membrane vapor permeation has the double benefit of increasing the achievable purity at the same time as reducing the energy consumption. The membrane module is placed as shown in Figure 5.6 and the main issue is related to the resistance of the membrane at a temperature of 130 °C, which is required to keep the water–ethanol mixture above the dew point.

Composite membranes were successfully applied by the authors. The membrane preferably permeates water that is reintroduced in the column. The energy saving in the integrated system was quantified as half of the requirement when only distillation was considered. Even higher savings were reported by Vane et al. [61] when the vapor permeation was coupled with a stripping column.

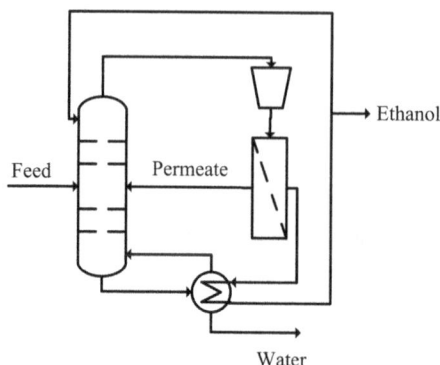

Feed

Permeate

→ Ethanol

Water

Fig. 5.6: Distillation/vapor permeation hybrid system.

5.2.2 Hybrid distillation/pervaporation processes for biobutanol purification

Butanol produced through fermentation processes is usually referred as biobutanol. Before 1950, almost two-thirds of global butanol supplies were produced in this way. Development of the petrochemical industry made synthetic butanol cheaper and biological processes were abandoned. Recently, the discovery of new strains able to tolerate higher concentrations of butanol, development of new separation options, and the possible use of biobutanol as biofuel are attracting the interest of academia and industry and stimulating reconsideration of the fermentation production route. In particular, biobutanol as biofuel has some advantages over bioethanol. It has lower vapor pressure, is not hygroscopic, is less corrosive, can be used pure or blended in any concentration with gasoline, and has a higher energy content [62]. The anaerobic fermentation of starchy substrates using different strains of *Clostridium acetobutylicum* or *Clostridium beijerinckii* produces a fermentation broth that is typically a mixture of acetone, butanol, and ethanol (ABE mixture) with a component ratio of 3 : 6 : 1 [63]. However, separation is challenging and energy intensive because of the diluted feed and presence of the ethanol–water homogeneous azeotrope and the butanol–water heterogeneous azeotrope.

When only distillation is considered as separation method, different alternatives have been reported in the literature, as reviewed by Liu et al. [64]. In general, the main possibility for decreasing the energy demand of the process is increasing the butanol concentration in the distillation feed. In contrast to bioethanol purification, where the pervaporation step is used to enhance product purity, in this case the optimal pervaporation position is before the separation train. Figure 5.7(a) shows a distillation-based sequence for the purification of ABE mixture similar to that proposed by Marlatt and Datta [65]. Figure 5.7(b) reports the hybrid process studied by Rom et al. [66]. For both configurations, it was assumed that acetone and ethanol had already been removed from the feed.

The main difference between the configurations is that, in Figure 5.7(a), the first column is fed with a diluted stream, then a decanter is used to separate an aqueous

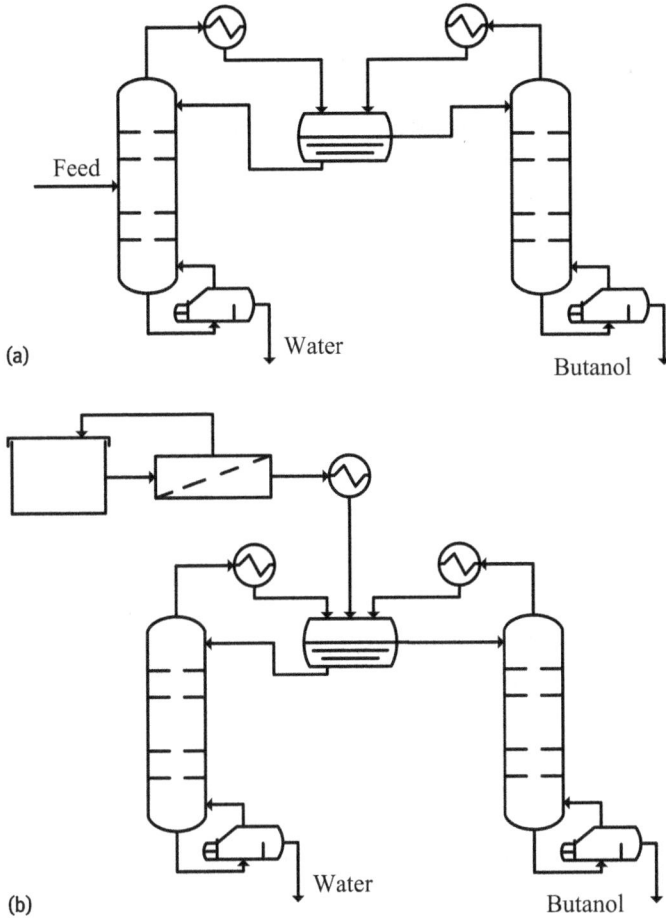

Fig. 5.7: Configurations for (a) distillation and (b) hybrid pervaporation-assisted distillation alternative.

phase and a butanol-rich phase. The second column is used to reach the required purity for the butanol. In the configuration of Figure 5.7(b), the pervaporation unit increases the concentration of butanol in the permeate to the immiscibility region. In this way, it is possible to directly connect the feed to the decanter. Rom et al. [66] compared the two alternatives, considering a feed stream with 0.5 wt% butanol at 35 °C, using a poli(dimethylsiloxane) organophilic membrane and a permeate side pressure of 0.004 bar. The energy consumption for the separation by distillation was quantified at 72 MJ/kg butanol. When the butanol concentration was increased to 9 wt% by means of the pervaporation unit, the energy demand was reduced by 50%. If it were possible to bring the concentration to 50%, the energy demand could be 90% less than in the distillation-based design. It is necessary to investigate membranes able to reach this high concentration. For the hybrid flowsheet, a butanol concentration of 5 wt% was identified as the lower limit for economic advantage.

Most of the possibilities explored in the literature on the use of pervaporation modules in biobutanol production focus on pervaporation-assisted fermentation. The objective of this coupling is to remove butanol from the fermentor to alleviate product inhibition, increase the butanol final concentration, and allow the use of more concentrated feedstocks. It is evident that a higher concentration of butanol corresponds to lower energy consumption in the final separation step. Similar results to the case discussed were reported, among others, by Cai et al. [67], Setlhaku et al. [68], and Van Hecke et al. [69].

5.2.3 Hybrid distillation/pervaporation processes: final remarks

Distillation and pervaporation are two distinct unit operations that can be interlinked in a more efficient hybrid process. The main benefits derive from the partial shift of some separation duties from distillation to pervaporation. In particular, this shift concerns cases of difficult separations such as azeotropic mixtures or compounds with a low relative volatility. The shift allows distillation to operate in its optimal region, avoiding the use of a high number of stages, high reflux, or external mass-separation agents. Pervaporation, not being limited by the vapor–liquid equilibria, can efficiently improve the overall separation economy. Bioalcohols, as produced by fermentation processes, are dilute aqueous solutions containing one or more azeotropes. Pervaporation is able to reduce the energy consumption, limiting the heat duty to that needed for the pervaporate. Moreover, no external mass-separating agents are required to overcome the azeotropic composition. In this way, the product has a higher market appeal; furthermore, the process is safer and has a low environmental impact. Some challenges remain in developing industrial hybrid pervaporation/distillation processes and are mainly related to membranes that are able to keep high selectivity at high flux.

Intensified hybrid pervaporation/distillation systems, where the two separations are combined into a single unit, have also been described in the literature [70]. An intensified system is shown in Figure 5.8.

Some of the column stages are replaced by a pervaporation module. The module can be a ceramic hollow fiber membrane such that the permeate is recovered from the inside lumen. The main advantages of this kind of integration are as follows:
- Energy required for pervaporation is provided by the distillation column vapor, removing the need for interstage heating.
- Mass and energy transfer between liquid and membrane is enhanced by turbulence induced by the vapor.
- The driving force is maximized because the liquid is close to its saturation point.

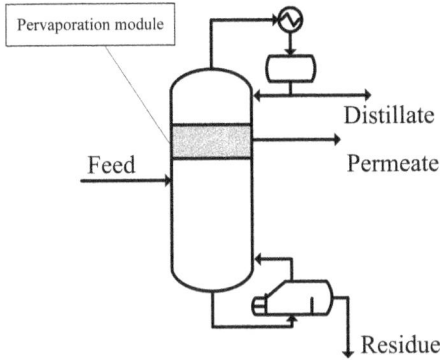

Fig. 5.8: Intensified single-unit pervaporation/distillation system.

- The membrane area necessary for separation is less than that needed for external pervaporation.

This single-unit intensified system is in its early stages of development and more experimental work is required to promote its application on an industrial level.

5.3 Liquid–liquid extraction-assisted distillation

Liquid–liquid extraction is the separation of one or more components included in a liquid mixture by contact with another immiscible or partly miscible liquid. The separation is achieved when the components in the feed distribute preferably in the second liquid. As for distillation or absorption, it is possible to repeat the separation in stages to reach the required purity.

In the process description, the following definitions apply [71]:

Feed: inlet stream in which the substance to be extracted is initially dissolved
Solute: substance transferred from the feed
Solvent: second liquid phase added to the process, in which the solute is dissolved
Extract: outlet stream containing solute-enriched solvent extracted from the feed
Raffinate: outlet stream containing solute-depleted feed

In general, extraction is favored over distillation for the following [72]:
- dissolved or complexed inorganic substances in organic or aqueous solutions
- removal of diluted contaminants
- removal of diluted high-boiling components
- recovery of heat-sensitive components
- separation of mixtures according to chemical type (e.g., removal of aliphatics from aromatics)
- separation of close-boiling liquids where solubility differences can be exploited
- separation of azeotropic mixtures

The choice of solvent is a crucial point and affects the economy and feasibility of the whole process. Different authors have reported selective criteria, some of which are listed as follows [73, 74]:

Selectivity: expresses how the solute is distributed between the solvent and feed

Partition ratio: related to the amount of solvent required for the separation

Density: large density difference between the feed and solvent is usually required to make separation easier

Miscibility: feed and solvent ideally should be immiscible

Safety: nontoxic and noninflammable solvents are preferred

Cost: cost of the solvent is part of the overall economic evaluation

The equipment used to bring the two liquid phases in contact and permit material transfer is normally classified into four types [75]: mixer-settlers, continuous counterflow extractors, continuous counterflow extractors with mechanical agitation, and centrifugal extractors.

The first type is the simplest technology. In mixer-settlers, the two liquids are mixed in a vessel using different types of impellers; then, the two phases are separated in a gravity decanter. Usually, to reach the required purity target, different mixers and settlers are connected in countercurrent flow. Continuous counterflow extractors are generally spray, packed, or tray columns. Continuous counterflow extractors with mechanical agitation are required for a low density difference between the liquid phases or for high viscosity liquids. In this case, the column is equipped with rotating agitators driven by an axial shaft.

The extractor design includes identification of the number of stages and the amount of solvent required to perform the target separation. Details on the design procedure for the different extractor types can be found in Henley et al. [72].

Liquid–liquid extraction could be considered intrinsically hybrid, because the solvent is usually recovered by distillation. An example of extraction followed by distillation solvent recovery is showed in Figure 5.9.

The extract obtained from the extraction column is fed to a distillation column and the solvent is recovered and recycled to the extractor. This configuration is not considered a liquid–liquid extraction-assisted distillation scheme because the solvent re-

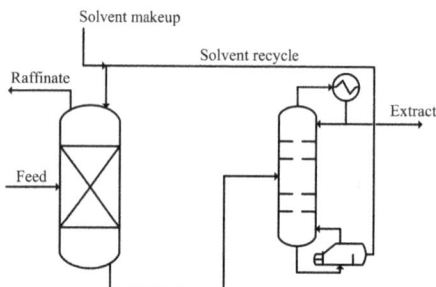

Fig. 5.9: Extraction and solvent recovery by distillation.

covery section is a necessary step for a reasonable and sustainable design. Integrated liquid–liquid extraction/distillation flowsheets are here intended as combined operations where extraction is used to perform part of the whole separation process in more efficient way.

5.3.1 Hybrid liquid–liquid extraction/distillation processes for bioethanol purification

As discussed in the Section 5.2.1, different distillation-based configurations are used to concentrate the bioethanol produced by fermentation processes. One of the most commonly used alternative includes a preconcentration column to approach the azeotropic composition, followed by an extractive distillation column to overcome the azeotropic composition, as reported in Figure 5.10. The overall configuration is composed of three columns. Because the extractive column uses an external entrainer, an additional column is required for its recovery. Several studies have used this configuration as a base case for comparison with different alternatives [76, 77].

Aviles Martinez et al. [78] proposed two hybrid flowsheets where liquid–liquid extraction was used to reduce the energy demand of distillation-based alternatives. The hybrid flowsheets are reported in Figure 5.11.

The authors compared the hybrid alternatives shown in Figure 5.11 with the reference case of Figure 5.10, considering the energy consumption, total annual cost (TAC), and carbon dioxide emission as performance criteria. The ethanol feed composition was set at 10 mol%, according to the typical yield obtainable from the fermentation of sugar cane bagasse. The purity targets were 99.99 mol% for water and ethanol and 99 mol% for the solvents to be recycled. For the extractive distillation, ethanol purity in the distillate stream of the prefractionator was set to 83.76 mol% and glycerol was used as solvent. In the liquid–liquid extractor, n-dodecane was selected for its ability

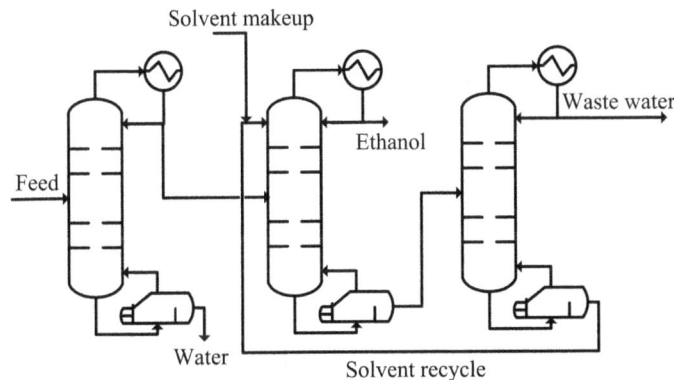

Fig. 5.10: Base case extractive distillation configuration for dehydration of diluted ethanol feeds.

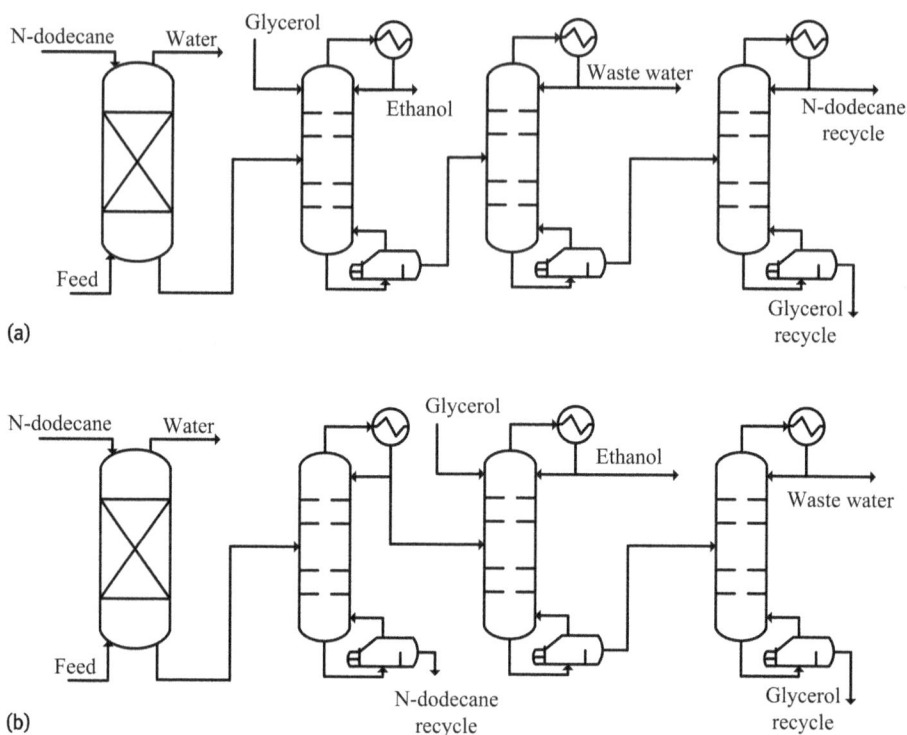

Fig. 5.11: Hybrid liquid–liquid extraction-assisted distillation configurations for ethanol dehydration.

to separate light alcohols from water. For the configuration shown in Figure 5.11(a), an almost pure water stream was obtained as raffinate in the liquid–liquid extractor. The flow rate of this stream was very similar to that of water separated as bottom stream in the prefractionator column of the base configuration in Figure 5.10. The extract stream containing *n*-dodecane, ethanol, and some water was fed to the extractive distillation column where, by adding glycerol, ethanol at the required purity was recovered as distillate. The bottom stream proceeds for the recovery of both solvents. The second hybrid configuration proposed, as reported in Figure 5.11(b), differs from the first in the separation order after the extractor. Although the authors did not report any synthesis procedure for generating alternatives, it is possible to suppose that this second hybrid alternative was generated by following the general rule to remove the mass-separation agent right after its introduction. It is possible to see from Figure 5.11(b) that *n*-dodecane is separated as bottom stream in the first distillation column; then, the distillate is sent to the extractive distillation column for ethanol purification. The last column performs glycerol recovery.

In all configurations, the wastewater stream is composed mainly of water (85 mol%), traces of solvents, and ethanol. Comparing the performances of the hybrid alternatives with the base case, for the configuration of Figure 5.11(a) the energy consumption and the carbon dioxide emissions were 37% higher than for the distillation-based alternative, while the TAC was almost 19% higher. These performances were expected because the solvent used in the extractor is recovered in the last distillation column, flowing through all the intermediate separation units. On the other hand, the configuration reported in Figure 5.11(b) exhibited excellent performance, demonstrating 30% reduction in energy consumption, 47% TAC reduction, and 30% reduction in carbon dioxide emissions. Although this hybrid alternative includes a higher number of units than the pure distillation case, the liquid–liquid extraction can save the amount of energy required by the prefractionator for separation of the water stream. The cost corresponding to the amount of energy saved is higher than the expected increase in annualized capital costs. Liquid–liquid extraction-assisted distillation processes are a valid alternative for reducing energy consumption in bioethanol separation plants.

The optimal sequence of Figure 5.11(b) was also considered by Vazquez-Ojeda et al. [79] and compared with the typical extractive distillation configuration. Their work focused on process optimization by means of a differential evolution algorithm with restrictions. Octanoic acid, octanol, and iso-octanol were considered as possible solvents for the extraction, while ethylene glycol was used in the extractive column. Different feed compositions were explored, ranging from 2 to 15 mol% of ethanol. The results obtained for a feed composition of 10 mol%, even if different solvents were employed, were in agreement with the results obtained by Aviles Martinez et al. [78]. When the ethanol content in the feed was reduced to 5 or 2 mol%, the hybrid configuration lost its convenience, mainly due to the need for a high number of stages in the liquid–liquid extractor.

5.3.2 Hybrid liquid–liquid extraction/distillation processes for biobutanol purification

As discussed in Section 5.2.2, the separation of butanol obtained by fermentation represents a bottleneck in the overall process economy. Because it is coproduced together with acetone and ethanol, the complexity of the mixture and the dilute fermentation broth make separation by distillation an energy-intensive process. One of the most competitive ways to produce pure biobutanol is by liquid–liquid extraction-assisted distillation. One of the first attempts to define liquid–liquid hybrid flowsheets for biobutanol production was carried out by Dadgar and Foutch [80] with the aim of promoting the fermentative over the synthetic process. Their example was then followed by other researchers. For example, Kraemer et al. [81] gathered the properties of 44 different solvents proposed for the ABE extraction and identified mesitylene as a novel extracting solvent, taking advantage of progress in solvent screening by computer-aided molecular design. The general hybrid configuration is reported in Figure 5.12.

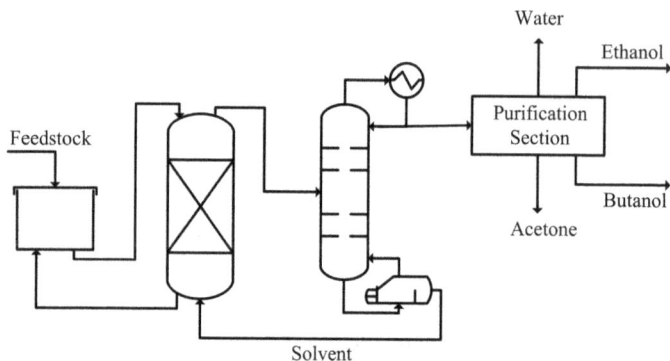

Fig. 5.12: Hybrid liquid–liquid extraction/distillation flowsheet for purification of acetone, butanol, and ethanol.

The authors demonstrated how the downstream separation sequence depends on the composition of the distillate stream obtained from the recovery column, which, in turn, depends on the solvent selected in the extraction column. The purification section, indicated as a box in Figure 5.12, was defined considering mesitylene and oleyl alcohol as solvents. The results were compared with those obtained using separation by distillation. The specific energy demand for the mesitylene process was 4.8 MJ/kg butanol produced versus 13.3 MJ/kg for the oleyl alcohol and 18.4 MJ/kg for the distillation. Although the main objective of the work was to prove the convenience of hybrid operations, another issue was unconsciously introduced. When multicomponent separations are considered, the combination of two different unit operation is not univocal. This means that, for the same separation, different hybrid configurations are possible. Therefore, it is necessity to define the synthesis procedure for the generation of a research space including all the alternatives. If the generation method is able to predict all the possible alternatives, then the optimal configuration can be identified. For the ABE separation, this issue was explored by Liu et al. [64], who used the following four-step approach:

1. The unit operations used for the separations were defined.
2. A P-graph representation of the operating units was performed.
3. The network was constructed including all the combinatorically feasible flowsheets.
4. A finite number of optimal and near-optimal flowsheets was generated.

The fermentation broth was fed to either a gas stripper or a liquid–liquid extractor. After these units, different combinations of distillation, gas stripping, and extraction were considered. Using the TAC as objective function, the first ten optimal flowsheets obtained were all composed of liquid–liquid extraction followed by different distillation column sequences. The optimal configuration selected is reported in Figure 5.13.

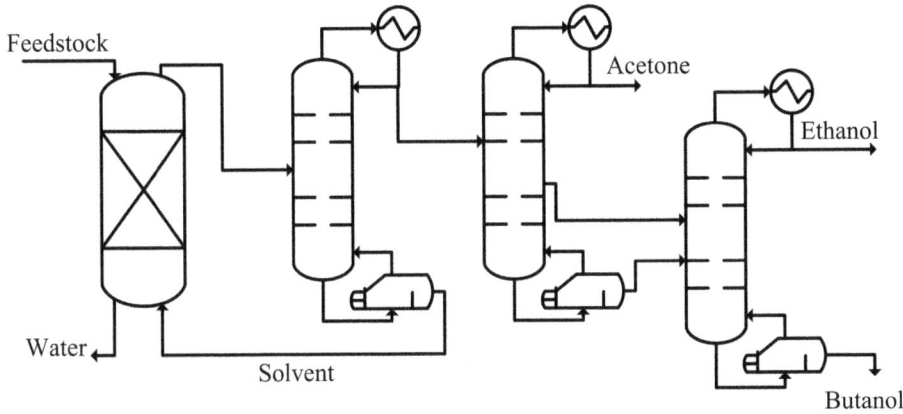

Fig. 5.13: Optimal hybrid configuration defined by Liu et al. [64].

The hybrid configuration is composed of a liquid–liquid extractor followed by three distillation columns. The first column performs solvent recovery, the second is used for acetone separation, and the third for butanol purification from ethanol. The distinctness of this configuration is represented by the acetone/ethanol butanol column. This column has a side stream and, therefore, classifies as a complex column. For the separation of ternary mixtures, this kind of sequence was also reported by Doherty and Malone [82] and classified as a "complex direct configuration."

Defining a hybrid configuration is not a simple combination of two or more unit operations. The combination can be realized in different ways and, for multicomponent separations, various alternatives are usually possible. In this case, process synthesis strengthens process intensification by generating a set of alternatives to be explored in order to define the optimal solution for the separation problem.

5.3.3 Hybrid liquid–liquid extraction/distillation processes: final remarks

The combination of liquid–liquid extraction and distillation is usually considered a natural consequence of the need to recover solvent from the extract stream. In the context of the definition of hybrid configurations, liquid–liquid extraction is considered a unit operation able to improve the separation economy of pure distillation alternatives. In particular, extraction is considered efficient in separating components that exhibit a strongly nonideal behavior, such as azeotropic mixtures. Both separation methods are mature, with well-defined design procedures, and are extensively applied on an industrial scale. In hybrid flowsheets, it is very common that liquid–liquid extraction is used first; then, the components separated in the extract are recovered by distillation. Feed composition and distillate-to-feed ratio are the main parameters that liquid–liquid extraction alters to make distillation more efficient. A fundamental point in developing such hybrid configurations is development of highly selective solvents with a low environmental impact. As shown by the two separation cases considered, for binary mixtures such as bioethanol–water, the process structure can be easily predicted. More alternatives are possible for a multicomponent feed such as biobutanol fermentation broth. The dimension of the problem can grow even more when complex configurations are included in the search space. For multicomponent distillation, process synthesis is a fundamental tool for generation of all possible alternatives. This aspect is considered in the following case study.

5.4 Synthesis, design, and optimization of alternative hybrid configurations for biobutanol separation

The hybrid flowsheet identified by Liu et al. [64], and reported in Figure 5.13, is here considered as a reference for generating alternative configurations. The synthesis procedure includes the following three main steps:

Introduction of thermal couplings: A thermal coupling is a bidirectional vapor–liquid stream used to replace one or more condensers and/or reboilers associated with nonproduct streams. Referring to Figure 5.13, there are two auxiliary exchangers that satisfy the conditions for substitution, the condenser of the first column and the reboiler of the second column. The exchangers can be eliminated individually or in a combinatorial way. Figure 5.14 reports the possible thermally coupled alternatives.

Section recombination: The introduction of one or more thermal couplings creates a structural degree of freedom to move the column section, providing the common reflux ratio and/or vapor boil-up between columns connected by thermal coupling. The configurations obtained are called thermodynamically equivalent. For the case con-

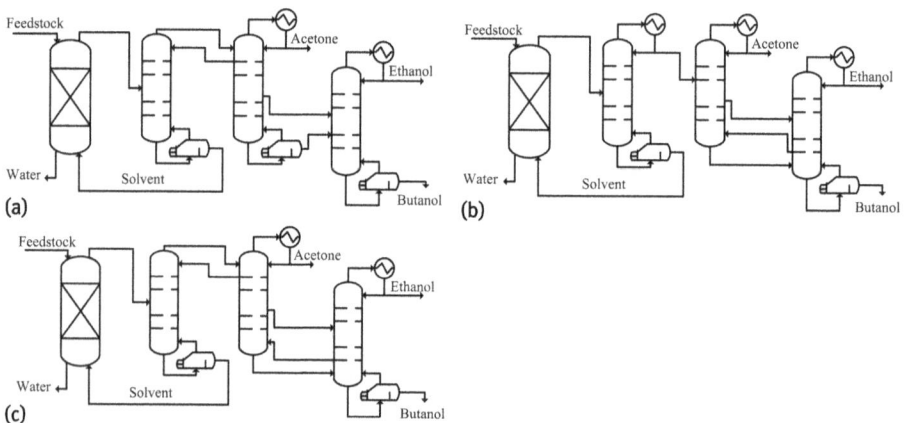

Fig. 5.14: Three possible thermally coupled configurations derived from Figure 5.13.

sidered, five configurations are possible. As an example, the possible configurations obtained from the thermally coupled alternative reported in Figure 5.14(c) are shown in Figure 5.15. The complete set of alternatives was reported by Errico et al. [83].

Intensification: The objective of intensification is a reduction in the number of columns compared with the reference case. The reduction is performed, starting from the thermodynamically equivalent configurations, by elimination of single column sections. For example, the second column of the configuration in Figure 5.15(a) is used mainly for transportation of the ethanol–butanol mixture, its elimination leading to the intensified configuration of Figure 5.16(a). Proceeding in the same way, the third column of the configuration in Figure 5.15(b) can be eliminated and the ethanol directly withdrawn from the second column, as depicted in Figure 5.16(b). Five possible intensified configurations are possible. Here, for the sake of brevity only three are reported in Figure 5.16. All the alternatives were reported by Errico et al. [83].

The synthesis procedure described here allows the designer to define different classes of alternatives, varying from thermally coupled to thermodynamically equivalent to intensified structures. All the configurations are structurally related to the reference used to initialize the procedure. This point proved to be very useful during design and optimization [84, 85].

To prove the potential of the configurations proposed, they were modeled by means of the process simulator Aspen Plus. The feed composition, as defined by Wu et al. [86], is reported in Table 5.1 together with its physical characterization.

The nonrandom two-liquid (NRTL) Hayden–O'Connell thermodynamic model was used and hexyl acetate was selected as solvent for the liquid–liquid extraction. The purity requirements were fixed as 99.5 wt% for butanol and acetone and 95 wt% for ethanol. The column pressure was optimized to use cooling water in the overhead condensers. Two different objective functions were chosen to compare the perfor-

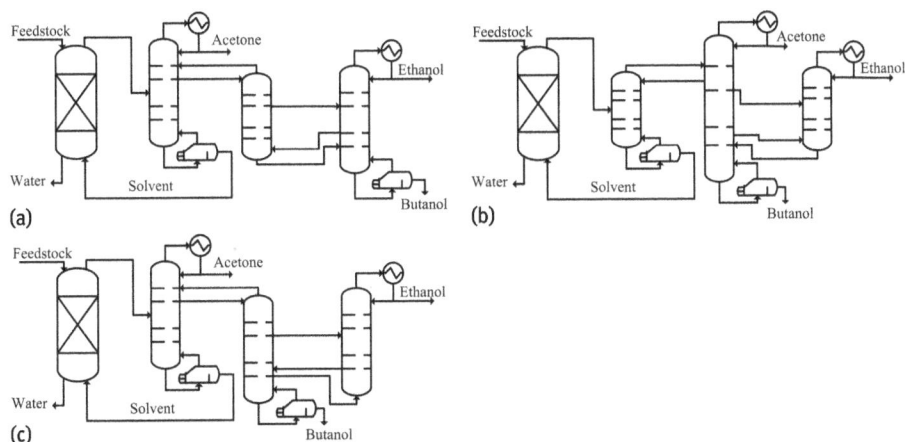

Fig. 5.15: Three thermodynamically equivalent configurations obtained from Figure 5.14(c).

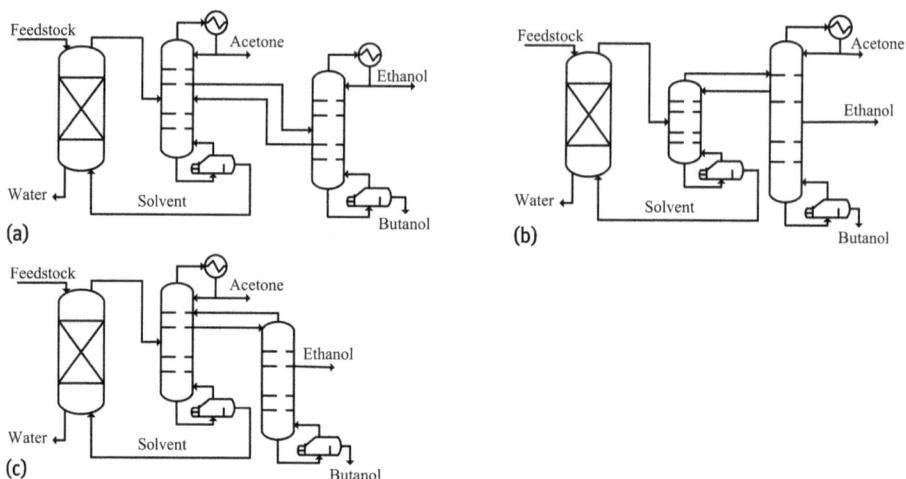

Fig. 5.16: Three intensified alternatives derived from Figure 5.15.

Tab. 5.1: Feed characterization.

Temperature [K]	322.04
Vapor fraction	0
Flow rate [kg/h]	45.36
Composition	
Butanol [wt%]	0.3018
Acetone [wt%]	0.1695
Ethanol [wt%]	0.0073

mances of the different alternatives. The first was the TAC, which is related to process economy. The second function was eco-indicator 99, which measures the environmental impact of the production based on life-cycle assessment methodology. Details on eco-indicator 99 evaluation have been published by Geodkoop and Spriensma [87]. The alternatives were optimized using a multi-objective optimization strategy based on the combination of differential evolution and tabu search. Differential evolution is based on the idea of evolution of populations of possible solutions, which occurs through operations of mutation, crossover, and selection. The tabu algorithm keeps a record of recently visited points to avoid further revisits of already explored areas. Differential evolution with tabu search was successfully applied for optimization of complex distillation configurations [88].

The design and the objective function values for the reference configuration of Figure 5.13 are reported in Table 5.2.

The design of the thermally coupled configurations of Figure 5.14 was obtained by considering the correspondence of the column sections in the reference case and

Tab. 5.2: Design, operative parameters, and comparison criteria for the reference configuration (Figure 5.13). C_1, C_2, C_3 indicate the column number in Figure 5.13.

Parameter	Extractor	C_1	C_2	C_3
Number of theoretical stages	5	26	46	20
Overall efficiency	0.654	0.766	0.721	0.834
Reflux ratio	–	0.905	6.034	14.836
Feed stage	1	13	32	5/15
Solvent feed stage	5	–	–	–
Side stream stage	–	–	44	–
Column diameter [m]	0.335	0.322	0.325	0.292
Operative pressure [kPa]	101.353	101.353	101.353	101.353
Distillate flow rate [kg/h]	–	21.687	7.694	0.333
Side stream flow rate [kg/h]	–	–	1.901	–
Solvent flow rate [kg/h]	708.549	–	–	–
Solvent makeup [kg/h]	0.709	–	–	–
Condenser duty [kW]	–	7.284	7.736	1.239
Reboiler duty [kW]	–	65.919	8.428	0.907
TAC [k$/year]	234.172			
Eco-indicator [points/year]	13,017			

each of the alternatives generated [84, 85]. The values of the objective functions are summarized in Table 5.3.

From the values obtained, it is clear that of all the thermally coupled alternatives the one shown in Figure 5.14(c) has the best performance criteria. Figure 5.17 shows the Pareto-optimal solutions, where the chosen solution is marked with a circle. It is evident how the two objective functions compete.

It has been extensively proven that there is a correspondence between alternatives included in the different subspaces [89, 90]. This means that, once the best configuration is identified in a specific subspace of alternatives, only the configurations derived from that one are expected to be promising. For this reason, because the best thermally

Tab. 5.3: Objective function values for all the alternatives of Figures 5.14–5.16.

Configuration	TAC [k$/year]	Eco-indicator 99 [points/year]
5.14a	214.280	12,462
5.14b	212.428	13,350
5.14c	188.143	11,642
5.15a	189.102	12,017
5.15b	188.471	11,571
5.15c	184.930	11,894
5.16a	198.160	19,684
5.16b	168.490	16,681
5.16c	163.631	15,595

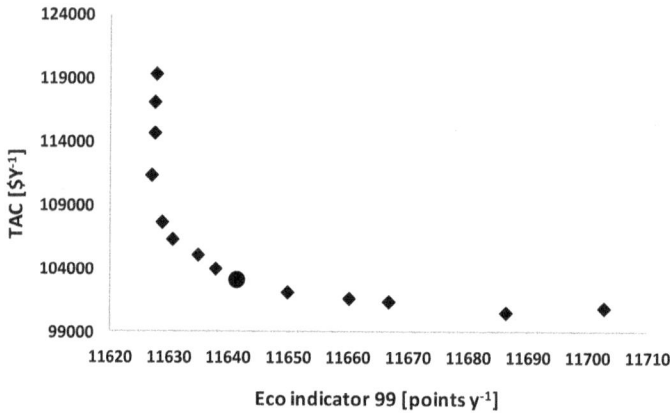

Fig. 5.17: Pareto-optimal solution for the best thermally coupled configuration, as shown in Figure 5.14(c).

coupled configuration has been identified, only the alternatives derived for that configuration are considered. Their performances are reported in Table 5.3. All the thermodynamically equivalent configurations have a better TAC than the reference case. The configuration in Figure 5.15(c) achieved the best economic performance, although the eco-indicator value was not as good as for the best thermally coupled configuration. Considering the intensified alternatives, the improvement in TAC value is evident. For the best case, the alternative shown in Figure 5.16(c), the TAC was reduced to 43% of the reference configuration. The corresponding design parameters are reported in Ta-

Tab. 5.4: Design and operational parameters for the best intensified configuration, as shown in Figure 5.16(c). C_1, C_2 indicate the column number in Figure 5.16.

Parameter	Extractor	C_1	C_2
Number of theoretical stages	5	58	20
Overall efficiency	0.654	0.783	0.718
Reflux ratio	–	27.182	–
Feed stage	1	45	–
Solvent feed stage	5	–	–
Side stream stage	–	–	12
Column diameter [m]]	0.335	0.323	0.324
Operative pressure [kPa]]	101.353	101.353	101.353
Distillate flow rate [kg/h]	–	7.711	–
Themal coupling flow rate [kg/h]	–	118.621	–
Side stream flow rate [kg/h]	–	–	0.336
Solvent flow rate [kg/h]	708.289	–	–
Solvent makeup [kg/h]	0.684	–	–
Condenser duty [kW]	–	31.094	0.000
Reboiler duty [kW]	–	65.642	24.517

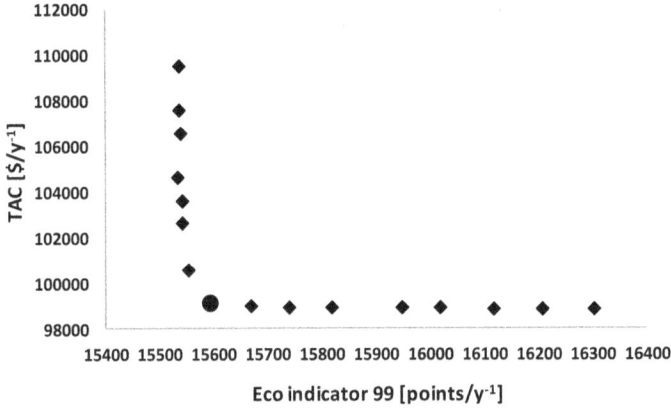

Fig. 5.18: Pareto-optimal solution for the best intensified configuration, as shown in Figure 5.16(c).

ble 5.4, while the Pareto-optimal solutions are in Figure 5.18. As before, a circle is used to mark the solution chosen.

Although the best intensified alternative realized a significant reduction in TAC, the opposite was observed for the eco-indicator. In this case, a 16.5% increase was observed compared with the reference case. This increase is a result of the higher reboiler duty of the intensified alternative and the consequent increase in carbon dioxide emission, corresponding to the production of the auxiliary fluid.

Recently, the study of alternative configurations was completed by including the possibility of waste heat recovery [91] and proving their controllability [92].

5.5 Conclusions

Bioethanol and biobutanol produced by fermentation processes are the frontrunner bioalcohols for the substitution of petro-derived gasoline. Compared with chemical and petrochemical processes, bioprocessing is characterized by lower capacity, dilution of the streams, and (normally) by the presence of azeotropes. For all these reasons, unit operations typically applied to final product purification result in high energy demand, reducing the overall competitiveness of the process.

The definition of alternative separation schemes remains an active research field and distillation one of the most studied processes. Different distillation-based processes have been proposed over the years, including complex configurations and divided-wall columns. Hybrid solutions are also emerging as viable alternatives for finally bringing biofuel production to an industrially competitive level. Hybrid flowsheets can be obtained by the combination of different unit operations to overcome their individual limitations. From the cases examined, it is clear that pervaporation and liquid–liquid-assisted distillation are valid alternatives for bioalcohol separation, but there are issues to be resolved. For the case of multicomponent separations, hybrid flowsheets are generated by intuition and there is seldom systematic prediction of all the possible alternatives. Moreover, the search space for the best alternative should include complex distillation columns such as thermally coupled or divided columns, because their convenience has been proven in many separation cases.

5.6 Bibliography

[1] Coyne CJ. Constitutions and crisis. Journal of Economic Behavior & Organization, 2011, 80, 351–357.

[2] Offe C. Crisis of crisis management: elements of a political crisis theory. International Journal of Politics, 1976, 6, 29–67.

[3] Garrison CE. The energy crisis: a process of social definition. Qualitative Sociology, 1981, 4, 312–322.

[4] Linnhoff B. Thermodynamic analysis in the design of process networks. PhD thesis, 1972, The University of Leeds.

[5] Linnhoff B, Flower JR. Synthesis of heat exchanger networks: systematic generation of energy optimal networks. AIChE Journal, 1978, 24, 633–642.

[6] Linnhoff B, Sahdev V. Pinch Technology. Ullmann's encyclopedia of industrial chemistry. Wiley, 2000

[7] Plesu V, Subirana Puigcasas J, Benet Surroca G et al. Process intensification in biodiesel production with energy reduction by pinch analysis. Energy, 2015, 79, 273–287.

[8] Fujimoto S, Yanagida T, Nakaiwa M, Tatsumi H, Minowa T. Pinch analysis for bioethanol production process from lignocellulosic biomass. Applied Thermal Engineering, 2011, 31, 3332–3336.

[9] Walmsley TG, Atkins MJ, Walmsley MRW, Neale JR. Appropriate placement of vapor recompression in ultra-low energy industrial milk evaporation systems using Pinch Analysis. Energy, 2016, 116, 1269–1281.

[10] Van Gerven T, Stankiewicz A. Structure, synergy, time – the fundamentals of process intensification. Industrial & Engineering Chemistry Research 2009, 48, 2465–2474.

[11] Stankiewicz A, Moulijn JA. Process intensification: transforming chemical engineering. Chemical Engineering Progress, 2000, 22–34.

[12] Yao X, Zhang Y, Du L, Liu J, Yao J. Review of the application of microreactors. Renewable and Sustainable Energy Reviews, 2015, 47, 519–639.

[13] Yildirim O, Kiss AA, Kenig EY. Dividing wall columns in chemical process industry: A review on current activities. Separation and Purification Technology, 2011, 3, 403–417.

[14] Segovia-Hernandez JG, Hernandez S, Bonilla Petriciolet A. Reactive distillation: a review of optimal design using deterministic and stochastic techniques. Chemical Engineering and Processing: Process Intensification, 2015, 97, 134–143.

[15] Grossmann IE, Westerberg AW. Research challenges in process systems engineering. AIChE Journal, 2000, 46, 1700–1703.

[16] Moulijn JA, Stankiewicz A, Grievink J, Gorak A. Process intensification and process systems engineering: A friendly symbiosis. Computers and Chemical Engineering, 2008, 32, 3–11.

[17] Van Gerven T, Stankiewicz A. Structure, energy, synergy, time – the fundamentals of process intensification. Industrial & Engineering Chemistry Research, 2009, 48, 2465–2474.

[18] Gundersen T. Implementing agreement on process integration – a process integration primer. International Energy Agency 2000.

[19] Babi DK, Sales Cruz M, Gani R. Fundamentals of process intensification: a process systems engineering view. In: Segovia Hernandez JG, Bonilla Petriciolet A (eds). Springer International Publishing, 2016, 7–33.

[20] Baldea M. From process integration to process intensification. Computers and Chemical Engineering, 2015, 81, 104–114.

[21] Edgar TF, Himmelblau DM, Lasdon LS. Optimization of Chemical Processes. McGraw Hill, 2001.

[22] Taylor R, Krishna R. Modelling reactive distillation. Chemical Engineering Science, 2000, 55, 5183–5229.

[23] Rong BG. Synthesis of dividing-wall columns (DWC) for multicomponent distillations-A systematic approach. Chemical Engineering Research and Design, 2011, 89, 1281–1294.

[24] Almeida–Rivera C, Swinkles P, Grievink J. Designing reactive distillation processes: present and future. Computers and Chemical Engineering, 2004, 28, 1997–2020.

[25] Nishida N, Stephanopoulos G, Westerberg AW. A review of process synthesis. AIChE Journal, 1981, 27, 321–351.

[26] Babi DK, Holtbruegge J, Lutze P, Gorak A. Sustainable process synthesis-intensification. Computers and Chemical Engineering, 2015, 81, 218–244.

[27] Babi DK, Lutze P, Woodley JM, Gani R. A process synthesis-intensification framework for the development of sustainable membrane-based operations. Chemical Engineering and Processing: Process Intensification, 2014, 86, 173–195.

[28] Lutze P, Roman–Martinez A, Woodley JM, Gani R. A systematic synthesis and design methodology to achieve process intensification in (bio) chemical processes. Computers and Chemical Engineering, 2012, 36, 189–207.

[29] Lutze P, Babi DK, Wooldley JM, Gani R. Phenomena based methodology for process synthesis incorporation process intensification. Industrial & Engineering Chemistry Research, 2013, 52, 7127–7144.

[30] Freund H, Sundmacher K. Towards a methodology for the systematic analysis and design of efficient chemical processes Part 1. From unit operations to elementary process functions. Chemical Engineering and Processing: Process Intensification, 2008, 47, 2051–2060.

[31] Kober PA. Pervaporation, perdistillation and percrystallization. Journal of the American Chemical Society, 1917, 39, 944–948.

[32] Wijmans JG, Baker RW. A simple predictive treatment of the permeation process in pervaporation. Journal of Membrane Science, 1993, 79, 101–113.

[33] Baker RW. Pervaporation. In: Membrane technology and applications, 3rd edn. Wiley, 2012, pp 379–416.

[34] Ong YK, Shi GM, Le NL, Tang YP, Zuo J, Nunes SP, Chung TS. Recent membrane development for pervaporation processes. Progress in Polymer Science, 2016, 57, 1–31.

[35] Brüchke HEA. State-of-the-art of pervaporation processes in the chemical industry. In: Pereira Nunes SP, Peinemann KV (eds). Membrane technology in the chemical industry. 2nd edn. Wiley, 2006, 151–202.

[36] Uragami T, Matsuoka Y, Miyata T. Permeation and separation characteristics in removal of dilute volatile organic compounds from aqueous solution through copolymer membranes consisted of poly(styrene) and poli (dimethylsiloxane) containing a hydrophobic ion liquid by pervaporation. Journal of Membrane Science, 2016, 506, 109–118.

[37] Smitha B, Suhanya D, Sridhar S, Ramakrishna M. Separation of organic-organic mixtures by pervaporation-a review. Journal of Membrane Science, 2004, 241, 1–21.

[38] Chapman PD, Oliveira T, Livingston AG, Li K. Membranes for the dehydration of solvents by pervaporation. Journal of Membrane Science, 2008, 318, 5–37.

[39] Sahin S. Principles of pervaporation for the recovery of aroma compounds and applications in the food and beverage industries. In: Rizvi S. (ed). Separation, extraction and concentration

process in the food, beverage and nutraceutical industries. 1st edn. Woodhead Publishing Limited, 2010, 219–243.

[40] Alshehri A, Lai Z. Attainability and minimum energy of single-stage membrane and membrane/ distillation hybrid processes. Journal of Membrane Science, 2014, 472, 272–280.

[41] Davis JC, Valus RJ, Eshraghi R, Velikoff AE. Facilitated transport membrane hybrid systems for olefin purification. Separation Science and Technology, 1993, 28, 463–476.

[42] Gottschlich DE, Roberts DL. Energy minimization of separation processes using conventional/ membrane hybrid systems. Technical Report for EG&G Idaho, Inc. and U.S. Department of Energy 1990, http://www.osti.gov/scitech/servlets/purl/6195331.

[43] Moganti S, Noble RD, Koval CA. Analysis of a membrane/distillation column hybrid process. Journal of Membrane Science, 1994, 93, 31–44.

[44] Stephan W, Noble RD, Koval CA. Design methodology for a membrane / distillation column hybrid process. Journal of Membrane Science, 1995, 99, 259–272.

[45] Errico M, Rong BG, Tola G, Spano M. Optimal synthesis of distillation systems for bioethanol separation. Part 1: Extractive distillation with simple columns. Industrial & Engineering Chemistry Research, 2013, 52, 1612–1619.

[46] Errico M, Rong BG, Tola G, Spano M. Optimal synthesis of distillation systems for bioethanol separation. Part 2: Extractive distillation with complex columns. Industrial & Engineering Chemistry Research, 2013, 52, 1620–1626.

[47] Verhoef A, Degreve J, Huybrechs B, van Veen H, Pex P, Van der Bruggen B. Simulation of a hybrid-distillation process. Computers and Chemical Engineering, 2008, 32, 1135–1146.

[48] Quintero JA, Cardona CA. Process simulation of fuel ethanol production from lignocellulosics using Aspen Plus. Industrial & Engineering Chemistry Research, 2011, 50, 6205–6212.

[49] Babalou AA, Rafia N, Ghasemzadeh K. Integrated systems involving pervaporation and applications. Basile A, Figoli A (eds). Pervaporation, vapour permeation and membrane distillation. Woodhead Publishing, 2015, 65–86.

[50] Chovau S, Gaykawad S, Straathof AJJ, Van der Bruggen B. Influence of fermentation by products on the purification of ethanol from water using pervaporation. Bioresource Technology, 2011, 102, 1669–1674.

[51] Vane LM. A review of pervaporation for product recovery from biomass fermentation processes. Journal of Chemical Technology and Biotechnology, 2005, 80, 603–629.

[52] Di Luccio M, Borges CP, Alves TLM. Economic analysis of ethanol and fructose production by selective fermentation coupled to pervaporation: effect of membrane costs on process economics. Desalination, 2002, 147, 161–166.

[53] Tusel G, Ballweg A. Method and apparatus for dehydrating mixtures of organic liquids and water. US Patent 4,405,409 Sep. 20, 1983.

[54] Sander U, Soukup P. Design and operation of a pervaporation plant for ethanol dehydration. Journal of Membrane Science, 1988, 36, 463–475.

[55] Nagy E, Mizsey P, Hancsok J, Boldyryev S, Varbanov P. Analysis of energy saving by combination of distillation and pervaporation for biofuel production. Chemical Engineering and Processing: Process Intensification, 2015, 98, 86–94.

[56] Luyben WL. Control of a column/pervaporation process for separating the ethanol/water azeotrope. Industrial & Engineering Chemistry Research, 2009, 48, 3484–3495.

[57] Gooding CH, Bahouth FJ. Membrane-aided distillation of azeotropic solutions. Chemical Engineering Communications, 1985, 35, 267–279.

[58] Pressly G, Ng KM. A break-even analysis of distillation-membrane hybrids. AIChE Journal, 1998, 44, 93–105.

[59] Brüschke HEA, Tusel GF. Economics of industrial pervaporation processes. In: Drioli E (ed). Membranes and Membrane Process. New York, Springer Science + Business Media, 1986, 581–586.

[60] Huang Y, Baker RW, Vane LM. Low-energy distillation-membrane separation process. Industrial & Engineering Chemistry Research, 2010, 49, 3760–3768.

[61] Vane LM, Alvarez FR, Huang Y, Baker RW. Experimental validation of hybrid distillation-vapor permeation processes for energy efficient ethanol-water separation. Journal of Chemical Technology and Biotechnology, 2010, 85, 502–511.

[62] Dürre P. Butanol: an attractive biofuel. Biotechnology Journal, 2007, 2, 1525–1534.

[63] Qureshi N, Blaschek HP. ABE production from corn: A recent economic evaluation. Journal of Industrial Microbiology & Biotechnology, 2001, 27, 292–297.

[64] Liu J, Fan LT, Seib P, Friedler F, Bertok B. Downstream process synthesis for biochemical production of butanol, ethanol, and acetone from grains: Generation of the optimal and near-optimal flowsheets with conventional operation units. Biotechnology Progress, 2004, 20, 1518–1527.

[65] Marlatt JA, Datta R. Acetone-butanol fermentation process development and economic evaluation. Biotechnology Progress, 1986, 2, 23–28.

[66] Rom A, Miltner A, Wukovits W, Friedl A. Energy saving potential of hybrid membrane and distillation process in butanol purification: Experiments, modelling and simulation. Chemical Engineering and Processing: Process Intensification, 2016, 104, 201–211.

[67] Cai D, Chen H, Chen C, Hu S, Wang Y, Chang Z, Miao Q, Qin P, Wang Z, Wang J, Tan T. Gas stripping-pervaporation hybrid process for energy-saving product recovery from acetone-butanol-ethanol (ABE) fermentation broth. Chemical Engineering Journal, 2016, 287, 1–10.

[68] Sethaku M, Heitmann S, Gorak A, Wichmann R. Investigation of gas stripping and pervaporation for improved feasibility of two-stage butanol production process. Bioresource Technology, 2013, 136, 102–108.

[69] Van Hecke W, Vandezande P, Claes S, Vangeel S, Beckers H, Diels L. Integrated bioprocess for long-term continuous cultivation of Clostridium acetobutylicum coupled to pervaporation with PDMS composite membranes. Bioresource Technology, 2012, 111, 368–377.

[70] Fontalvo J, Keurentjes JTF. A hybrid distillation-pervaporation system in a single unit for breaking distillation boundaries in multicomponent mixtures. Chemical Engineering Research and Design, 2015, 99, 158–164.

[71] Koch J, Shiveler G. Design principles for liquid-liquid extraction. CEP Magazine 2015, November, 22–30.

[72] Henley H, Seader JD, Roper DK. Liquid-liquid extraction. In: Separation Process Principles, John Wiley & Sons, Inc., 2011, 323–388.

[73] Sinnott RK. Separation columns (distillation, absorption and extraction). In: Chemical engineering design, Vol. 6, 4th ed, Elsevier Butterworth–Heinemann, 2005, 493–633.

[74] Gmehling J, Schedemann A. Selection of solvents or solvent mixtures for liquid-liquid extraction using predictive thermodynamic models or access to the Dortmund data bank. Industrial & Engineering Chemistry Research, 2014, 53, 17794–17805.

[75] Peters MS, Timmerhaus KD, West RE. Separation equipment – design and cost In: Plant design and economics for chemical engineers. McGraw-Hill, 2003, 754–876.

[76] Errico M, Ramirez Marquez C, Torres Ortega C, Rong BG, Segovia Hernandez JG. Design and control of an alternative distillation sequence for bioethanol purification. Journal of Chemical Technology and Biotechnology, 2015, 90, 2180–2185.

[77] Taylor M, Wankat PC. Increasing the energy efficiency of extractive distillation. Separation Science and Technology, 2004, 39, 1–17.

[78] Aviles Martinez A, Saucedo-Luna J, Segovia Hernandez JG, Hernandez S, Gomez-Castro F, Castro-Montoya AJ. Dehydration of bioethanol by hybrid process liquid-liquid extraction/extractive distillation. Industrial & Engineering Chemistry Research, 2012, 51, 5847–5855.

[79] Vazquez-Ojeda M, Segovia-Hernandez JG, Hernandez S, Hernandez-Aguirre A, Kiss AA. Design and optimization of an ethanol dehydration process using stochastic method. Separation and Purification Technology, 2013, 105, 90–97.

[80] Dadgar AM, Foutch GL. Improving the acetone-butanol fermentation process with liquid-liquid extraction. Biotechnology Progress, 1988, 4, 36–39.

[81] Kraemer K, Harwardt A, Bronneberg R, Marquardt W. Separation of butanol from acetone-butanol-ethanol fermentation by a hybrid extraction-distillation process. Computers and Chemical Engineering, 2011, 35, 949–963.

[82] Doherty MF, Malone MF. Column sequencing and system synthesis. In: Conceptual design of distillation systems. McGraw-Hill, 2001, 289–346.

[83] Errico M, Sanchez-Ramirez E, Quiroz-Ramirez JJ, Segovia-Hernandez JG, Rong BG. Synthesis and design of new hybrid configurations for biobutanol purification. Computers and Chemical Engineering, 2016, 84, 482–492.

[84] Errico M, Rong BG, Torres-Ortega CE, Segovia-Hernandez JG. The importance of the sequential synthesis methodology in the optimal distillation sequences design. Computers and Chemical Engineering, 2014, 62, 1–9.

[85] Errico M, Pirellas P, Torres-Ortega CE, Rong BG, Segovia-Hernandez JG. A combined method for the design and optimization of intensified distillation systems. Chemical Engineering and Processing: Process Intensification, 2014, 85, 69–76.

[86] Wu M, Wang M, Liu J, Huo H. Life-cycle assessment of corn-based biobutanol as a potential transportation fuel. ANL/ESD/07–10, 2007.

[87] Geodkoop M, Spriensma R. The eco-indicator 99. A damage oriented for life cycle impact assessment. Methodology report and manual for designers. Technical report. Amersfoort, The Netherlands: Pré Consultants, 2001.

[88] Torres-Ortega CE, Errico M, Rong BG. Design and optimization of modified non-sharp column configurations for quaternary distillations. Computers and Chemical Engineering, 2015, 74, 15–27.

[89] Rong BG. Synthesis of dividing-wall columns (DWC) for multicomponent distillations-A systematic approach. Chemical Engineering Research and Design, 2011, 89, 1281–1294.

[90] Smith R, Linnhoff B. The design of separators in the context of overall processes. Chemical Engineering Research and Design, 1988, 66, 195–228.

[91] Gonzalez-Bravo R, Sanchez-Ramirez E, Quiroz-Ramirez JJ, Segovia-Hernandez JG, Lira-Barragan LF, Ponce-Ortega JM. Total heat integration in the biobutanol separation process. Industrial & Engineering Chemistry Research, 2016, 55, 3000–3012.

[92] Sánchez-Ramírez E, Alcocer-García1 H, Quiroz-Ramírez JJ, Ramírez-Márquez C, Segovia-Hernández JG, Hernández S, Errico M, Castro-Montoya AJ. Control properties of hybrid distillation processes for the separation of biobutanol. Journal of Chemical Technology and Biotechnology, 2016, doi:10.1002/jctb.5020.

Petri Uusi-Kyyny, Saeed Mardani, and Ville Alopaeus

6 Process intensification for microdistillation using the equipment miniaturization approach

Abstract: Process development has gradually been moving to smaller scale test facilities for individual testing of chemical reactions, catalysts, physical properties, phase equilibria, and other aspects of industrial processes. This information has typically been combined with integrated process models developed with process flowsheet simulators and other unified modeling platforms. Despite this progress, pilot plants are often needed to obtain experimental confirmation of longer test runs with recycle streams similar to those in real plant operation. Miniaturized pilots can be used to reduce the high costs of typical pilot operations. These pilots can be smaller than traditional bench-scale test facilities, yet provide similar data quality. Because such a pilot is small and safe to operate, it can be built earlier during the process development project, and several pilots can be used for parallel testing of several process concepts. This can lead to significantly reduced time-to-market for new processes. Miniaturized process units have been available for several years for many operations, such as reactors, mixers, and pumps. However, distillation, perhaps the most often used large-scale separation process, has not been miniaturized successfully to other unit scales. This chapter discusses progress in miniaturization of distillation and integration of distillation as a unit operation in a continuous milliscale pilot with recycle streams.

Keywords: Distillation, process development, miniaturization, pilot, metal foam, 3D printing

6.1 Introduction

New processes are needed for introducing biobased feedstocks into the chemical industry on the basis of demand. In addition, situations frequently arise where alternative designs for existing processes are proposed or the demand for a novel product requires process modification. Process development has to be as fast as possible while keeping costs as low as possible [1]. Thus, the design of novel chemical processes or upgrading of old processes are generally based on extensive use of modeling and simulation. Mathematical modeling and simulation techniques enable evaluation of several process concepts, thus reducing the number of pilot plant experiments and speeding up process design. Despite major advances in modeling-based process development, there are still many instances where piloting cannot be totally omitted. There are three reasons for this: (1) Small amounts of impurities are present that are not necessarily detected in a single-run reactor tests. These may originate from impurities in

https://doi.org/10.1515/9783110465068-006

the raw materials or potentially unknown side reactions and only become evident in prolonged runs with recycle streams. (2) An integrated process may be needed to increase trust in the process concept, and longer continuous runs are necessary for this. (3) A sufficient volume of product that is close to the expected final industrial plant quality is needed for testing product properties in various applications or subsequent process units. The main benefit of miniaturized pilots lies in the first of these reasons, but such a pilot can also benefit the process development project for the second and third reasons. A miniaturized pilot is sufficient if the subsequent process units are also piloted at the same scale. Furthermore, final product testing can, in some cases, be done at relatively small scales, although volumes exceeding miniaturized pilot capabilities are often needed.

Typically, piloting is relatively expensive and only the most promising process setups can be tested. This can lead to a situation where the decision to build an expensive pilot, sometimes as big as one-tenth of the full-scale plant, is not made because of a lack of proper integrated unit information from smaller scales. A miniaturized pilot could generate such information and build confidence. Furthermore, several process concepts can be tested in parallel at a small scale, especially if the miniaturized pilot fits within a regular fume hood and can be run partially unmanned because only very small volumes of dangerous chemicals are used. This is not possible with bench-scale units, which typically require full-time operator attendance, leading to much higher operating costs.

Fast process development is a prerequisite for minimizing development costs and being first on the market with a novel chemical product. Miniaturized testing of catalysts, reactors, and relevant physical and mass transfer properties, all simultaneously, is already common at the very beginning of process development. Furthermore, there has been extensive use of modeling tools in the chemical, oil refining, and petrochemical industries for decades, and increasingly in other fields. If this relatively well-established practice is augmented with miniaturized pilots, the so-called "valley of death" in process development cash flow occurs earlier, with earlier time to market, and earlier break-even point for cash flow. This is illustrated in Figure 6.1.

Naturally, there are also limitation the fluid streams. For handling solids, the natural limitation is that all void parts must be larger than the largest solid size, typically by at least an order of magnitude, to avoid blockage.

A summary of the properties and capacities of various pilot plant sizes are compared in Figure 6.2. Please note that although the terms "micro" and "milli" refer directly to volumes, they are not necessarily always used in a strictly defined manner. The volumes may refer to the feed volumes per (undefined) time, total volume of the pilot, or typical volumes of individual units. Sometimes, this terminology is used simply to separate different scales without precise reference to volumes. Furthermore, the concepts of minipilot and bench scale may sometimes be indefinite, and even the full-scale pilot size depends on the volume of typical final commercial production for each individual product. Figure 6.2 defines the scales used in this comparison.

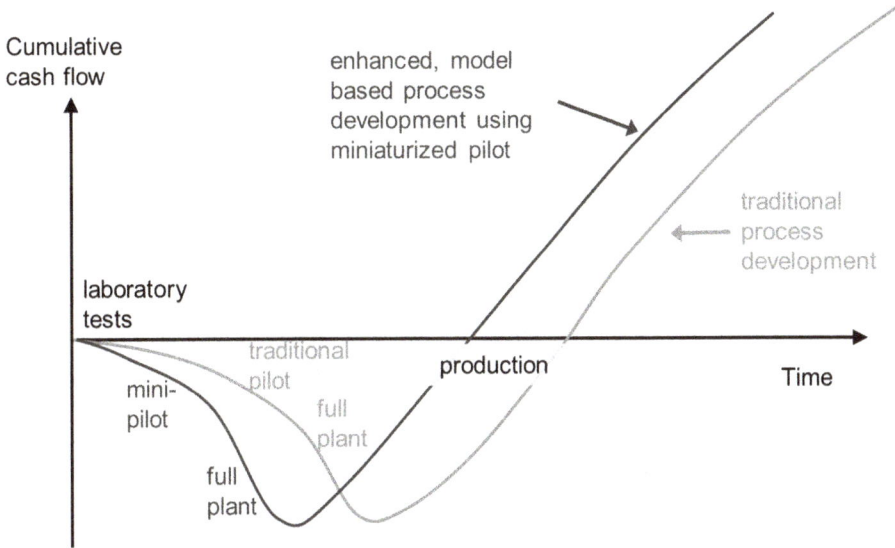

Fig. 6.1: Cash flow as a function of time in a typical process development project. The lowest point is called the "valley of death" because the maximum amount of money is spent without any revenue. With extensive parallel laboratory testing and use of modeling tools and miniaturized pilots, the valley of death occurs earlier, and the break-even point where the investments are paid back occurs earlier. This is usually of crucial importance to companies.

In milliscale piloting, the amount of chemicals needed is small, resulting in increased safety and reduced labor costs. This is also an advantage because, in the initial process development phase, industrial feeds may be in short supply because a plant producing the feed chemicals may not yet be in use. Reduced labor costs are achieved

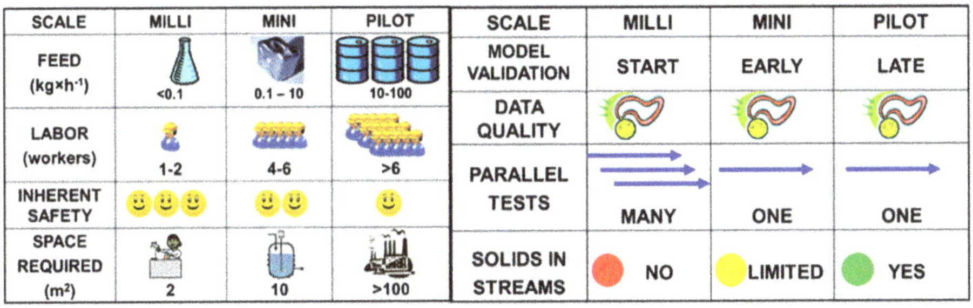

SCALE	MILLI	MINI	PILOT	SCALE	MILLI	MINI	PILOT
FEED (kg×h⁻¹)	<0.1	0.1 – 10	10-100	MODEL VALIDATION	START	EARLY	LATE
LABOR (workers)	1-2	4-6	>6	DATA QUALITY			
INHERENT SAFETY				PARALLEL TESTS	MANY	ONE	ONE
SPACE REQUIRED (m²)	2	10	>100	SOLIDS IN STREAMS	NO	LIMITED	YES

Fig. 6.2: Properties of different sizes of milli/micro plants. Distillation is used as the principal separation method in the majority of industrial processes. Small-scale continuous distillation devices with more than one theoretical stage and a small hold-up are extremely important for miniaturization of a miniplant into a microplant. The main reason for this is the need to reduce the recycle cumulative residence time, which should be less than 24 h according to Wörz [2] (figure by D.Sc. Aarne Sundberg).

because process monitoring can be relaxed (as the chemical amount is very small). Also, changes to the process and maintenance are rapid (replacing small-scale tubing, filters, etc. during the run). Furthermore, the space needed for the laboratory and personnel is drastically reduced. Thus, several process options can be built and screened simultaneously from the very beginning of the process development cycle. Consequently, the effort allocations for modeling and determination of the kinetics and physical properties can be shared correctly. The idea of using miniaturized pilots in process development is presented in Figure 6.3. To obtain data from a milliplant with the same quality as from larger units, the instrumentation and analytics should be equally extensive. Sampling for process analysis can be a problem in large plants, especially if costs are being minimized by keeping the sampling instrumentation limited. However, at the small scale, the sampling system can be much simpler because the whole process stream can be directed through the analyzers. Milli/micro plants are also more feasible due to their easy tailoring possibilities and adaptation to desired changes. The agility they offer is especially needed when considering design tools for the chemical industry. As mentioned earlier, the inherent drawback of very small-scale pilots is their inability to handle solids in the process.

Figure 6.3 presents a simple conceptual diagram of the intentions behind pilot miniaturization. At the laboratory scale (test-tube scale), many alternatives can be screened to find potentially useful solutions. In many cases, information is only semi-quantitative, but allows most unfeasible solutions to be discarded. At a slightly larger scale, individual small-scale reactors and other equipment are used to test ideas from the laboratory. Typically, communication between chemists and engineers should be guaranteed at this step. At the second step, as much quantitative information as possible is obtained about the system, but traditionally from different sources. After that, the traditional approach is to use various larger scale pilots (bench and larger sizes) that aim to mimic the industrial process operation. Process modeling has also been used for some decades to aid planning during this phase, although the whole development path was more or less based on heuristics and the experience of development engineers. In the framework of pilot miniaturization, it is sometimes possible to omit

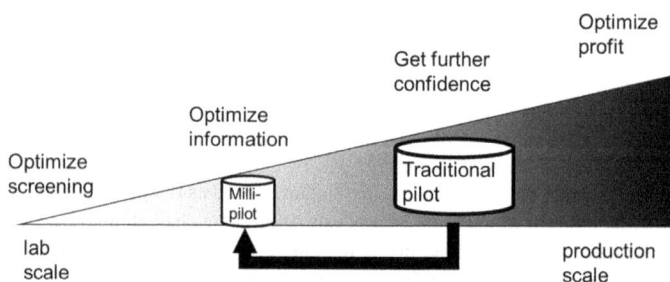

Fig. 6.3: The benefit of using microplants in process development, compared with a traditional pilot plant. Further confidence in the process can be obtained at an earlier stage.

completely the large-scale pilot, which is sometimes built just to gain confidence before full-scale plant investment. Information from small-scale individual experiments, such as reaction kinetics, catalyst deactivation, phase equilibria, and thermophysical properties, can be combined in a unified process flowsheet simulator, which is used to design large-scale plants. The small-scale pilot would then bring confidence, especially related to long trial runs, recycle streams, and validation of the simulation model. This approach can save considerable money and time, as illustrated in Figure 6.1.

6.2 Development of small-scale distillation units

Distillation is the main separation method used in chemical industries [3]. Separation is achieved by controlled transfer of heat to and from the distillation column to form vapor and liquid phases. Phase equilibrium behavior is the key phenomenon for consideration.

Because one equilibrium stage separation seldom fulfils the separation requirements, a distillation column inevitable requires counter-current flow of the contacting phases, leading to several ideal stages per unit. This principle of counter-current contact is essential in developing a successful microdistillation column.

Although microprocess technology has been applied in microscale reactors, mixers, extraction units, etc. for several years, and some can be considered as mature technology, there has been a lack of miniaturized distillation columns that can match the other unit operations. One of the main reasons for this is difficult temperature control. Multistage distillation works very well at large scales, when the process is very close to adiabatic (except in the reboiler and condenser). Various side condensers, side strippers, and other more complex column arrangements have been applied, especially in the oil refining industry, but the temperature profile of the column remains relatively straightforward to maintain because operation of the trays is close to adiabatic. The main heat flow along the column is due to liquid and vapor flows. Temperature changes result from the vapor pressures and compositions of the components that are being distilled. By contrast, at very small scales, heat losses become significant compared with heat flows. In many cases, this is precisely the benefit of miniaturization, because temperature can be controlled more easily and processes are thus inherently safer than their larger scale adiabatic counterparts. For distillation, this could be problematic because the temperature profile could have a minimum in the middle of the column, especially if the condenser operates above ambient temperature. This is illustrated in Figure 6.4.

This kind of behavior is highly problematic, because it causes column internal vapors to condense in the middle of the column instead of in the condenser. For traditional distillation columns, this would cause complete failure of the process because either there would be no vapor to enter the condenser or the condenser might not be

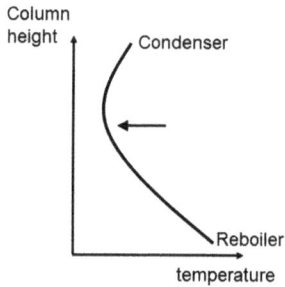

Fig. 6.4: Potential problems in the temperature profile of very small-scale distillation columns as a result of uncontrolled heat losses.

able to condense the vapor that enters. Thus, there would be no distillate product. For horizontal distillation columns (as described in Section 6.3.2), the situation is not as detrimental because both vapor and liquid can pass the minimum temperature location. Even then, most of the column height would be unused because the lowest boiling components would be enriched in the middle of the column instead of in the top distillate.

One way to avoid this is to design a column with appropriate heat conduction properties so that heat from the reboiler dissipates along the column, and the lowest temperature is obtained only near the condenser. Efficient insulation should also be used and intermediate heaters could be added to compensate for heat losses.

Table 6.1 summarizes developments in counter-current small-scale vapor–liquid separation. The maximum number of ideal stages is reported to be 18. It is clear that

Tab. 6.1: Summary of separation methods with countercurrent flow.

Reference	N/HETP [m]	Capacity [mol/s]	Type
Seok and Hwang [4]	4–5/0.06–0.16	0.013–0.1	Heat pipe
Ziogas et al. [5]	12/0.01	NA	Plate
Fink and Hampe [6]	2.5/0.016	NA	Collimator
Tegrotenhuis and Stenkamp [7]	NA/0.018	1.5×10^{-4}	Plate
Cypes et al. [8]	NA	0.7–32×10^{-4}	Stripper
Tonkovich et al. [9]	NA/0.018	NA	Falling film
Sundberg et al. [10]	2.5/0.05	Total reflux	Plate
	5.7/0.1	1.4×10^{-4}	
Sundberg et al. [11]	18/0.016	Total reflux	Plate
	5/0.058	6.3×10^{-5}	
MacInnes et al. [12]	6.7/0.053	2.7×10^{-5}	Stripper
Lam et al. [13]	4/0.02	2.0×10^{-6}	Chip
Cvetković et al. [14]	NA/NA	3.0×10^{-7}	Chip
Jang and Kim [15]	NA	0.007–0.022	Plate
Mardani et al. [16]	4.6/0.070	Total reflux	Packed
	3.5/0.095	1.3×10^{-4}	

N is the number of ideal stages, HETP is height equivalent of a theoretical plate, and capacity is the feed flow rate.

further plates are needed for widening the possibilities of use in process development. Note that some of the studies do not involve such challenging temperature control issues as distillation because the feed of noncondensable gases helps maintain the vapor. In traditional distillation, these gases are unwanted because they lower the capacity of the column and condenser in particular.

6.2.1 Reflux ratio control

A small distillation column needs reflux control and controlled energy input. This requirement for accurate reflux control is specific to distillation (among separation processes operating in counter-current mode) and constitutes a further challenge.

The flow rates of the streams have to be measured to give meaningful results for the reflux ratio determination. It is beneficial to use Coriolis-type mass flow controllers because of their insensitivity to the medium, compared with mass flow controllers based on the principle of differential pressure or thermal conductivity. A schematic setup for one potential robust small-scale reflux control system, as used by Sundberg et al. [17], is presented in Figure 6.5. The distillate is first condensed in a heat exchanger. The noncondensing gases are vented from the distillate and the condensate level in the tube acting as a condensate tank is detected with a differential pressure transmitter. The output of the differential pressure transducer is used to control the pump. The ratio of condensate and reflux streams is controlled using mass flow controllers. Thus, it is possible to operate both in constant flow rate and in flow ratio-controlled modes. This setup using an external condenser system enables small-scale columns to be switched for testing in a very short time. Also, heat losses can be conveniently determined with this column in full reflux mode, because the distillate flow rate is measured.

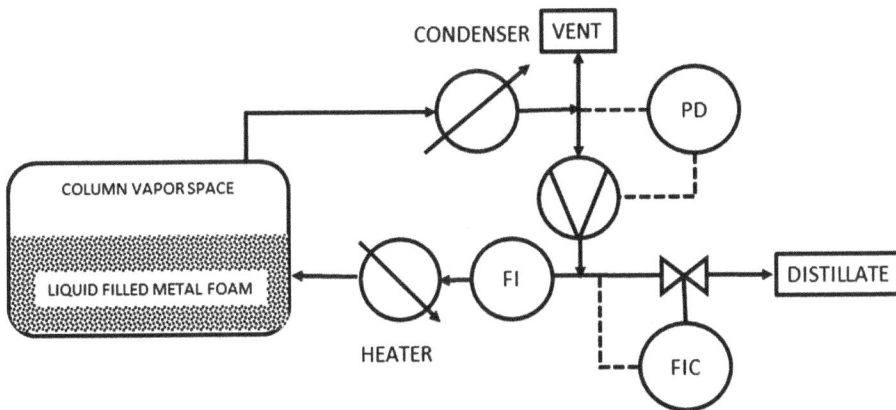

Fig. 6.5: Reflux control, as described by Sundberg et al. [17].

6.2.2 Reboiler types

Over the past few years, our research group has tested three principal reboiler types for application in small-scale distillation columns. The benefits and drawbacks of each type are as follows:

Heat transfer through conduction of the column wall: This type of reboiler was used by Sundberg et al. [10, 11] and in most of the other units presented in Table 6.1. This type is essentially used in the heat pipe type of column. The drawback is that the amount of energy used for the boil-up cannot be measured and, thus, the internal flows cannot be fully known. Some energy is lost to the surroundings via conduction and convection before heating the liquid. However, the benefit is that external connections to the column can be minimized into just one outlet connection. This type of reboiler is presented in Figure 6.6 (left).

External reboiler: Energy input control is achieved using pumps, Coriolis mass flow controllers, and evaporation of the whole stream flowing through the reboiler. This type of reboiler is presented in Figure 6.6 (middle). The temperature of the vapor stream is measured and visual observation is used to assure that the feed is completely vaporized. The benefit is that the energy input can be determined from the heat of vaporization of the mixture, stream temperatures, and heat capacities. A drawback is the possibility of high-boiling impurities not evaporating and causing accumulation and blockage.

Internal electrical reboiler: The reboiler can be located in the reboiler part of the column, as presented in Figure 6.6 (right). The energy input is measured from the electrical energy input, which can be precisely controlled. In small-scale systems, the challenge is controlling the liquid level of the reboiler. If this is too low, the result is overheating of the heater cartridge and possible electrical failure if the heater is operating in a completely dry environment. A possible solution for local

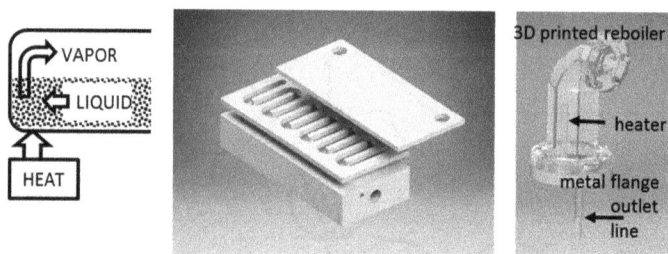

Fig. 6.6: Three tested reboiler types. Left: heat transfer using conduction through the column wall. Middle: external reboiler structure consisting of sandwiched metal plates. The plates are assembled and welded by shielded metal arc welding. Volume of this particular reboiler channel is 5.7 cm^3. Right: internal electrical reboiler.

overheating is attachment of high surface area metal foam or a wick to the heater element to dissipate heat more evenly. Another option is to measure heater wire resistance so that a warning is given before overheating, in analogy to hot wire anemometry. This is still to be tested.

6.3 Distillation column structures

Development of the column structures presented here was mainly based on work done by our research group during recent years. The development was naturally based on general knowledge of manufacturing techniques, distillation column operation, multiphase flow, and phase equilibria. However, as the development was largely hand crafted, it was also based on trial and error. Observed features during test runs of each design were used in the development of subsequent structures. The idea of using metal foams as packing in a plate-type horizontal column originated from interest in the materials themselves and their availability. Three-dimensional (3D) printing, on the other hand, gives freedom of design and the possibility of discovering various designs in a short time frame, because complex products can be printed rapidly. Thus, it was possible to use and test very complex structures not possible to realize with normal manufacturing techniques such as drilling and milling.

6.3.1 Brass column with heat pipe type of operation

The main issues in the development of small-scale distillation units are control of heat loss, temperature profile, flow rate, and flow ratio. As the size of the device decreases, the surface-to-volume ratio increases, which increases the significance of heat losses. The idea of using metal foam as packing was first tested by manufacturing and running a brass column with heat pipe type of operation, as presented in Figure 6.7. The metal foam manufacturer was Recemat International (the Netherlands). The reboiler was based on heat transfer through conduction of the column wall and was relatively easy to manufacture. It was possible to take samples from several locations. Lines for introducing feed and removing distillate and bottom product were available. Visual observation of boiling, flow, and condensing was possible through a window, by lifting part of the insulation. This enabled detecting the effect of different metal foam thicknesses. Condensation of the liquid was initially obtained by air cooling (i.e., using natural heat loss from the unit). Air cooling was found to be less stable than water cooling, which was used in later experiments. Experiments with total reflux and continuous flow were conducted after solving initial challenges related to sealing of the window. Details of the experiments are described by Sundberg et al. [11]. One drawback of the column was that it was not possible to assess quantitatively the flow rates

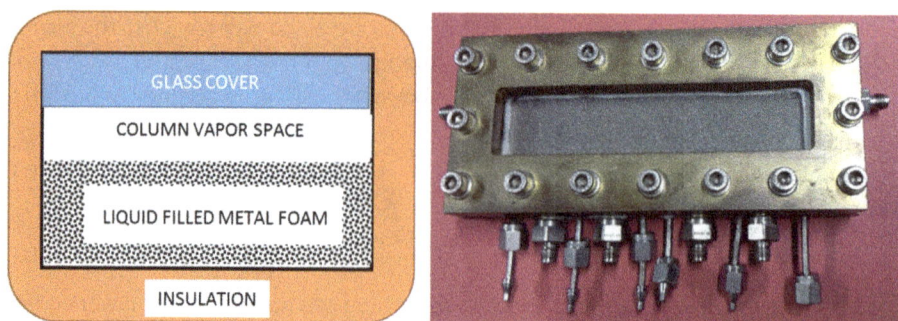

Fig. 6.7: Schematic of planar metal foam column (left) and realized column sampling points (right).

in the column due to lack of information on boil-up and reflux ratio. Control of heat loss was also challenging.

6.3.2 Stainless steel plate type of column

The column wall was manufactured out of stainless steel. A scheme of the column is presented in Figure 6.8. The shape of the distillation chamber was planar because this shape was easy to manufacture. External dimensions of the column were height 20 mm, width 50 mm, and length 300 mm. The internal dimensions of the vapor–liquid contacting channel (VLCC) were height 5 mm, width 30 mm, and length 290 mm. A KF40 flange was welded to the cold end of the column, allowing maintenance and connection of the column to a condenser. Metal foam (3 mm thick) was placed on the bottom of the VLCC and secured in place using a metal plate of 2 mm height, 30 mm width, and 290 mm length. Experiments were also carried out with 2 mm of vapor space above the metal foam [10]. Less theoretical stages were obtained in the experiments than in setups without vapor space.

The column was heated by heat transfer through the column wall using an electrically heated aluminum plate. Another electrically heated plate was attached on top of the column. Heat loss could thus be controlled to some extent and the temperature was very stable compared with systems heated using electrical tracing wires, as commonly used. Feed was introduced with a syringe pump without further heating. The bottom product was drawn by gravitation, and flow rate control was achieved using a needle valve. The bottom product line was also the sampling line near the end of the column. The distillate was drawn using the principle of communicating vessels. For the analytics, a refractive index detector was used. The temperature profile was measured throughout the column length using temperature probes.

Test distillations were performed with a mixture of *n*-hexane and cyclohexane. Determination of the composition profile enabled observation of the separation efficiency along the column. Relatively inefficient separation was observed near the

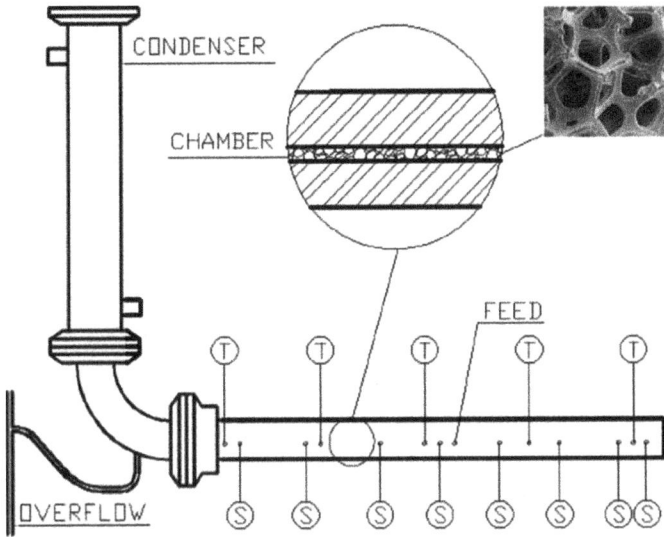

Fig. 6.8: Schematic of a microdistillation column [18]. The vapor–liquid contact chamber is filled with metal foam (Recemat International); *S* sampling points, *T* temperature probes.

condenser end of the column. This may have been caused by condenser back-mixing and/or low vapor flow rates. Heat losses in the column were considered high, even with the aluminum plates.

The maximum number of theoretical stages in these experiments was found to be 18. Almost pure cyclohexane was obtained from the bottom of the column. This result required optimization of the aluminum plate temperatures. The continuous flow results showed that the separation efficiency of the column decreased rapidly when feed flow rate increased from 0.25 to 2 mL/min, indicating low column capacity. On the other hand, good stability of the column concentration profiles was observed in runs lasting up to 3 weeks. The main improvement over the brass column was the use of communicating vessels for withdrawing distillate. This design also produced distillation results for a column that had no separate space for vapor flow on top of the column. In this design, the separation efficiency was higher without the vapor space for 2 mm of metal foam. However, the separation efficiency came at the expense of reduced column capacity.

6.3.3 Modular copper column

The modular copper column was only tested in a continuous mode of operation. The column had an open space of 2 mm above metal foam. The feed rate was 2 cm^3/min of *n*-hexane/cyclohexane mixture and both distillate and bottom product flow rates were approximately 1 cm^3/min. *n*-Hexane concentrations at the ends of the column were 0.67 ± 0.03 and 0.26 ± 0.01 in mole fractions. The separation efficiency was at best 4.7 ± 0.5 ideal stages within 440 mm of metal foam, giving the height equivalent of a theoretical plate (HETP) as 93 mm. This corresponds to a similar separation efficiency obtained for the steel column with 2 mm of empty vapor space above the metal foam. Stability of the column was not as good as using the steel column. The reason was probably due to insufficient control of the liquid level. The parts of the modular column are shown in Figure 6.9 [10].

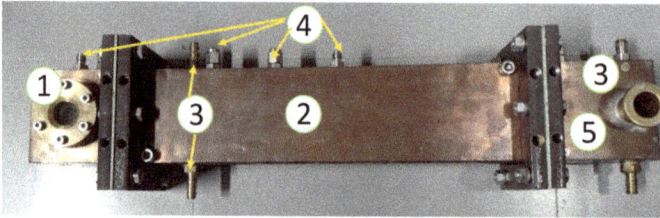

Fig. 6.9: Modular distillation column consisting of a reboiler section (1), column (2), fittings for column temperature control (3), sample ports and feed and product line fittings (4), and condenser section (5).

6.3.4 Laser-welded square column

Laser-welded square distillation column parts and the metal foam packing were laser cut and welded at the Lappeenranta University of Technology. Parts of the column are presented in Figure 6.10. The internal cross-section of the column was a square with

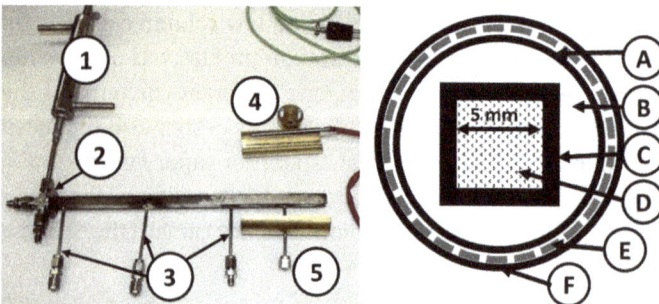

Fig. 6.10: Left: Laser welded microdistillation column (adapted from Hirvimäki et al. [19]) showing condenser (1), distillate outlet and return lines (2), feed lines (3), reboiler heating element consisting of a brass block with a PID controlled cartridge heater and a temperature probe well (4), and bottom product outlet line (5). Right: Schematic cross-section of the distillation column indicating the stainless steel cover (A), insulation (B), column wall (C), metal foam (D; Recemat International, RCM-NC-2733.03), temperature controlled electrical tracing (E), and insulation (F).

side lengths of 5 mm. The column was completely filled with metal foam. The reboiler was based on heat conduction through the column.

The maximum number of plates obtained was approximately four. The column was only operated in continuous mode. This experiment also showed that filling the column completely caused a decrease in separation efficiency when 5 mm of foam was used instead of 2 mm (stainless steel column). The reduced efficiency could be because the high foam thickness resulted in channeling of vapor flow, even though the column was completely filled.

6.3.5 3D-printed coiled compact distillation column

The idea of 3D printing a distillation column arose from the fact that novel structures and freedom of design was needed to realize ideas for new column geometries. Selection of the polymeric printing material was based on availability of the printers and cost of printing. 3D printing of polymeric materials is inexpensive so multiple designs can be tested rapidly. For final designs, the material of choice is stainless steel or titanium because they have better chemical compatibility than polymers. Unfortunately, 3D printers for producing metallic parts are expensive and involve substantial printing and service costs. Although it is possible to obtain printing services from third-party companies, there is always a lead time in delivery, printing is costly, and the design–printing–redesign cycle time is prolonged.

The idea for the column was based on coiling the column to a compact spiral shape in an attempt to lower heat loss and decrease the space that the column occupied. The column is presented in Figure 6.11. The column length (number of stages) could potentially be increased, with a minimal increase in external size compared with the plate-type column of Sundberg et al. [11]. In the first design, the cross-sectional area for fluid flow inside the column proved to be too small, and flooding was observed for the flow rates enabled by the external reboilers. One reason for this may be the additive manufacturing technique itself. When printing resin material, part of the resin may have unintentionally hardened inside the channels. Autodesk Inventor® 2016 was used for the 3D model design in this work. EnvisionTEC Perfactory® III Mini SXGA+ 3D printer

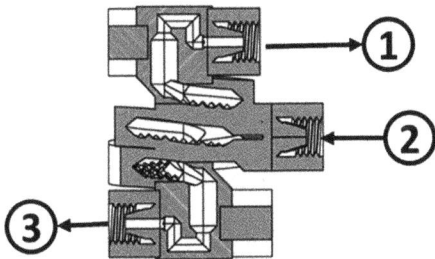

Fig. 6.11: Cut-out of the coiled distillation column: distillate connector (1), feed connector (2), and bottom product outlet (3).

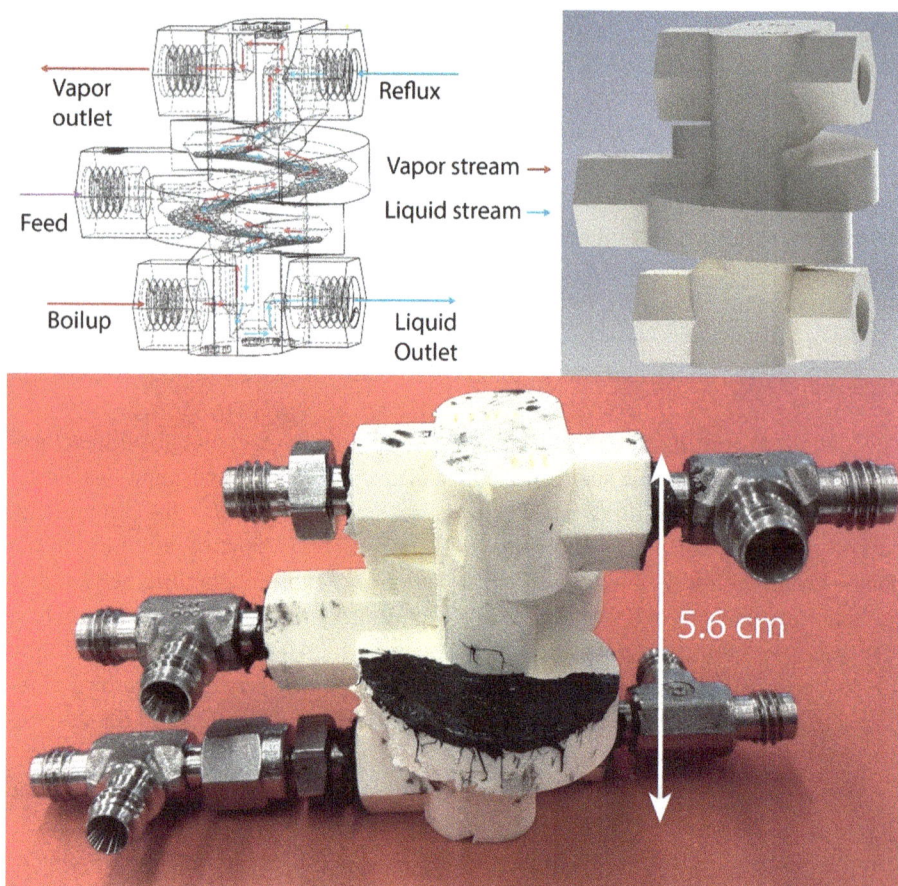

Fig. 6.12: Distillation column: illustration of the flows in the column (top left), external view obtained from the software tool (top right), and realized column with fittings installed (bottom).

was used for the stereolithography (SLA). The printing material was HTM140IV provided by EnvisionTEC.

One interesting feature was that it was possible to print the fittings for connecting the column to the external reboiler. The mechanical durability of these fittings was unfortunately low, but good sealing was achieved by using an additional sealant, as can be seen from Figure 6.12. This column is described in a report by Mardani et al. [16].

6.3.6 3D-printed modular coiled distillation column

A coiled modular 3D-printed distillation column is presented in Figure 6.13. It was designed, manufactured, and tested as described by Mardani et al. [16]. The 3D model

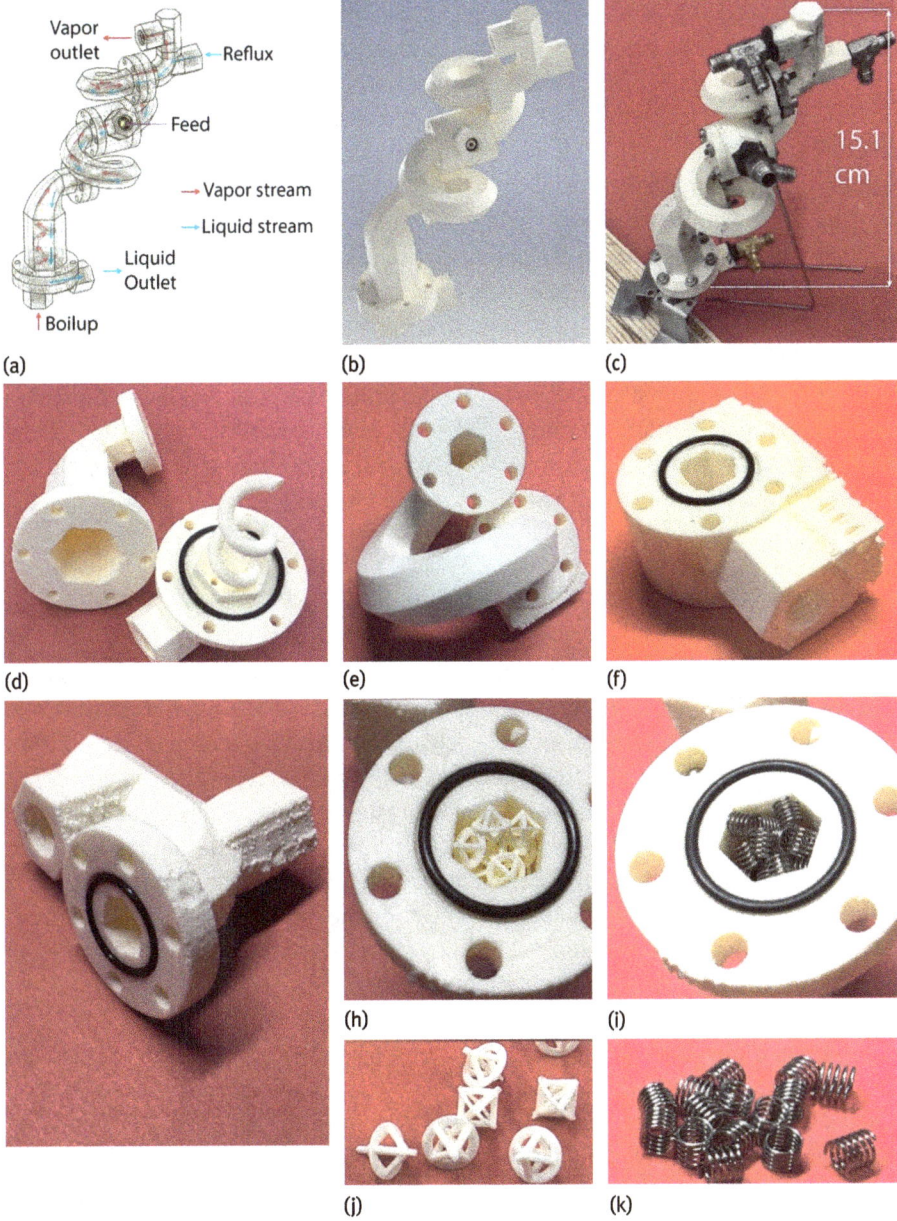

Fig. 6.13: (a) Modular coiled distillation column showing the interior and stream flow directions of the column, (b) 3D model design of the modular column, (c) final printed version with fittings connected, (d) bottom inlet and outlet, (e) main body of the column, (f) column feed inlet, (g) top inlet and outlet section, (h) 3D-printed packing in the column, (i) metal spring type packing in the column, (j) 3D-printed packing, and (k) metal spring packing.

design software, SLA printer, and printing material were the same as for the coiled compact column case (Section 6.3.5). The internal width was slightly larger than for the first coiled column and the column was packed. This construction proved that distillation columns can be designed and manufactured with exotic forms and still be suitable for distillation. The column was modular in structure and used an external reboiler and reflux ratio control.

The sections of the column, such as feed inlet, main body of the column, and top and bottom sections were designed and manufactured separately. The separate parts were then connected together using the flange connections. This type of modular structure made the design and manufacturing tasks much easier. The approach also enabled modification of individual sections, without affecting the main structure of the column. Additionally, because individual printed parts are smaller and simpler, the failure probability of the whole unit was reduced. Moreover, the flange connections enabled sections to be added or removed to achieve the desired separation efficiency. This could be a desirable feature for columns used for different separations.

6.3.7 Conclusion of distillation column structure review

Small-scale distillation columns contain similar main functional parts as larger columns (i.e., reboiler, column, packing, and condenser). The applicable flow rate ranges and the separation efficiencies tested and obtained are presented in Table 6.1. Flow rate limits arise not only from the effect of flow rate on separation efficiency but also because of the performance of pumps and other peripheral devices. Design and manufacture of the structures were based on ideas generated during research and were improved on the basis of experimental results. In these works, manufacturing capability and challenges also had major effects on the design. On the other hand, additive manufacturing allows quick testing of new designs, which can be very different from the usual perception of a distillation column.

Visualization and discussion related to system behavior were the main design methods, in addition to rapid manufacture and trial. This method was selected because rigorous simulation of the flow and composition patterns before manufacturing and testing would limit the spectrum of units tested within a given project time span. This is due to the relatively complex structures tested, for both the distillation apparatus and its internal details, and the requirement for independent validation of physical closure models, such as capillary forces within the unit. However, once a good design is found then the application of more elaborate and elegant methods, such as computational fluid dynamics (CFD) for analysis of the operation of the units, are useful for further improvements.

6.4 Metal foam as a packing material

Open cell metal foams were found to be good packing materials for horizontal distillation columns [10]. They are widely available industrially in a multitude of materials, pore sizes, and dimensions. Mechanical modifications such as cutting, bending, and drilling of the metal foams are relatively convenient. The effects of foam properties on the distillation performance in a plate-type distillation column were discussed by Sundberg [20]. The main conclusions are presented in Table 6.2.

One important aspect of metal foam selection is that they must be constructed with connected pores to allow liquid flow. This is not necessarily the case for all available foams. The foam connections should also be isotropic (i.e., flow should be allowed in all directions). In the length direction, an open structure is needed for the liquid to flow in general. In the height direction, an open structure enhances mass transfer because vertically isolated liquid paths would lead to saturation of the surface and bypassing of liquid at the bottom of the foam. Sideways, an open structure is especially required for feeding and removing products from the column in a uniform manner. Some of the available foams are shown in Figure 6.14.

To evaluate mass transfer performance of the metal foams, mass transfer measurement methods, such as those of Hoffman et al. [23] and Last and Stichlmair [24], were applied with the CO_2 and NaOH-solution system and air–nitrogen system described by Ojala et al. [25].

A schematic view of the apparatus for mass transfer measurements is presented in Figure 6.15. Composition of the gas phase was analyzed using a mass spectrometer and composition of the liquid phase either by conductivity or using a dissolved oxygen

Tab. 6.2: The effect of foam properties on distillation column operation.

Variable	Observed effect
Foam pore size	Capillary forces weaken
Foam porosity	Permeability increases [21]
Level of interconnections	Permeability increases [22]
Foam thickness	Mass flux and HETP increase
Specific surface area	None

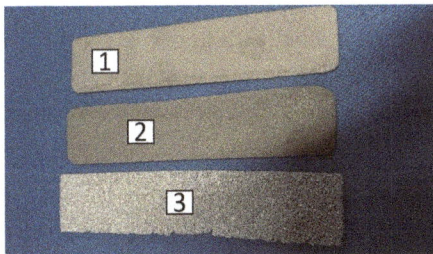

Fig. 6.14: Foam 1 has isolated open pores, which do not permit interpore transport and was not usable as a distillation column packing material. Open cell metal foams 2 and 3 have interconnected pores and can be used as a distillation packing.

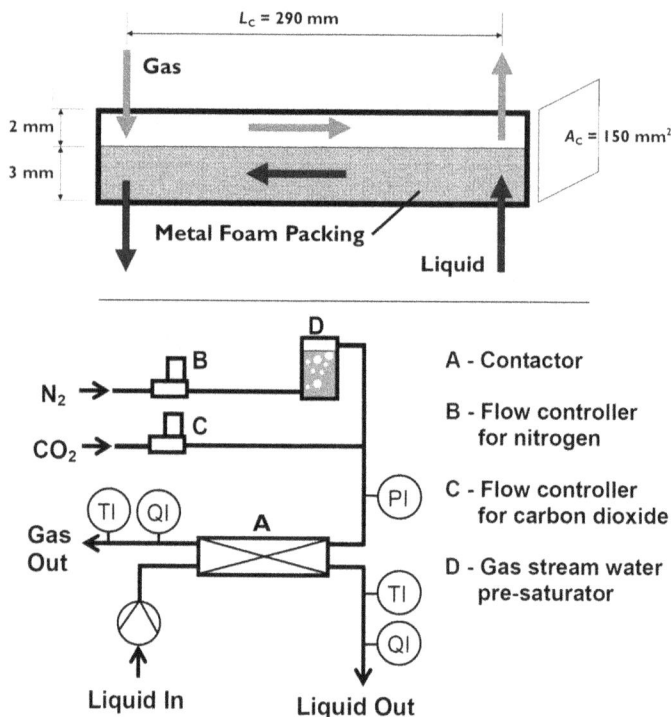

Fig. 6.15: Contactor for mass transfer experiments (top) and equipment setup for mass transfer measurements (bottom) [25].

meter, depending on the test system. It was found that the gas–liquid contact area is closely related to the top surface area of the foam, not the full specific area inside the foam pores [20]. This is to be expected because the same capillary forces enabling good wetting properties and flow of liquid (even in horizontal units) also tend to fill the foam with liquid, provided the surface properties are such that liquid wets the packing material as it should. In columns where the cross-sectional area is full of metal foam, vapor inevitably occupies part of the foam; however, vapor probably tends to form a separate layer on top of the liquid. In all cases, for the vapor to flow it must form a continuous vapor space in horizontal installations. Experimental results also showed that the volumetric mass transfer coefficient does not depend on the gas flow rate. This implies that the capillary forces keep the metal foam filled with liquid, even at high gas flow rates. On the other hand, the volumetric mass transfer coefficient seems to depend linearly on the liquid flow rate. The foam porosity and interconnections are important for the flow properties of the packing.

6.5 3D-printed packings

Besides additive manufacturing of the column itself, 3D printing can be used for packing materials that potentially could be used instead of metal foams or other packing materials. It could be possible to manufacture expensive laboratory distillation column packing using 3D printing. Students could design and print their own packing and test its performance in a normal laboratory column. Additive manufacturing (3D printing) offers freedom in the manufacture and selection of the packing structure. Both random packing and structured packing can be designed and manufactured. Additionally, the column and packing can be printed simultaneously to produce an integrated column with a built-in packing. One packing is shown in Figure 6.16 [16]. The main objective of that work was to test additive manufacture of the packing elements and their structural strength. Full mass transfer analysis of printed packings remains to be done. The wetting behavior of the packing manufactured from polymeric materials can be improved by coating the polymer using electroless plating or by printing it directly with a metal printer.

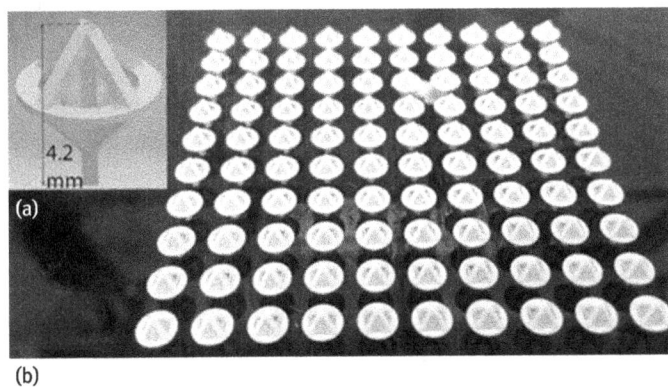

Fig. 6.16: 3D-printed packing. (a) Single packing demonstrating packing height. (b) 100 printed packings on the base plate of the 3D printer.

6.6 Application of microscale distillation for small-scale piloting

A small-scale trial plant was developed and its usability demonstrated by making experimental runs. A system with known reaction equilibrium, kinetics, and phase equilibria was selected in order to focus on the pilot equipment performance and reduce degrees of freedom in the analysis [20].

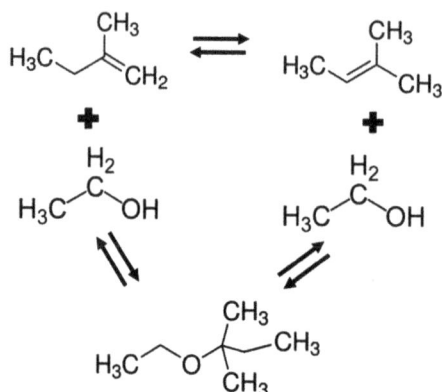

Fig. 6.17: Main reactions in the synthesis of 2-ethoxy-2-methylbutane from ethanol and isoamylenes and the isomerization of isoamylenes.

The chemical system studied was the synthesis of 2-ethoxy-2-methylbutane (ETAE) from ethanol and isoamylenes over an acidic ion-exchange resin in the liquid phase. ETAE is a gasoline additive with favorable vapor pressure and octane number. Reaction equilibria and kinetic expressions have been determined in earlier works by Rihko and Krause [26] and Linnekoski et al. [27, 28]. The modeling approach took into account the activities of the chemical systems modeled using original UNIFAC [29]. The relevant main reactions are given in Figure 6.17. Subsequent separation of a fuel ether with recycle of unreacted alkene and alcohol is demonstrated in Figure 6.18. The reactor was modeled with an in-house simulator, FLOWBAT, in steady state operation.

The phase equilibria for the test case were obtained from literature sources, as presented in Table 6.3, and were modeled according to the gamma-phi approach. The Wilson liquid activity coefficient model [30] and the Soave modification of the Redlich–Kwong equation of state [31] were used. These systems have high positive deviations from Raoult's law and also exhibit azeotropic behavior. The ideal liquid assumption was used for the hydrocarbon + hydrocarbon and hydrocarbon + ether binary pairs/systems. The parameters of the models and further details are given by Sundberg et al. [32].

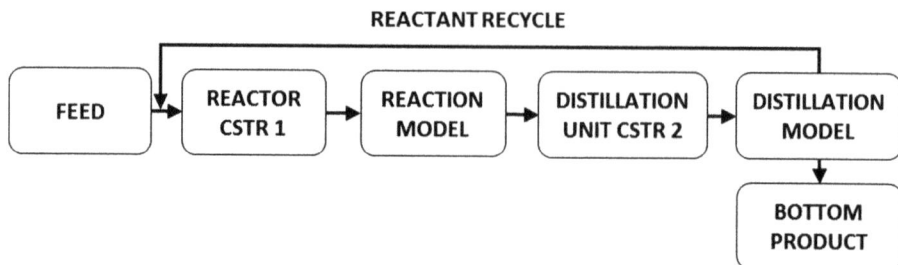

Fig. 6.18: Process model used in the simulations of tertiary amyl ethyl ether production in the microplant. Volume of CSTR1 was 10 cm³ and that of CSTR2 was 40 cm³.

Tab. 6.3: Phase equilibria (vapor–liquid equilibria, azeotropic data, and excess enthalpy) used for modeling of the distillation column. Reality of the liquid phase was only taken into account for the ethanol binary systems.

System, ethanol +	Literature source
2-Methyl-1-butene	Verrazi et al. [33]
	Gmehling et al. [34]
2-Ethoxy-2-methylbutane	Arce et al. [35]
	Gmehling et al. [34]
	Heine et al. [36]
Pentane	Ishii [37]
	Campbell et al. [38]
	Collins et al. [39]
	Gmehling et al. [34]

6.6.1 Distillation model

The distillation model was based on the distillations of several mixtures of known vapor–liquid equilibria, as presented by Sundberg et al. [40]. The HETP was determined as a function of the flooding factor. The number of theoretical stages was then calculated from the obtained HETP values and the length of the column. The feed was located in the middle of the column. The column simulations were performed as a steady-state distillation using the previously presented vapor–liquid equilibrium model and the FLOWBAT simulator.

6.6.2 Process model

The reactor and the distillation column were modeled as steady-state processes. The physical liquid hold-up was taken into account by combining the modules with continuous-flow stirred tank models (CSTR), as presented in Figure 6.18. The liquid volume in the reactor was obtained from the total reactor volume and the void fraction of the packed catalyst. The distillation column hold-up was experimentally determined from the measured amount of liquid injected for obtaining steady-state operating conditions. The total hold-up of the systems was 50 cm^3 including pumps, piping, and other spaces in the system. The residence time in the process varied from 2 to 4 h depending on the feed flow rate.

Process model validation was achieved by comparing stream composition results from the model against experimental results. Impurity accumulation was not observed during the experiments and, in general, very good agreement was seen.

6.6.3 Apparatus and instrumentation

The piping and instrumentation diagram of the microplant is presented in Figure 6.19 and a photograph of the equipment in a fume hood in Figure 6.20.

The microplant consisted of a reactor and a distillation column with the possibility of returning the unreacted and separated feed to the reactor. The feed to the reactor

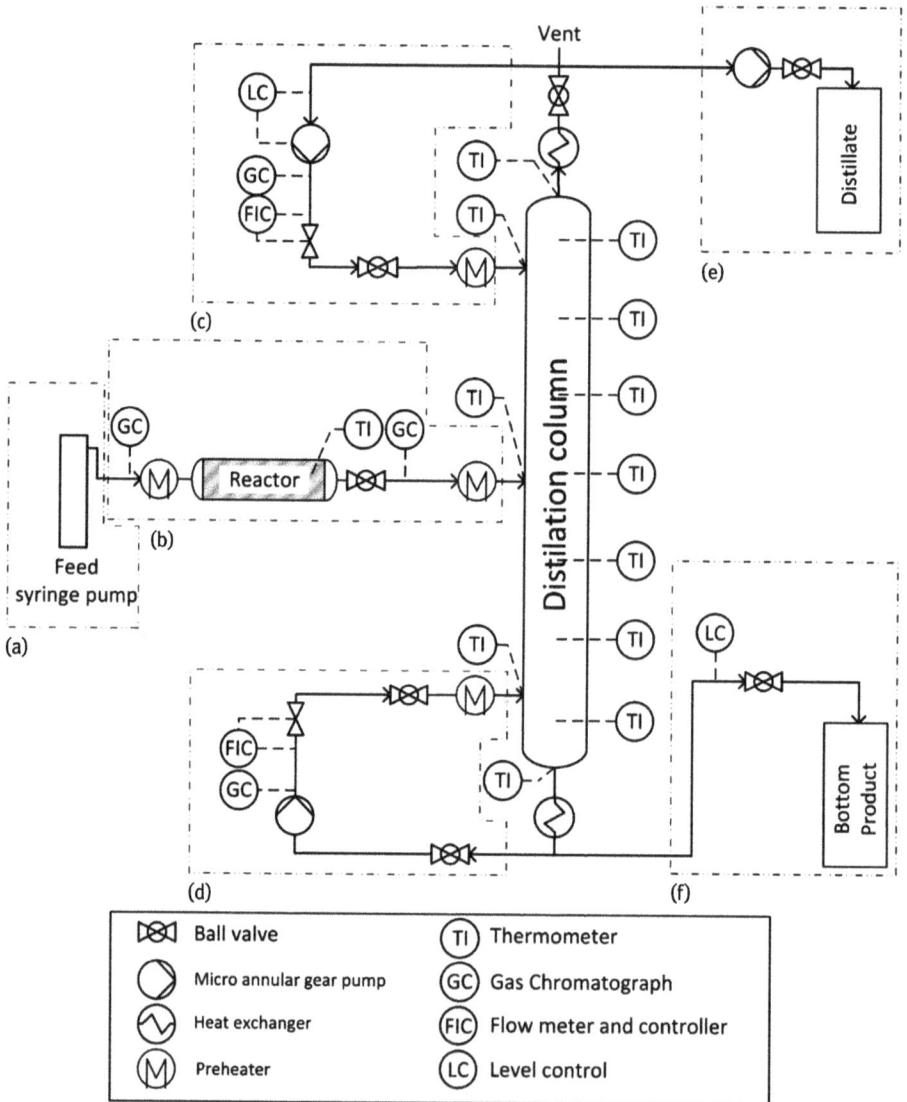

Fig. 6.19: Microplant used in the work of Sundberg [32]: (a) feed syringe pump, (b) reactor, (c) reflux setup, (d) reboiler setup, (e) distillate collection, and (f) bottom product collection.

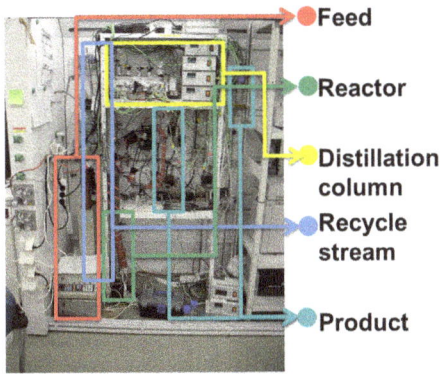

Fig. 6.20: Microplant containing feed pump, tube reactor, distillation column, recycle stream, and product collection.

was analyzed using an online gas chromatograph (GC). After analysis of the feed, the stream was fed to the reactor. For temperature control, the reactor setup was either located in a thermostatic water bath or electrically heated. The composition of the reactor product was analyzed using on-line GC. Before entering the distillation column, the feed could be preheated to a set temperature using a small-scale electrical heater. Subsequently, the product stream was directed to a distillation column to separate the product from reactants. At the top of the column, the vapor exiting the column was condensed and cooled using a heat exchanger. Condensate was collected in a small container (reflux drum). The container had a vent on the top to keep the distillation column at atmospheric pressure. The collected liquid was split into the reflux stream and the distillate stream. The distillate stream was collected in a flask at a constant flow rate. The reflux stream was returned back with a level-controlled mechanism to adjust the level in the liquid container. The reflux stream was directed through the on-line GC for analysis and through a flowmeter to check the flow rate. The reflux ratio could thus be measured. The temperature of the reflux stream was adjusted before entering the distillation column using a preheater and, finally, the stream was injected into the top of the column. The bottom product of the distillation column was collected in a tubular container. The stream was cooled with a heat exchanger and divided into the bottom product stream and the boil-up stream. The boil-up stream was pumped at a constant flow rate through the on-line GC system and analyzed. The constant flow rate was achieved by adjusting the flow rate with a combination of pump and mass flow controller. The bottom product stream was taken out by controlling the liquid level. To make it simple, the communicating vessels principle was used to take the liquid out. Additional details can be found in publications by Mardani et al. [16] and Sundberg et al. [20].

6.6.4 Impurity accumulation test

The reactor model was verified with runs performed without recycle (i.e., with once-through experiments). The mass balances were very good (within 1% after collector vessel cooling). The reactor performed well in prolonged runs and catalyst deactivation was not observed. The reactor product stream was directed subsequently to the distillation column and the column model could describe the experiments with good accuracy. Once-through experiments gave the confidence to use the reactor and distillation model separately to describe the dynamic behavior of the system.

Etherification with recycling of the light components showed that long-term runs were possible and that the models could follow the behavior of the system very well, as presented in Figure 6.21.

The suitability of the microplant for impurity accumulation trials was tested by adding an inert component, in this case n-pentane, to the feed mixture (1:1 molar ratio of ethanol to isoamylene). The boiling point of n-pentane is lower than that of the product ETAE. n-Pentane was expected to accumulate in the recycled distillate stream, based on the vapor pressures of the components. This was verified experimentally. The results of the runs are presented in Figure 6.22.

The results of the impurity accumulation run compared favorably with both the process model and the extrapolation model presented by Wörz [2]. The results showed that it is possible to use the microplant for accumulation studies and that the results can be interpreted using common process simulation tools. This result showed that the use of very small-scale continuous distillation devices for process development is possible.

6.6.5 Conclusions from the small-scale pilot test runs

The results showed that the scale of the apparatus and the method of operation are applicable for processes having a reactor and distillation setup with a recycle. It was also possible to model the accumulation of an impurity in the recycle stream accurately. This case validates the usefulness of this approach and shows that the distillation process is consistent with larger scale operation. The test process (manufacture of ETAE) was selected as a base case because the reaction kinetic models, catalyst behavior, and vapor–liquid equilibria were known, based on prior research on the system.

It is, however, evident that further developments are still needed for wider application of the miniaturized distillation column and integrated micropilot. Some necessary improvements include better reliability of pumps and instrumentation, improved analytic capability, and development of the distillation column itself with additional equilibrium stages. Process control issues may arise if the systems are made more complex by the addition of additional unit processes and recycles.

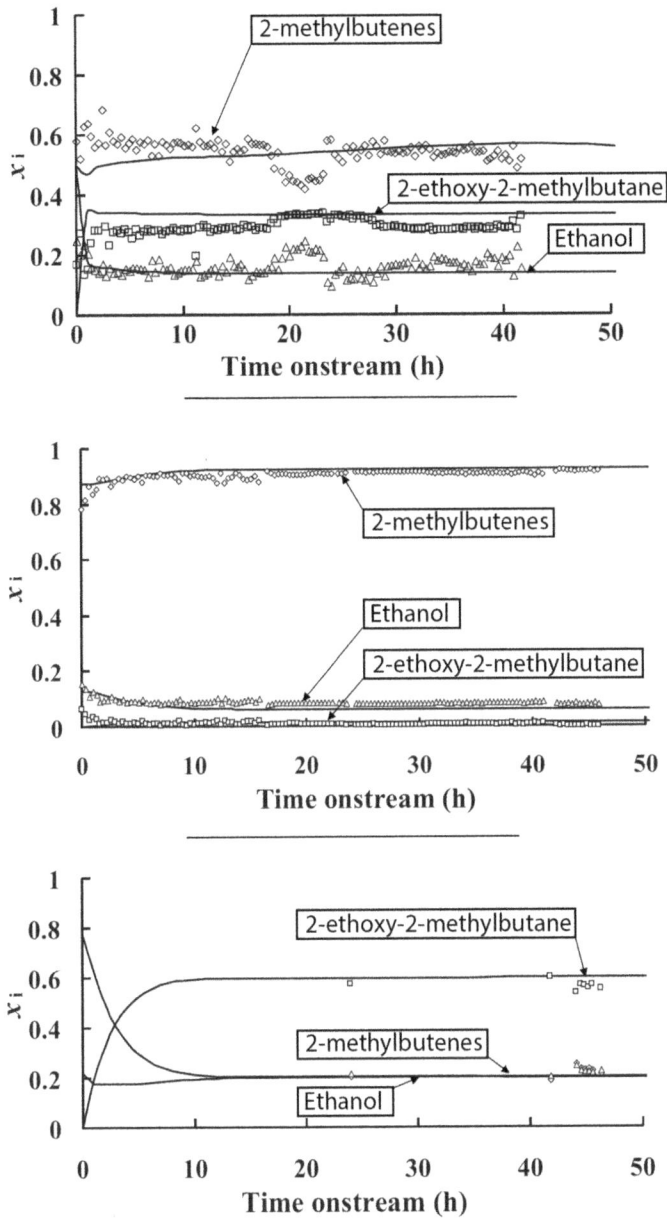

Fig. 6.21: Test runs for the etherification microplant: reactor product (top), reflux composition (middle), and boil-up for the etherification with recycle (bottom). Lines represent simulated values. Adapted from Sundberg et al. [32].

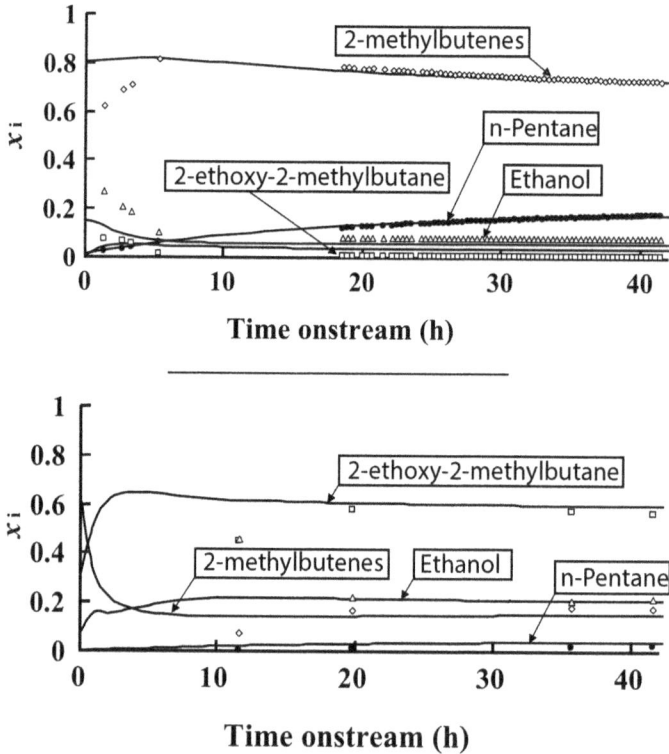

Fig. 6.22: Accumulation tests for the etherification microplant: reflux stream composition with clear accumulation of n-pentane (top) and bottom product with small increase in the n-pentane level (bottom). Adapted from Sundberg et al. [32].

6.7 Conclusions

The role of miniaturized pilot operations and rapid prototyping methods in the construction of small-scale pilot plants cannot be overlooked. Because each pilot plant can be considered unique for its application, the dimensions of process units of the plant also have to vary. This can be taken into account by using, for example, 3D printing for the manufacture of different parts of the plants. It is possible to print almost the entire process with currently available printers and materials. The ultimate simplified vision (now within reach) is the design of a process flow sheet, which then can be almost directly manufactured at a small scale by rapid prototyping methods. Rapid prototyping/manufacturing methods have developed considerably in recent years, enabling cost-effective utilization of a wide variety of materials, including many metals, alloys, and polymers.

Several challenges arose during the development of small-scale distillation columns. First, although temperature control issues have been partly solved, they can still be improved. Second, durable long lasting and easily serviceable pumps are desperately needed for continuous runs lasting several weeks. Third, information on polymer types and chemical compatibility, needed for evaluating the chemical compatibility of polymers used in 3D printing, is not readily available even from manufacturers of the printers.

It is clear that, despite all the challenges, miniaturized pilots and additive manufacturing of component parts hold promise. There are huge opportunities in this approach, especially in the framework of enhanced process development, that justify further research on this topic.

6.8 Bibliography

[1] Vogel GH. Process development: From initial idea to the chemical production plant. Wiley, 2006.

[2] Wörz O. Process development via a miniplant. Chem Eng Process Process Intensif, 1995, 34, 261–268.

[3] Fair JR. Distillation: King in Separations. Chem Processing, 1990, 9, 23–31.

[4] Seok DR, Hwang ST. Zero-gravity distillation utilizing the heat pipe principle (micro-distillation). AIChE J, 1985, 31, 2059–2065.

[5] Ziogas A, Cominos V, Kolb G, Kost H-J, Werner B, Hessel V. Development of a Microrectification Apparatus for Analytical and Preparative Applications. Chem Eng Tech, 2012, 35, 58–71.

[6] Fink H, Hampe MJ. In: Ehrfeld W (ed(s)) Microreaction Technology: Industrial Prospects, IMRET 3. Springer-Verlag, Berlin, Germany, pp 664–673.

[7] Tegrotenhuis WE, Stenkamp V. Conditions for fluid separations in microchannels, capillary-driven fluid separations, and laminated devices capable of separating fluids, V.S. 2005 US Patent 6875247B2.

[8] Cypes S, Bergh SH, Hajduk D. Microscale flash separation of fluid mixtures, 2007, WO 2007/033335.

[9] Tonkovich AL, Simmons WW, Silva LJ, Qiu D, Perry ST, Yuschak T, Hickey TP, Arora R, Smith A, Litt RD, Neagle P. Distillation process using microchannel technology, 2009, US Patent 7610775 B2.

[10] Sundberg A, Uusi-Kyyny P, Jakobsson K, Alopaeus V. Modular, Horizontal micro-distillation column. AIChE Annual Meeting, Nashville, USA, 2009.

[11] Sundberg A, Uusi-Kyyny P, Alopaeus V. Novel Micro-Distillation Column for Process Development. Chemical Engineering Research and Design, 2009, 87, 705–710.

[12] MacInnes JM, Ortiz-Osorio J, Jordan PJ, Priestman GH, Allen RWK. Experimental demonstration of rotating spiral microchannel distillation. Chem Eng J, 2010, 159, 159–169.

[13] Lam KF, Cao E, Sorensen E, Gavriilidis A. Development of multistage distillation in a microfluidic chip, LAB CHIP, 2011, 11, 1311–1317.

[14] Cvetković BZ., Lade O, Marra L, Arima V, Rinaldi R, Dittrich PS. Nitrogen supported solvent evaporation using continuous-flow microfluidics, RSC Adv, 2012, 2, 11117–11122.

[15] Jang DJ, Kim YH. A new horizontal distillation for energy saving with a diabatic rectangular column. Korean J Chem Eng, 2015, 32, 2181–2186.

[16] Mardani S, Ojala L, Uusi-Kyyny P, Alopaeus V. Development of a unique modular distillation column using 3D printing. Chem Eng Process Process Intensif, 2016, 109, 136–148.

[17] Sundberg A, Uusi-Kyyny P, Jakobsson K, Alopaeus V. Control of reflux and reboil flowrates for milli and micro distillation. Chem Eng Res Des, 2013, 91, 753–760.

[18] Sundberg A, Uusi-Kyyny P, Jakobsson K, Alopaeus V. Development and modeling of micro distillation column. Proceedings Distillation and Absorption 2010, Eindhoven, the Netherlands, pp 217–222.

[19] Hirvimäki M, Manninen M, Sundberg A, Uusi-Kyyny P, Salminen A. Feasibility of laser material processing in the design and manufacture of small scale devices. MECHANIKA, 2013, 19, 336–343.

[20] Sundberg A. Micro-Scale distillation and microplants in process development, Dissertation, Aalto University, 2015.

[21] Sundberg A. The use of Micro Technology in process development. M.Sc. Thesis, Aalto University Finland, 2007.

[22] Khayargoli P, Loya V, Lefebvre LP, Medra M. The impact of microstructure on the permeability of metal foams. CSME Forum, 2004, 220–228.

[23] Hoffmann A, Mackowiak JF, Górak A, Löning J-M, Runowski T, Hallenberger K. Standardization of mass transfer measurements. Basis for the description of absorption processes. Trans IChemE, Part A, Chem Eng Res Des, 2007, 85(A1), 40–49.

[24] Last W, Stichlmair J. Determination of Mass Transfer Parameters by Means of Chemical Absorption, Chem Eng Technol, 2002, 25, 385–391.

[25] Ojala MS, Sundberg AT, Uusi-Kyyny P, Alopaeus V. Characterization of mass-transfer of a small-scale distillation column. Paper presented at the CHISA 2012 – 20th International Congress of Chemical and Process Engineering and PRES 2012 – 15th Conference PRES.

[26] Rihko LK, Krause AO. Reactivity of isoamylenes with ethanol. Appl Catal A: Gen, 1993, 101, 283–295.

[27] Linnekoski JA, Krause AO, Rihko LK. Kinetics of the heterogeneously catalyzed formation of tert-amyl ethyl ether. Ind Eng Chem Res, 1997, 36, 310–316.

[28] Linnekoski JA, Paakkonen PK, Krause AO, Rihko-Struckmann LK. Simultaneous isomerization and etherification of isoamylenes. Ind Eng Chem Res, 1999, 38, 4563–4570.

[29] Fredenslund Aa. Vapor-liquid Equilibria Using UNIFAC, Elsevier, Amsterdam, 1977, ISBN: 978-0-444-41621-6.

[30] Wilson GM. Vapor liquid Equilibrium. XI. A new expression for the excess free energy of mixing. J Am Chem Soc, 1964, 86, 127–130.

[31] Soave G. Equilibrium constants form a modified Redlich-Kwong equation of state. Chem Eng Sci, 1972, 27, 1197–1203.

[32] Sundberg A, Uusi-Kyyny P, Alopaeus V. The use of Microplants in process development – Case study of the etherification of 2-ethoxy-2-methylbutane. Chem Eng Process, 2013, 74, 75–82.

[33] Verrazi A, Kikic I, Garbers P, Barreau D, Le Roux. Vapour-liquid equilibrium in binary systems ethanol+C4 and C5 hydrocarbons. J Chem Eng Data, 1998, 43, 949–953.

[34] Gmehling J, Menke J, Krafczyk J, Fischer K. A data bank for azeotropic data—status and applications. Fluid Phase Equil, 1995, 103, 51–76.

[35] Arce A, Arce A Jr, Rodil E, Soto A. Isobaric vapor–liquid equilibria for systems composed by 2-ethoxy-2-methylbutane, methanol or ethanol and water at 101.32 kPa. Fluid Phase Equilib, 2005, 233, 9–18.

[36] Heine A, Fischer K, Gmehling J. Various thermodynamic properties for binary systems with tertiary ethers. J Chem Eng Data, 1999, 44, 373–378.

[37] Ishii N. The volatility of fuels containing ethyl alcohol. VII. Total and partial partial vapor pressures of mixtures of ethyl alcohol and pentane. J Soc Chem Ind Jpn, 1935, 38, 705–707.

[38] Campbell SW, Wilsak RA, Thodos GJ. (Vapor + liquid) equilibrium behavior of (n-pentane + ethanol) at 372.7, 397.7, and 422.6 K. J Chem Thermodyn, 1987, 19, 449–460.

[39] Collins SG, Christensen JJ, Izatt RM, Hanks RW. The excess enthalpies of 10 (n-pentane + an n-alkanol) mixtures at 298.15 K. J Chem Thermodyn, 1980, 12, 609–614.

[40] Sundberg A, Uusi-Kyyny P, Jakobsson K, Alopaeus V. Control of reflux and reboil flowrates for milli and micro distillation. Chem Eng Res Des, 2013, 91, 753–760.

Carlo Edgar Torres-Ortega and Ben-Guang Rong

7 Integrated biofuels process synthesis: integration between bioethanol and biodiesel processes

Abstract: Second and third generation bioethanol and biodiesel are more environmentally friendly fuels than gasoline and petrodiesel, and more sustainable than first generation biofuels. However, their production processes are more complex and more expensive. In this chapter, we describe a two-stage synthesis methodology for integrating both biodiesel and bioethanol processes. In the first stage, to minimize unit production costs we screened different technological paths by formulating a mixed integer nonlinear problem superstructure solved in GAMS. In the second stage, we intensified one portion of the optimal technological path. We used the concept of column section recombination and employed Aspen Plus V8.8 and its economic evaluation tool to evaluate the structural changes. The first stage identified the optimal technological routes and the integration of bioethanol (30% used for biodiesel process), glycerol (10% used for bioethanol process), and steam and electricity from combustion (54% used as electricity) in the bioethanol and biodiesel processes. In the second stage, we saved about 5% in equipment costs and 12% in utility costs for bioethanol separation. This dual synthesis methodology, consisting of a top-level screening task followed by a down-level intensification task, proved to be an efficient methodology for integrated biofuel process synthesis. The case study illustrates and provides important insights into the optimal synthesis and intensification of biofuel production processes with the proposed synthesis methodology.

Keywords: Bioethanol, biodiesel, MINLP, intensification, synthesis, integration, optimization

7.1 Introduction

7.1.1 Energy world consumption projections

According to different projections, an increasing population and development of the world economy, mainly of non-OECD countries, will increase the worldwide energy demand by 48% from 2012 to 2040 [1] (Figure 7.1). Of the primary energy sources, fossil fuels will remain the dominant source of energy for powering the global economy, providing about 60% of the growth in energy. Gas will be the fastest growing fossil fuel, in contrast to coal, which will suffer a sharp drop in use. Of the nonfossil fuels, it is expected that renewables, including biofuels, will grow rapidly, increasing from 3% today to 9% by 2035 [2].

The global demand for liquid fuels (oil, biofuels, and other liquids) is projected to increase roughly six times by 2035. This increased demand will come from emerg-

https://doi.org/10.1515/9783110465068-007

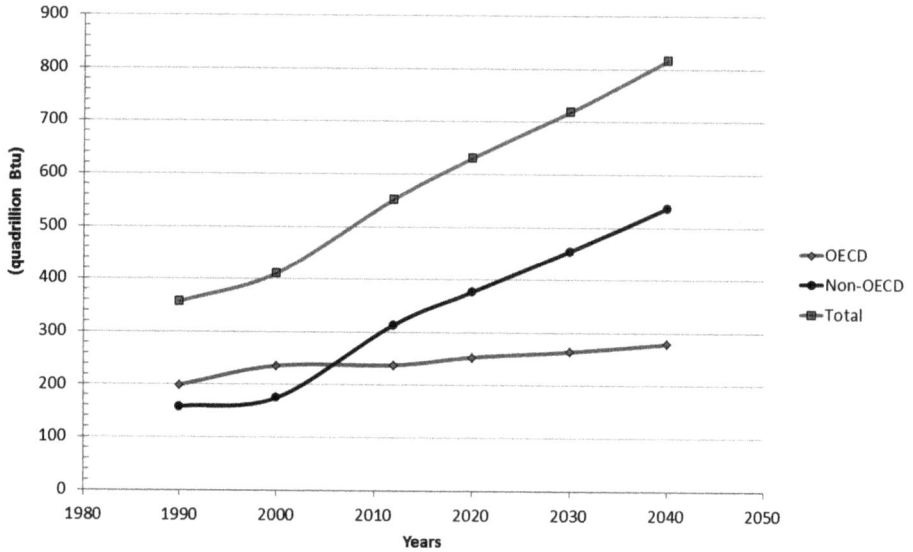

Fig. 7.1: Projection of worldwide energy consumption until 2040 (data taken from *EIA International Energy Outlook 2016* [1]).

ing economies, with China and India accounting for over half of the increase [2]. The growth in global consumption of liquid fuels is driven by transportation and industry, with transportation accounting for almost two-thirds of the increase.

7.1.2 Worldwide transportation sector

In the non-OECD regions, where 80% of the world's population resides, the demand for transportation energy is forecast to almost double between 2012 and 2040 [1]. According to the US Energy Information Administration's projections on the consumption of liquid transport fuels between 2012 and 2040, diesel (including biodiesel) will show the largest gain, followed by jet fuel and motor gasoline (including bioethanol). Motor gasoline will still account for the largest consumption of transportation fuel. Passenger transportation, in particular light-duty vehicles, will consume the most transportation energy, with light-duty vehicles consuming more energy than all modes of freight transportation, including heavy trucks, marine, and rail combined. Figure 7.2 depicts the 2012 consumption of liquid transport fuels by mode. The transportation sector accounted for 25% of the total global delivered energy consumption in 2012 [1].

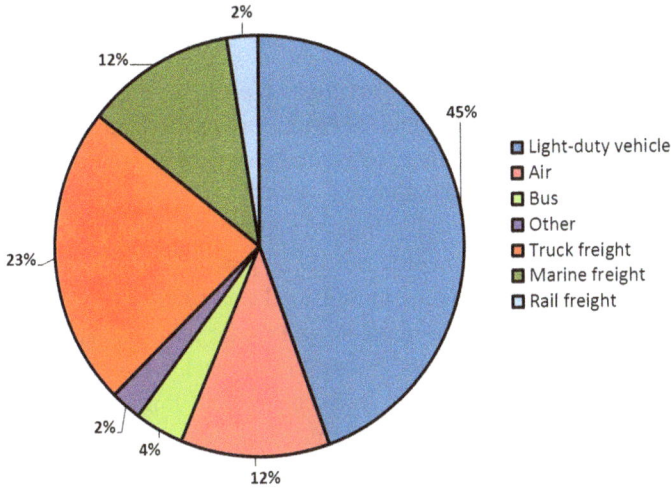

Fig. 7.2: World transport energy consumption by mode in 2012 (data taken from *EIA International Energy Outlook 2016* [1]).

7.1.3 Biofuel potential

Growth of the bioindustry fuel sector is driven by the needs to reduce reliance on fossil fuels, decelerate climate change, increase fuel security, and develop a wider range of bioproducts for a growing global population [3]. One sustainable alternative for addressing these needs is to convert an abundant natural resource such as biomass into liquid transportation fuels.

Technology development has reached either production scale or demonstration scale for the following four types of biomass:
–   energy crops
–   forest products
–   agricultural residues
–   animal manure

Considering that these types of resources are more widely distributed than fossil fuels, their use as raw material could have dual potential as substitutes for the following:
–   fossil fuels in all end-use sectors and in power and district heat generation
–   biomass in traditional usages (e.g., wood and agricultural residues for cooking and heating in the residential sector)

7.1.4 Rural and industrial market and development

From a social point of view, biofuel production can create a positive social synergy. According to a recent study by the US Department of Agriculture, rural counties with bioethanol plants can attribute 32% of countywide employment growth from 2000 to 2008 to the development and operation of ethanol refineries [4].

The largest producers of bioethanol are the USA and Brazil. The USA produced about 60% of the global total in 2012 from corn, and Brazil produced about a quarter from sugar cane. Global biodiesel production grew 47-fold between 2000 and 2013, with Europe leading the growth. The biofuels mandate and the large use of light-duty diesel vehicles in Europe have driven this rapid growth in biodiesel. Rapeseed is the feedstock for more than half the global biodiesel production. Regarding biofuel consumption, bioethanol consumption grew from 272 PJ in 2000 to 1426 PJ in 2010 (18% increase per year); meanwhile, biodiesel grew from 18 to 616 PJ (42% increase per year).

Advanced biofuels based on lignocellulosic crops (wood and straw) produce ethanol and diesel substitutes from the woody parts of existing food crops and from crops that thrive in land that is unsuitable for food crops. Production of advanced biofuels is just taking off, with high investment and production prices and only a few plants in operation worldwide. About 0.2% of total biofuel production in 2012 was of this type [5].

Some of the main strategic challenges in the biomass supply chain are listed below:

– Differences in the natural environment and in human activity mean that biomass resources are distributed unevenly across regions.
– Large-scale international trade in bioenergy requires the transport of high-density commodities at low cost and using efficient logistic systems that consider seasonal fluctuations in supply.
– The difference between the pace of food demand and yield growth defines land availability.
– Optimal extraction of agricultural and forest residues is needed. An increase in residue use could limit land expansion and reduce the problem of whether land should be used for food or energy. However, if extraction is too high, degradation of the soil could occur.
– Logistics and biomass price affect the supply chain. Every feedstock supply chain system depends on locally available feedstock and is therefore affected by its price, regional characteristics, logistic radius defining the biorefinery capacity, ethanol selling price, global warning impact, etc. [6].
– Seasonality of biomass supply could necessitate multifeedstock biorefineries. A big challenge is the uncertainty of a continuous supply of biomass feedstock throughout the year and the whole plant lifetime. Thus, the possibility of a biore-

finery that can work with different feedstocks should be considered and evaluated from the conceptual stage. The effect of different feedstocks on the products should also be studied [7].

7.1.5 Environmental situation

The annual rate of growth of carbon dioxide emissions is now less than half that of the past 20 years, reflecting gains in energy efficiency and small changes to fuels. However, emissions continue to rise, suggesting the need for further action. Despite the slowdown in emissions growth, the level of carbon dioxide emissions continues to grow and is estimated to increase by 20% between 2014 and 2035.

With bioenergy demand estimated to double between 2010 and 2030, there are important environmental issues that have to be considered. Plants convert CO_2 from the atmosphere into biomass. The carbon stored in biomass in called biogenic carbon. When the biomass of the plants is combusted, the biogenic carbon is released into the atmosphere to complete the cycle. If this biogenic carbon is sequestered by plants, then the system is in balance. This means that the amount of CO_2 in the atmosphere does not increase. When short-rotation energy crops or agricultural residues are used as fuel, a balanced carbon cycle is obtained because the crops grow/renew themselves annually. Moreover, energy crops could also contribute to emission reductions if they are sustainably cultivated on surplus land or land unsuitable for food crops. However, not all scenarios present a positive balance. The use of forest residues could result in negative effects because the rate of carbon sequestration into biomass is slower than the rate of combustion of forest residue. However, sustainable forest management strategies such as thinning and prevention of fires could contribute to accelerated forest growth. Another negative scenario is the change of forestland into agricultural land for bioenergy crops. This change implies the harvest of biomass, which stores less carbon than forests, potentially increasing CO_2 emissions in the atmosphere.

In addition to greenhouse gas (GHG) emissions, there are a number of other environmental issues related to sustainability of bioenergy. Some of the most relevant are listed below:

- Changes in land use have direct effects when bioenergy crops are grown on land not previously used for cropland or farming.
- Changes in land use can have indirect effects such as an increase in agricultural commodity prices, food security, land use change in other regions, etc.
- Expanding bioenergy use can increase water stress because the water for biomass production does not necessarily go back into the system.
- Changes in biodiversity can occur.
- Soil quality is affected by the use of fertilizers, high agricultural residue extraction rates, etc.

7.1.6 Fuel properties of bioethanol and biodiesel

Analysis of the physical and chemical properties of bioethanol and biodiesel is necessary when they are considered as either fuel or fuel additives. Table 7.1 presents an overview of the fuels properties. We need to identify their limitations and explore new possibilities to overcome these disadvantages. The main advantages and disadvantages of bioethanol and biodiesel are listed next.

Main advantages of bioethanol:
- higher octane number and oxygen content
- lower soot, carbon oxide, and unburned hydrocarbon emissions compared with fossil fuels
- broader flammability limits, higher flame speed, higher heat of vaporization, higher compression ratio, and shorter ignition timing
- biodegradable and does not contain toxic substances such as aromatic compounds

Main disadvantages of bioethanol:
- miscible with water and therefore has a corrosive effect on engine components
- undesirable effect on electric fuel pumps because it increases internal wear and generates sparks
- lower heating value and lower energy density than conventional fossil fuels
- pure ethanol is difficult to vaporize, which leads to difficult vehicle start-up in cold weather
- higher tendency to reach the crankcase and contaminate engine oil

Main advantages of biodiesel:
- produces less emissions, specifically CO_2, sulfur compounds, and particulate matter
- better lubricity properties, which prolong engine life
- agriculture-orientated, nontoxic, biodegradable, and renewable
- high cetane number

Main disadvantages of biodiesel:
- increased emissions of nitrogen oxides, which could result in formation of smog and acid rain
- lower heating value than petrodiesel
- use of farm land to grow biodiesel crops could result in increased cost of food, and food scarcity
- higher cloud and pour points, which represents a challenge for use in cold regions
- oxidative instability due to exposure of unsaturated fatty acid chains to oxygen

Tab. 7.1: Fuel properties of bioethanol, biodiesel, and their fossil fuel counterparts (data retrieved from [8]).

Fuel property	Units	Bioethanol	Gasoline	Biodiesel	Diesel
Density at 15 °C	$kg\,m^{-3}$	790	737	880	837.3
Kinematic viscosity at 40 °C	$mm^3\,s^{-1}$	1.13	0.593	6	2780–3180
Cetane number	–	5.8	10.0–15.0	47–65	47.64–53.90
Oxygen	mass %	34.7	0–4	11	–
Octane number	–	110	86–94	−25	10.0–30.0
Latent heat of vaporization	$kJ\,g^{-1}$	921.1	289	330	370
Calorific value	$MJ\,kg^{-1}$	25.22–26.70	34.84	39.4	43.8
Flash point	°C	13	−43	170	64.74
Auto-ignition temperature	°C	332.8–420.0	257	225	230
Water content	$mg\,kg^{-1}$	2024	–	–	50

In this chapter, we discuss how to implement process synthesis and process intensification for the conceptual design of processes for simultaneous production of bioethanol and biodiesel. As described above, bioethanol and biodiesel are two of the most promising biofuels for partial replacement of fossil fuels, and are already produced industrially. Moreover, we consider environmental impact aspects such as production of GHG emissions, land use, and competition with food production. Biomass for advanced (second and third generation) biofuels could provide enough raw material to fulfill production requirements and minimize environment impact. Therefore, we focus on advanced bioethanol and biodiesel for further analysis.

We first describe the most relevant production processes for bioethanol and biodiesel. Representative nonedible crops, forest residues, and agricultural residues are selected as raw materials. We describe the possibilities for integration between both biofuel processes, considering mass and energy integration. Next, we describe the synthesis and intensification methodologies for two consecutive stages. The first stage performs an optimal synthesis over different production technological routes or paths. The second stage examines the selected technological routes in detail to determine relevant process variables for process integration and intensification. We close the chapter with a case study, in which the two-stage methodology was used to solve the problem of synthesis and intensification of integrated bioethanol and biodiesel production processes.

7.2 Lignocellulosic bioethanol production process

A sustainable production process for bioethanol involves the implementation of second-generation raw materials (i.e., lignocellulosic materials) because they are cheap, abundant, renewable, and do not negatively affect the human food supply [9]. However, these convenient traits are compromised because of the complex nature of these raw materials, meaning that processing is more expensive than for their starch counterparts. Lignocelluloses are composed of cellulose, hemicelluloses, and lignin; in other words, they have an intricate structure, which is recalcitrant to decomposition. Typical compositions of various lignocelluloses are depicted in Table 7.2.

The total production process for lignocellulosic bioethanol is depicted in Figure 7.3. After dealing with the distribution logistics of lignocellulosic biomass, there are six major production process steps: biomass handling, pretreatment, hydrolysis, fermentation, separation or recovery, and dehydration of bioethanol. Brief descriptions of the state of the art of these process steps are given in Sections 7.2.1 to 7.2.4.

7.2.1 Biomass handling

The transformation of biomass into fuels or chemicals at industrial level implies the need for robust and reliable handling of bulk solids. Processes involving solids are more challenging than those involving fluids. Processes involving solids typically have significantly longer start-up times and larger start-up costs.

The particular case of biomass is more complicated in the sense that there may be relevant differences between plant species in their composition, physical and chemical properties, moisture content, contaminants, etc. In addition to the inherent feedstock properties, pretreatments (milling, densification, thermal treatment, etc.) before starting the chemical or biochemical conversion can significantly change the physical properties of biomass solids, again altering the handling. The differences in solid properties between species and stages in the same process make biomass more unpredictable than conventional granular materials processed by industry, especially regarding flow reliability and control [11]. Therefore, this kind of material often re-

Tab. 7.2: Content of main carbohydrates in popular lignocellulosic feedstocks (adapted from [10]).

Feedstock	Content (wt% dry)		
	Cellulose	Hemicellulose	Lignin
Corn stover	35.1–39.5	20.7–24.6	11.0–19.10
Hardwood stems	40–55	24–40	18–25
Rice straw	29.2–34.7	23–25.9	17–19
Rice husk	28.7–35.6	11.96–29.3	15.4–20
Wheat straw	35–39	22–30	12.0–16
Newspaper	40–55	24–39	18–30
Pine	42–49	13–25	23–29
Switchgrass	35–40	25–30	15–20
Softwood stems	45–50	24–40	18–25

quires semi-empirical approaches for correlating discharge and flow with particle size distribution.

Key developments in biomass solid handling are robust and reliable characterization methods for flow properties (offline) and procedures (preferably nondestructive) for characterization of biomass structure and composition (online). The relevance of robust and reliable characterization methods for biomass solids is related to the lack of standardized equipment for handling solid biomass. Most conventional solid-handling equipment is designed for common granular solids, such as coal. Conventional granular material is nonfibrous, rigid, homogenous in particle shape, and with particle sizes of less than a few millimeters. However, few biomass materials meet those criteria. For example, technical problems regarding biomass with a tendency to arch or bridge across hopper outlets compromise the storage and flow of material in silos or in conical or wedge-shaped hoppers [12].

Safety and dust generation are also key factors during biomass handling [13]. Fires (self-heating during storage) and dust explosions are latent issues that can result in worker injuries, loss of life, and economic or environmental problems. Thus, venting devices to release pressure, modification of the operating pressure of the vessels, and other issues are of special concern.

In conclusion, factors such as a better understanding of biomass particles (characterization), the relationship between vessel size, fill height, and diameter on the arching behavior, and venting devices are crucial for the design of reliable and safe procedures.

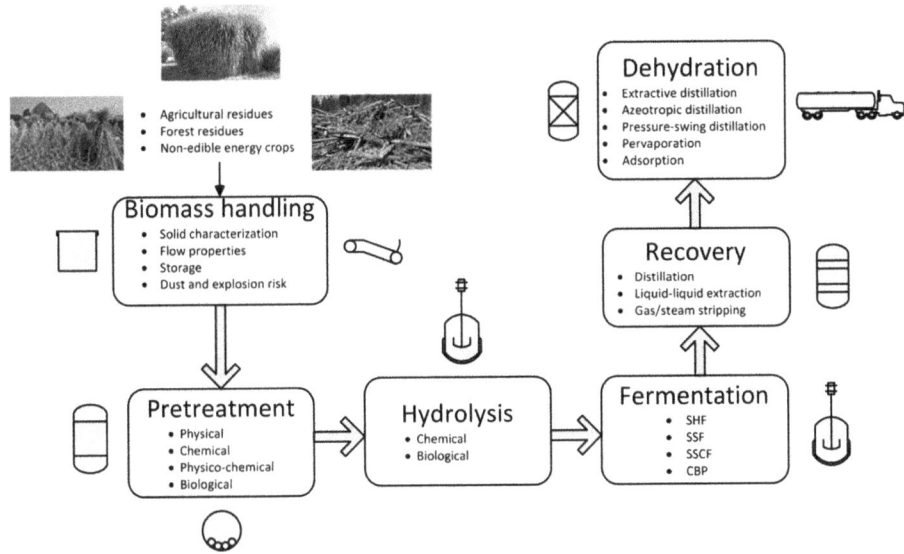

Fig. 7.3: Flowchart of the total production process of bioethanol from lignocellulosic materials.

7.2.2 Pretreatment of lignocellulosic materials

The main target of pretreatment is to break down the lignin seal to make cellulose and hemicellulose more accessible to enzymatic hydrolysis for conversion. This step is crucial for the entire production process and is costly. Suitable pretreatment must disrupt the hydrogen bonds in crystalline cellulose, break down the crosslinked matrix of hemicelluloses and lignin, and raise the porosity and accessible surface area of cellulose for saccharification [14]. The choice of a suitable pretreatment process depends on the feedstock and its economic assessment and environmental impact.

We can classify pretreatments as follows, according to the principles used to disrupt lignocellulose structures:

Physical pretreatment: Methods include grinding and milling, microwave irradiation, and extrusion. These methods increase the surface area and reduce the particle size of lignocellulosic materials. Usually, they are employed in combination with other pretreatments.

Chemical pretreatment: Alkalis, acids, organosolvs, ozonolysis, and ionic liquids (green solvents) can be used. This is the most studied pretreatment technique, especially dilute acid pretreatment.

Physico-chemical pretreatment: Procedures include steam explosion, liquid hot water treatment, ammonia fiber explosion, wet oxidation, and CO_2 explosion. These pretreatments combine chemical and physical processes.

Biological pretreatment: Biomass can be treated with white or brown fungi and bacteria. This method employs microorganisms that degrade lignocellulose to modify the biomass. Biological pretreatment is promising because does not require chemicals, needs low energy input, operates at mild conditions, and is environmentally friendly. However, its main disadvantages are a very slow process speed and the need for intense control [10].

A negative side effect of pretreatment is the formation of inhibitory compounds, mainly during chemical pretreatment, which inhibit further saccharification and the fermentation process. These compounds are grouped into three categories: weak acids (acetic and formic acids, etc.), furan derivatives (hydroxymethylfurfural and furfural), and phenolic compounds (aldehydes, benzoic acids, etc.)

As the reader will guess, it is possible to combine pretreatment methods to overcome their stand-alone limitations and drawbacks. Examples are pretreatments using alkalis and acids, alkalis and ionic liquids, and dilute acids and steam explosion.

7.2.3 Hydrolysis of cellulose and hemicellulose, and fermentation strategies

Cellulases and hemicellulases hydrolyze cellulose and hemicellulose to generate soluble fermentation sugars. However, the lignin interferes with hydrolysis by blocking the access of enzymes to cellulose and hemicellulose. Despite intensive research over the past few decades, the enzyme hydrolysis step remains a major techno-economic bottleneck in lignocellulosic ethanol processes.

Fermentation can be performed in a variety of configurations, such as separate hydrolysis and fermentation (SHF), simultaneous saccharification and fermentation (SSF), simultaneous saccharification and cofermentation (SSCF), and consolidated biomass processing (CBP). The following is a brief description of these fermentation strategies:

- SHF is the conventional two-step process in which lignocellulose is first hydrolyzed in a vessel by cellulases to form the reduced sugars; then, these sugars are fermented to bioethanol in a second vessel using various yeasts. The main advantage of this process is that each step can operate at its optimum conditions, commonly at different operating conditions.
- SSF is an intensification step in which saccharification and the subsequent fermentation of sugars to bioethanol by yeasts such as Saccharomyces or Zymomonas take place in the same vessel. On the one hand, it is challenging to find appropriate operating conditions that are convenient for both processes. On the other hand, there might be certain synergy by increasing the hydrolysis rate of cellulose and hemicellulose because of removal of end-product inhibition.
- SSCF takes SSF into a further intensification step by including the fermentation of pentoses, which usually represent a relevant proportion of the total reduced sugars formed. Techno-economic studies such as that conducted by Sassner et al. [15] demonstrate the importance of utilizing the pentose fraction for bioethanol production to obtain good process economy and high bioethanol yield. The total xylose uptake could be increased from 40% to as much as 80% by controlling the enzyme feed [16].
- CBP represents the highest degree of intensification of hydrolysis and fermentation. One single process includes the production of cellulases and hemicellulases, hydrolysis of cellulose and hemicellulose to monomeric sugars, and the fermentation of hexose and pentose sugars. CPB avoids the capital costs, substrate and other raw material costs, and utility costs associated with cellulase production. However, CBP requires a microbial culture that can handle all these tasks, including substrate utilization and product formation; so far, there is no microbial culture able to do this at a sufficient level.

7.2.4 Separation and dehydration of bioethanol

After fermenting monomeric sugars into bioethanol, subsequent steps involve the separation and dehydration of bioethanol for engine use. The first step is separation of bioethanol from the fermentation broth. Commonly, this broth contains 5–10 wt% of bioethanol together with gas, a large amount of water, soluble organics, and solids.

The gas is mainly CO_2 generated after fermentation and the soluble organics are compounds such as unconverted sugars, side products of fermentations, inhibitors, etc. The solids fraction contains unconverted lignocellulosic material, ash, and some side products of pretreatments.

Gas can be vented during fermentation and/or further separated by flashes or distillation columns. Solids can be separated either by filters, especially if they are valuable and the flow rates are not high, or by centrifuges. Usually, the organic soluble compounds are separated together with most of the water. Typically, bioethanol is concentrated up to 95.63 wt% purity through distillation, close to the ethanol–water azeotrope, but it also is possible to recover bioethanol using gas/steam stripping or liquid–liquid extraction. However, distillation is preferred when bioethanol is considered as a bulk chemical for massive production. The presence of an azeotrope implies a thermodynamic limitation when changing the composition of the components in both liquid and vapor phases, meaning that it is not possible to achieve higher bioethanol purity through conventional distillation.

After obtaining a concentrated stream of bioethanol and water, there are several technological alternatives for dehydrating the bioethanol and achieving 99.5 wt% purity, as specified for fuel for car engines. Among the different possibilities are pervaporation and adsorption as energy-saving alternatives, but with inconvenient capital costs and separation factor levels when large scale processing is required [17]. Pressure-swing distillation is not convenient because the bioethanol–water azeotrope has low sensitivity to pressure changes. As alternatives to ordinary distillation, azeotropic distillation and extractive distillation can break the azeotrope and handle large-scale flows, but at the expense of low energy efficiency. A brief description of these alternatives is given below:

Gas/steam stripping: The removal of alcohol from fermentation broth by transferring the alcohol into a gas/steam stream is relatively simple, and can operate at mild conditions. However, the alcohol transferred to the gas/steam phase must later be recovered by condensation and membrane separation or adsorption.

Extraction: The liquid–liquid extraction process can be coupled to the fermenter to extract the bioethanol and increase the fermentation yield by reducing bioethanol inhibition. However, the operation faces the technical challenges of finding a solvent with low solubility in the aqueous phase, short separation time, chemical inertness, and environmental friendliness.

Pervaporation: This technique uses membranes to separate bioethanol from water. Energy is only required for evaporation of the permeate. However, the separation capacity is low for most organic membranes, requiring a reduced concentration of water at the entrance.

Adsorption: This method operates under two general strategies: liquid-phase or vapor-phase with adsorbents such as zeolites and silica gel. Implementation requires a simple process design, with long service life. However, it is efficient only when there is a low concentration of water in the feed stream.

Pressure-swing distillation: This system typically consists of two distillation columns and applies changes of pressure to displace the azeotrope of the mixture. However, the bioethanol–water azeotrope is not very sensitive to changes in pressure.

Azeotropic distillation: In this distillation system, an external substance, forming new homogeneous or heterogeneous azeotropes with the components of the initial mixture, is introduced to change the original volatilities. It is important that the resulting azeotrope should be either easy to separate or reusable without being separated. For the bioethanol–water azeotrope, benzene, cyclohexane, or toluene are typically used [18].

Extractive distillation: Here, an external agent is added to change the relative volatilities of the original components without forming new azeotropes. With an appropriate high-boiling separating agent, this separation method is not energy intensive because the separation agent is not evaporated. Typical liquid external agents for the bioethanol–water azeotrope are ethylene glycol and glycerol. Dissolved salts, mixtures of separation agents and salts, ionic liquids, and highly branched polymers have been also proposed as separating agents [19, 20].

7.3 Fatty acid ethyl esters: biodiesel production process

The use of nonedible oils such as algae oil, toxic vegetable oils, and waste oil can help to reduce competition with food production. From 2004 to 2007, 34% of edible oil produced worldwide was employed to produce biodiesel; it is projected that from 2005 to 2017 more than a third of edible oil production will be used for biodiesel production [21].

Algae are one of the most promising sources of nonedible oils. Laboratory studies have shown appealing characteristics such as fast growth and higher oil content (up to 80 wt% dry based) than any oil crop. They have the potential to yield 100 times more biodiesel than soybeans [22]. In addition, algae can grow in open ponds, creating zero demand for arable land. However, harvesting algae is highly energy consuming and estimated to account for 20–30% of the total production cost.

Of the different possible nonedible oil sources, those with a high percentage of monounsaturated fatty acids are the best options for biodiesel production [23]. Some nonedible oils and fats with a high content of monounsaturated fatty acids are presented in Table 7.3.

The use of vegetable oils and their blends directly as fuel in diesel engines is impractical because of their high viscosity and acidity, in addition to the formation of gum during storage and combustion. One way to address these disadvantages is the reaction between alcohols and oils/fats to form alkyl esters. Methanol is the preferred alcohol for study and application because of its low price and zeotropic nature with water. However, bioethanol has attracted recent interest because of advantages such as low toxicity and derivation from renewable resources. Moreover, certain physical

Tab. 7.3: Main fatty acid components of common oils and fats rich in monounsaturated fatty acids (adapted from [24]).

Fatty acid	Fatty acid content (wt%)						
	Jatropha curcas oil	Neem oil	Caper spurge	Rice bran seed oil	Honge oil	Tallow oil	Poultry fat
Palmitic	14.20	18.10	6.80	17.70	10.60	23.30	22.20
Stearic	7.00	18.10	2.20	2.20	6.80	19.40	5.10
Oleic	44.70	44.50	81.46	40.60	49.40	42.40	42.30
Linoleic	32.80	18.30	34.60	35.60	19.00	2.90	19.30
Saturated	21.60	37.00	8.78	22.00	29.20	45.60	27.30
Monounsaturated	45.40	44.50	82.16	40.83	51.80	42.40	50.70
Polyunsaturated	33.00	18.50	6.49	37.40	19.00	3.80	20.30

properties of fatty acid ethyl esters (FAEEs) derived from bioethanol are better than those of fatty acid methyl esters (FAMEs) derived from methanol, such as cloud and pour points, heat content, and cetane number [25].The methoxide ion is more reactive than the ethoxide ion, thus making the reaction rate of methanolysis higher than that of ethanolysis [26]. This is mainly attributed to the greater steric hindrance effect of (bio)ethanol compared with methanol. However, some reaction technologies seem to minimize this difference, such as supercritical transesterification, ultrasonic irradiation, and microwave irradiation.

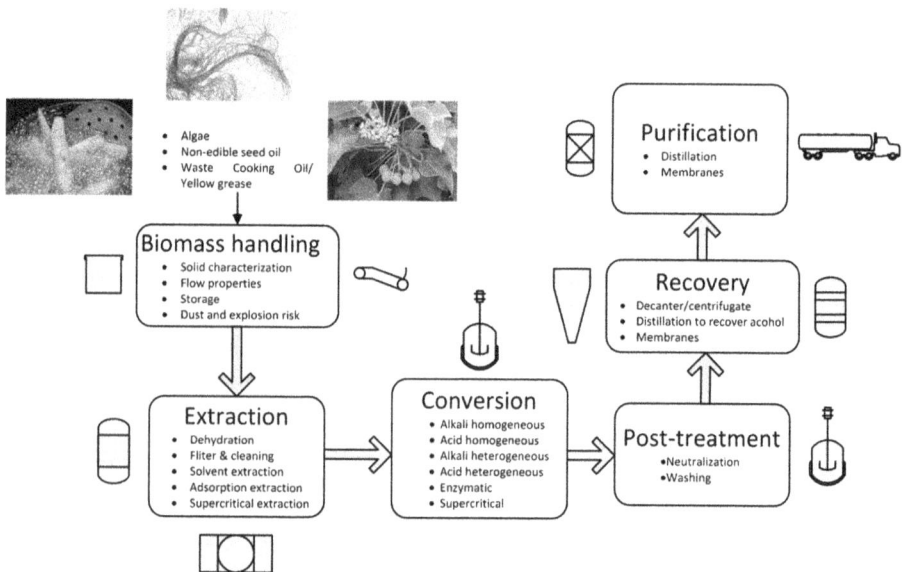

Fig. 7.4: Flowchart of the total production process for biodiesel from nonedible oils/fats, with ethanol as acyl acceptor.

An overview of the general biodiesel production process is depicted in Figure 7.4. Second and third (algae) generation raw materials are selected. Briefly, the oil first has to be extracted and mechanically cleaned from the raw material. Then, it is converted together with a surplus of bioethanol to FAEEs through different technological alternatives. After that, and depending on the type of conversion selected, further post-treatment occurs before recovery and purification. More details of the process are given in Sections 7.3.1 to 7.3.3.

7.3.1 Extraction and conversion of oils

Extraction of seed oils usually involves a combination of processes such as pressing and solvent extraction. First, seeds usually undergo a preheating step and then are mechanically pressed in expellers, which extracts about 70–75% of the oil. Seeds then undergo a counter-current extraction process in a multistage unit with an organic solvent such as chloroform, hexane, cyclohexane, acetone, or benzene until most of the oil is recovered (about 99%).

In the case of waste cooking oil, filtration to remove solid impurities is required to obtain clear oil that consists of a mixture of triglycerides (TGs), diglycerides (DGs), monoglycerides (MGs), and free fatty acids (FFAs). The composition varies according to the effects of temperature during use of the oil.

For extraction of algae oil, concentration and dehydration are indispensable pretreatment steps before oil extraction. Techniques such as centrifugation, flocculation, sedimentation, and filtration are first employed to concentrate the algal biomass. After that, drying methods such as spray-drying, drum-drying, freeze-drying, and sun-drying using solar drying devices are used [27].

7.3.2 Conversion of oil into alkyl esters

The first and most common industrially used reaction for conversion of oil into alkyl esters is transesterification. The mechanism of this reaction is summarized in Figure 7.5.

The conversion of oils into biodiesel can be carried out using homogeneous catalysis, heterogeneous catalysis, or supercritical fluids, as described next.

Homogeneous catalysis

Acid conditions (esterification): Acid homogeneous conditions are suitable for feedstocks with high content of FFAs. The most common acid used is sulfuric acid. However, this method is sensitive to the presence of water, which reduces ester yield and completely inhibits it at 50 g kg^{-1} water concentration [28]. In general,

Desired reaction path   TG + R'OH ⬌ DG + FAEE ⬌ MG + FAEE ⬌ Glycerol + FAEE

Mechanism in every reversible step
- 1.- Production of tetrahedral
- 2.- Breakdown of tetrahedral into FAEE and DG/MG/Glycerol
- 3.- Recovery of catalyst by proton transfer

Undesired reaction path: water   TG + water ⟶ DG + FFA

Undesired reaction path: FFA   FFA + alkali catalyst ⟶ Soap + Water

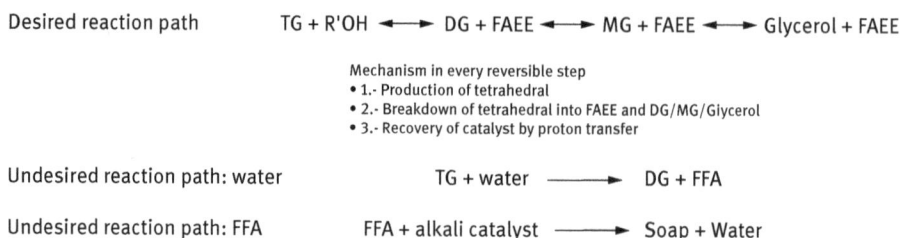

Fig. 7.5: Summary of desired and undesired reactions occurring during transesterification.

the acid reaction is much slower than the alkaline reaction and requires higher oil-to-alcohol molar ratio.

Alkali conditions (transesterification): Common catalysts are alkaline metal hydroxides, alkoxides, and sodium and potassium carbonates. These catalysts show high conversion of triglycerides when the amount of FFAs and water is low. However, the presence of water and/or FFAs leads to saponification reactions. A technical problem related to alkali transesterification using (bio)ethanol is related to the degree of emulsion, which is higher than for methanol. This higher emulsion hinders product separation and downstream processing. One way to address the problem of high FFA content in the feedstock is to combine esterification with acid catalysis to convert FFAs to alkali esters, followed by transesterification with alkali catalyst.

Heterogeneous catalysis

Heterogeneous catalysts can be recycled and re-used several times, with simpler and downsized separation processes for the final product. Techno-economic studies at different production capacities, such as reported by Sakai et al. [29], have shown that heterogeneous processes are economically superior because they allow processing of lower cost raw materials, give higher quality products and byproducts, and require simpler downstream processing.

Acid heterogeneous transesterification: This system can perform transesterification and esterification reactions simultaneously and is therefore suitable for treatment of low quality oil feedstocks. Acid heterogeneous catalysis can operate at low (45–75 °C) or high (200–300 °C) temperatures, giving over 90% yield [30].

Alkali heterogeneous transesterification: The activity of a heterogeneous catalyst resembles its homogeneous counterpart at the same operating conditions. However, Lee and Saka [31], among others, reported low tolerance to high contents of FFA and water in the oil feedstocks. In addition to industrial alkali catalysts, the use of natural renewable and low cost sources has been reported. Examples of these natural sources are materials rich in calcium, such as eggshell, limestone calcite, dolomite, and fly ash-based catalysts, yielding over 85% conversion to biodiesel [32].

Enzymatic transesterification: This system employs enzymes generated from microorganisms, animals, and plants. In general, this kind of process is highly selective and gives a high yield of biodiesel, while operating under mild conditions (30–40 °C). The recovery of glycerol is simple (immobilized enzymes) and gives high purity product. In addition, this biocatalysis can be employed on raw materials with a high FFA content (up to 80%) [33] and high water content (40–300 g kg^{-1}). Nevertheless, its commercial application is very limited because of the high cost of lipase catalyst, long reaction times, and limited re-use of the lipases.

Supercritical fluid technology

Conventional processing methods have to deal with technical drawbacks such as the presence of water in the mixture, catalyst consumption and recovery, low purity glycerol, and generation of wastewater. When the reaction is carried out under supercritical conditions, the mixture becomes homogeneous and esterification and transesterification occur simultaneously without the presence of catalyst. Small amounts of water can actually enhance biodiesel yield [34]. Regarding FFA content, some processes intentionally hydrolyze the raw materials to preform FFAs before the supercritical transesterification process to avoid side reactions and isomerization of the esters. However, the challenges of this technology are related to the severe conditions (280–400 °C and 10–30 MPa) and high alcohol-to-triglyceride ratio [35].

7.3.3 Separation and purification of biodiesel

The basic separation and purification steps are recovery of unreacted alcohol, purification of fatty acid alkyl esters from the catalyst, and separation of glycerol as main byproduct. Conventionally, the unreacted alcohol is recovered and sent back just after reaction to minimize the presence of alcohol in the wastewater effluent. The next separation step is recovery of biodiesel from glycerol, employing simple techniques such as gravitational settling or centrifugation. After transesterification, the addition of water can enhance the separation of alkyl esters and glycerol. Moreover, a water wash, with addition of either alkali or acid, can function as a means to neutralize the catalyst. After washing, the alkyl ester must be dried to achieve the required biodiesel specifications.

In addition to the units already mentioned, hybrid units have been proposed for enhancing biodiesel separation. The most known are membrane reactors based on inorganic microporous ceramic membranes. A membrane reactor removes unreacted oil, shifting the reaction equilibrium to the product side. Reactive distillation is another hybrid technology, which consists of a multifunctional reactor where chemical reaction and thermodynamic separation are combined. A benefit of this technology is a reduction in the alcohol-to-oil ratio as the forward shift of the reaction equilibrium is driven by the continuous removal of byproducts. In consequence, shorter reaction times and higher productivities are reported, together with simpler downstream processing because solid acids are used instead of homogenous catalyst [31].

7.3.4 New uses for glycerol

Glycerol is the major byproduct in biodiesel production, with a yield of at least 10 wt%. Thus, it is necessary to develop convenient uses for glycerol, with the aim of reducing the total production costs for biodiesel. Zhou [36], among others, has proposed different alternatives that range from syngas to more ester production, including several intermediate commodities. In practice, there are two main strategies for using glycerol in large quantities: obtaining commodities from glycerol and producing oxygenated additives for fuels. Among the possible commodities, epichlorohydrin and acrolein are intermediates in production of polymers such as epoxide resins, vinyl, and styrene. The most promising fuel additive is glycerol tertiary butyl ether (GTBE) for biodiesel, and gasoline. GTBE can lead to a reduction in emissions, mainly of NOx and unburned hydrocarbons. In addition, in the case of biodiesel, GTBE could reduce viscosity and improve cloud and pour points.

7.4 Integration between bioethanol and biodiesel processes

The degree of success of biofuel production and its inherent high cost of processing are deeply connected to the possibilities for integration, in situ and with other processes, to generate valuable coproducts simultaneously.

Integration strategies contemplate the integration of different production lines to take advantage of potential synergies. For instance, in bioethanol production, the coproduction of lignocellulosic fermentation processes with other well-established processes used for starchy feedstock can help to develop lignocellulosic processes, such as the case of corn kernel with corn fiber production, and grain wheat with wheat fiber production. Specific studies showed promising results for integration of bioethanol production from softwood thinning with an existing biomass power plan [37]. Grassi [38] proposed the theoretical integration of a bioethanol production plant with coproduction of methanol via catalytic hydrogenation of CO_2.

Another integration strategy involves cogeneration and formation of coproducts. In the case of bioethanol, it is common to consider the thermal conversion of lignin to provide vapor and electricity for the entire production process, or even to generate a surplus of electricity that can be commercialized. Some studies show that this strategy can provide all the steam and electricity for the biomass-to-bioethanol process [39]. Carrocci and James [40] proposed a two-step cogeneration system to increase the overall efficiency of bioethanol production. The first step is use of stillage (byproduct from bioethanol separation) to produce biogas that can be burned to generate electricity or used as transportation fuel. The second step is use of the lignin residues for conventional cogeneration.

Regarding the conversion of coproducts and wastes into valuable and marketable products, there are several alternatives already described in the literature, but their use depends on the type of feedstock employed. In the case of bioethanol, coproducts such as yeast, bagasse, fructose, corn fiber, ruminant food, 2,3-butanediol, xylitol, furfural, and unaltered lignin can be commercialized to increase the value of the biorefinery. From waste lignin, it is possible to produce fuel additives, adhesives, concrete plasticizers, asphalt antioxidants, and coatings [41]. Process synthesis can potentially evaluate the point at which diversification of biochemical products in biofuel production makes the entire process more profitable (biorefinery concept).

Several authors have already analyzed the concept of partial integration between bioethanol and biodiesel production processes. Gutierrez et al. [42] studied the possibility of applying in-situ bioethanol production to reduce biodiesel production costs from a single feedstock (oil palm) through process simulation. In that work, the oil extracted from fresh fruit bunches was transesterified to biodiesel, with in-situ production of bioethanol from two lignocellulosic palm industry byproducts (empty fruit brunches and palm press fiber), reducing biodiesel production costs by up to 40%. The integration consisted of recirculation of material streams to purify bioethanol and exchange of heat stream unit separations.

Martín and Grossmann [43] studied the optimization of algae composition to simultaneously produce bioethanol and biodiesel. They considered several transesterification routes to transform algae oil into biodiesel using bioethanol produced from the algae starch. In addition to optimizing the production of biofuels, the authors performed heat integration and minimized water consumption. They found that an algae composition (dry weight) of 60% oil, 30% starch, and 10% protein, together with the integrated processes, reduced the production costs to 20% as a result of savings in raw materials (bioethanol). In a more recent work, Martin and Grossmann [44] optimized the simultaneous production of biodiesel (FAEEs) and bioethanol from switchgrass. They proposed a superstructure with a set of alternatives, formulated as a mixed integer nonlinear program (MINLP), to achieve optimization and heat integration, considering costs and water consumption. They successfully identified key parameters that could make the whole process profitable.

A recent strategy called "lignocellulosic biodiesel" simultaneously produces bioethanol and biodiesel from the same lignocellulosic biomass using oleaginous microorganisms. Oleaginous microorganisms such as microalgae, bacilli, fungi, and yeasts, with a content of at least 20% lipids, act as lipid-accumulating agents that can be used to digest lignocellulosic biomass while generating lipids for biodiesel production. However, the entire process requires pretreatment and hydrolysis of the biomass. Furthermore. aerobic fungal fermentation has a high energy demand because of the need for agitation, aeration, and cooling.

Kiss [45] proposed the use of hydrous bioethanol and a solid catalyst with low sensitivity to water and FFAs to treat waste vegetable oil in an integrated reaction–separation unit. This process avoids the step of bioethanol dehydration, which is an energy-intensive process. Regarding the esterification/transesterification stage, a specific type of solid catalyst that tolerates the presence of water and FFAs was required to make use of this hydrous bioethanol. The flexibility in the type of raw material, use of hydrous ethanol, and use of reactive distillation to save equipment costs were translated into 40% reduction in total investment costs.

When we analyze the production processes for bioethanol and biodiesel, it is possible to classify the integration possibilities according to three different perspectives: mass integration, energy integration, and integration between units → intensification. A brief description of important aspects of these integration categories is given in Sections 7.4.1 to 7.4.3.

7.4.1 Mass integration

Feedstocks: Integration of feedstocks can be carried out when at least one of the biomass feedstocks is used as raw material for both biodiesel and bioethanol production. An example of this is the use of algae for simultaneous production of algae oil and starch.

Solvents: During reaction or separation processes in either bioethanol or biodiesel production, solvents are usually employed to enhance the conversion or improve separation factors. Moreover, the products, byproducts, and chemicals obtained or recycled in a process can be potentially used in another process for solvent purposes. As an example, glycerol is a byproduct of biodiesel production and can be employed as a solvent in extractive distillation for bioethanol dehydration.

Water: Bioprocesses are highly water consuming. For instance, during lignocellulosic bioethanol processing, water use is high during pretreatment and fermentation. Likewise, conventional homogeneous transesterification/esterification requires several water washes and neutralization steps. The collection of this wastewater and its subsequent treatment and redistribution are essential in achieving a sustainable operation.

Byproducts/coproducts: In bioethanol and biodiesel production, several substances can be interchangeable and coproducts can increase the general value of the biorefinery. For instance, some of the bioethanol produced can be used for production of FAEEs; some of the glycerol can be employed for bioethanol dehydration; and several coproducts from sugars and lignin can be commercialized to increase the profitability of the production processes.

7.4.2 Energy integration

Heating/cooling: It is possible to synthesize heat exchanger networks (HENs) between process units. Crucial steps such as pretreatments and separations by distillation are usually highly energy consuming; thus, energy savings in these critical areas can improve the efficiency of simultaneous processes. We can use superstructure optimization or pinch technology analysis to synthesize HENs. In addition to saving heating and/or cooling through HENs, it is possible to use alternative technologies such as heat pumps.

Generally, the reduction of heating and/or cooling in a process can be achieved by two strategies: first, by utilizing all the process streams (including residues) to provide energy for the system and, second, by increasing the efficiency of the system by means of process integration. For example, in the specific case of lignocellulosic bioethanol, an important amount of heat can be provided by burning the lignin-rich process residues and avoiding or reducing the use of external fossil fuels [46]. Pinch technology is one of the most common conventional methods for increasing process efficiency by means of process integration and can be used in different ways to reduce the use of utilities. Pinch technology can improve the core of the process design, as reported by Fujimoto et al. [47], who found that the use of heat pumps improved the efficiency of a lignocellulosic bioethanol process. Another strategy for increasing process efficiency by means of pinch technology is to optimize the combined heat and power (CHP) or polygeneration (simultaneous heating, cooling, and electricity production) system associated with utility allocation [48].

Papoulias and Grossmann [49] developed mathematical programming-based methods for heat integration and design of HENs. Since then, a large amount of work has been done on this topic. One of the main advantages of using a mathematical approach is the possibility to carry out heat integration and process optimization simultaneously, instead of using a sequential approach. In the sequential approach, heat integration is not considered during process optimization; consequently, the whole process may not be fully optimized and could overestimate the utility cost.

Chen et al. [50] recently suggested a new approach that combines rigorous simulation with mathematical heat integration. Here, the heat integration subproblem was formulated as a linear problem model and solved simultaneously during optimization of the flowsheet in a commercial process simulator to update the minimum utility and heat exchanger area targets.

If heat integration is done either after process optimization or simultaneously, it has the potential to modify the whole process design, such as changing the design of an individual piece of equipment, temperature and pressure operations, and so on, to provide important savings after designing the resulting HEN.

Electricity: The electricity demand of both bioethanol and biodiesel production process can be satisfied by thermal conversion of lignin (the nonfermentable fraction of lignocellulose), followed by mechanical conversion using turbines to generate electricity by using steam as work fluid. By doing so, the whole production system could not only satisfy its own energy demand, but also produce a commercializable surplus of electricity. Moreover, the implementation of heat pumps using electricity could reduce steam consumption in unit operations.

7.4.3 Integration between units → intensification

The integration between unit operations, indirectly related with intensification, implies two levels of interconnection. A high interconnection level is represented by the fusion of different unit operations, such as two or more reaction steps in one single unit, two or more separation steps, or hybrid reaction–separation steps merged into one single step. On the other hand, a low interconnection level implies the coupling of two or more unit operations by sharing recycle streams, without the operations occurring in the same equipment. A higher interconnection may increase the synergy of the system, improving the performance of the process, but simultaneously involves a higher level of compromise during the operation.

We consider intensification to be the development of technology that implies a more efficient use of energy, is cheaper to implement, and uses less plant space, or any combination of these features. Regarding the possibilities of intensification, strategies for reaction–reaction steps, separation–separation, and reaction–separation for bioethanol and biodiesel production have all been reported. For the case of integration of the reaction–reaction steps of bioethanol production, proposals such as cofermentation of lignocellulosic hydrolyzates, simultaneous saccharification and fermentation, simultaneous saccharification and cofermentation, and consolidated bioprocessing have been proposed for intensification of the reaction systems. For biodiesel production, the use of heterogeneous acid catalysts for low quality fatty sources can perform both transesterification and esterification in order to save extra equipment and utilities.

In the field of separation–separation intensification for bioethanol and biodiesel production, hybrid systems such as extractive distillation and membranes, distillation and molecular sieves, and distillation and pervaporation have been suggested in order to reduce costs. Distillation columns are the most common and energy-demanding separation units in industrial use. Fusion of distillation columns, distillation columns with flashes, or distillation columns with absorbers have been recently proposed [51] for further reduction in energy and capital costs. Energy consumption is especially high for bioethanol dehydration.

In the case of reaction–separation intensification, it is possible to highlight some examples for bioethanol and biodiesel processes. In the case of bioethanol, one strategy is fusion of enzymatic hydrolysis and membrane separation to separate the enzyme inhibitors and increase hydrolysis yield. Similarly, bioethanol removal from the culture broth (to reduce inhibition of yeast by bioethanol) can be accomplished by vacuum, gas stripping, membranes, and biocompatible liquid extraction, either inside or coupled to the fermentation stage. In the case of biodiesel, the integration of reaction and separation using reactive distillation is a highly promising intensification approach for improving biodiesel production technology.

7.5 Methodological framework for synthesis and intensification

Process synthesis determines the best interconnection of processing units and the type and design of these process units within a system. Any synthesis procedure must include the following steps:
1. definition of a product/service(s)
2. definition of the performance indices for evaluating the system
3. decomposition and evaluation of the problem into interconnected tasks
4. synthesis of promising technologies for the product/service of interest
5. identification of relevant design parameters for each technology
6. quick localization of the better alternatives without totally enumerating all the options

A continuous challenge for process synthesis is related to the changing of performance indices or criteria. Conventionally, indices such as economics, energy consumption, controllability, and (more recently) environmental impact have shaped most synthesis solutions into one or sometimes two simultaneous objectives. However, social factors such as sustainability, life cycle impact, atom economy, and labor utilization now have a higher relevance.

When we face a synthesis problem, we usually encounter two possibilities: a starting configuration to be improved or no starting configuration, in which case we need to find and select the best candidate configuration. Moreover, in both cases, systematic generation of alternatives based on human-implemented or algorithmic methods is essential to guarantee that the optimal alternative is present in the search space.

If we are starting with an initial system, we can apply either evolutionary methods or superstructure approaches. In evolutionary methods, we propose feasible modification to the previous system, leading to an improved system. If the new modification improves the system's performance evaluation, we use it; otherwise, we go back to the previous system and try another feasible modification. However, this method is highly sensitive to the initial system and we may need to generate this initial system through another method such as heuristics. In the superstructure case, we can consider all possible alternatives and their interconnections into one integrated superstructure. Here, material flows, scaling, operating conditions, and design parameters can be determined in one enormous simultaneous program.

For the second scenario, where alternative candidates have to be generated without a previous starting point, the literature mention techniques such as breadth and depth-first, bounding, heuristics, and decomposition [52]. Breadth and depth-first is a tree-search method. This kind of method uses nodes interconnected by a set of links or paths between nodes, where each node represents a particular state of the problem. To avoid the exploration of all possible nodes, several authors have combined tree-search methods with heuristic or bounding methods. Combinations of tree-search and bounding methods are usually referred to as branch and bound methods. In each level of nodes, a bound is set on the basis of the best solution obtained. Those nodes whose bounds are inferior to the best solution already obtained are not analyzed further. The nodes with promising bounds are expanded one at a time, depending on their values with respect to the best solution.

Another way to avoid examining all possible alternatives (i.e., screening the search space) is based on rules-of-thumb, also named heuristics. These heuristics consist of older methods based on process engineering experience in designing similar equipment and on insights from the physical-chemical phenomena related to the specific unit operation. This method is fast, but there is no guarantee of optimality. Furthermore, heuristics have to be continually re-evaluated in the light of changes in practice, constraints, and economics. The decomposition method consists of sequential decomposition of the original problem into smaller subproblems that are simpler to solve.

One way to overcome the disadvantages of individual methodologies is through combination of methods already reported in the literature, such as heuristic and evolutionary, branch and bound heuristics, and evolutionary bounding search.

The synthesis of an integrated biofuel production process can be addressed from a diversity of perspectives. One way to proceed is to separate the synthesis problem into two detailed-level stages. In the first stage, we define a superstructure optimization problem, including the most promising technological routes, based on heuristics and literature research. For the first stage, simplified models are used to represent different technological routes. Once the superstructure solution is defined and specific technological routes are selected, we can continue with the second stage. The second stage uses the previous solution as starting point and then uses rigorous models to evaluate and confirm the feasibility and profitability of the solutions.

In addition to performing a rough evaluation followed by rigorous confirmation, we can couple specific synthesis tools into each stage. For instance, in the first stage, besides discriminating between technological routes and raw materials, we can include possibilities for mass and energy integration between the biofuel production processes simultaneously. By doing so, the superstructure explores the integrated and nonintegrated technological routes, because this integration can significantly change the performance. In the second stage, we can examine the possibilities of integration between units (intensification), that is, integration of reaction–reaction, separation–separation, and reaction–separation, knowing that rigorous models will be used to confirm the feasibility of any structural change in the processes and to evaluate any real improvement in the system.

Before describing this synthesis approach in the integrated bioethanol–biodiesel case study, we briefly explain the methodology framework for each stage, with special emphasis on mixed integer nonlinear programming problems (MINLP) for synthesis purposes in the first stage, and systematic generation of intensified reaction–separation and separation–separation systems for the second stage by recombination of column sections.

7.5.1 First stage: formulating and solving the superstructure synthesis problem through MINLP

The solution-orientated synthesis problem with MINLP consists of giving the specifications for input (raw materials, flow rates, etc.) and outputs (desired products and purities) streams and then integrating the process alternatives that convert the inputs into desired outputs, meeting design specifications while maximizing or minimizing a given objective function(s).

The synthesis problem deals with the selection of technological routes, and how they should be interconnected, as well as design of their parameters (dimensions and operating conditions). The first group of tasks implies discrete decisions, whereas the second search is in a continuous space. The equations used to represent mass and energy balances, thermodynamic properties, economics, etc. are usually a set of linear and nonlinear equations. Therefore, the optimal synthesis problem involves a discrete–continuous nonlinear problem. A typical mathematical representation is given in Equations (7.1)–(7.4):

$$\min Z = f(x, y) \tag{7.1}$$
$$\text{s.t. } h(x, y) = 0 \tag{7.2}$$
$$g(x, y) \le 0 \tag{7.3}$$
$$x \in X \quad \text{and} \quad y \in Y \tag{7.4}$$

where Y is an integer binary vector of $0-1$ variables standing for the existence (1) or not (0) of unit operations. X stands for the vector of continuous variables such as material/energy flows, size of equipment, operating conditions, etc. Z is the objective function to minimize the subject to constraints in the form of equalities and inequalities.

A schematic description of superstructure formulation for a process synthesis is depicted in Figure 7.6. In this framework, several feedstocks, reactions paths, and separation paths are explored for generation of certain products or outputs. Moreover, recycles and interconnections between processes and unit operations can be explored at this level.

In addition to the superstructure representation, where simplified representations of unit operations and streams are explicitly considered, it is possible to use aggregate models for a higher level of representation, where the synthesis problem is simplified

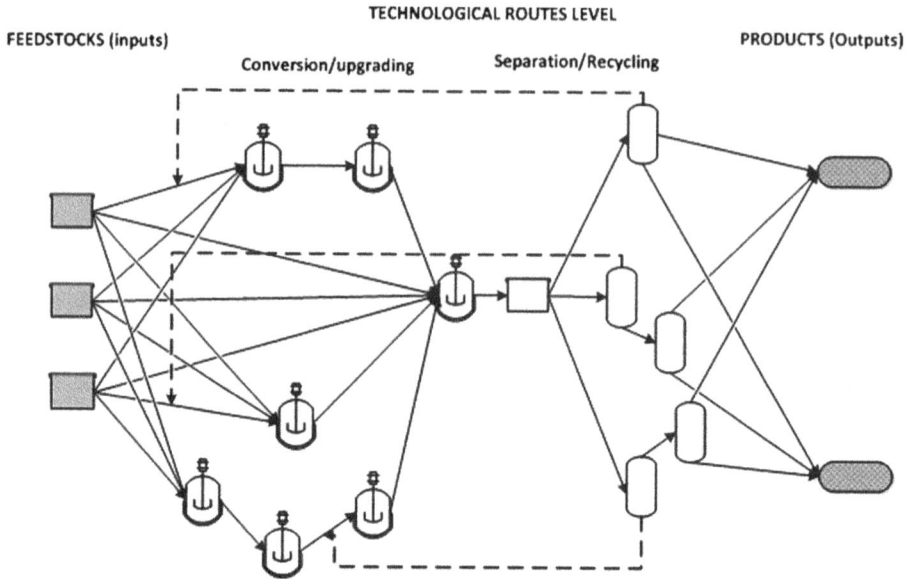

Fig. 7.6: Conceptual framework of a superstructure optimization applied to process synthesis.

by the use of design targets. In this way, the original MINLP problem can be simplified to nonlinear programming (NLP), mixed-integer linear programming (MILP), or linear programming (LP) problems.

The first step in formulating and solving an MINLP problem is to deal with the representation of alternatives. The representation of the superstructure can be explicit or implicit. The first type is related to the construction of networks, whereas the second is related to construction of trees. Tree representations lend themselves to decomposition, where the alternatives are analyzed through a branch and bound search. For networks, different alternative systems can be represented by the same subset of nodes. The advantage of this representation is that it is more compact for modeling the problem explicitly as an MINLP.

A crucial point related to MINLP modeling is related to the discrete 0–1 variables. In most cases, these variables are assigned to the components of the superstructure and then the interconnection between them is activated or deactivated. However, there are cases where the 0–1 variables are directly assigned to the interconnections. Handling 0–1 variables involves multiple-choice constraints for selecting among a subset of components. The use of 0–1 variables can also be used to activate or deactivate continuous variables, inequalities, or equations related to an activated or deactivated unit.

In general, there are the following three major empirical guidelines for a "good" MINLP formulation [53]:

- Try to keep the problem as linear as possible.
- Try to develop a formulation that is as tight an NLP relaxation as possible.
- If possible, reformulate the MINLP as a convex programming problem (highest probability for finding global optimal).

Common algorithms to solve MINLP problems include branch and bound, generalized Benders decomposition, and outer-approximation.

Applications of MINLP are diverse and include biomedical and biological engineering, communication engineering, computational geometry, finance, psychology, robotics, and transportation. In the specific field of process networks, there are studies in batch processes, heat and power plants, crude oil scheduling, data reconciliation, distillation sequences, HENs, water networks, multicommodity flow, pooling, reverse osmosis, and some direct applications related to biofuel production [54, 55].

7.5.2 Second stage: intensification through recombination of column sections

Once the MINLP problem associated with the synthesis process is solved, the next stage is to select the optimal path. Focusing on the separation train and its interaction with reactive sections, we can employ further systematic synthesis procedures through the definition of column sections. A column section is a portion of the column where there is no mass or heat flow entering or exiting that section. Column sections are used in trays or packing columns operating under distillation, reactive distillation, extraction, or absorption principles. The aim of defining column sections is to ease the generation of intensified sequences based on simple columns. Broadly speaking, after identifying the respective column sections in a column or sequence, we can manipulate the sections by transfer, substitution, or combination of sections to save energy and/or equipment costs.

When analyzing a separation train, the definition and use of column sections allows us to change the structural configuration of separation trains and generate complete new subspaces of separation trains with intensification potential. According to the study presented by Errico et al. [56], simple column sequences can be intensified using a systematic four-step generic procedure based on the following steps:

1. Simple column sequences (SCSs) to thermally coupled sequences (TCSs): Replace one or more condenser(s)/reboiler(s) not associated with the product stream by bidirectional vapor–liquid connection(s) [57–59].
2. Thermally coupled sequences (TCSs) to thermodynamically equivalent sequences (TESs): Move a column section related to a condenser/reboiler providing a common reflux ratio/vapor boil-up between two consecutive columns. The sequences are thermodynamically equivalent because they keep the same level of energy consumption, whatever the structural change [57, 60, 61].

3. Thermodynamically equivalent sequences (TESs) to intensified sequences (ISs) with less than $N-1$ columns: Remove side column sections that no longer have a separation task.

Similarly, the study presented by Rong [62] discusses the first two steps related to SCSs and TCSs. The third step is different in that, instead of generating ISs by removing column sections, Rong suggests the synthesis of divided-wall columns (DWCs), as explained below:

3* TESs to DWCs: Sometimes, the side column sections purify a product or an intermediate substance inside the separation trains; thus, their removal compromises the design specifications. However, these side column sections can be incorporated into the column through their thermal links, generating DWCs.

Recently, Torres-Ortega et al. [63] combined both systematic procedures in a nonsharp distillation configuration with more than $N-1$ columns. This study first showed that the systematic procedure could generate intensified sequences with less than $N-1$ columns, presenting savings in energy and fixed investment. Second, a more generalized synthesis procedure was proposed, adding a bifurcation part between simple ISs and DWCs as follows:

3** TESs to either ISs or DWCs: The criterion for deciding how to proceed with the side column sections is based on evaluation of the functionality of the side column section. If the side column section performs a separation task, then we should synthesized a DWC. However, if it performs a transportation task, an IS can be proposed without hindering the separation. Use of a process simulator is a way to evaluate the functionality of the side column section through analysis of composition profiles.

As an example, Figure 7.7 depicts the complete synthesis procedure described in Torres-Ortega et al. [63] for a nonsharp separation train of a conceptual quaternary mixture ABCD. In this case, bidirectional liquid–vapor streams were only considered in the first distillation column. Therefore, the entire subspaces of SCS, TCS, TES, IS, and DWC systems are not presented here.

It can be inferred from Figure 7.7 and previous studies [57, 58, 64] that there are structural correspondences between the different separation systems. This correspondence is translated into resemblance between the design parameters for the separation trains. In other words, if one wants to design an intensified separation system whose design procedure is complex or not clear, the design parameters can be obtained from a simpler separation system with a well-known and straightforward design procedure.

After selection of a technological route or path from the superstructure, and a further synthesis-intensification procedure in the separation or reaction–separation section,

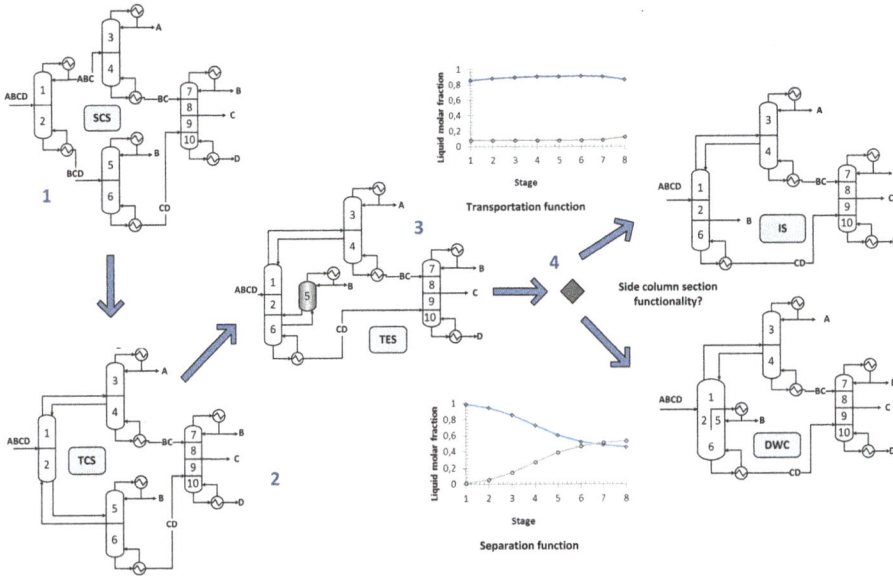

Fig. 7.7: Synthesis and intensification of conceptual structures from a nonsharp separation train for a four-component ABCD mixture.

it is necessary to use rigorous simulators to validate the feasibility and evaluate performance indices for the selected production train by considering kinetics, thermodynamic equations, mass and energy balances, etc.

7.6 Case study: integrated lignocellulosic bioethanol and biodiesel process synthesis

Lignocellulosic bioethanol and second and third generation biodiesel are promising biofuels that can be produced on a large scale for the transportation sector while minimizing environmental effects and social inconvenience. However, their production implies complex production trains, intensive use of energy and water, and use of chemicals and solvents that increase environmental stress and compromise the profitability of these processes. Process synthesis can search for new production processes that exploit available transformation and separation technologies, together with mass and energy integration tools to improve the performance of these processes and lower production costs. The methodological framework was described in the previous section; here, we present a case study that exemplifies these synthesis methods in biofuel production. We analyzed the synthesis production problem for bioethanol and biodiesel through two steps: superstructure optimization (MINLP) for screening purposes and analysis of column sections for intensifying the separation sections.

7.6.1 Problem description for superstructure optimization

Assumptions
Mass balances:
- Lignocellulosic raw material: softwood (Scots pine), wheat straw, and switchgrass consist only of cellulose, hemicellulose, and lignin with fixed compositions based on the literature [10, 65, 66].
- Oil raw materials: microalgae oil, waste cooking oil, and jatropha oil consist only of triolein, tripalmitin, oleic acid, and palmitic acid with fixed compositions based on the literature [24, 67].
- Yields, conversions, and recovery factors (98% for key components) were fixed according to the literature [67–76].

Energy balances:
- Energy consumption for each equipment unit was calculated through rigorous simulation in Aspen Plus V8.8 using the nonrandom two liquid (NRTL) model and UNIQUAC functional-group activity coefficients (UNIFAC) for missing parameters to predict thermodynamic properties.
- Constant heats of combustions were used to determine the cogeneration capacity of the combustor and turbine under constant efficiency coefficients.
- All units were assumed to operate under adiabatic conditions without thermal losses.

Costing:
- Aspen's Economic Analyzer V8.8 was used for fixed flowrates as base cost units.
- Prices of raw materials, chemicals, solvents, and catalysts were fixed according to the literature.

Given
We defined the superstructure of lignocellulosic bioethanol and the second to third generation biodiesel production processing network that considers different raw materials, technological routes, mass integration, and cogeneration (steam and electricity). Moreover, we defined the flow rate and composition of the feedstock, operating conditions and yields for each of the operation units, and market prices for raw materials, utilities, solvents, and intermediate chemicals.

Determine
The optimal processing route that minimizes the total manufacturing cost per unit of gasoline gallon equivalent, (bioethanol → 1.5 and biodiesel → 0.96 units of gasoline by volume). The mathematical model considers technological selection constraints, mass balance constraints, and techno-economic evaluation constraints. The major decision variables are the following:

- technological route selection
- mass flow rate and composition
- energy consumption of each technology
- overall capital and operational costs per equipment unit
- glycerol and bioethanol integration between bioprocesses
- associated cost for waste disposal
- split of cogeneration into steam and electricity production
- use of glycerol for cogeneration
- recycle fraction of key components

7.6.2 Superstructure setting and MINLP solution

General equations describing the superstructure are given next.
Objective function

$$\text{Minimize unit TCOM} = \frac{\sum (C_{UT} + C_{DW} + C_{RM} + 2.215 C_{OL} + 0.190 \, \text{COM} + 0.246 \, \text{FCI})}{gge}$$

(7.5)

FCI stands for the fixed cost investment, with 10% depreciation allowance; C_{UT}, C_{DM}, and C_{RM} stand for the costs of utilities, disposal of wastes, and raw materials; C_{OL} stands for operating labor costs; COM stands for manufacturing costs; TCOM stands for total costs of manufacturing; and gge stands for gallon of gasoline equivalent.

TCOM consists of direct manufacturing costs (DMC), which include raw materials, waste disposal, direct supervisory and clerical labor, maintenance and repairs, operating supplies, laboratory charges, and patents and royalties. Fixed manufacturing costs (FMC) include depreciation, local taxes and insurance, and plant overhead costs. General expenses (GE) include management, distribution and sales, financing, and research and development.

Subject to the next constraints
Technological (binary variables) selection constraints (Eq. 7.6–7.12)

$$y_1 + y_2 = 1 \tag{7.6}$$
$$y_3 + y_4 = 1 \tag{7.7}$$
$$y_5 + y_6 = 1 \tag{7.8}$$
$$y_7 + y_8 + y_9 = 1 \tag{7.9}$$
$$y_{10} + y_{11} = 1 \tag{7.10}$$
$$y_{12} + y_{13} + y_{14} = 1 \tag{7.11}$$
$$y_{15} + y_{16} = 1 \tag{7.12}$$

Mass constraints (Equations 7.13–7.19)

$$m_{i,j,k}^{\text{upstream}} + m_{i,j,k}^{\text{new}} + m_{i,j,k}^{\text{recyc}} = m_{i,j,k}^{\text{in}} \tag{7.13}$$

$$m_{i,j,k}^{\text{out\_a}} + m_{i,j,k}^{\text{out\_b}} = m_{i,j,k}^{\text{in}} \tag{7.14}$$

$$m_{i,j,k}^{\text{product}} + m_{i,j,k}^{\text{downstream}} + m_{i,j,k}^{\text{recyc}} = m_{i,j,k}^{\text{out\_a}} \tag{7.15}$$

$$m_{i,j,k}^{\text{upstream}} = m_{i,j,k}^{\text{downstream}} \tag{7.16}$$

The term $m_{i,j,k}^{\text{upstream}}$ represents the upstream mass flow (i, j) of component (k) per hour. The superscripts indicate the type of mass flow (i, j, k): "new," "recyc," "product," "in," and "downstream" stand for new or make-up, recycled component, product (biofuel), and input and downstream component mass flows, respectively. A schematic representation of the component mass balance used in the superstructure section is depicted in Figure 7.8.

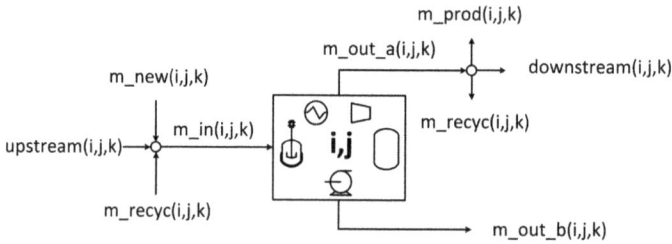

Fig. 7.8: Representative unit for the general component mass balance.

Dependency on new or make-up raw material using the factor $F_{i,j,k}^{\text{chemical}}$

$$m_{i,j,k}^{\text{new}} = F_{i,j,k}^{\text{new}} \cdot m_{i,j,k}^{\text{upstream}} - m_{i,j,k}^{\text{recyc}} \tag{7.17}$$

The term $F_{i,j,k}^{\text{new}}$ stands for the new or make-up factor for the respective component mass flow.

Separation of streams using the factor $F_{i,j,k}^{\text{separation}}$

$$m_{i,j,k}^{\text{out\_a}} = F_{i,j,k}^{\text{separation}} \cdot m_{i,j,k}^{\text{in}} \tag{7.18}$$

The term $F_{i,j,k}^{\text{separation}}$ stands for the recovery factor for the respective component mass flow.

The change in composition by chemical reaction is represented by

$$m_{i,j,k}^{\text{out\_a}} = m_{i,j,k}^{\text{in}} \cdot \left(1 + F_{i,j,k}^{\text{reaction}}\right) \tag{7.19}$$

The term $F_{i,j,k}^{\text{reaction}}$ stands for the reaction factor for the respective component mass flow.

7.6.2.1 Techno-economic constraints (Equation 7.20–7.24)

The cost factor $FC_{i,j}$ for each unit at the base-flow rate m_b is calculated in Aspen's Economic Analyzer as the equipment cost coefficient. The module cost $C_{Bi,j}$ of each unit is calculated by Equation (7.20), which considers the cost factor and the component flow mass input $m_{i,j,k}^{in}$.

$$C_{Bi,j} = CF_{i,j} \left(\frac{\sum_{k \in K} m_{i,j,k}^{in}}{m_b} \right)^{0.8} \tag{7.20}$$

FCI is calculated as a function of module cost $C_{Bi,j}$, which represents the cost of a completely new facility.

$$FCI = 1.18 \sum_{i \in I} \sum_{j \in J} C_{Bi,j} \tag{7.21}$$

The energy cost factor $ECF_{i,j}$ for each unit at m_b is calculated in Aspen's Economic Analyzer as the utility cost coefficient. The final utility cost $C_{UTi,j}$ of each unit is described by Equation (7.22).

$$C_{UTi,j} = ECF_{i,j} \left(\frac{\sum_{k \in K} m_{i,j,k}^{in}}{m_b} \right)^{0.8} \tag{7.22}$$

COM accounts for depreciation, costs of operating labor C_{OL}, utility costs C_{UT}, cost of waste disposal C_{DW}, and raw material costs C_{RM}.

$$COM = 0.280 FCI + 2.73 C_{OL} + 1.23 \left(C_{UT} + C_{DW} + C_{RM} \right) \tag{7.23}$$

The operating labor cost C_{OL} necessary to run the processes can be estimated using Equation (7.24). The number of units No_{units} accounts for the number of pieces of equipment (excluding pumps and vessels) that the process contains.

$$C_{OL} = 236,700 \left(6.29 + 0.23 \sum No_{units} \right)^{0.5} \tag{7.24}$$

The superstructure for the given problem is depicted in Figure 7.9. The main products are bioethanol (ethanol) and biodiesel (FAEE-C16 → ethyl palmitate and FAEE-C18 → ethyl oleate). We considered $2\,t \cdot h^{-1}$ each of Scots pine, wheat straw, and switchgrass, and $1\,t$ each of microalgae oil, waste cooking oil, and jatropha oil as fixed feedstock flow rates. The total generation of direct (directly produced) and indirect (produced by the steam and electricity supplied to the process) CO_2 equivalents (CO_2Eq) is collected in $CO_2(1,10)$. As self-utilities, electricity and steam are produced by a combustor–boiler–turbine system represented by COMB(1,11). Collection and disposal of wastes is represented by WWT(1,12). The blocks RXN and SEP represent conversion and separation units, which can contain equipment such as reactors, heaters, coolers, compressors, pumps, and vessels. We integrated the bioethanol and biodiesel processes by sharing bioethanol (green lines), glycerol (purple lines), steam and electricity self-produced (red lines), and disposal of wastes (blue lines). The dotted red line implies energy connection rather than mass connection. Raw materials

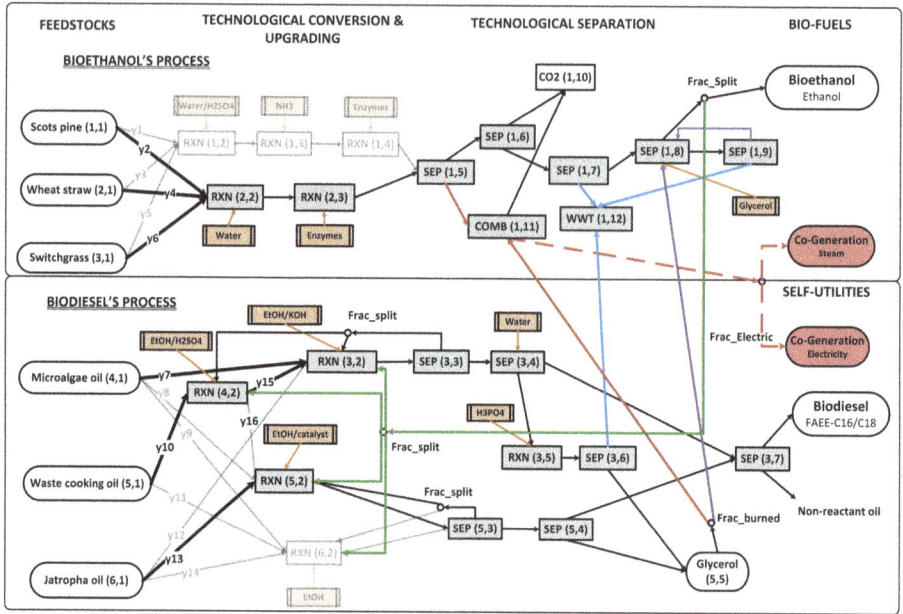

Fig. 7.9: *Bioethanol process*: RXN(1,2) dilute acid pretreatment, RXN(1,3) dilute acid neutralization, RXN(1,4) dilute acid fermentation, RXN(2,2) liquid hot water pretreatment, RXN(2,3) liquid hot water fermentation, SEP(1,5) centrifuge filter, SEP(1,6) flash, SEP(1,7) distillation to recover EtOH, SEP(1,8) extractive distillation to purify EtOH, SEP(1,9) distillation recovery of solvent, COMB(1,11) co-generation unit, WWT (1,12) disposal of wastes, CO_2(1,10) direct and indirect CO_2 equivalent units. *Biodiesel process*: RXN(3,2) homogenous–alkaline transesterification, RXN(4,2) homogeneous–acid esterification, RXN(5,2) heterogeneous–alkaline transesterification, RXN(6,2) supercritical transesterification, SEP(3,3) distillation to recycle EtOH, SEP(3,4) decanter, RXN(3,5) neutralization, SEP(3,6) distillation to purify glycerol, SEP(3,7) distillation to purify FAEEs, SEP(5,3) distillation to recycle EtOH, SEP(5,4) decanter. See text for an explanation of the colored lines and boxes.

are depicted in orange boxes. Integer binary variables related to the selection or not of technologies are depicted as variables $y1$, $y2$, etc. Fractional variables related to split of streams are depicted as Frac_split, Frac_Electric, and Frac_burned.

The problem was solved on a hp EliteBook laptop with Intel(R) Core(TM) i5-4200, 2.30 GHz CPU, 8 GB RAM, and Windows 7 64-bit. The model was coded in GAMS 24.4.6. The MINLP solver utilized was BARON 14.4, which found the global optimal solution after 20,704 iterations.

The minimum value of the unit production cost was 1.938 US\$/gge, generating 92.793 kg CO_2Eq/gge. The global optimal technological routes are highlighted in gray blocks in Figure 7.9; meaning that the blocks in white were not selected. Technological paths selected are highlighted with thicker lines.

Regarding the cogeneration section, 53.9% of the material burned in the combustor was used to produce electricity and the rest to produce steam. The bioethanol and biodiesel system produced enough steam and electricity to cover 100% of its utility requirements.

The recycle of bioethanol from distillation unit SEP(3,3) was only to unit RXN(3,2) (homogeneous–alkaline transesterification) because it has a higher ethanol demand. Recycle from distillation unit SEP(5,3) was only to unit RXN(5,2) (heterogeneous–alkaline transesterification) because unit RXN(6,2) was not selected. A minimum percentage of 10% (lower bound) of glycerol was integrated back to the bioethanol process as solvent and the majority burned in the combustor to produce steam and electricity (90%). Of the bioethanol produced and purified in distillation unit SEP(1,8), 70% was selected as final product (upper bound) to increase biofuel production and the rest was recycled back to the biodiesel process. From this bioethanol fed to the biodiesel process, 7.5% was sent to unit RXN(3,2), 0.7% to RXN(4,2) (homogeneous–acid esterification), and 21.8% to unit RXN(5,2). These values are in congruence with both the selection of technological routes by binary variables and the required bioethanol flow rates for RXN(3,2), RXN(4,2), and RXN(6,2).

Figure 7.10(a) demonstrates the relevance of the fixed costs of the equipment, which accounts for about two-thirds of the total cost. The next important costing factor is represented by the purchase of raw materials such as feedstocks, solvents, chemicals, and catalysts, which account for one-fifth of the total costs. Figure 7.10(b) shows the high contribution of the separation section to the total cost of network production processes, accounting for almost 60%, followed by the costs related to raw materials and conversion. Optimization via binary variables in the conversion section helped reduce its cost in comparison with the separation section.

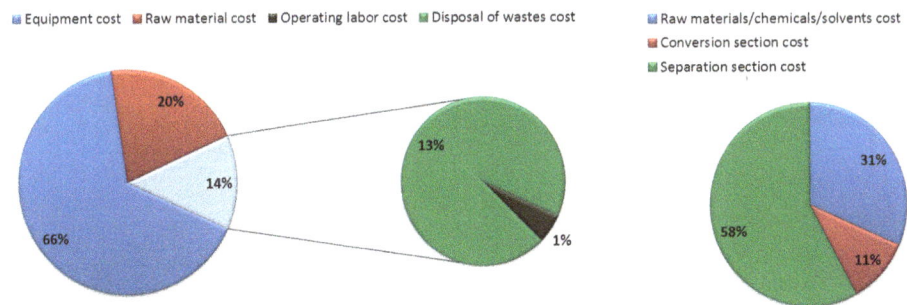

Fig. 7.10: Main results from the overall synthesis optimization: (left) contribution (%) of different factors in total bioethanol and biodiesel production and (right) contribution (%) of main sections of the total production process.

7.6.3 Synthesis-intensification: column section methodology

After determining the global optimal technology paths for simultaneously producing bioethanol and biodiesel by the MINLP approach, the next step is to analyze in detail the units involved in the optimal paths. It is possible to study one of each unit in the paths to explore intensification possibilities. For practical reasons, we focused on one specific section. As observed in Figure 7.10(b), the separation section is responsible for a high proportion of the production costs. Considering both biofuels, intensification of bioethanol separation is probably one of the most challenging tasks for the following reasons: (1) presence of solids, liquids, and gas; (2) very low concentration of bioethanol in the fermentation broth; and (3) presence of the bioethanol–water azeotrope, which prevents straightforward separation between them. Therefore, we focused on the separation section of the bioethanol process, highlighted by a black rectangle in Figure 7.11.

Torres-Ortega and Rong [77] have previously studied the nature of a lignocellulosic fermentation broth, and explored different separation sequences for a production process incorporating dilute acid pretreatment. By adjusting minor changes in the fermentation broth, we can use similar separation sequences for liquid hot water (LHW) pretreatment of fermentation broth. Mainly, we need to remove inhibitors

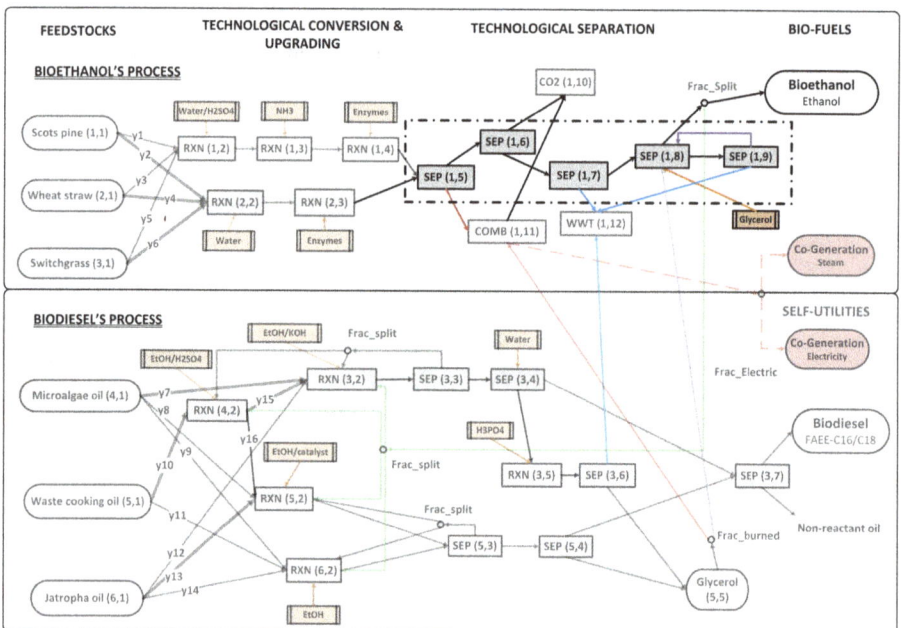

Fig. 7.11: Superstructure, with bioethanol separation section highlighted. SEP(1,5) centrifuge filter, SEP(1,6) flash, SEP(1,7) distillation to recover EtOH, SEP(1,8) extractive distillation to purify EtOH, SEP(1,9) distillation to recover solvent.

(hydroxymethylfurfural and furfural) and salts (ammonium sulfate and acetate) and slightly increase the concentration of water. We assumed that the sequence suggested by Torres-Ortega and Rong [77], depicted in Figure 7.12, was a good starting point for defining bioethanol separation under a bioethanol and biodiesel integration scenario. In Figure 7.12, the separation units employed to change the composition of the fermentation broth are highlighted in gray blocks and labeled. The fermentation broth is first conditioned to enter the centrifuge filter CF-1, where the solids components are removed and sent to the combustor–boiler–turbine system. The fluid fraction is fed to a distillation column DC-2, where the CO_2, bioethanol, and some water are separated at the top. Most of the water is separated at the bottom as stillage and sent for disposal or treatment. The top product of DC-2 is separated by a set of flashes, F-3 and F-4, to separate most of the CO_2. Any amount of bioethanol that can be lost in F-3 and F-4 is recovered in the absorption column AC-5 using water and then sent back to DC-2. A mixture of approximately 90 wt% bioethanol and 10% water is sent to extractive distillation column DC-6, using glycerol as solvent, to obtain bioethanol with a purity of 99.5 wt%. Finally, the glycerol used is recovered by combination of flash F-7 and stripping column SC-8 and sent back to DC-6. The starting setups and designs for each equipment unit were taken from Torres-Ortega and Rong [77].

Based on this starting sequence, different column sections can be identified in DC-2 (three column sections), DC-6 (three column sections), as well as F-3, F-4, F-7, and AC-5. SC-8 can be also treated as a column section [51]. To ease the description of the present synthesis and intensification methodology, the column sections of interest are labeled in Figure 7.12. Notice that column section number three accounts for both flashes F-3 and F-4.

Considering these eight column sections, there are multiple ways to combine them to achieve different levels of intensification. Torres-Ortega and Rong [51] made an extensive study of the synthesis tools described in Figure 7.7, as well as new synthesis

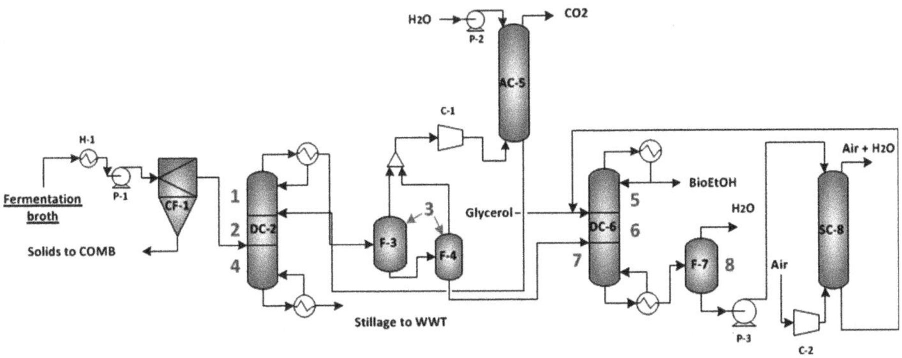

Fig. 7.12: Starting separation sequence for LHW-pretreated fermentation broth. H heater, P pump, C compressor, CF centrifuge filter, F flash, AC absorption column, DC distillation column, and SC stripping column.

tools such as hybridizing unit operations by reformulating column sections, replacing flashes by column sections, and relocating column sections. One of the most promising intensified sequences is depicted in Figure 7.13. Here, a distillation column consisting of eight column sections and two vertical walls (i.e., a multipartition DWC) replaces DC-2, F-3, F-4, DC-6, and F-7. This DWC system has two reboilers, two partial condensers, one total condenser, and four feed streams. In this way, the separation sequence was reduced from eight to four columns, potentially saving capital costs and energy consumption. However, design of the DWC is more complex and requires a more elaborate synthesis strategy. Details of the synthesis procedure to obtain the sequence in Figure 7.13 is beyond the scope of this chapter, but are extensively discussed by Torres-Ortega and Rong [51]. However, Figure 7.14 gives a brief description of the systhesis methodology applied to this sequence.

First, the flashes F-3 and F-4 were replaced by the single column section number three. This column section was stacked on one side of the original unit DC-2. By doing so, we obtained a DWC column where the partition wall was between column sections one, two, and three, as shown in Figure 7.14(a).We next stacked column DC-6 (i.e., its column sections five, six, and seven) into the DWC unit just obtained, givin the configuration shown in Figure 7.14(b). There might be different ways to combine the column sections of DC-6 into the initial DWC, but we chose the option that satisfied the best of the specifications for the separation unit. This option stacked column sections five and six next to three and four; column section seven was set as new bottom section of the multipartition DWC because it separates the heaviest component in the

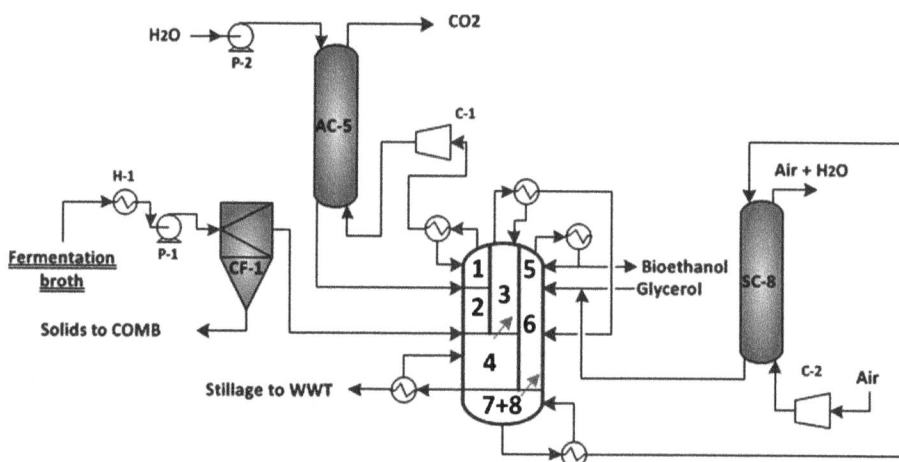

Fig. 7.13: Intensified separation sequence for LHW-pretreated fermentation broth; H heater, P pump, C compressor, CF centrifuge filter, AC absorption column, DC distillation column, and SC stripping column. The arrows in red describe the fraction split of the vapor moving up; thicker vertical lines in the column indicate dividing walls.

mixture. Finally, as shown in Figure 7.14(c), flash F-7 was replaced by another column section; however, this replacement is not necessary because it performs a separation with much lower flow rate than in the cases of flashes F-3 and F-4. Therefore, column section seven in the new configuration could take over the functionality of F-7 in addition to its original separation function, thus simplifying the separation sequence.

7.6.4 Evaluation with the process simulator

To obtain a numerical evaluation and compare the intensified sequence with the original separation sequence shown in Figures 7.12 and 7.13, we did rigorous simulations in Aspen Plus V8.8. We used Henry components and National Renewable Energy Laboratory (NREL) physical property data to predict the thermodynamics of the separation problem. Inside the simulator, we used RadFrac to model the separation units. The composition of the fermentation broth used in this case is depicted in Table 7.4. The content of the main components such as bioethanol, water, and CO_2 are within the range of values used for the superstructure optimization described in Section 7.6.2. The precise values for these main components and the content of other minor components present during LHW pretreatment of lignocellulosic feedstock were taken from the literature [68–70, 78]. We considered a flow rate of 30,000 kg h^{-1}.

As mentioned, the setups and designs from previous works [51, 77] were taken as starting points. Then, because of changes in the composition and flow rates, we made appropriate adjustments to the design parameters to achieve the recoveries and purities specified for the new LHW-pretreated fermentation broth (bioethanol recovery >97 wt% and bioethanol final purity of 99.5 wt%).We focused on the fluid-separation units, where intensification occurs, and did not consider the centrifuge filter in the

Tab. 7.4: Separation of a fermentation broth pretreated with liquid hot water.

Component	Content (%)
O_2	0.01
CO_2	4.29
Ethanol	4.49
H_2O	81.10
Glucose	0.53
Xylose	7.27
Extractive	1.35
Soluble lignin	0.67
Lactic acid	0.14
Xylitol	0.06
Glycerol	0.01
Succinic acid	0.02
DAP	0.06

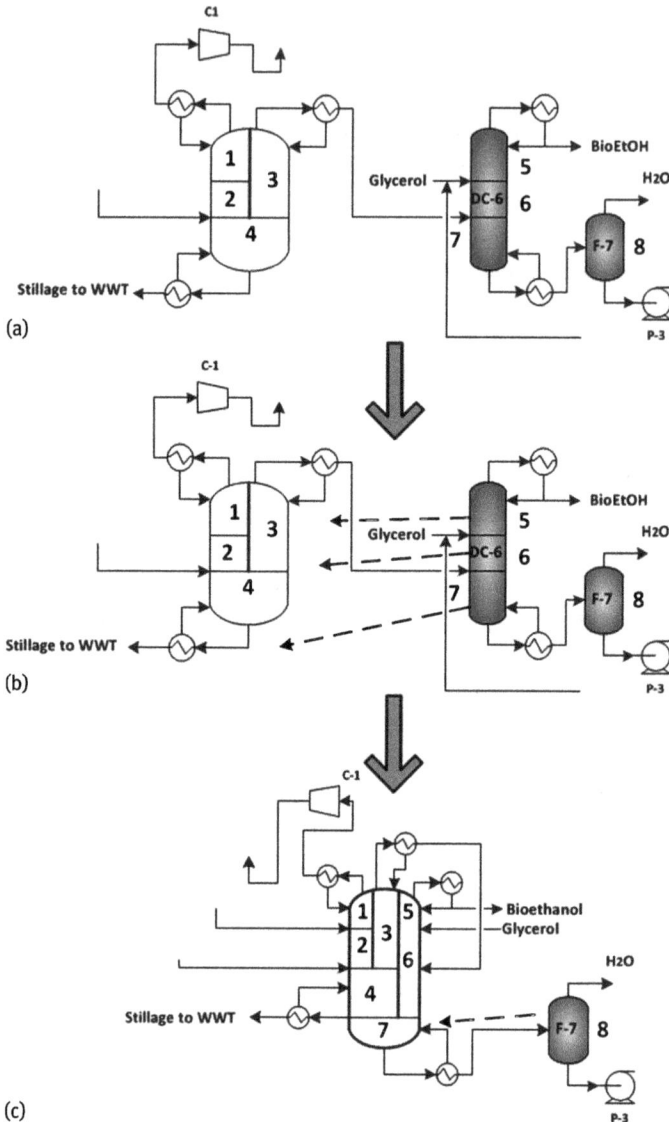

Fig. 7.14: Brief synthesis procedure to obtain the intensified separation sequence for LHW-pre-treated fermentation broth; (a) Replacement of Flashes F-3 and F-4 by column section 3 forming a new DWC on the left, (b) Incorporation of DC-6 in the DWC, (c) Simplification of flash F-7 by the column section 7 inside the DWC. H heater, P pump, C compressor, CF centrifuge filter, AC absorption column, DC distillation column, and SC stripping column.

simulator. To model DWC in Aspen Plus V8.8, we used an approximation based on interconnecting column sections, as depicted in Figure 7.15. The mass and energy balances were properly modeled by this method. Suitable DWC equipment costing was case-based [51].

After achieving the required recovery and purity specifications for bioethanol in the simulations, we used Economic Evaluation V8.8 tool to calculate the utility and equipment costs for both sequences. We assumed that calculation of the equipment cost of the multipartition column could be done directly in the way shown in Figure 7.15. The economic results obtained for both separation sequences are presented in Table 7.5. For reproducibility purposes, the design parameters of the two separation sequences are given in Table 7.6.

Table 7.5 shows the moderated savings obtained when intensification was used in the presented separation problem. The major benefits in the intensified sequence are reflected in utility costs, implying higher thermodynamic efficiency of the sequence.

Entries in bold refer to the separation units shown in Figures 7.12 and 7.13.

Note that superscripts 1, 2, and 3 stand for the regions divided by vertical walls, labeled from left to right in Figure 7.13. For instance, in Figure 7.13, 0.04^1 corresponds to the rr of column section one (first region at the left), 1.99^2 to column section three (second region in the middle), and 0.23^3 to column section five (third region at the right). In the particular case of stages, superscript 1 stands for the total number of

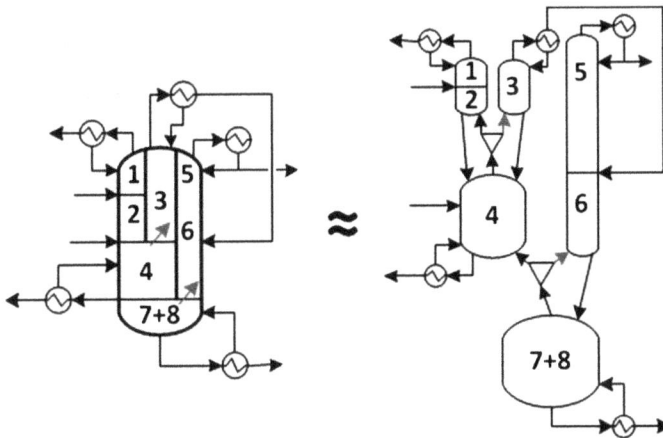

Fig. 7.15: Approximation of a DWC system by column sections in Aspen Plus V8.8.

Tab. 7.5: Economic evaluation of the reference and intensified sequences for separation of LHW-pretreated fermentation broth.

	Equipment costs (US$)	Utility costs (US$/year)
Reference sequence of Figure 7.12	1,650,900	1,099,380
Intensified sequence of Figure 7.13	1,732,300	958,052
Percentage improvement [%]	4.70	12.86

Tab. 7.6: Main design parameters for the reference and the intensified sequences for separation of LHW-pretreated fermentation broth.

	Reference case (Figure 7.12)			Intensified sequence (Figure 7.13)			
Parameter	**Separation units**			**Separation units**			**Parameter**
	DC-2	**EDC-6**		**Multipartition DWC**			
Number of stages	28	26		35[1]			Number of stages (total)
Feed stage from CF-1	9			11[2]			Number of stages (intermediate section)
Feed stage from AC-5	10			19[3]			Number of stages (final section)
Feed stage solvent		2		3[1]			Feed stage from CF-1
Feed stage from F-4		19		11[1]			Feed stage from AC-5
Solvent/feed ratio		1.02		2[3]			Feed stage solvent
Distillation flow rate [kg h^{-1}]	3100	1325		12[3]			Feed stage from column section three
rr (mass reflux ratio)	0.98	0.11		1.02			Solvent/feed ratio
Pressure [atm]	1.0	1.0		1348[1], 1510[2], 1325[3]			Distillation flow rate [kg h^{-1}]
Split fraction				0.04[1], 1.99[2], 0.23[3]			rr (mass reflux ratio)
	F-3	**F-4**	**F-7**	1.0			Pressure [atm]
Temperature [K]	316.15	325.29	385.56	0.5[1] (to 3), 0.81[2] (to 6)			Split fraction (arrow direction)
Pressure [atm]	1.0	0.60	0.10				
	AC-5			**AC-5**			
Number of stages	10			10			Number of stages
Pressure [atm]	3.0			3.0			Pressure [atm]
Water flow rate [kg h^{-1}]	1721			662.45			Water flow rate [kg h^{-1}]
	SC-8			**SC-8**			
Number of stages	12			12			Number of stages
Pressure [atm]	1.0			1.0			Pressure [atm]
Air flow rate [kg h^{-1}]	439.03			1465.07			Air flow rate [kg h^{-1}]
	P-1	**P-2**	**P-3**	**P-1**	**P-2**	**P-3**	
Discharge pressure [atm]	1.1	3.1	1.1	1.1	3.1	1.1	Discharge pressure [atm]
	Comp-1	**Comp-2**		**Comp-1**	**Comp-2**		
Discharge pressure [atm]	3.1	1.1		3.1	1.0		Discharge pressure [atm]

stages, 2 for the stages of the first vertical wall (from left to right) and 3 for the stages of the second vertical wall (last from left to right). For instance, the DWC system has 35 stages in total, with a first vertical wall of 11 stages and a second vertical wall of 19 stages; in the same way, a feed stage such as 2^3 stands for a feed stream entering the second stage of the third region of the column (third from left to right).

7.7 Discussions

Biomass, the source of biofuels, is abundant and better distributed throughout the world than fossil fuels. Moreover, the use of biomass to produce biofuels can potentially boost other sectors, such as the rural sector, generating a synergic effect. However, not all kinds of biomass are suitable for biofuel production because of possible negative environmental effects. This is especially true for first-generation fuels that are produced from food crops such as wheat, sugar cane, and oilseed rape. Use of second and third generation biofuels could be an important solution to the problem. Production of these fuels uses biomass such as agricultural, forestry, and household residues (usually not exploited), energy crops grown on land not suitable for edible crops, or algae grown in open pools or photoreactors not occupying croplands.

The crucial problem regarding the use of second and third generation bioethanol and biodiesel is related to the nature of their molecular structure and their processing. For instance, lignocellulosic feedstocks contain mainly xylan, cellulose, and lignin plus other substances, mainly pentoses and hexoses. Breaking down their complex structure, hydrolyzing their complex sugars into simple ones, and fermenting them into bioethanol are more complicated and expensive processes than the popular corn- and cane-based bioethanol production processes. Moreover, the bioethanol produced is of low concentration and must be separated and purified for engine use. In the case of biodiesel, the process is complicated by the need for an appropriate oil extraction technique and treatment for the high content of FFAs. Furthermore, to increase the productivity of the process, byproduced glycerol has to be used somehow to bring extra value to biodiesel processing.

Second and third generation bioethanol and biodiesel production processes are relevant from both research and industrial viewpoints, but they involve high demands for energy, water, and chemicals. It is possible, conceptually, to integrate both processes to increase the efficiency and productivity of both biofuels, with the aim of reducing production costs. The integration can be carried out according to different perspectives. Energy in the form of steam and electricity can be shared by burning mainly lignin and glycerol in a combustor–boiler–turbine system. Mass can be shared by using some of the bioethanol produced to replace the make-up of bioethanol in the biodiesel process, and glycerol byproduct can replace the make-up of glycerol used as a solvent in the bioethanol purification step. There is also a third level of integration, which focuses on the pieces of equipment rather than individual processes. Here, two or more operation units can be integrated or intensified to achieve savings in energy consumption, equipment cost, and land space.

In this chapter, we proposed a two-stage methodology for synthesis of a more efficient production process for the integrated production of bioethanol and biodiesel. In the first stage, we defined a superstructure, considering different reactions and separation units to produce and purify bioethanol and biodiesel, as well as different possibilities for mass and energy integration between them. We employed GAMS 24.4.6 and the solver BARON 14.4 globally to minimize the unit production cost of both biofuels per gallon of gasoline equivalent. In the second stage, we used column section methodology to intensify the unit operations. From the superstructure optimization, we identified a critical cost-intensive section of the total process for further intensification. This section corresponded to the separation and purification of fermentation broth under LHW pretreatment (the optimal pretreatment, found during superstructure optimization). We proposed a starting separation sequence, based on other works, that was suitable for the separation problem. We identified and labeled relevant column sections in the sequence and briefly described the synthesis procedure by replacing flashes and moving column sections to neighboring positions until we obtained an intensified sequence. We used conceptual synthesis and then the rigorous simulator Aspen Plus V8.8 and its Economic Evaluation V8.8 tool to confirm design specifications and perform a numerical comparison between systems.

By applying this combined methodology to current biofuel production synthesis, we were able to screen the different conversion-and-upgrading technological routes and choose those that reduced the total cost of manufacture. Moreover, split fractions of key process variables were chosen, such as the cogeneration section and recycling of solvents and chemicals. Finally, after identifying the optimal technological routes, under a specific scenario of interest, we focused on intensifying critical sections of the optimal paths. By applying the method of column section recombination, we obtained moderate savings for the separation and purification sections of bioethanol production in both equipment cost and utility cost.

7.8 Conclusions

We as a society need to face the challenge of a continuously increasing world demand for energy, where the need for more energy directly reflects the progress and economic growth of a country. One of the most important energy sectors is related to transportation, which demands liquid fuels, namely, fossil fuels such as gasoline, diesel, and, to a lesser degree, natural gas.

Bioethanol and biodiesel, produced from second or third generation biomass, represent a potential sustainable solution to the demand for liquid fuels. However, their processing is of high complexity and requires high energy consumption. A dual synthesis methodology, as presented in this chapter, evaluated the synthesis problem of reducing unit manufacturing costs for integrated bioethanol and biodiesel production processes. The first part of the methodology considered the screening process between reaction and separation units, as well as integration (mass and energy) between biofuel production processes through MINLP superstructure optimization. The second part applied column section methodology to intensify critical sections of the optimal technological paths.

We conclude that this two-stage synthesis methodology can be used to exploit the advantages of both methods: the screening process by superstructure optimization (selecting technologies and defining mass and energy integration) and the intensification process by column section recombination on the selected optimal path (reduction in the number of unit operations and in energy consumption).

7.9 Bibliography

[1] US Energy Information Administration. International Energy Outlook 2016, vol. 484, 2016. Retrieved from http://www.eia.gov/forecasts/ieo/world.cfm.

[2] Melorose J, Perroy R, Careas S. Bp Energy Outlook, 2016 Edition. Statewide Agricultural Land Use Baseline, 2016, 1. doi:10.1017/CBO9781107415324.004.

[3] European Biofuels TP. Global biofuels – an overview. EBTP-SABS, 2015. Retrieved January 26, 2016, from http://www.biofuelstp.eu/global_overview.html.

[4] Brown JP, Weber JG, Wojan TR. Emerging Energy Industries and Rural Growth, USDA, 2013.

[5] International Renewable Energy Agency. Global Bioenergy Supply and Demand Projections: A working paper for REmap 2030, 2014. Retrieved from http://www.igc.int/en/downloads/grainsupdate/igc_5yrprojections.pdf.

[6] Kim S, Dale BE. All biomass is local: The cost, volume produced, and global warming impact of cellulosic biofuels depend strongly on logistics and local conditions. Biofuels, Bioproducts and Biorefining, 2015, 9, 422–434. doi:10.1002/bbb.

[7] Giuliano A, Poletto M, Barletta D. Process optimization of a multi-product biorefinery: The effect of biomass seasonality. Chemical Engineering Research and Design, 2016, 107, 236–252. doi:10.1016/j.cherd.2015.12.011.

[8] Khuong LS, Zulkifli NWM, Masjuki HH, Mohamad EN, Arslan A, Mosarof MH, Azham A. A review on the effect of bioethanol dilution on the properties and performance of automotive lubricants in gasoline engines. RSC Adv, 2016, 6(71), 66847–66869. doi:10.1039/C6RA10003A.

[9] Alvira P, Tomas-Pejo E, Ballesteros M, Negro MJ. Pretreatment technologies for an efficient bioethanol production process based on enzymatic hydrolysis: A review. Bioresource Technology, 2010, 101(13), 4851–4861. doi:10.1016/j.biortech.2009.11.093.

[10] Menon V, Rao M. Trends in bioconversion of lignocellulose: Biofuels, platform chemicals & biorefinery concept. Progress in Energy and Combustion Science, 2012, 38(4), 522–550. doi:10.1016/j.pecs.2012.02.002.

[11] Falk J, Berry RJ, Broström M, Larsson SH. Mass flow and variability in screw feeding of biomass powders – Relations to particle and bulk properties. Powder Technology, 2015, 276, 80–88.

[12] Barletta D, Berry RJ, Larsson SH, Lestander TA, Poletto M, Ramírez-Gómez Á. Assessment on bulk solids best practice techniques for flow characterization and storage/handling equipment design for biomass materials of different classes. Fuel Processing Technology, 2015, 138, 540–554. doi:10.1016/j.fuproc.2015.06.034.

[13] Saleh K, Jaoude MMA, Morgeneyer M, Lefrancois E, Bihan OLe, Bouillard J. Dust generation from powders: A characterization test based on stirred fluidization. Powder Technology, 2014, 255, 141–148.

[14] Mosier N, Wyman C, Dale B, Elander R, Lee YY, Holtzapple M, Ladisch M. Features of promising technologies for pretreatment of lignocellulosic biomass. Bioresource Technology, 2005, 96(6), 673–686. doi:10.1016/j.biortech.2004.06.025.

[15] Sassner P, Galbe M, Zacchi G. Techno-economic evaluation of bioethanol production from three different lignocellulosic materials. Biomass and Bioenergy, 2008, 32(5), 422–430. doi:10.1016/j.biombioe.2007.10.014.

[16] Olofsson K, Palmqvist B, Lidén G. Improving simultaneous saccharification and co-fermentation of pretreated wheat straw using both enzyme and substrate feeding. Biotechnology for biofuels, 2010, 3, 17. doi:10.1186/1754-6834-3-17.

[17] Vane LM. Separation technologies for the recovery and dehydration of alcohols from fermentation broths. Biofuels Bioproducts & Biorefining, 2008, 553–588. doi:10.1002/bbb.

[18] Offeman RD, Franqui-Espiet D, Cline JL, Robertson GH, Orts WJ. Extraction of ethanol with higher carboxylic acid solvents and their toxicity to yeast. Separation and Purification Technology, 2010, 72(2), 180–185. doi:10.1016/j.seppur.2010.02.004.

[19] Huang H-J, Ramaswamy S, Tschirner UW, Ramarao BV. A review of separation technologies in current and future biorefineries. Separation and Purification Technology, 2008, 62(1), 1–21. doi:10.1016/j.seppur.2007.12.011.

[20] Seiler M, Ko D, Arlt W. Hyperbranched polymers: new selective solvents for extractive distillation and solvent extraction. Separation and Purification Technology, 2002, 29, 245–263.

[21] Balat M. Potential alternatives to edible oils for biodiesel production – A review of current work.Energy Conversion and Management, 2011, 52(2), 1479–1492. doi:10.1016/j.enconman.2010.10.011.

[22] Moazami N, Ashori A, Ranjbar R, Tangestani M, Eghtesadi R, Nejad AS. Large-scale biodiesel production using microalgae biomass of Nannochloropsis. Biomass and Bioenergy, 2012, 39, 449–453. doi:10.1016/j.biombioe.2012.01.046.

[23] Ef A, Betiku E, Dio I, Tv O. Production of biodiesel from crude neem oil feedstock and its emissions from internal combustion engines. African Journal of Biotechnology, 2012, 11(22), 6178–6186. doi:10.5897/AJB11.2301.

[24] Aransiola EF, Ojumu TV, Oyekola OO, Madzimbamuto TF. ScienceDirect A review of current technology for biodiesel production: State of the art. Biomass and Bioenergy, 2013, 61, 276–297.

[25] Černoch M, Hájek M, Skopal F. Ethanolysis of rapeseed oil – Distribution of ethyl esters, glycerides and glycerol between ester and glycerol phases. Bioresource Technology, 2010, 101(7), 2071–2075. doi:10.1016/j.biortech.2009.11.035.

[26] Cavalcante KSB, Penha MNC, Mendonça KKM, Louzeiro HC, Vasconcelos ACS, Maciel AP, Silva FC. Optimization of transesterification of castor oil with ethanol using a central composite rotatable design (CCRD). Fuel, 2010, 89(5), 1172–1176. doi:10.1016/j.fuel.2009.10.029.

[27] Behera S, Singh R, Arora R, Sharma NK, Shukla M, Kumar S. Scope of Algae as Third Generation Biofuels. Frontiers in Bioengineering and Biotechnology, 2015, 2(February), 1–13. doi:10.3389/fbioe.2014.00090.

[28] Canakci M, Gerpen J van. Biodiesel Production Via Acid Catalysis. Transactions of the ASAE (American Society of Agricultural Engineers), 1999, 42(1984), 1203–1210.

[29] Sakai T, Kawashima A, Koshikawa T. Economic assessment of batch biodiesel production processes using homogeneous and heterogeneous alkali catalysts. Bioresource Technology, 2009, 100(13), 3268–3276. doi:10.1016/j.biortech.2009.02.010.

[30] Narasimhara K, Brown DR, Lee AF, Newman AD, Siril PF, Tavener SJ, Wilson K. Structure-activity relations in Cs-doped heteropolyacid catalysts for biodiesel production. Journal of Catalysis, 2007, 248(2), 226–234. doi:10.1016/j.jcat.2007.02.016.

[31] Lee JS, Saka S. Biodiesel production by heterogeneous catalysts and supercritical technologies. Bioresource Technology, 2010, 101(19), 7191–7200. doi:10.1016/j.biortech.2010.04.071.

[32] Ngamcharussrivichai C, Nunthasanti P, Tanachai S, Bunyakiat K. Biodiesel production through transesterification over natural calciums. Fuel Processing Technology, 2010, 91(11), 1409–1415. doi:10.1016/j.fuproc.2010.05.014.

[33] Gog A, Roman M, Toşa M, Paizs C, Irimie FD. Biodiesel production using enzymatic transesterification – Current state and perspectives. Renewable Energy, 2012, 39(1), 10–16. doi:10.1016/j.renene.2011.08.007.

[34] Niza NM, Tan KT, Lee KTA. Biodiesel production by non-catalytic supercritical methyl acetate:thermal stability study. Appl Energy, 2012, 101, 198–202. doi:10.1016/J.apenergy.2012.03.033.

[35] Marulanda VF, Anitescu G, Tavlarides LL. Investigations on supercritical transesterification of chicken fat for biodiesel production from low-cost lipid feedstocks. Journal of Supercritical Fluids, 2010, 54(1), 53–60. doi:10.1016/j.supflu.2010.04.001.

[36] Zhou CH, Beltramini JN, Fan YX, Lu GQ. Chemoselective catalytic conversion of glycerol as a biorenewable source to valuable commodity chemicals. Chemical Society Reviews, 2008, 37(3), 527–549. doi:10.1039/b707343g.

[37] Kadam KL, Wooley RJ, Aden A, Nguyen QA, Yancey MA, Ferraro FM. Softwood forest thinnings as a biomass source for ethanol production: A feasibility study for California. Biotechnology Progress, 2000, 16(6), 947–957. doi:10.1021/bp000127s.

[38] Grassi G. Modern bioenergy in the European Union. Renewable Energy, 1999, 16, 985–990. doi:10.1016/S0960-1481(98)00347-4.

[39] Reith JH, Uil H Den, Veen H Van, Niessen JJ, Jong E De, Elbersen HW, Raamsdonk L. Co-production of bio-ethanol, electricity and heat from biomass residues. Twelfth European Conference and Technology Exhibition on Biomass for Energy, Industry and Climate Protection, (July), 17–21, 2002. Retrieved from http://www.ecn.nl/publicaties/PdfFetch.aspx?nr=ECN-RX--02-030.

[40] Carrocci LR, James EH. Cogeneration at alcohol production plants in Brazil. Energy, 1991, 16(8), 1147–1151. doi:10.1016/0360-5442(91)90147-E.

[41] Arato C, Pye EK, Gjennestad G. The Lignol Approach to Biorefining of Woody Biomass to Produce Ethanol and Chemicals. Applied Biochemestry and Biotechnology, 2005, 121–124, 871–882.

[42] Gutiérrez LF, Sánchez OJ, Cardona Ca. Process integration possibilities for biodiesel production from palm oil using ethanol obtained from lignocellulosic residues of oil palm industry. Bioresource technology, 2009, 100(3), 1227–1237. doi:10.1016/j.biortech.2008.09.001.

[43] Martín M, Grossmann IE. Optimal Engineered Algae Composition for the Integrated Simultaneous Production of Bioethanol and Biodiesel, 2013, 59(8). doi:10.1002/aic.

[44] Martín M, Grossmann IE. Optimal simultaneous production of biodiesel (FAEE) and bioethanol from switchgrass. Industrial and Engineering Chemistry Research, 2015, 54(16), 4337–4346. doi:10.1021/ie5038648.

[45] Kiss Aa. Separative reactors for integrated production of bioethanol and biodiesel. Computers & Chemical Engineering, 2010, 34(5), 812–820. doi:10.1016/j.compchemeng.2009.09.005.

[46] Sassner P, Zacchi G. Integration options for high energy efficiency and improved economics in a wood-to-ethanol process. Biotechnology for Biofuels, 2008, 11, 1–11. doi:10.1186/1754-6834-1-4.

[47] Fujimoto S, Yanagida T, Nakaiwa M, Tatsumi H, Minowa T. Pinch analysis for bioethanol production process from lignocellulosic biomass. Applied Thermal Engineering, 2011, 31(16), 3332–3336. doi:10.1016/j.applthermaleng.2011.06.013.

[48] Kravanja P, Modarresi A, Friedl A. Heat integration of biochemical ethanol production from straw – A case study. Applied Energy, 2013, 102, 32–43. doi:10.1016/j.apenergy.2012.08.014.

[49] Papoulias SA, Grossmann IE. A structural optimization approach in process synthesis. Computers & Chemical Engineering, 1983, 7(6), 707–721.

[50] Chen Y, Eslick JC, Grossmann IE, Miller DC. Simultaneous process optimization and heat integration based on rigorous process simulations. Computers and Chemical Engineering, 2015, 81, 180–199.

[51] Torres-Ortega CE, Rong BG. Synthesis and Simulation of Efficient Divided Wall Column Sequences for Bioethanol Recovery and Purification from an Actual Lignocellulosic Fermentation Broth. Industrial and Engineering Chemistry Research, 2016, 55(1), 7411–7430. doi:10.1021/acs.iecr.5b02773.

[52] Nishida N, Stephanopoulos G, Westerberg AW. A review of process synthesis. AIChE Journal, 1981, 27(3), 321–351. doi:10.1002/aic.690270302.

[53] Grossmann IE, Kravanja Z. Mixed-integer nonlinear programming techniques for process systems engineering. Computers Chemical Engineering, 1995, 19(95), 189–204. doi:10.1016/0098-1354(95)87036-9.

[54] Gong J, You F. Global Optimization for Sustainable Design and Synthesis of Algae Processing Network for CO_2 Mitigation and Biofuel Production Using Life Cycle Optimization. AIChE Journal, 2014, 60, 3195–3210. doi:10.1002/aic.

[55] Garcia DJ, You F. Multiobjective Optimization of Product and Process Networks: General Modeling Framework, Efficient Global Optimization Algorithm, and Case Studies on Bioconversion. AIChE Journal, 2015, 61, 530–554. doi:10.1002/aic.

[56] Errico M, Rong B-G, Tola G, Turunen I. A method for systematic synthesis of multicomponent distillation systems with less than N-1 columns. Chemical Engineering and Processing: Process Intensification, 2009, 48(4), 907–920. doi:10.1016/j.cep.2008.12.005.

[57] Rong B-G, Kraslawski A. Partially thermally coupled distillation systems for multicomponent separations. AIChE Journal, 2003, 49(5), 1340–1347. doi:10.1002/aic.690490525.

[58] Rong BG, Kraslawski A, Turunen I. Synthesis of functionally distinct thermally coupled configurations for quaternary distillations. Industrial & Engineering Chemistry Research, 2003, 42(6), 1204–1214. doi:10.1021/ie0207562.

[59] Rong B-G, Rong B-G, Kraslawski A, Kraslawski A, Turunen I, Turunen I. Synthesis of Heat-Integrated Thermally Coupled Distillation Systems for Multicomponent Separations. Industrial & Engineering Chemistry Research, 2003, 42(19), 4329–4339. doi:10.1021/ie030302k.

[60] Rong B-G, Kraslawski A, Turunen I. Synthesis and Optimal Design of Thermodynamically Equivalent Thermally Coupled Distillation Systems †. Industrial & Engineering Chemistry Research, 2004, 43(18), 5904–5915. doi:10.1021/ie0341193.

[61] Rong BG, Turunen I. A new method for synthesis of thermodynamically equivalent structures for Petlyuk arrangements. Chemical Engineering Research & Design, 2006, 84(A12), 1095–1116. doi:10.1205/cherd06012.

[62] Rong BG. Synthesis of dividing-wall columns (DWC) for multicomponent distillations-A systematic approach. Chemical Engineering Research and Design, 2011, 89(8), 1281–1294. doi:10.1016/j.cherd.2011.03.014.

[63] Torres-Ortega CE, Errico M, Rong B-G. Design and optimization of modified non-sharp column configurations for quaternary distillations. Computers & Chemical Engineering, 2015, 74, 15–27. doi:10.1016/j.compchemeng.2014.12.006.

[64] Rong B-G, Kraslawski A. Optimal Design of Distillation Flowsheets with a Lower Number of Thermal Couplings for Multicomponent Separations. Industrial & Engineering Chemistry Research, 2002, 41(23), 5716–5726. doi:10.1021/ie0107136.

[65] Sun Y, Cheng J. Hydrolysis of lignocellulosic materials for ethanol production: a review. Bioresource Technology, 2002, 83(1), 1–11. doi:10.1016/S0960-8524(01)00212-7.

[66] Haghighi Mood S, Hossein Golfeshan A, Tabatabaei M, Salehi Jouzani G, Najafi GH, Gholami M, Ardjmand M. Lignocellulosic biomass to bioethanol, a comprehensive review with a focus on pretreatment. Renewable and Sustainable Energy Reviews, 2013, 27, 77–93. doi:10.1016/j.rser.2013.06.033.

[67] Martín M, Grossmann I. Simultaneous optimization and heat integration for biodiesel production from cooking oil and algae. Industrial & Engineering Chemistry, 2012, 51, 7998–8014. Retrieved from http://pubs.acs.org/doi/abs/10.1021/ie2024596.

[68] Garlock RJ, Balan V, Dale BE, Ramesh Pallapolu V, Lee YY, Kim Y, Warner RE. Comparative material balances around pretreatment technologies for the conversion of switchgrass to soluble sugars. Bioresource Technology, 2011, 102, 11063–11071. doi:10.1016/j.biortech.2011.04.002.

[69] Pérez Ja, Ballesteros I, Ballesteros M, Sáez F, Negro MJ, Manzanares P. Optimizing Liquid Hot Water pretreatment conditions to enhance sugar recovery from wheat straw for fuel-ethanol production. Fuel, 2008, 87(17–18), 3640–3647. doi:10.1016/j.fuel.2008.06.009.

[70] Sun TTW. Impact of Pretreatment Methods on Enzymatic Hydrolysis of Softwood. University of Toronto, 2013.

[71] Yat SC, Berger A, Shonnard DR. Kinetic characterization for dilute sulfuric acid hydrolysis of timber varieties and switchgrass. Bioresource technology, 2008, 99(9), 3855–3863. doi:10.1016/j.biortech.2007.06.046.

[72] Normark M, Winestrand S, Lestander Ta, Jönsson LJ. Analysis, pretreatment and enzymatic saccharification of different fractions of Scots pine. BMC biotechnology, 2014, 14, 20. doi:10.1186/1472-6750-14-20.

[73] Saha BC, Nichols NN, Qureshi N, Kennedy GJ, Iten LB, Cotta Ma. Pilot scale conversion of wheat straw to ethanol via simultaneous saccharification and fermentation. Bioresource technology, 2015, 175, 17–22. doi:10.1016/j.biortech.2014.10.060.

[74] Li E, Xu ZP, Rudolph V. MgCoAl-LDH derived heterogeneous catalysts for the ethanol transesterification of canola oil to biodiesel. Applied Catalysis B: Environmental, 2009, 88(1–2), 42–49. doi:10.1016/j.apcatb.2008.09.022.

[75] Severson K, Martín M, Grossmann I. Optimal integration for biodiesel production using bioethanol. AIChE Journal, 2013, 59(3), 834–844. doi:10.1002/aic.

[76] Valle P, Velez A, Hegel P, Mabe G, Brignole EA. Biodiesel production using supercritical alcohols with a non-edible vegetable oil in a batch reactor. Journal of Supercritical Fluids, 2010, 54(1), 61–70. doi:10.1016/j.supflu.2010.03.009.

[77] Torres-Ortega CE, Rong B-G. Synthesis, Design, and Rigorous Simulation of the Bioethanol Recovery and Dehydration from an Actual Lignocellulosic Fermentation Broth. Industrial & Engineering Chemistry Research, 2015, 55, 210–225. doi:10.1021/acs.iecr.5b02773.

[78] Humbird D, Davis R, Tao L, Kinchin C, Hsu D, Aden A, Dudgeon D. Process design and economics for biochemical conversion of lignocellulosic biomass to ethanol. Technical Report NREL, (May), 2011. Retrieved from http://www.researchgate.net/publication/229843699_Process_Design_and_Economics_for_Biochemical_Conversion_of_Lignocellulosic_Biomass_to_Ethanol_Dilute-Acid_Pretreatment_and_Enzymatic_Hydrolysis_of_Corn_Stover/file/9fcfd5011638d5a2af.pdf.

Chandrakant R. Malwade, Haiyan Qu, Ben-Guang Rong, and
Lars P. Christensen

8 Process synthesis for natural products from plants based on PAT methodology

Abstract: Natural products are defined as secondary metabolites produced by plants and form a vast pool of compounds with unlimited chemical and functional diversity. Many of these secondary metabolites are high-value-added chemicals that are frequently used as ingredients in food, cosmetics, pharmaceuticals, and other consumer products. Therefore, process technology toward industrial-scale production of such high value chemicals from plants has significant importance. This chapter discusses a process synthesis methodology for recovery of natural products from plants at the conceptual level. The methodology generates different process flowsheet alternatives consisting of multiple separation techniques. Decision making is supported by heuristics as well as basic process information already available from previous studies. A process analytical technology (PAT) framework, part of the quality by design approach, has been included at various steps to obtain molecular-level information on process streams and, thereby, support rational decision making. The formulated methodology has been used to isolate and purify artemisinin, an antimalarial drug, from dried leaves of the plant *Artemisia annua*. A process flowsheet is generated consisting of maceration, flash column chromatography, and crystallization unit operations for extraction, partial purification, and final purification of artemisinin, respectively. PAT tools such as high-performance liquid chromatography, liquid chromatography–mass spectroscopy, and chemometric methods have been used extensively to characterize the process streams at the molecular level. The generated process information is used to further optimize the process flowsheet.

Keywords: Process synthesis, natural products, process analytical technology, artemisinin, chromatography, crystallization

8.1 Introduction

8.1.1 Natural products from plants

Nature has provided humankind with tremendous wealth in the form of plants, which humans started to use thousands of years ago as food and as remedies for various ailments [1, 2]. Over the centuries, this knowledge of plants and their use to alleviate pain and cure various ailments has trickled down through the generations, with addition of invaluable experience from successive generations. Ultimately, this resulted in the development of sophisticated traditional medicinal systems, which are still practiced in many parts of the world. The earliest written records of such medicinal systems in-

https://doi.org/10.1515/9783110465068-008

clude documents written in cuneiform on clay tablets in Mesopotamia, Ebers Papyrus from Egypt, Chinese Materia Medica, Ayurvedic remedies from India, De Materia Medica from Greek botanist Pedanious Dioscorides, and many more [3–5]. These records contain valuable information about plants with medicinal properties, their intended uses, and preparation of their extracts. Even today, the traditional medicinal systems serve as a major source of lead drug molecules for the pharmaceutical industry [6–9]. Current drugs are mainly synthesized, including those of natural origin. However, some natural products have very complicated chemical structures and stereochemistry, making their synthesis extremely difficult and very expensive. Consequently, isolation of drugs from plants still constitutes an important source of several important drugs, such as the anticancer drugs vincristine and vinblastine and the antimalarial drug artemisinin (Figure 8.1). The procedures for isolating drugs from plants and other natural sources are often very tedious, expensive, and sometimes not very sustainable; therefore, there is a need to develop sustainable large-scale isolation processes for drugs of natural origin. This chapter describes the application of process analytical technology (PAT) to assist sustainable development of such isolation processes.

Plants produce a vast and diverse assortment of natural products, of which the great majority do not take part in growth and development. These compounds are traditionally referred to as secondary metabolites and are differently distributed in the plant kingdom. Some secondary metabolites are restricted to a particular taxonomic group, such as a family, genus, or even species, whereas others are widely distributed. Secondary metabolites play a central role in defense mechanisms, signaling pathways, attracting pollinators, etc. [9], although the function of many secondary metabolites is still unknown. Secondary metabolites constitute, however, an important source of bioactive natural products and are still the most important source of new drugs for the pharmaceutical industry, either in their native form or as lead compounds [8]. In contrast, primary metabolites such as phytosterols, acyl lipids, sugars, nucleotides, amino acids, and organic acids are found in all plants and perform metabolic roles that are essential for the growth and development of plants. They also act as precursors and/or building blocks in the biosynthesis of secondary metabolites [10]. Based on their biosynthetic origins, plant natural products (primary and secondary metabolites) can be divided into four major groups [10]:

- fatty acids and polyketides derived from the acetate pathway
- aromatic amino acids and phenylpropanoids from the shikimate pathway
- terpenoids and steroids derived from the mevalonate and methylerythritol phosphate pathways
- alkaloids containing nitrogen and derived from aliphatic and aromatic amino acids

Secondary metabolites are usually synthesized in specialized cells and are accumulated in plant cells in smaller quantities than primary metabolites; thus, the extraction and purification of most secondary metabolites can be challenging. In Table 8.1,

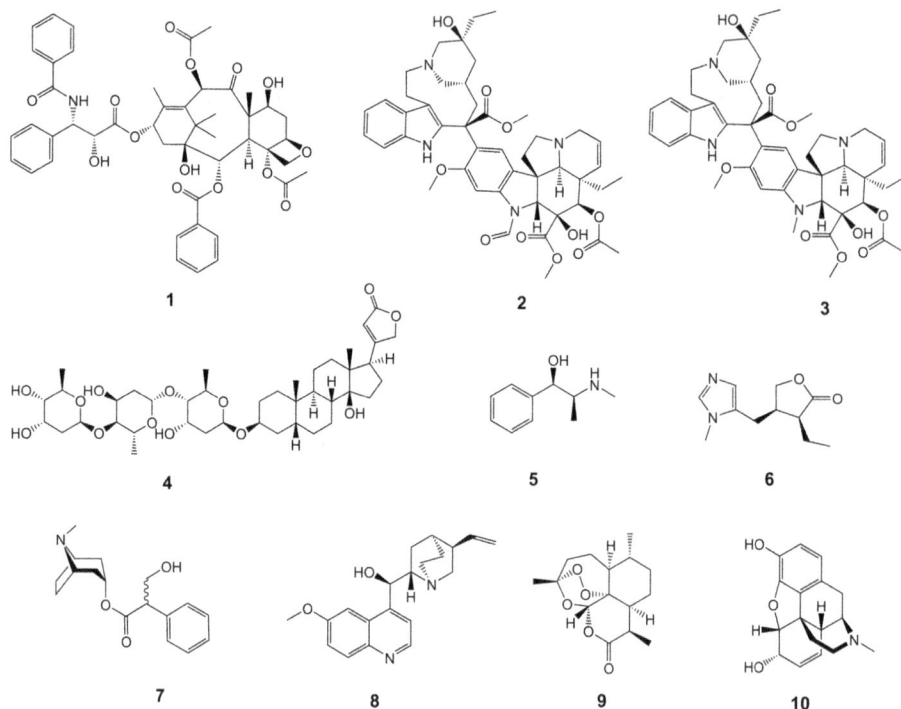

Fig. 8.1: Chemical structures of natural products used as drugs. Paclitaxel (taxol) (1) is a diterpenoid anticancer drug from the bark of *Taxus brevifolia*. Vincristine (2) and vinblastine (3) are terpenoid indole alkaloid anticancer drugs from the leaves of Madagascar periwinkle (*Catharanthus roseus*). Digitoxin (4) is a steroidal saponin and a cardiotonic drug from the leaves of red foxglove (*Digitalis purpurea*). Ephedrine (5) is the main alkaloid from *Ephedra* species and an adrenergic drug. Pilocarpine (6) is an imidazole alkaloid and an antiglaucoma agent from the leaves of *Pilocarpus jaborandi*. The tropane alkaloid atropine (7) is the racemic form of hyoscyamine and an anticholinergic drug from deadly nightshade (*Atropa belladonna*). The antimalarial drug quinine (8) is a quinolone alkaloid, from the bark of *Cinchona* species and the sesquiterpene lactone artemisinin (9) is from leaves of *Artemisia annua*. An analgesic drug morphine (10) is from the seedpods of *Papaver somniferum* [11].

important types of secondary metabolites are listed according to their main biosynthetic pathways, some of which have been used for the development of drugs. Several biosynthetic pathways are often involved in the biosynthesis of secondary metabolites, which also explains the structural diversity among secondary metabolites and the large number of natural products isolated from plants. *Dictionary of Natural Products* has already documented around 139,000 of such natural products along with their chemical and physical properties, with supplements being in progress [10]. Scientists believe that this is just the tip of the iceberg as only about 10% of the 250,000 described species of flowering plants have been studied chemically [11]. Therefore, the database of natural products from plants may grow significantly in future, thereby offering a

Tab. 8.1: Important types of secondary metabolites from plants, listed according to their main biosynthetic pathways, some of which have been used for the development of drugs.

Biosynthetic origin	Type of secondary metabolites
Acetate pathway	Unsaturated fatty acids Polyacetylenes Alkylamides Anthrones Anthraquinones Cannabinoids
Shikimate pathway	Lignans Phenylpropenes Coumarins Furanocoumarins Styrylpyrones Diarylheptanoids Benzoic acid derivatives, including tannins Cinnamic acid derivatives Flavonoids Isoflavonoids Stilbenes Quinones, including terpenoid quinones
Mevalonate and methylerythritol phosphate pathways	Iridoids Sesquiterpenoids Diterpenoids Triterpenoids and triterpenoid saponins Steroids and steroid saponins
Alkaloids	Pyrrolidine and tropane alkaloids Phenylethylamine and tetrahydroisoquinoline alkaloids Modified benzyltetrahydroisoquinoline alkaloids Phenylethylisoquinoline alkaloids Terpenoid tetrahydroisoquinoline alkaloids Terpenoid indole alkaloids Quinoline alkaloids Imidazole alkaloids

huge variety of natural products with interesting chemical structures and bioactivities that surpass the human imagination. Natural products have widespread applications as pharmaceuticals, food ingredients, dietary supplements, flavors and fragrances, pesticides, and cosmetics [11–14]. According to a technical market report published by BCC Research, the global market for botanical and plant-derived drugs was valued at 22.1 billion US$ in 2012. The forecasts show that the total market value is expected to reach 26.6 billion US$ in 2017 after increasing at a five-year compound annual growth rate of 3.7% [15]. Recent market surveys clearly indicate the growing preference of consumers for dietary supplements, food ingredients, and cosmetics containing natural

Tab. 8.2: Examples of natural products from plants used as dietary supplements.

Ingredient	Botanical source	Activity
Terpenoids, flavonoids	Gingko biloba	Enhancement of memory
Allicin, *S*-allylcysteine	Garlic	Cardiovascular diseases
Alkylamides, flavonoids	Echinacea	Improve immune system
Hypericin, hyperforin	St John's Wort	Against depression
Ginsenosides	Ginseng	Anticancer
Diarylheptanoids, gingerols	Ginger	Anti-inflammatory, anticancer
Iridoids	Valerian	Sedative
Acylglycerides	Saw palmetto	Chronic prostatitis

products. This preference is expected to fuel the growth in the market value of these products [16, 17]. A few examples of natural products from plants used as pharmaceuticals and dietary supplements are shown in Figure 8.1 and Table 8.2, respectively.

8.1.2 Need for recovery of natural products from plants

Today's global pharmaceutical industry, a major application area for natural products from plants, is highly regulated in terms of the standardization of products. Standardization of products is related to their consistent quality and efficacy. Natural products from plants in crude forms such as concoctions, extracts, or raw plants are almost impossible to standardize, mainly because of their natural origin. The standard qualities of such products depend on several factors that are difficult to control, for example, environmental conditions in which the plants are grown. The medicinal plants grown in one part of the world may not have the same medicinal value as those grown in another part of the world. Thus, it is almost impossible for such products to pass the stringent requirements set by regulatory authorities such as the US Food and Drug Administration (USFDA) and the European Medicine Agency. These regulatory authorities require the active principles responsible for therapeutic activity of the crude natural products to be obtained in pure form, which can be then used to formulate different dosage forms [18]. Although natural products such as dietary supplements and cosmetics are not regulated under such stringent requirements, concerns have been raised recently regarding their adulteration and the deceptive and questionable marketing of dietary supplements in the USA. In addition to this, lack of information about the interactions of dietary supplements with drugs is a cause of concern for regulatory authorities, which may force such products to be regulated similarly to pharmaceuticals from plants [19]. Taking into consideration the regulatory concerns and the additional advantages of pure natural products (compactness, reduced side effects, and enhanced activity), recovery of natural products from plants assumes great importance.

Several other approaches, such as metabolic engineering, plant tissue culture, hairy root culture, genetic manipulation, and total synthesis, have been developed for the manufacture of natural products from plants [20–22]. Although some natural products have been manufactured successfully using these techniques, the applications are very limited and the methods are not economically feasible for most natural products. Therefore, processes involving isolation and purification remain the methods of choice for recovery of natural products from plants, which is the focus of this chapter. Recent developments in the field of agronomy regarding cultivation of medicinal plants, combined with efficient design of processes for recovery of natural products from plants, offer great potential to deliver natural products in pure form economically.

8.1.3 Challenges in recovery of natural products from plants

Most commonly encountered separation problems in the chemical industry involve purification of a product from a relatively well-defined reaction mixture in terms of product concentration and identity of side products. However, the purification of natural products from plants (Figure 8.2) is a very different scenario and encounters various challenges, as explained briefly in Sections 8.1.3.1 to 8.1.3.4.

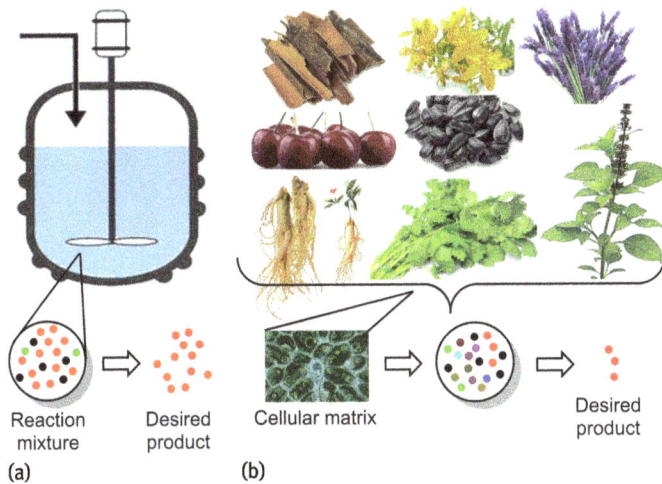

Reaction mixture   Desired product   Cellular matrix   Desired product

(a)   (b)

Fig. 8.2: Problem specification for recovery of natural products from plants: (a) purification of a desired product from reaction mixture and (b) purification of natural product from plants. Red dots indicate the desired product to be recovered, other dots indicate impurities. The number of dots indicates the representative concentration of each component.

8.1.3.1 Raw material variation

Consistent quality of raw material is a primary requirement for the design of a robust process that converts raw material into the final product. In the case of isolation and purification of natural products from plants, the raw material could be flowers, stems, fruits, roots, seeds, leaves, and sometimes the whole plant obtained from different sources. Bioactive secondary metabolites are biosynthesized and stored in the cellular matrix of these parts of the plants, as shown in Figure 8.2(b). Biosynthetic pathways are very sensitive to various factors, such as environmental conditions (high and low temperature, drought, alkalinity, salinity, UV stress, and pathogen infection) and nutrient availability in the plant-growing region [23, 24]. These kind of stresses, which vary from region to region, can influence the biosynthetic pathways producing secondary metabolites and thereby affect their concentration level [25]. This can lead to either positive or negative changes in the levels of desired secondary metabolites. In both cases, inconsistent quality of raw material can be problematic from the process design point of view as it has a direct connection with process parameters such as equipment, analytical techniques, and operating conditions.

8.1.3.2 Low concentration of product in the raw material

The concentration of secondary metabolites produced in plants is often very low. Paclitaxel, an anticancer drug, is a very good example of such a natural product. It is obtained from the bark of *Taxus brevifolia* and, during its development phase, about 1200 kg of bark was collected and processed to obtain enough paclitaxel ($\sim 10\,g$) for clinical trials [26]. This certainly contrasts with the generally encountered purification problems in the chemical industry, which involve purification of the desired product, often in high concentrations; furthermore, scale up of the process is generally straightforward.

8.1.3.3 Complicated mixture containing many impurities

The desired natural products from plants are often embedded in the cellular matrix along with other cometabolites. The presence of several cometabolites having similar chemical structure and physicochemical properties to the desired natural product can significantly influence the performance of stand-alone separation processes. For instance, cometabolites having similar chromatographic properties to those of the desired natural product could be difficult to separate by chromatographic separation. Similarly, cometabolites influencing the solubility or boiling point of the desired natural product could influence the crystallization or distillation significantly. Therefore, isolation and purification of natural products from plants demand the use of multiple separation techniques arranged in an optimal sequence.

8.1.3.4 Lack of process information

The availability of fundamental process data such as physicochemical properties of the desired product, impurities present in the mixture, and thermodynamic data (solid–liquid, liquid–liquid, and liquid–vapor equilibria) is crucial for the design of an efficient separation process. In the case of recovery of natural products, the physicochemical properties of desired product and, to some extent, the impurities could be available from pharmacognostic and agronomic studies. However, there is always a dearth of thermodynamic data that needs to be either predicted theoretically or determined experimentally during design of the separation process.

8.1.4 Process synthesis for separation of multicomponent mixtures

Most chemical processes are associated with some kind of separation tasks for the overall process synthesis and design, for example, separation of a product from the reaction mixture or separation of multicomponent mixtures into different pure products. Therefore, separation process synthesis has received considerable attention in the chemical industry and has progressively evolved to deal with challenging separation problems involving multicomponent mixtures. Process synthesis is a complicated task involving multitask attention, spanning different disciplines and points of view, and with different focus and levels of detail for conversion of raw material into finished product. Several methodologies have been developed to design separation processes for purification of multicomponent mixtures, as shown in Figure 8.3.

The heuristic approach is totally empirical and based on experience collected by engineers and researchers. It is widely used in the chemical process industry for rapid decision making and there are many well-established heuristics, developed over time, especially regarding separation problems involving distillation [27]. The evolutionary design approach involves making a series of modifications to a previously syn-

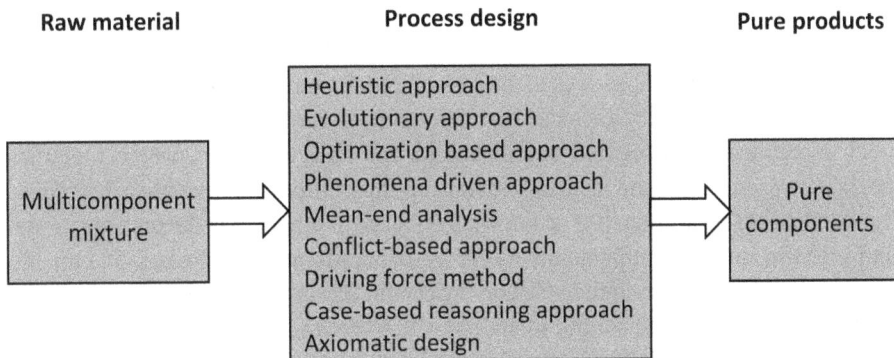

Fig. 8.3: Different methodologies developed for separation process design.

thesized process, leading to an improved design. Thus, evolutionary approaches for design of processes depend very much on the initially synthesized process flowsheet. Stephanopoulos and Westerberg [28] describe the steps involved in evolutionary process synthesis as follows: an initial flowsheet, rules to make systematic and small changes to the initial flowsheet generating an alternative flowsheet, effective strategy to apply these rules, and a means to compare the original flowsheet with any of its alternative flowsheets. Douglas [29] proposed a hierarchical heuristic procedure for chemical process design, which is a combination of heuristic and evolutionary approaches.

Optimization-based approaches for process synthesis involve use of algorithmic methods, such as mixed integer nonlinear programming (MINLP) to optimize the given objective function, which could be economic, environmental, or safety related [30]. An important drawback of optimization-based methods is the need for huge computational effort and the fact that the optimality of the solution can be guaranteed only with respect to the alternatives that have been considered a priori. In addition, this approach encounters great difficulties when dealing with the optimization of underdefined design problems and uncertainties that result from the multi-objective requirements of the design problem.

Phenomena-driven design proposes that reasoning should not start at the level of building blocks but at a low level of aggregation (i.e., at the level of the phenomena that occur in those building blocks). Jaksland et al. [31] studied separation process design and synthesis based on thermodynamic phenomena. They explored the relationships between the physicochemical properties, separation techniques, and conditions of operation. The number of alternatives for each separation task was reduced by systematically analyzing these relationships. Then, the possible flowsheets were produced with a list of alternatives for the separation tasks.

The mean end analysis method is based on application of various operations in such a sequence that the differences in properties between the raw materials and products are systematically eliminated, resulting in transformation of the raw materials into the desired products. Siirola [32] describes the hierarchy of property difference reduction as identity, amount, concentration, temperature, pressure, and form. A major disadvantage of this approach is the lack of understanding about undesirable side effects from an increase or decrease in the differences of other properties.

The conflict-based method is founded on the TRIZ approach, which identifies the system's conflicts and contradictions for the solution of inventive problems. The conflict-based approach is a knowledge-based method that decomposes a design problem into subproblems instead of applying hierarchical design. It is an efficient method for modifying the solution space and screening alternatives at an early stage [33]. Case-based reasoning methods utilize the solutions that were applied to past similar problems to find a solution for an existing problem. It deals with very specific data from previous situations and reuses results and experience to fit new problem situations. During the first step, retrieval, a new problem is matched against problems of previous cases by calculating the value of the similarity functions in order to find the most similar problem and its solution. If the proposed solution does not meet the necessary requirements of the new problem, case-based reasoning proceeds to the next step, adaptation, and creates a new solution.

Sauar et al. [34] have proposed a new principle of process design based on the equipartition of the driving forces. They claimed that process design should be optimized by the equal distribution of driving forces throughout the process by assuming that the rates of entropy production are proportional to the square of the driving forces. However, the basic assumption that entropy production rates are proportional to the square of the driving forces is not valid for many important chemical processes. The axiomatic approach to design is used to define both a design methodology and a set of rational criteria for decision making. The first principle is the independence axiom, which states that a good design maintains the independence of the functional requirements. The second principle is the information axiom, which claims that the information content is minimized in a good design. It establishes the information content as a criterion for the evaluation of design alternatives.

8.2 Process synthesis for recovery of natural products from plants

The process synthesis approaches mentioned above are suitable for general chemical products and could be applicable in the case of clearly defined separation problems. As mentioned earlier, the recovery of natural products from plants is a challenging task and represents a very different problem from the separation problems commonly encountered in the chemical industry. In addition, systematic process design for recovery of natural products from plants has received less attention in the past and only a handful of reports are available. A heuristic approach has been discussed by Harjo et al. [35] for design of a manufacturing process for phytochemicals. The approach consists of five steps: product specification and feed characterization, selection of product recovery techniques, construction of flowsheet alternatives, selection of operating conditions, and evaluation of flowsheet alternatives. Gertenbach [36] has described thermodynamic and kinetic aspects of solid–liquid extraction (SLE) technologies for manufacture of nutraceuticals, including fundamental aspects such as solid–liquid equilibrium, mass transfer, and the effect of process parameters (solvent composition, temperature, particle size, and liquid-to-solid ratio). The engineering aspects of equipment selection and design for extraction processes have also been discussed. A new approach has been proposed by Kassing et al. [37, 38] and Ndocko et al. [39] that combines the design of experiments and rigorous process modeling for developing plant-based extraction processes. Intensified collaboration between different disciplines such as process engineering, botany, and analytical chemistry has also been emphasized for addressing shortcomings in the existing processes.

Strube et al. [40] described the process synthesis for recovery of natural products from plants at the conceptual level. A process consisting of multiple unit operations such as SLE followed by liquid–liquid extraction (LLE) and chromatography, and their combination, has been suggested for recovery of natural products. Additionally, guidelines have been provided for optimal selection of solvents for SLE, LLE, and chromatography and for selection of equipment. Most of the literature describing process development for recovery of natural products is limited to the extraction process and related process parameters. Process development approaches that consider multidisciplinary collaboration and rigorous process modeling for recovery of natural products from plants are more realistic. However, the fundamental process information required for rigorous process modeling is rarely available. In addition, these approaches do not consider the interconnections between multiple unit operations at the molecular level, which can be very important in developing an optimal process for recovery of a natural product.

Therefore, there is a need for the development of a general methodology that can systematically generate and evaluate different process alternatives and allow the synergistic effect between different separation techniques to be determined at the conceptual level. The following two major tasks should be considered in the methodology for recovery of natural products from plants:

1. selection of suitable separation methods for isolation and purification of the target compound
2. characterization and analysis of the required data and information concerning samples in each of the processing steps, as well as the separation performance in terms of concentration, yield, purity, and productivity

Considering the complex nature of the problem, the first task can be supported by the heuristics approach, whereas the second needs process analytical technology (PAT) to obtain the required data and information. Thus, we have formulated a methodology based on PAT for recovery of natural products from plants. The application of PAT is crucial for generation of process information at the molecular level, which can play an important role in decision making and in determining the synergic effect between multiple unit operations. Before describing the details of PAT-based methodology for recovery of natural products from plants, we give a brief introduction to the PAT framework and the encompassing tools.

8.2.1 Process analytical technology

The genesis of PAT lies in the development of process analytical chemistry, which has been used in the petrochemical industry for many decades to improve understanding of processes and thereby enable their control [41]. Control of continuous processes in the petrochemical industry is effected by collecting real-time data on critical process parameters. Originally, PAT was a part of the quality by design (QbD) approach developed by Juran [42] with the aim of designing quality into products and bringing discipline into process development in order to remove quality failure at source. The International Council for Harmonization (ICH) has adopted the QbD approach in the development and manufacture of pharmaceuticals, as described in the ICH guidelines. According to the QbD approach, a quality target product profile (QTPP) containing the product specifications is defined and the design space adjusted such that the final product specifications are met by monitoring the critical quality attributes of raw materials, in-process materials, and final product. For example, the QTPP for a drug obtained from plants contains specifications such as purity, particle size distribution, crystal polymorph, and optical purity, which depend on the intended application of the final product. Recently, the USFDA also released a draft guidance for the pharmaceutical industry, which recommends use of PAT to improve the efficiencies of both manufacturing and regulatory processes through improved understanding of process [43]. Process understanding is generally measured in terms of identification and explanation of all critical sources of variability, variability management by the process, and accurate and reliable prediction of product quality attributes over the design space. Thus, the application of PAT in the early stage of process design assumes great significance in realizing the basic tenet of QbD. In addition, it can open the door for innovation and improvement of process through generation and application of knowledge during the development phase. PAT is emerging as an important process systems engineering tool in the manufacture of pharmaceuticals and chemicals [44–47]. Therefore, we include PAT in the process synthesis methodology for recovery of natural products from plants, where lack of process information is always a bottleneck. As stated in the USFDA draft guidance, PAT comprises multivariate tools for design, data acquisition, and analysis; process analyzers; process control tools; and continuous improvement and knowledge management tools. However, in the context of process synthesis for recovery of natural products at the conceptual level, we consider process analyzers and multivariate tools for design and data analysis.

8.2.1.1 Process analyzers

Process analyzers play an important role in monitoring the critical quality attributes of products and process streams. Modern process analytical chemistry offers a number of techniques for measuring different physical and chemical properties of simple and complicated samples at the atomic, molecular, and bulk levels. A comprehensive review of modern process analyzers and their applications is provided by Workman et al. [41]. A major development in the area of process analyzers is the possibility of on-line/in-line measurements, which give a real-time snapshot of the process and, thereby, a better understanding of the process. On-line measurements are performed on a diverted process stream, whereas in-line measurements are performed directly on the process stream, either invasively or noninvasively. At-line measurement requires the sample to be removed from the process stream and analyzed separately. In the context of process design for recovery of natural products from plants, the most commonly encountered analyses include the characterization of raw materials, process streams, and finished products. Very often, analyses are performed on complex mixtures containing many components. Therefore, techniques such as high-performance liquid chromatography (HPLC), high-performance liquid chromatography combined with diode-array detection (HPLC-DAD), gas chromatography (GC), liquid chromatography coupled with mass spectrometry (LC-MS), and gas chromatography coupled with mass spectrometry (GC-MS), which are capable of achieving the separation of

Tab. 8.3: Commonly used process analysis techniques, their applications, and their mode of operation [41].

Process analysis technique	Applications	Mode of operation
High-performance liquid chromatography (HPLC)	Separation, quantification, identification of solid/liquid samples	At-line
Gas chromatography (GC)	Separation, quantification, identification of volatiles or samples that can be volatilized	At-line
Liquid chromatography–mass spectrometry (LC-MS)	Separation, quantification, identification of solid/liquid samples	At-line
Gas chromatography–mass spectrometry (GC-MS)	Separation, quantification, identification of volatiles or samples that can be volatilized	At-line
UV–Vis spectroscopy	Quantification and identification	On-line, at-line
Raman spectroscopy	Quantification, identification of polymorphs, solvates, cocrystals	On-line, in-line, at-line
FTIR spectroscopy	Quantification and identification of solid/liquid samples	On-line, in-line, at-line
Nuclear magnetic resonance (NMR)	Quantification, identification, structure elucidation	At-line
X-ray powder diffraction (XRPD)	Quantification, crystal structure determination, polymorph identification	At-line
Focused beam resonance measurement (FBRM)	Particle size, metastable zone width	On-line, in-line, at-line
Polarimetry	Determination of optical rotation	On-line, at-line
Thermogravimetric analysis (TGA)	Moisture content, degradation, determination of inorganics	At-line
Differential scanning calorimetry (DSC)	Melting enthalpy, melting point, solubility, purity	At-line

chemical components from samples and subsequent qualitative and quantitative analysis, are widely used in research on natural products. Table 8.3 shows the commonly used process analysis techniques, their possible applications, and their mode of operation. Several analyzers used for measurement of critical quality attributes of finished products, such as purity, melting enthalpy, crystal polymorph, particle size, chemical structure, and optical rotation, are also mentioned in Table 8.3.

8.2.1.2 Multivariate tools for design, data acquisition, and analysis

Similar to general chemical products and processes, natural products and the processes for their recovery from plants are complex multifactorial systems. The traditional approach of one-factor-at-a-time experiments is laborious and does not provide adequate insights into the interactions between product qualities and process variables. In addition, the knowledge base for most natural product recovery processes is almost negligible and involves many unknown variables. The PAT framework offers several multivariate tools for methodological experiments based on statistical principles of orthogonality, reference distribution, and randomization, which provide effective means of identifying the effect and interaction of product qualities and process variables. Examples of multivariate tools include statistical design of experiments, response surface methodologies, artificial neural network methods, and pattern recognition tools. Detailed descriptions of various multivariate tools for design are given elsewhere [48, 49].

Application of advanced process analyzers on-line, at-line, or in-line to acquire molecular level information of process streams very often generates an enormous amount of data. Manual extraction of relevant chemical information from such databases can be tedious and time consuming. Multivariate or multiway data analysis techniques offer the possibility of rapidly mining relevant chemical information from such databases. From a chemometric perspective, process analyzers and data are classified as univariate, multivariate, or multiway on the basis of the dimensionality of the data set obtained. Process analyzers such as pH meters or single-channel photometers represent the simplest type and produce univariate data (zeroth-order data tensor), whereas HPLC produces multivariate data (first-order data tensor). In the case of combined techniques such HPLC-DAD, LC-MS, and GC-MS, the data generated is often multiway (second- or higher-order data tensor). In the context of process design for recovery of natural products from plants, application of advanced process analyzers such as HPLC-DAD, LC-MS, and GC-MS to acquire molecular level information of process streams is very common. Chemometrics offers different multivariate and multiway data analysis methods, such as principal component analysis (PCA), multiway curve resolution (MCR), PARAFAC, and Tucker models for analysis of such datasets [50–52]. Mathematical background, development, and application of chemometric methods for data analysis have been reported in several reviews and books [52–54]. Thus, the knowledge base acquired through the application of process analyzers, multivariate design, and data analysis tools serves as a foundation for robust design of processes for recovery of natural products from plants. It can also support the development of process simulation models, which can contribute to continuous learning and help to reduce overall development time.

8.2.2 PAT-based methodology for recovery of natural products from plants

Most of the natural products from plants evolve through the different stages, as shown in Figure 8.4. First, plants with medicinal or health benefits are identified through the references in traditional medicinal systems or from empirical knowledge. This is followed by pharmacognostic investigation and study of the agronomic aspects of cultivation of medicinal plants. Finally, the raw material (biomass) is converted into finished natural products such as pharmaceuticals, food, cosmetics, dietary supplements, and chemicals. Thus, basic process information about the raw material and the natural product to be recovered, as shown in Table 8.4, is mostly available from empirical traditional medicinal systems and pharmacognostic and agronomic investigations. A scheme for recovery of natural products from plants based on PAT methodology is depicted in Figure 8.5. In the first step, an initial process flowsheet is generated with the help of knowledge about the raw material and target products, which is available from previous studies, process information generated through PAT tools, and heuristics. In the next step, underlying equilibrium data for target compounds and associated impurities are determined either experimentally or using prediction methods. Finally, the initial process flowsheet is modified for further improvement on the basis of thermodynamic data.

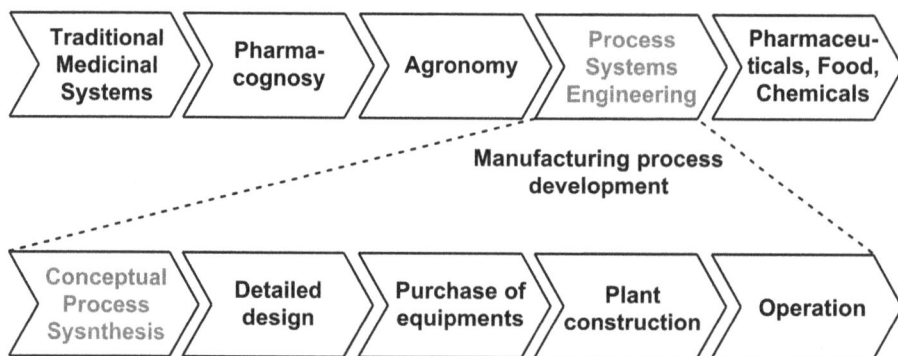

Fig. 8.4: Different stages involved in the evolution of natural products from plants.

Tab. 8.4: Basic information on the desired natural product and its biological source that is available from previous investigations.

Traditional medicinal systems	Pharmacognostic studies	Agronomic studies
– plants with medicinal values – morphological description of plants – application to cure various ailments – parts of the plant being used for applications – preparation of dosage	– botanical information on medicinal plants – anatomy of useful parts of the plants – pharmacological actions of bioactive principles – chemistry of secondary metabolites – physicochemical properties of bioactive molecules	– cultivation of medicinal plants – effect of environmental conditions on quality of medicinal plants – nutritional requirements for cultivation – harvesting time – storage conditions

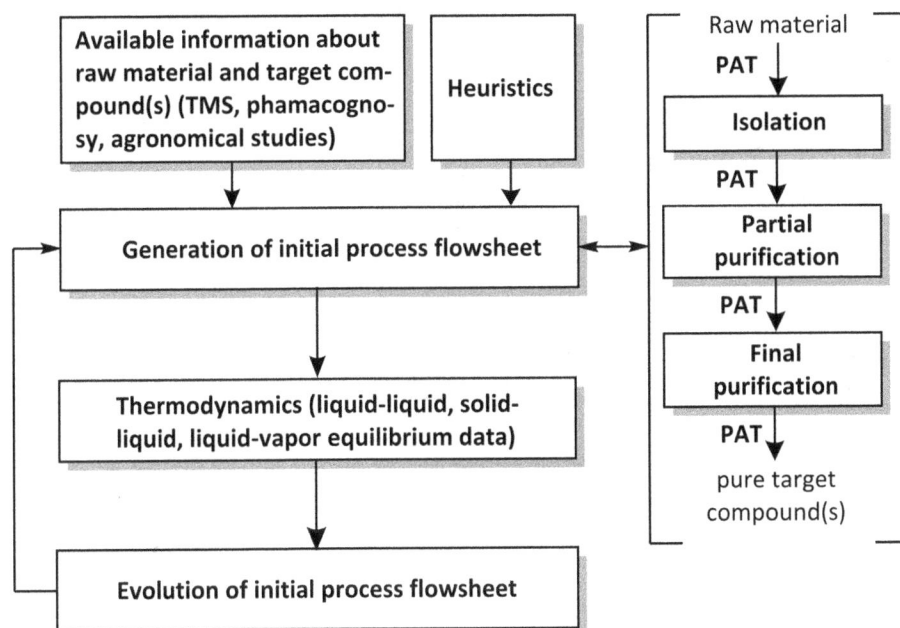

Fig. 8.5: PAT-based process synthesis methodology for recovery of natural products from plants.

8.2.2.1 Generation of initial process flowsheet

As shown in Figure 8.5, the task of generating an initial process flowsheet is decomposed into isolation of target compound from raw material (biomass) and purification of target compound from the extract. Depending on the complexity of the extract, the purification task is further decomposed into partial and final purification. However, decomposition of the task into different steps depends solely on the complexity of the problem. Thus, for some problems, it may be possible to obtain the desired natural

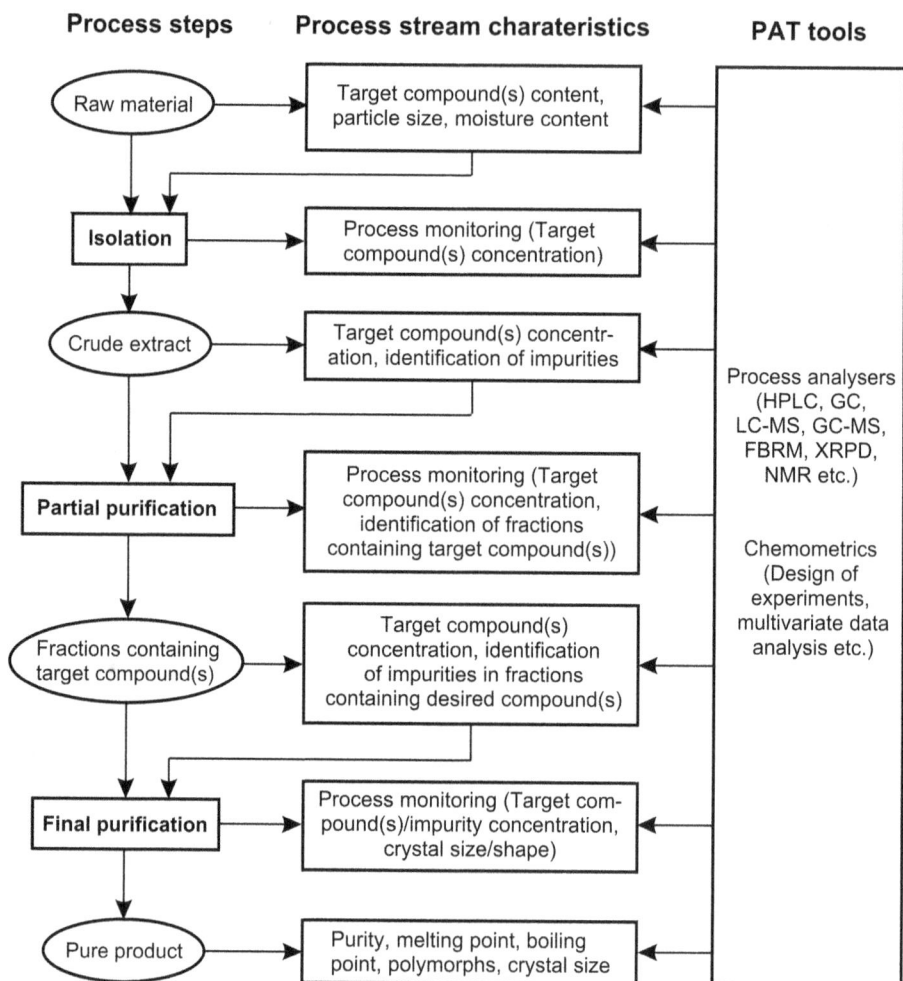

Fig. 8.6: Application of PAT tools for generation of process stream information in recovery of natural products from plants.

product in a single step, whereas other problems require several steps to obtain a high purity natural product. The major tasks involved during generation of an initial process flowsheet are pretreatment of raw material and selection of separation methods, solvents, and operating conditions. The basic process information on the raw material and target compound, as shown in Table 8.4, can be employed in decision making at every step of initial process flowsheet generation. In addition, general heuristics, which are principles based on the experience of previous designers, can be employed for selection of separation techniques, equipment, and operating conditions for recovery of natural products from plants.

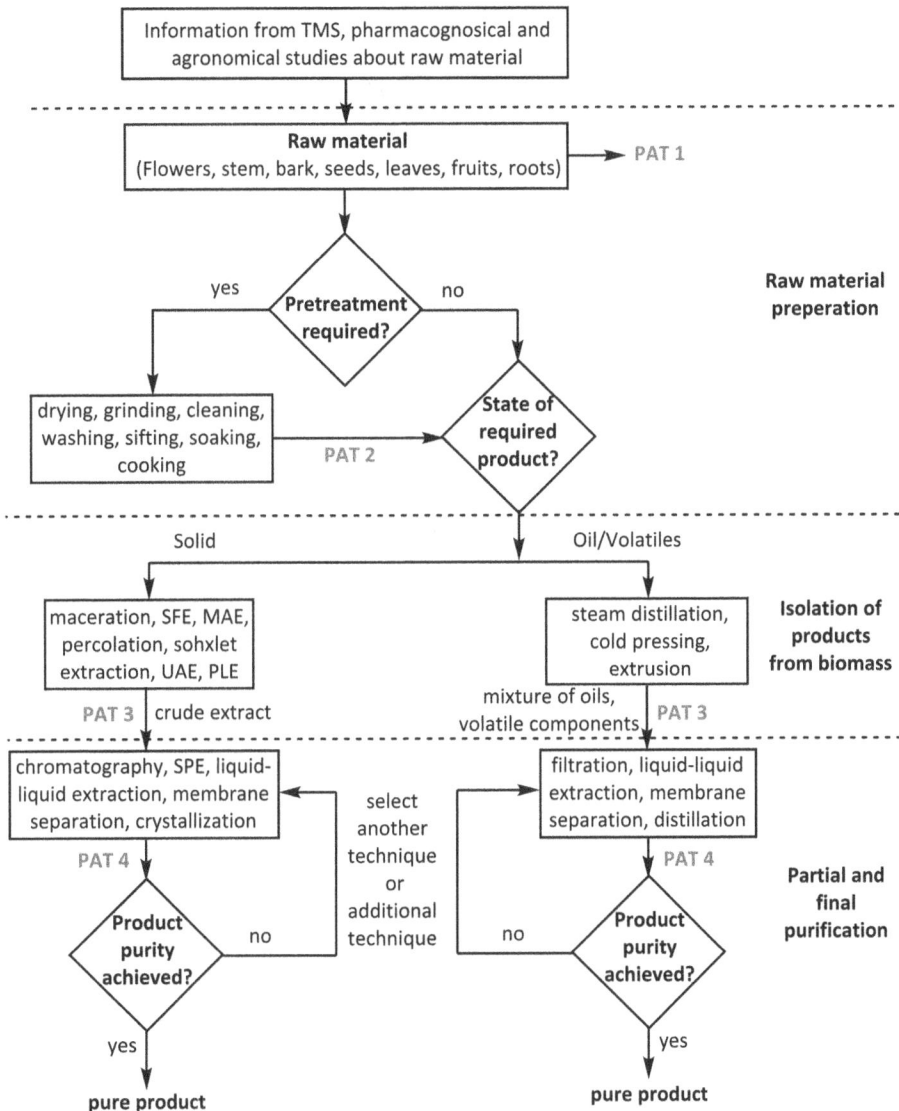

Fig. 8.7: Decision tree for generation of initial process flowsheet for recovery of natural products from plants. SFE supercritical fluid extraction, MAE microwave-assisted extraction, UAE ultrasound-assisted extraction, PLE pressurized liquid extraction, SPE solid phase extraction.

A general set of heuristics specific to the isolation and purification of natural products from plants have been outlined by Malwade et al. [55] and Harjo et al. [35] together with commonly employed separation methods. However, sometimes it is not possible to make a reasonable decision during generation of an initial flowsheet due to lack of specific heuristics, previous experience, or basic process information. In such situations, the application of PAT tools for simultaneous generation and consumption of information about process streams for rational decision making at different steps of initial flowsheet generation can significantly improve the quality of the flowsheet. As shown in Figure 8.6, the process stream characteristics at various steps can be obtained with the help of suitable PAT tools. For example, characteristics of the raw material (biomass) such as content of target compound, moisture content, particle size, and density can be used to select the separation method, operating conditions, solvents, equipment, etc. in the isolation step of initial flowsheet generation. Generation of a suitable initial process flowsheet is very important to the methodology; in fact, the better it is, the more quickly it will evolve into an efficient alternative, which can then be considered for detailed engineering design. A decision tree can be designed and used as a guide for generation of an initial flowsheet for converting raw materials into pure natural products, as shown in Figure 8.7.

8.2.2.2 Optimization of initial process flowsheet

Similar to any optimization problem where the starting point is crucial for quickly finding the optimal solution, the generation of a reasonable initial flowsheet is important for its fast evolution into an optimal flowsheet. Generally, process flowsheets generated only on the basis of heuristics do not guarantee an optimal flowsheet. However, inclusion of the PAT framework to generate process stream information can guide the generation of a near-optimal initial flowsheet. Improvement of an initial flowsheet firstly demands evaluation of the existing flowsheet. Frequently, process flowsheets are evaluated on the basis of technical parameters such as yield, product purity, and productivity and economic parameters such as volume of solvents, number of equipment units, and operating cost. Additionally, environmental parameters such as sustainability, health influence, and climate impact are increasingly taken into consideration. However, in the case of natural products from plants, process flowsheets are evaluated on the basis of product purity and yield. The reasons for this are that the process synthesis for recovery of natural products from plants is done at the conceptual level and that the products are of high value and therefore purity is crucial.

Algorithm-based optimization is used widely for optimization of a superstructure consisting of predefined flowsheet alternatives. In this approach, a superstructure consisting of all process alternatives is mathematically modeled as a general formulation that considers various process variables; equality constraints denoting total mass and energy balances, equilibrium relationships, etc.; inequality constraints corresponding to design specifications; and an objective function. The mathematical model is studied with respect to the theoretical properties that the objective function and constraints may satisfy. Finally, efficient algorithms are used to find the solution, which extracts the optimal flowsheets from the superstructure. However, the major drawback of this approach is that it considers only the process alternatives included in the superstructure and has limited ability to solve the resulting complicated mathematical models. Because process synthesis for recovery of natural products is being considered at the conceptual level, application of this approach is unsuitable because of lack of process information such as basic mass and energy balance modeling, equilibrium relationships, and availability of short-cut methods.

Therefore, an evolutionary approach is used, whereby small changes are made to an existing flowsheet to generate a better "neighboring" flowsheet and the process repeated until the optimal flowsheet is found. The changes made to the existing flowsheet may be based on the physical properties and measured sample stream information, distribution of the target compound and its neighboring compounds, and, more importantly, accurate thermodynamic data and information such as solubility and impurity effects involved in the relevant unit operations.

8.3 Recovery of artemisinin from *Artemisia annua* – a case study

The PAT-based process synthesis methodology has been applied to design a process for recovery of artemisinin, an antimalarial drug, from the dried leaves of *Artemisia annua*. The plant *A. annua*, a member of the Asteraceae family, is a vigorous, erect, annual (sometimes biannual), aromatic herbaceous plant reaching 120–200 cm in height and 100 cm in width, as shown in Figure 8.8 (left) [56]. The plant is native to China; however, it is cultivated in many parts of the world today. Historically, the aerial parts of *A. annua* have been used for treatment of fever and several other ailments in China for thousands of years. Youyou Tu, a Chinese scientist, was awarded a Nobel Prize in 2015 for the discovery that artemisinin can be used for treatment of drug-resistant malaria [57].

Artemisinin is an important antimalarial drug, recommended by the World Health Organization (WHO) for use in combination with other drugs against resistant *Plasmodium falciparum*-induced malaria [56]. Chemically, artemisinin is a sesquiterpene lactone bearing a peroxide bridge, as shown in Figure 8.8 (right). The potency of artemisinin against *P. falciparum*-induced malaria lies in this unique peroxide bridge. The latest report by WHO suggests a significant decrease in mortality rate, especially

Fig. 8.8: Fully grown *A. annua* plant (left) and structure of artemisinin (right).

among children, due to malaria and indicates the major role of medicines like artemisinin in the fight against malaria [58, 59]. Due to its importance as a major source of artemisinin, the plant *A. annua* has been investigated thoroughly in terms of pharmacognostic and agronomic aspects. Basic information about *A. annua* and artemisinin available from such investigations and traditional medicinal systems is shown in Table 8.5.

Tab. 8.5: Basic information about the plant *Artemisia annua* and artemisinin available from previous studies [55].

Traditional medicinal systems	Pharmacognostic studies	Agronomic studies
– identification of wild plant *A. annua* – morphology of *A. annua* – application to cure fever, malaria – preparation of *Artemisia* tea for treatment	– classification of *A. annua* to Asteraceae family – identification of artemisinin as major active principle – location of artemisinin in glandular trichomes on surface of leaves – about 600 secondary metabolites from *A. annua* identified – physicochemical properties of artemisinin	– development of high-yield cultivars of *A. annua* – optimal time of *A. annua* harvest – effect of growing conditions on artemisinin content – pretreatment and storage of *A. annua* leaves – artemisinin content range 0.2–2 wt%

8.3.1 Generation of initial process flowsheet

8.3.1.1 Isolation of artemisinin from *Artemisia annua* leaves

Prior to the isolation of artemisinin from dried leaves of *A. annua*, the raw material was characterized in terms of artemisinin content. HPLC analysis of the extract obtained from exhaustive extraction of dried leaves of *A. annua* showed an artemisinin content of 2.05 wt% [55]. The basic information about artemisinin and *A. annua* shown in Table 8.5 and the heuristics mentioned by Malwade et al. [55] were used to select the extraction method for isolation of artemisinin from dried leaves of *A. annua*. Due to the localization of artemisinin on the surface of leaves in glandular trichomes [60], the simple technique of maceration was chosen for artemisinin extraction. To avoid damage to the cellular matrix and consequent release of intracellular components, there was no pretreatment such as grinding or milling of leaves. Maceration involves immersion of biomass in extraction solvent, with intermittent shaking to facilitate the diffusion of dissolved artemisinin in the bulk solvent. Organic solvents of varying polarity were screened for the extraction of artemisinin in bench-scale experiments. The procedure included immersion of 5 g of dried leaves in 50 mL of solvent at room temperature for 16 h. Figure 8.9 shows the extracts obtained using different solvents, ar-

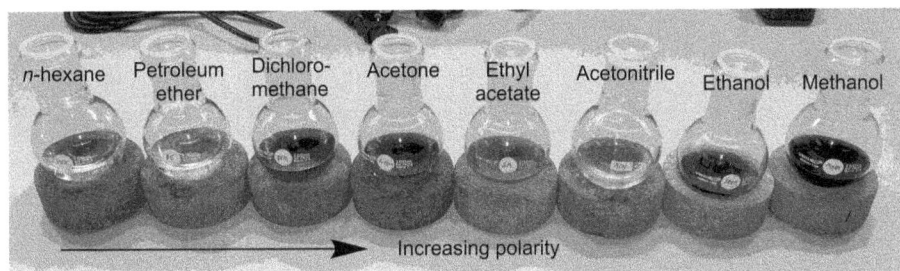

Fig. 8.9: Extracts of *A. annua* leaves obtained from solvents of varying polarity.

ranged in increasing polarity from left to right. It is evident that the nonpolar solvents hexane and petroleum ether yielded clean extracts, whereas solvents with higher polarity yielded relatively darker extracts, a clear indication of coextraction of waxes and chlorophylls in the latter.

Application of PAT tools for isolation of artemisinin

The yield of a target compound in SLE processes depends on several factors, such as temperature, biomass particle size, extraction time, and solvent-to-biomass ratio. Experimental evaluation of all process parameters is very demanding and time consuming. Therefore, PAT tools such as statistical design of experiments can be very useful in developing process models that can predict the impact of process parameters on the efficiency of SLE. Pilkington et al. [61] used a central composite design to assess the impact of extraction temperature, extraction duration, and solvent-to-biomass ratio on artemisinin recovery during SLE. A response surface methodology model and an artificial neural network model were generated by using the experimental results to predict artemisinin recovery and find the optimal operating conditions. However, the models do not predict the impact of process parameters on the molecular composition of extracts, which is also an important factor in the downstream purification of target compound from extracts. Therefore, HPLC-DAD was used to determine the composition of extracts obtained from solvents of varying polarity. The HPLC method employed earlier by Malwade et al. [55] was used for the analysis. The chromatograms were recorded at wavelengths of 190–600 nm to detect the maximum possible number of compounds in the extracts. However, analysis of extracts generated enormous amount of data, 410 chromatograms for each of 8 extracts. Manual extraction of process information such as artemisinin yield, UV profile, and relative concentration of impurities from the dataset was laborious as well as time consuming. Therefore, a multivariate data analysis technique called parallel factorization (PARAFAC) was used to extract information from the chromatograms. Prior to application of PARAFAC, the chromatograms were preprocessed to remove baseline drift and to align the retention time shift due to matrix effects. Figure 8.10 shows the preprocessed chromatograms

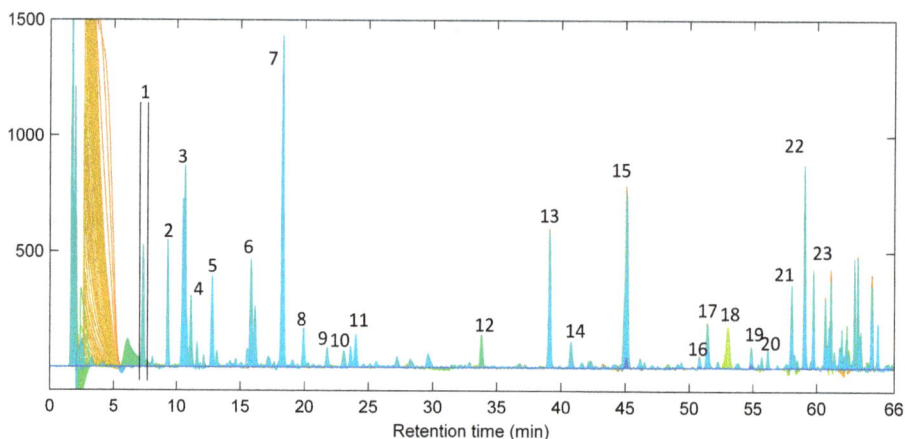

Fig. 8.10: Preprocessed chromatograms of *A. annua* extracts from acetonitrile, acetone, dichloromethane, ethyl acetate, ethanol, *n*-hexane, methanol, and petroleum ether.

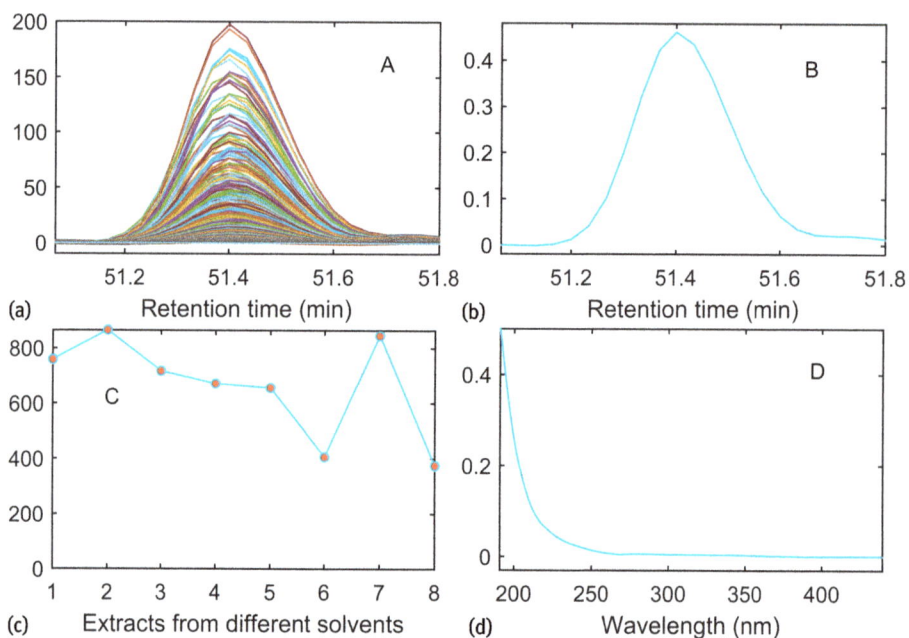

Fig. 8.11: Results from PARAFAC analysis of peak 17 (artemisinin) showing original interval data (a), retention time mode loadings (b), sample mode loadings (c), and UV mode loadings (d): 1 acetonitrile, 2 acetone, 3 dichloromethane, 4 ethyl acetate, 5 ethanol, 6 hexane, 7 methanol, and 8 petroleum ether.

(410×8) of *A. annua* extracts divided into 23 intervals corresponding to individual peaks. PARAFAC analysis as described by Malwade et al. [62] was performed on each interval to obtain the underlying UV profile and relative concentration in each extract. The results are shown in Figure 8.11.

The signal for the chemical entity (artemisinin), underlying UV profile, and relative concentration of artemisinin in all extracts were obtained from PARAFAC analysis of peak 17. It is obvious from Figure 8.11(c) that the yield of artemisinin is higher in acetone and methanol extracts, but lower in hexane and petroleum ether extracts.

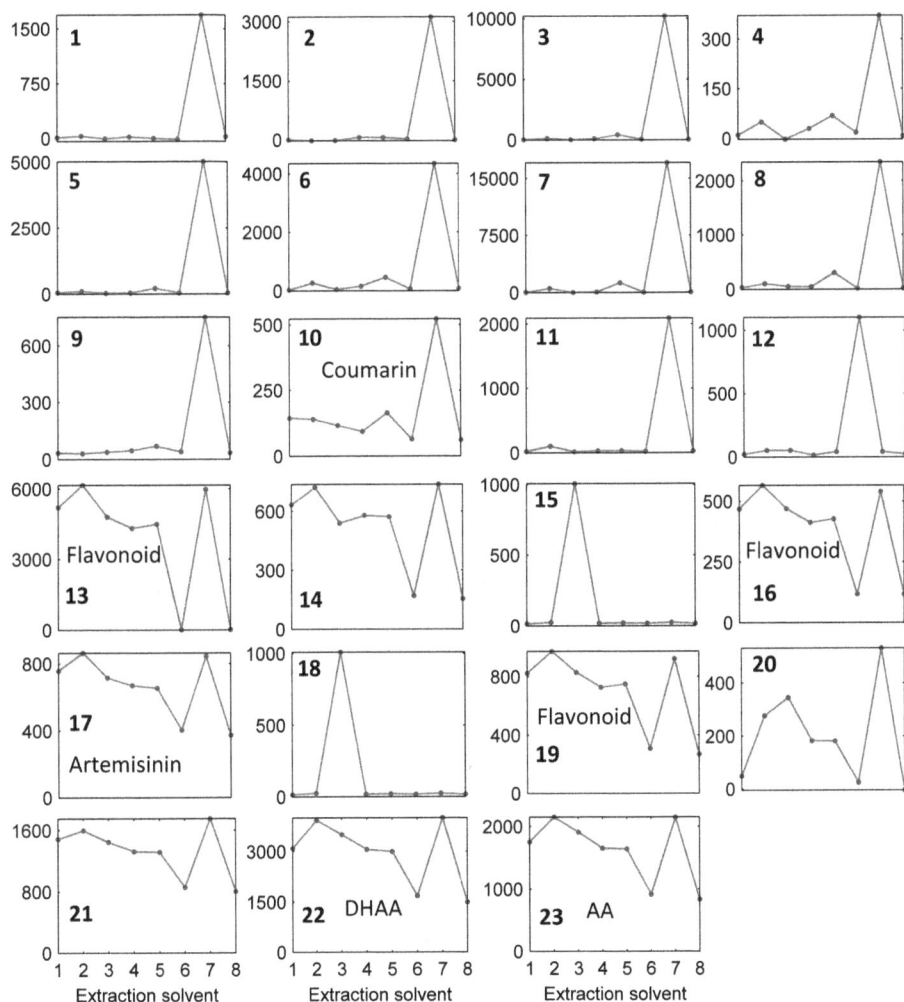

Fig. 8.12: Sample mode loadings (relative concentration profiles) of individual intervals from chromatograms of extracts: 1 acetonitrile, 2 acetone, 3 dichloromethane, 4 ethyl acetate, 5 ethanol, 6 hexane, 7 methanol, and 8 petroleum ether.

Likewise, PARAFAC analysis of all remaining peak intervals was performed and the sample mode loadings(i.e., the relative concentration of remaining peaks in all extracts) are shown in Figure 8.12. Peaks 10, 17, 22, and 23 were identified on the basis of UV profiles and standard samples as coumarin, artemisinin, dihydroartemisinic acid (DHAA), and artemisinic acid (AA), respectively. It is evident from Figure 8.12 that the impurities represented by peaks 1, 2, 3, and 5 are exclusively present in the methanol extract, whereas peaks 15 and 18 are exclusively present in the dichloromethane extract. This kind of process information can be crucial for the design of downstream purification of extracts. In this way, it was possible to obtain the chemical fingerprints of different extracts quickly and easily through the application of multivariate data analysis techniques. The dichloromethane extract of *A. annua* leaves was used to generate the initial process flowsheet.

8.3.1.2 Partial purification of *Artemisia annua* extract

Flash column chromatography (flash CC) is a powerful separation technique for purification of complicated mixtures and was selected to partially purify the dichloromethane extract of *A. annua* leaves. Considering the nonpolar to moderately polar nature of components present in the crude extract, normal phase silica was chosen as an adsorbent for flash CC according to heuristic principles. A mixture of hexane and ethyl acetate was selected as mobile phase based on solvent screening with thin layer chromatography (TLC). According to the heuristics provided by Marston et al. [63] for the design of chromatography, a column adsorbent-to-solute ratio of 20 : 1 is considered sufficient for separation of up to 10 g of sample loading. In the present work, the aim of flash CC was partial purification of the extract and removal of most of the impurities by a cost-effective technique such as crystallization.

An adsorbent-to-solute ratio of 20 : 1 was selected for partial purification of artemisinin from a 15 g extract of *A. annua*. The detailed procedure for operation of flash CC is described in the literature [55]. The flash CC fractions containing artemisinin were analyzed using HPLC-DAD and the chromatograms at 254 nm are shown in Figure 8.13. The multivariate data analysis technique PARAFAC was used to extract chemical information from HPLC data, such as relative concentrations of artemisinin and impurities in the fractions and their UV profiles, as described by Malwade et al. [62]. The extracted information was used to construct the elution sequence of artemisinin and impurities from the chromatography column. Thus, application of PAT tools made it possible to obtain a chemical fingerprint of the chromatography unit operation specific to the operating conditions used. The impurities identified on the basis of the UV profiles and LC-MS analyses are shown in Table 8.6.

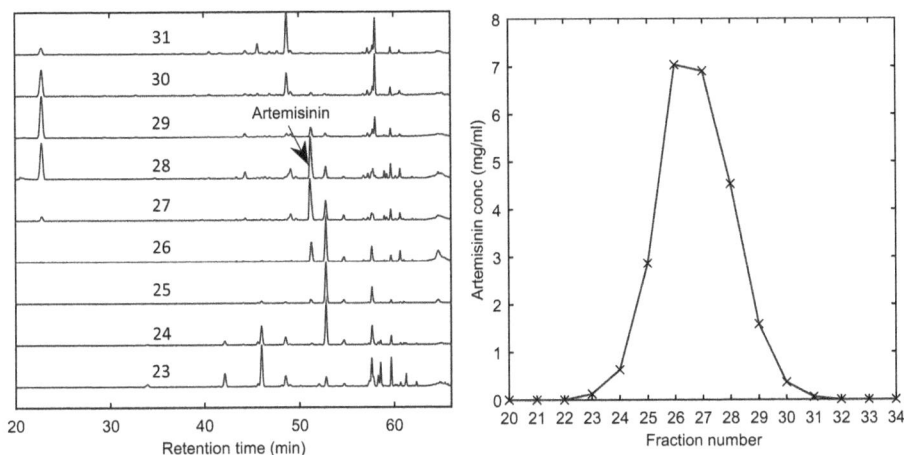

Fig. 8.13: Chromatograms of flash column chromatography fractions containing artemisinin measured at 254 nm (left) and concentration profile of artemisinin in the fractions (right).

Tab. 8.6: Identified impurities and their distribution in flash CC fractions.

Interval number	Retention time range (min)	Fraction number	Type of compound
1	22.40–23.40	25–30	Coumarin
2	31.80–32.60	23–29	–
3	38.80–39.50	25	Flavonoid
4	39.66–40.13	26–28	–
5	41.30–41.86	23–30	Artemisinin derivative
6	41.90–42.43	23–27	–
7	44.80–45.33	25	Casticin
8	50.76–51.63	23–31	Artemisinin
9	51.80–52.50	23–27	Artemisinin derivative
10	52.70–53.20	23–24	–
11	55.26–56.10	23–25	Polyacetylene
12	56.76–57.10	23–24	Artemisinin derivative
13	58.83–59.20	23–31	Dihydroartemisinic acid
14	59.56–59.80	23–31	Artemisinic acid

8.3.1.3 Final purification of artemisinin from flash CC fractions

It is known from pharmacognostic investigation of *A. annua* that the active principle, artemisinin, is crystalline in nature and able to form two different polymorphs, an orthorhombic form (QNGHSU02) and triclinic form (QNGHSU01) [64, 65]. Therefore, crystallization was used to obtain pure artemisinin from the chromatography fractions obtained in the previous step. Previously, Qu et al. [66] crystallized artemisinin from flash CC fractions of the dichloromethane extract of *A. annua* using two antisolvent crystallization steps. However, the operating conditions used for flash CC were different in terms of mobile phase and adsorbent-to-solute ratio and, therefore, the chemical composition of the fractions obtained were different. In this work, the same antisolvent crystallization steps were used to assess the effect of the chemical composition of fractions on the performance of antisolvent crystallization steps. An additional cooling crystallization step was included to crystallize remaining artemisinin from the mother liquor of antisolvent crystallization step 2. Crystallization results were significantly different from those of the previous work, which clearly indicates the influence of the chemical composition of fractions on the crystallization of artemisinin. In the previous work, 1-nonadecanol was crystallized in the first antisolvent crystallization step, whereas in this work no crystals were observed. The yield of artemisinin crystallized in this work (47%) was significantly higher than in the previous work (30%). Malwade et al. [55] have described the detailed procedure, mass balance, and analytical results for artemisinin crystals and mother liquor for individual crystallization steps.

8.3.2 Evaluation of initial process flowsheet

Evaluation of a process involves consideration of many factors, such as yield, purity, cost, environmental impact, and productivity. However, the process flowsheet generated at the conceptual design stage to recover artemisinin from *A. annua* (Figure 8.14) was evaluated on the basis of yield and purity of artemisinin. The concentration of artemisinin was monitored throughout the processing steps. The mass balance and yield of artemisinin for individual unit operations and the overall process were deduced from the analytical results and are shown in Table 8.7.

Tab. 8.7: Mass balance and yield of artemisinin for individual process steps and the overall process.

Unit operation	Mass of artemisinin in feed (mg)	Mass of artemisinin in output stream (mg)	Yield (%)
Maceration	3075	2767.5	90
Flash column chromatography	2767.5	2410	87
Crystallization	2410	1146	47
Overall process	3075	1146	37

It is clear from Table 8.7 that the yield of artemisinin in maceration and flash CC is significantly high; however, the efficiency of the overall process is affected by the poor performance of the crystallization step. Low yield of artemisinin during the crystallization step might be due to the presence of impurities in the fractions, which some reports indicate can suppress the crystallization of artemisinin [67–69]. The results also suggest that the thermodynamic behavior of the fractions strongly depends on the associated impurities, which directly affect the crystallization of artemisinin. Therefore,

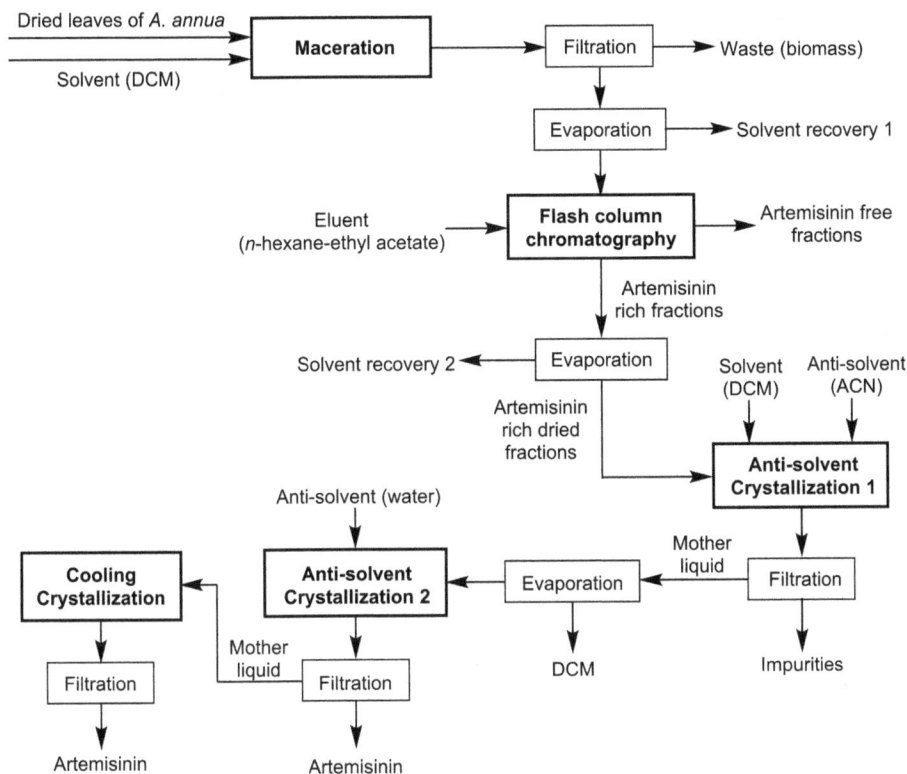

Fig. 8.14: Initial process flowsheet for recovery of artemisinin from *A. annua*.

identification of such impurities and their impact on the solid–liquid equilibrium of artemisinin is crucial in determining the synergy between flash CC and crystallization unit operations.

8.3.3 Measurement of solid–liquid equilibrium of artemisinin and impact of impurities

Rational design of crystallization processes requires knowledge about the solid–liquid equilibrium in a given solvent system and the effect of impurities on the equilibrium. The crystallization step in the initial process flowsheet for recovery of artemisinin was not based on thermodynamic data for artemisinin in the mobile phase. Therefore, the solubility of artemisinin in the mobile phase used for flash CC purification of the *A. annua* extract (i.e., *n*-hexane-ethyl acetate mixture) was measured at 5, 15, and 25 °C according to the procedure described by Malwade et al. [70]. The solubility of artemisinin was measured in *n*-hexane-ethyl acetate mixtures of composition 75 : 25, 80 : 20, 85 : 15, 90 : 10, and 95 : 5 (v/v). The impurities present in the fractions were identified and quantified according to artemisinin calibration [71]. As mentioned earlier, the presence of impurities can alter the crystallization process significantly, thereby demanding the identification of individual impurities affecting the solid–liquid equilibrium of artemisinin. However, most of the impurities present in the chromatography fraction containing artemisinin are not available commercially, which is the case for the majority of natural product purification problems. Therefore, the collective effect of these impurities on the solid-liquid equilibrium of artemisinin was determined by measuring the solubility of artemisinin in the combined chromatography fraction. The solubility of artemisinin in n-hexane–ethyl acetate mixtures is shown in Figure 8.15.

It is evident that the solubility of artemisinin decreases with decreasing temperature as well as with increasing amount of n-hexane in the solvent mixture. This clearly indicates the possibility of using antisolvent crystallization combined with cooling crystallization, as shown in Figure 8.15 by the path 1–5, to crystallize artemisinin from chromatography fractions. The solubility of artemisinin in chromatography fractions reconstituted in n-hexane–ethyl acetate mixtures is shown in Figure 8.16. It is clear that the impurities present in the fraction increase the solubility of artemisinin, confirming the phenomenon of cosolvency, where the solubility of a compound increases in the presence of impurities.

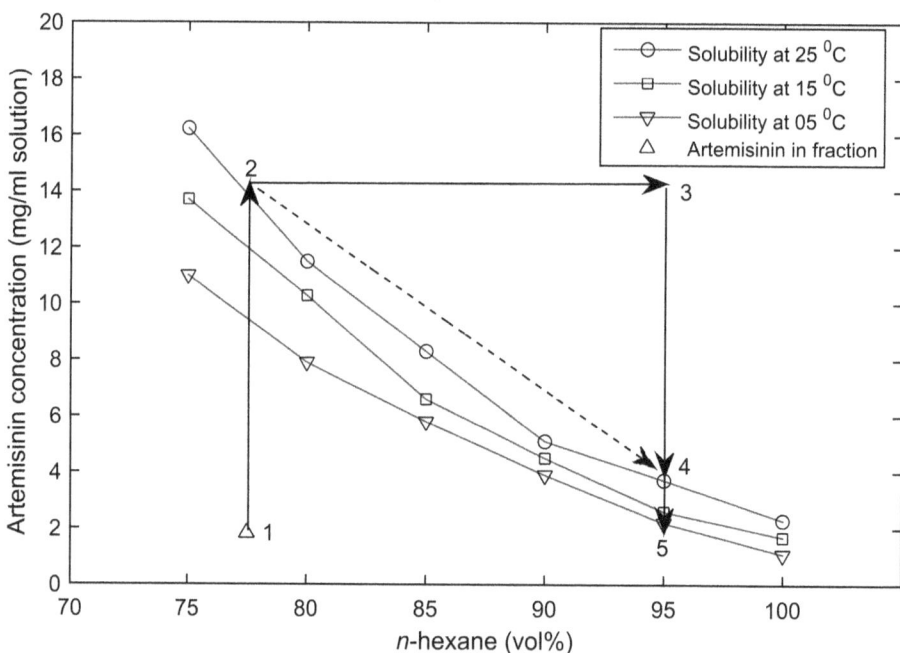

Fig. 8.15: Solubility of artemisinin in n-hexane–ethyl acetate mixtures and possible plan for crystallization of artemisinin. Solid lines are drawn for visual guidance.

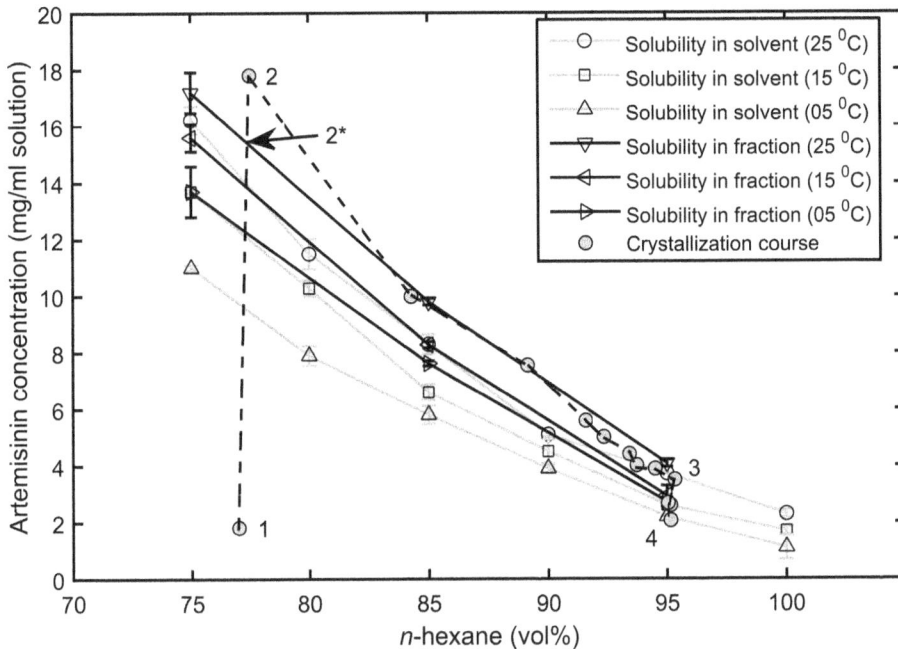

Fig. 8.16: Crystallization profile of artemisinin from flash column chromatography fraction along with solubility of artemisinin in *n*-hexane–ethyl acetate mixtures and chromatography fraction. Solid and dotted lines are drawn for visual guidance.

8.3.4 Generation of improved process flowsheet

Due to suboptimal performance of the crystallization step in the initial process flow-sheet, efforts to generate an improved flowsheet were directed toward redesign of the crystallization step. Based on thermodynamic data for artemisinin, crystallization of artemisinin was carried out in two steps, antisolvent crystallization by addition of *n*-hexane at 25 °C and cooling crystallization by cooling the solution from 25 °C to 5 °C, as suggested by Figure 8.15. The crystallization profile of artemisinin is given in Figure 8.16 and shows that the profile followed the expected path. Thus, an initial process flowsheet will take shape of an improved flowsheet, as shown in Figure 8.17, after re-designing the crystallization step according to thermodynamic data.

The new process flowsheet was better than initial process flowsheet in two ways: First, the yield of artemisinin obtained during the crystallization step designed ac-cording to equilibrium data (49%) was slightly higher than crystallization step in the initial process flowsheet (47%). Second, the complicated antisolvent crystallization steps employed in initial process flowsheet were avoided. Antisolvent crystallization step 1 in the initial flowsheet required complete evaporation of the mobile phase from chromatography fractions and dissolution in dichloromethane, followed by addition of acetonitrile as an antisolvent. In antisolvent crystallization step 2, dichloromethane

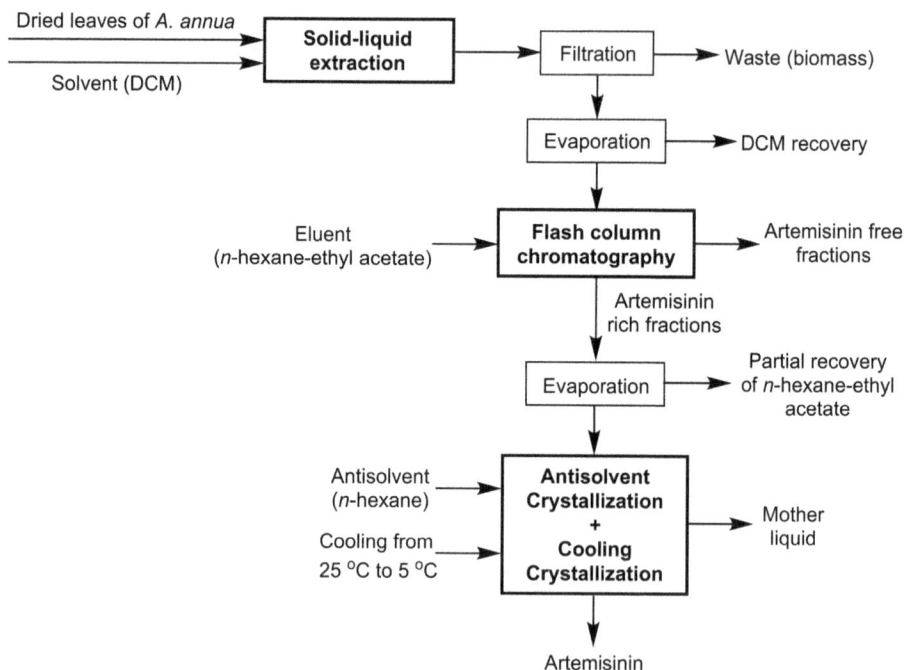

Fig. 8.17: Improved process flowsheet for recovery of artemisinin from dried leaves of *A. annua*.

was completely evaporated from the mother liquor of step 1, followed by addition of water as an antisolvent. In the improved flowsheet, antisolvent *n*-hexane was added directly to the chromatography fraction, followed by cooling crystallization. Thus, the crystallization step in the improved flowsheet was performed in a single equipment unit compared with several equipment units required in the crystallization step of the initial flowsheet.

8.4 Conclusions

Separation process design for recovery of natural products from plants is a challenging task for many reasons. First, the desired natural products are often embedded in the solid matrix of the source at very low concentrations. Second, there are many other secondary metabolites in the matrix that can have a strong influence on the properties of the desired natural products (boiling point, solubility, etc.) to be exploited for the separations. Third, there is always a lack of process information such as the identity of impurities, equilibrium data for desired products, and the influence of impurities on the equilibrium of desired natural products. Thus, the mixture from which desired natural products need to be recovered is very complex and underdefined in terms of chemical composition; therefore, it is almost impossible to achieve the task of recovering natural products from plants using standalone separation techniques.

In this work, PAT-based process synthesis methodology was developed for recovery of natural products from plants at the conceptual level. The formulated methodology employed the principle of problem decomposition into two subproblems to deal with the complexity of the problem. Thus, the task was divided into (1) isolation of natural products from the biomass to obtain a crude extract containing desired products along with impurities and (2) subsequent purification from the extract. The PAT framework was included in the methodology to obtain molecular level information about process streams required for the optimal interconnection of unit operations. Heuristics were provided for selection of separation techniques, solvents, and operating conditions.

The formulated methodology was applied to isolate and purify artemisinin from the dried leaves of the plant *A. annua*. An initial process flowsheet consisting of maceration, flash CC, and crystallization was generated and evaluated in terms of artemisinin yield. The performance of the overall process flowsheet was significantly affected by the low yield of artemisinin in the crystallization step. The low yield of artemisinin in the crystallization step was attributed to the presence of impurities and lack of thermodynamic data. The solid–liquid equilibrium of artemisinin and the impact of impurities on it were evaluated. The initial process flowsheet was modified by redesigning the crystallization step on the basis of thermodynamic data on artemisinin. The chemometric method PARAFAC was used to retrieve the number of impurities, their relative concentrations, and UV profiles from a large dataset obtained from HPLC-DAD measurements of extracts obtained from different solvents and flash CC fractions. UV profiles and LC-MS/MS of flash CC fractions confirmed that the impurities were coumarin, casticin, flavonoids, polyacetylene, and artemisinin-related compounds. Thus, the application of PARAFAC enabled process information to be retrieved from the complex dataset rapidly. The retrieved process information helped in understanding the fractionation sequence of components during flash CC.

Although the formulated methodology was only applied for the recovery of artemisinin from *A. annua*, it should be applicable for the recovery of the majority of natural products from plants. Chromatography and crystallization are powerful separation techniques that should be able to purify practically all natural products existing in solid form. Therefore, synergy between these two operations, which can be determined by revealing the process stream information at the molecular level, is crucial for the overall process in terms of yield and purity of the product as well as economy of the process. Thus, the formulated methodology can be applied to generate and evaluate initial process flowsheet alternatives, which can serve as a robust foundation for detailed process design for the recovery of natural products from plants.

8.5 Bibliography

[1] Hardy K, Buckley S, Collins MJ et al. Neanderthal medics? Evidence for food, cooking, and medicinal plants entrapped in dental calculus. Naturwissenschaften, 2012, 99(8), 617–626.

[2] Wood RE, Higham TFG, De Torres T et al. A New Date for the Neanderthals From El Sidrón Cave (Asturias, Northern Spain)*. Archaeometry, 2013, 55(1), 148–158.

[3] Spainhour CB. Natural products. In: Gad SC (ed). Drug Discovery Handbook. Hoboken: John Wiley & Sons, Inc., 2005, pp 11–72.

[4] Douillard J. Ayurveda – For optimal Health and Well-Being. Asia Pac Biotech News, 2004, 8(23), 1285–1289.

[5] Bart H. Extraction of Natural Products from Plants–An Introduction. In: Bart H-J, Pilz S (eds). Industrial Scale Natural Products Extraction, 2011, pp 1–25.

[6] Dias DA, Urban S, Roessner U. A Historical Overview of Natural Products in Drug Discovery. Metabolites, 2012, 2(4), 303–336.

[7] Bauer A, Brönstrup M. Industrial natural product chemistry for drug discovery and development. Nat Prod Rep, 2014, 31(1), 35–60.

[8] Newman DJ, Cragg GM. Natural products as sources of new drugs over the 30 years from 1981 to 2010. J Nat Prod, 2012, 75(3), 311–335.

[9] Harborne JB. Introduction to Ecological Biochemistry. Academic Press Inc., 1988.

[10] Dewick PM. Medicinal Natural Products: A Biosynthetic Approach. John Wiley & Sons, Ltd., 2009.

[11] Ross IA. Medicinal Plants of the World. Totowa, New Jersey, Humana press, 2005.

[12] Aburjai T, Natsheh FM. Plants used in cosmetics. Phytother Res, 2003, 17(9), 987–1000.

[13] Buchwald-Werner S, Bischoff F. Natural Products–Market Development and Potentials. In: Bart H-J, Pilz S (eds). Industrial Scale Natural Products Extraction, 2011, pp 247–267.

[14] Prasain J, Barnes S. Recent Advances in Traditional Medicines and Dietary Supplements. In: Osbourn AE, Lanzotti V (eds). Plant-derived Natural Products. Springer Science, 2009, pp 533–546.

[15] BCC Research. Botanical and plant-derived drugs: Global markets, 2012.

[16] Grand View Research, Inc. Organic Personal Care Market Worth $15.98 Billion By 2020, 2015.

[17] Persistence Market Research. Global Market Study on Dietary Supplements: Botanical Supplements To Be The Largest Market by 2020, 2015.

[18] Directive 2004/24/EC of the european parliament and of the council, 2004.

[19] United States Government Accountability Office. Herbal Dietary Supplements: Examples of Deceptive Or Questionable Marketing Practices and Potentially Dangerous Advice, 2010.

[20] Hussain MS, Fareed S, Ansari S, Ahmad MARIZ, Saeed M. Current approaches toward production of secondary plant metabolites. J Pharm Bioallied Sci, 2012, 4(1), 10–20.

[21] Bourgaud F, Gravot A, Milesi S, Gontier E. Production of plant secondary metabolites: A historical perspective. Plant Sci, 2001, 161(5), 839–851.

[22] Verpoorte R, Contin A, Memelink J. Biotechnology for the production of plant secondary metabolites. Phytochem Rev, 2002, 1(1), 13–25.

[23] Peñuelas J, Estiarte M. Can elevated CO_2 affect secondary metabolism and ecosystem function? Trends Ecol Evol, 1998, 13(1), 20–24.

[24] Akula R, Gokare AR. Influence of abiotic stress signals on secondary metabolites in plants. Plant Signal Behav, 2011, 6(11), 1720–1731.

[25] Gershenzon J. Changes in the levels of plant secondary metabolites under water and nutrient stress. In: Timmermann BN, Steelink C, Loewus FA (eds). Phytochemical adaptations to stress, 1984, pp 273–320.

[26] Miller H. The story of taxol: Nature and politics in the pursuit of an anti-cancer drug. Nat Med, 2001, 7(2), 148.

[27] Branan C. Rules of thumb for chemical engineers. Gulf Professional Pub., 2005.

[28] Stephanopoulos G, Westerberg AW. Studies in process synthesis—II: Evolutionary synthesis of optimal process flowsheets. Chem Eng Sci, 1976, 31(3), 195–204.

[29] Douglas J. Synthesis of separation system flowsheels. AIChE J, 1995, 41(12), 2522–2536.

[30] Grossmann I, Daichendt M. New Trends in Optimization-Based Approaches to Process Synthesis. Comput Chem Eng, 1996, 20, 665–683.

[31] Jaksland CA, Gani R, Lien KM. Separation process design and synthesis based on thermodynamic insights. Chem Eng Sci, 1995, 50(3), 511–530.

[32] Siirola JJ. Strategic process synthesis: Advances in the hierarchical approach. Comput Chem Eng, 1996, 20(96), S1637–1643.

[33] Li X. Conflict-based Method for Conceptual Process Synthesis, 2004.

[34] Sauar E, Ratkje SK, Lien KM. Equipartition of Forces: A New Principle for Process Design and Optimization. Ind Eng Chem Res, 1996, 35(11), 4147–4153.

[35] Harjo B, Wibowo C, Ng KM. Development of natural product manufacturing processes: Phytochemicals. Chem Eng Res Des, 2004, 82, 1010–1028.

[36] Gertenbach D. Solid–liquid extraction technologies for manufacturing nutraceuticals from botanicals. Funct foods Biochem Process Asp, 2001, 331–365.

[37] Kassing M, Jenelten U, Schenk J, Hänsch R, Strube J. Combination of Rigorous and Statistical Modeling for Process Development of Plant-Based Extractions Based on Mass Balances and Biological Aspects. Chem Eng Technol, 2012, 35(1), 109–132.

[38] Kassing M, Jenelten U, Schenk J, Strube J. A New Approach for Process Development of Plant-Based Extraction Processes. Chem Eng Technol, 2010, 33(3), 377–387.

[39] Ndocko Ndocko E, Bäcker W, Strube J. Process Design Method for Manufacturing of Natural Compounds and Related Molecules. Sep Sci Technol, 2008, 43(3), 642–670.

[40] Strube J, Werner B, Schulte M. Process Engineering and Mini – Plant Technology. In: Bart H-J, Pilz S, (eds). Industrial Scale Natural Products Extraction. Wiley-VCH Verlag GmbH & Co. KGaA., 2011, pp 123–179.

[41] Workman J, Koch M, Veltkamp D. Process analytical chemistry. Anal Chem, 2007, 79(12), 4345–4363.

[42] Juran JM. Juran on Quality by Design: The New Steps for Planning Quality into Goods and Services. The Free Press, 1992.

[43] United States Food & Drug Administration. PAT-A Framework for Innovative Pharmaceutical Development, Manufacturing and Quality Assurance, 2004.

[44] Gernaey KV, Cervera-Padrell AE, Woodley JM. A perspective on PSE in pharmaceutical process development and innovation. Comput Chem Eng, 2012, 42, 15–29.

[45] Klimkiewicz A, Mortensen PP, Zachariassen CB, van den Berg FWJ. Monitoring an enzyme purification process using on-line and in-line NIR measurements. Chemom Intell Lab Syst, 2014, 132, 30–38.

[46] Helmdach L, Feth MP, Ulrich J. Application of Ultrasound Measurements as PAT Tools for Industrial Crystallization Process Development of Pharmaceutical Compounds. Org Process Res Dev, 2015, 19(1), 110–121.

[47] Gurden SP, Martin EB, Morris AJ. The introduction of process chemometrics into an industrial pilot plant laboratory. Chemom Intell Lab Syst, 1998, 44, 319–330.

[48] Mason RL, Gunst RF, Hess JL. Statistical Design and Analysis of Experiments: With Applications to Engineering and Science, Second Edition. John Wiley & Sons, Inc., 2003.

[49] Dreiseitl S, Ohno-Machado L. Logistic regression and artificial neural network classification models: A methodology review. J Biomed Inform, 2002, 35(5–6), 352–359.

[50] Amigo JM, Skov T, Bro R. ChroMATHography: Solving chromatographic issues with mathematical models and intuitive graphics. Chem Rev, 2010, 110(8), 4582–4605.

[51] Acar E, Bro R, Schmidt B. New exploratory clustering tool. J Chemom, 2008, 22(1), 91–100.

[52] Rencher AC. Methods of Multivariate Analysis, Second Edition. John Wiley & Sons, Inc., 2012.

[53] Duarte AC, Capelo S. Application of Chemometrics in Separation Science. J Liq Chromatogr Relat Technol, 2006, 29, 1143–1176.

[54] Kourti T. The Process Analytical Technology initiative and multivariate process analysis, monitoring and control. Anal Bioanal Chem, 2006, 384(5), 1043–1048.

[55] Malwade CR, Qu H, Rong B-G, Christensen LP. Conceptual Process Synthesis for Recovery of Natural Products from Plants: A Case Study of Artemisinin from *Artemisia annua*. Ind Eng Chem Res, 2013, 52(22), 7157–7169.

[56] World Health Organization. WHO Monograph on Good Agricultural and Collection Practices (GACP) for *Artemisia Annua* L., 2006.

[57] Callaway E, Cyranoski D. China celebrates first Nobel. Nature, 2015, 174–175.

[58] World Health Organization. Global Malaria Programme, UNICEF, World Health Organization. Achieving the malaria MDG target: reversing the incidence of malaria 2000–2015, 2015.

[59] World Health Organization. Global technical strategy for malaria 2016–2030, 2015.

[60] Duke S, Paul R. Development and fine structure of the glandular trichomes of *Artemisia annua* L. Int J Plant Sci, 1993, 154(1), 107–118.

[61] Pilkington JL, Preston C, Gomes RL. Comparison of response surface methodology (RSM) and artificial neural networks (ANN) towards efficient extraction of artemisinin from *Artemisia annua*. Ind Crops Prod, 2014, 58, 15–24.

[62] Malwade CR, Qu H, Rong B-G, Christensen LP. Chemometrics for Analytical Data Mining in Separation Process Design for Recovery of Artemisinin from *Artemisia annua*. Ind Eng Chem Res, 2014, 53(13), 5582–5589.

[63] Marston A, Hostettmann K. Modern separation methods. Nat Prod Rep, 1991, 8(4), 391–413.

[64] Chan K, Yuen K, Takayanagi H. Polymorphism of artemisinin from *Artemisia annua*. Phytochemistry, 1997, 46(7), 1209–1214.

[65] Horosanskaia E, Seidel-Morgenstern A, Lorenz H. Investigation of drug polymorphism: Case of artemisinin. Thermochim Acta, 2014, 578(63968), 74–81.

[66] Qu H, Christensen KB, Fretté XC, Tian F, Rantanen J, Christensen LP. A Novel Hybrid Chromatography-Crystallization Process for the Isolation and Purification of a Natural Pharmaceutical Ingredient from a Medicinal Herb. Org Process Res Dev, 2010, 14(1), 585–591.

[67] Malwade CR, Qu H, Rong B-G, Christensen LP. Purification of artemisinin from quercetin by anti-solvent crystallization. Front Chem Sci Eng, 2013, 7(1), 72–78.

[68] Suberu JO, Yamin P, Leonhard K et al. The effect of O-methylated flavonoids and other co-metabolites on the crystallization and purification of artemisinin. J Biotechnol, 2013, 171C, 25–33.

[69] Lapkin AA, Peters M, Greiner L et al. Screening of new solvents for artemisinin extraction process using ab initio methodology. Green Chem, 2010, 12(2), 241–251.

[70] Malwade CR, Christensen LP. Simple multipurpose apparatus for solubility measurement of solid solutes in liquids. Educ Chem Eng, 2016, 16, 29–38.

[71] Malwade CR, Buchholz H, Rong BG et al. Crystallization of Artemisinin from Chromatography Fractions of *Artemisia annua* Extract. Org Process Res Dev, 2016, 20(3), 646–652.

Yufei Wang and Xiao Feng

9 Process synthesis for energy efficiency based on the pinch analysis approach

Abstract: The pinch analysis approach is widely used in process synthesis to guide the design and retrofitting of heat exchanger networks. Using this approach, the energy target clearly indicates the maximum energy recovery for a certain minimum temperature difference. The rules proposed in the pinch analysis approach can be used to find bottlenecks in a retrofitted heat exchanger network. Pinch analysis can be extended to the analysis of other energy systems. Grand composite curves derived using this approach are used to analyze utility systems, heat integration of reactors, and distillation columns. The total site pinch approach can be used to determine the energy target for multiple plants. This chapter introduces the applications of pinch analysis and illustrates the approach with an industrial case.

Keywords: Process synthesis, energy saving, pinch approach, heat exchanger network, utility system, separation system

9.1 The hierarchy of process synthesis

A process to transform a feed into products usually requires reactors, and isolation of products at the required purity requires a separation system. In the process, streams normally need to be heated or cooled to the desired temperatures. When heating and cooling requirements cannot be satisfied by heat recovery, external utilities (e.g., steam and cooling water) must provide heating and cooling. To use energy efficiently is to minimize utility consumption.

A process can be represented symbolically by an "onion diagram," as shown in Figure 9.1, in which each layer gives a hierarchy of the synthesis [1, 2].

For the synthesis of a new plant or grassroots design, the synthesis starts with the center and moves outward, that is, from the reactor to the separation system, the heat exchanger network (HEN), utilities, and so on. A reactor is likely to be the only place in the process where raw materials are converted into products. The chosen reactor design produces a mixture of unreacted feed materials, products, and byproducts that need separating. Thus, design of the separation system follows the reactor design. The reactor and separation system designs together define the process for heating and cooling duties. Thus, HEN design comes next. Those heating and cooling requirements that cannot be satisfied by heat recovery, dictate the need for external heating and cooling utilities.

When synthesis is carried out to modify an existing plant in retrofit, the reactor normally remains unchanged and the synthesis starts with the HEN. The pinch posi-

https://doi.org/10.1515/9783110465068-009

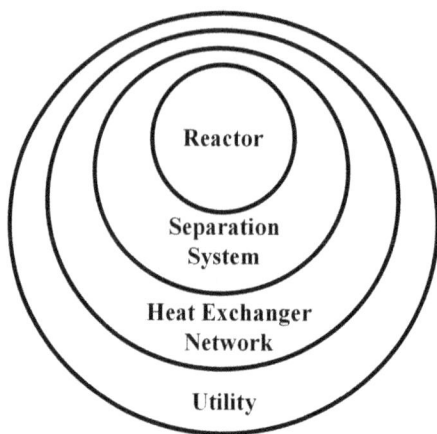

Fig. 9.1: The onion model.

tion of the HEN is determined first, because the subsequent analysis and optimization is based on the pinch position. Then, the separation system is considered, including single columns and the column system. Next to be considered is the heat integration of utilities. After that, coordination between the separation subsystem and the HEN subsystem is considered. The retrofit of the separation subsystem affects the HEN subsystem, causing a change in stream matches, so that a new heat integration for the HEN may be needed. In addition, coordination between the utility subsystem and the HEN subsystem can have an impact on the HEN, so that heat integration of the HEN is considered again [3].

9.2 Heat exchanger networks

9.2.1 Pinch and energy targets

The task of a HEN is to heat or cool all streams from their supply temperatures to their target temperatures. A stream that will be heated is termed a cold stream, and one that will be cooled is named a hot stream. The thermal characteristic of a stream can be represented by a curve (or a line) in the temperature–enthalpy ($T–H$) diagram, as shown in Figure 9.2. The stream is heated or cooled from its supply temperature T_S to its target temperature T_T. The enthalpy change refers to the heat duty released by a hot stream or received by a cold stream. The relative position of the curve can be changed by moving it horizontally, because the reference enthalpy for the stream can be changed.

In a process, there are usually several hot and cold streams. The overall behavior of these hot (or cold) streams can be quantified by combining them in the $T–H$ diagram. The temperature ranges are defined by the supply and target temperatures of

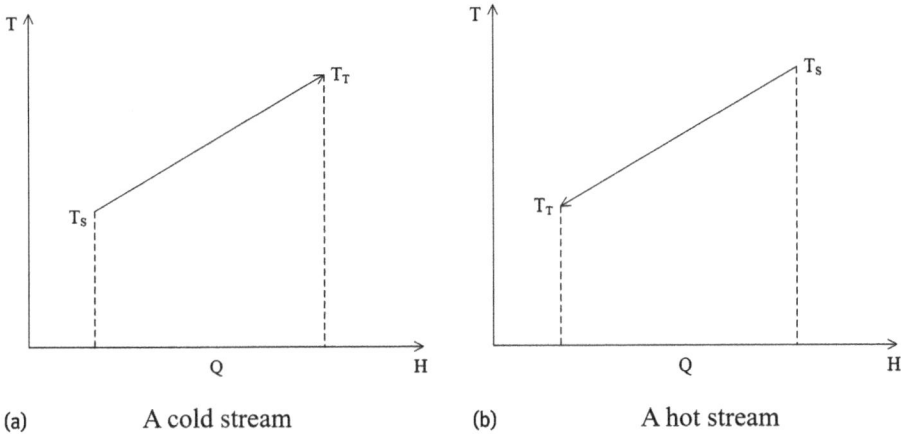

Fig. 9.2: A stream in a temperature–enthalpy diagram.

the streams. Figure 9.3 shows an example of three hot streams being combined into one composite curve. The heat capacity flowrates of the three streams are A, B, and C, as shown in Figure 9.3(a). From the supply and target temperatures of each stream, the temperature axis is divided into four ranges: temperature range 1, from T_1 to T_2; temperature range 2, from T_2 to T_3; temperature range 3, from T_3 to T_4; and temperature range 4, from T_4 to T_5. The temperature range can also be defined by changes in the overall rate of change of enthalpy with temperature, as shown in Figure 9.3(b). Within each temperature range, the streams are combined to produce a composite hot stream, with the enthalpy change determined by Equation (9.1).

$$\Delta H_i = \sum_j CP_j(T_i - T_{i+1}) \tag{9.1}$$

where H is enthalpy (in kW), CP is heat capacity flow rate (kW/°C), T is temperature (°C), subscript i is the temperature range i, and j refers to stream j.

The cold composite curve can be produced similarly. In this way, the hot/cold composite curve is like a single stream that is equivalent to the individual hot/cold streams in terms of temperature and enthalpy.

The hot and cold composite curves can then be plotted in the same $T–H$ diagram, as shown in Figure 9.4. For feasible heat exchange between the hot and cold streams, the hot composite curve must be above the cold composite curve at all points. The relative position of the two curves can be changed by moving them horizontally relative to each other, until a specified minimum temperature difference (ΔT_{min}) is reached. Heat can be recovered vertically from the hot streams that comprise the hot composite curve into the cold streams that comprise the cold composite curve in the overlap region in Figure 9.4. The maximum overlap between the curves corresponds to the maximum heat recovery and, thereby, the minimum external utility requirements for heating and cooling duties. Hot utility is required for the part of the cold stream that

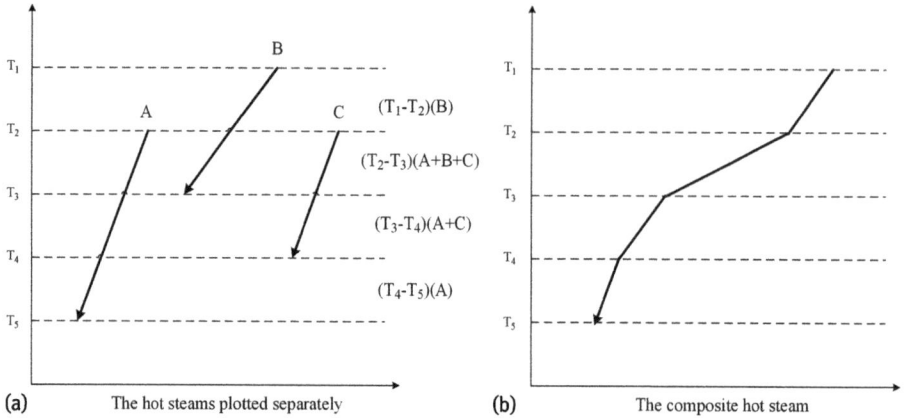

Fig. 9.3: The hot composite curve.

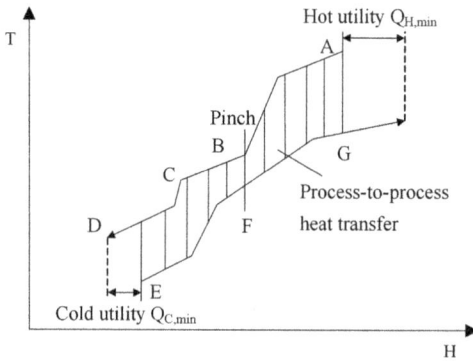

Fig. 9.4: Composite curves for targeting.

extends beyond the start of the hot stream in Figure 9.4, and cold utility is required for the part of the hot stream that extends beyond the start of the cold stream.

In this way, we can determine the pinch position where the temperature difference is equal to ΔT_{min} and the energy targets (the minimum hot and cold utilities). Obviously, the energy targets will be different for a different ΔT_{min}. The importance of ΔT_{min} is that it sets the relative location of the hot and cold composite curves and, therefore, the amount of heat recovery; as ΔT_{min} increases, the energy targets increase.

Choice of ΔT_{min} is the result of a trade-off between energy and capital cost. Figure 9.5 illustrates how the cost of the system changes with ΔT_{min}. At the limit $\Delta T_{min} = 0$, there is no driving force for heat transfer (at least at one point in the process), which would require infinite heat transfer area and hence infinite capital cost. As ΔT_{min} increases, the energy target (and hence energy cost) increases. At the same time, the capital cost decreases, which is due to the decrease in heat transfer area by increased temperature differences throughout the process.

The point where ΔT_{min} occurs between the hot and the cold composite curves is called the pinch point. The pinch point divides the process into two parts, as shown in Figure 9.6. Above the pinch (in temperature terms), in addition to heat recovery between process streams, heat is received from the hot utility and no heat is rejected. The part acts as a net heat sink. Below the pinch, in addition to heat recovery between process streams, no heat is received and heat is rejected to the cold utility. The part acts as a net heat source.

Above the pinch, only hot utility is needed to satisfy the heating demand of the system. If cold utility XH is used for cooling hot streams above the pinch, the heat of such hot streams cannot be recovered for heating cold streams above the pinch. To satisfy the enthalpy imbalance above the pinch, import of $(Q_{H\,min} + XH)$ heat from the hot utility is required. As a consequence, $(Q_{C\,min} + XH)$ of cold utility is used. Similarly, below the pinch, only cold utility is needed to satisfy the cooling demand of the system. If hot utility XH is used for heating cold streams below the pinch, the heating duty of such cold streams cannot be used to recover the heat of hot streams below the pinch. To satisfy the enthalpy imbalance below the pinch, $(Q_{C\,min} + XH)$ of cold utility is required. As a consequence, $(Q_{H\,min} + XH)$ of hot utility is used.

In addition to appropriate use of utilities, heat transfer between process streams should avoid heat transfer across the pinch. If an amount of heat XH is transferred from the system above the pinch to the system below the pinch (i.e., hot streams above the pinch exchange heat with cold streams below the pinch), it creates a deficit of heat XH above the pinch and an additional surplus of heat XH below the pinch. An extra

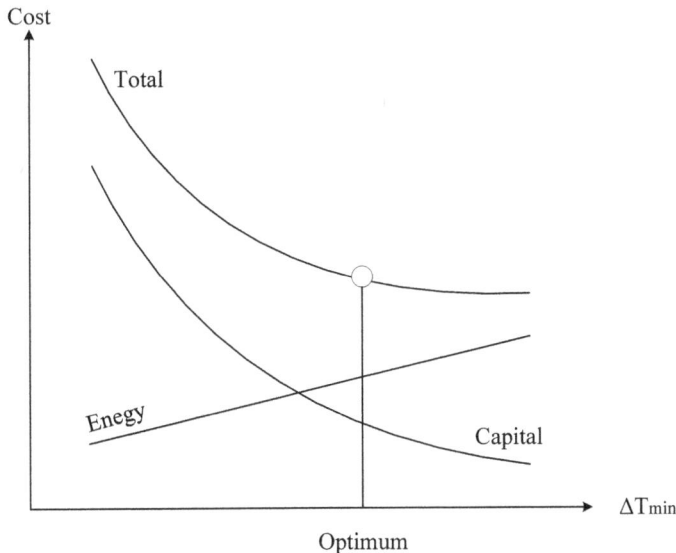

Fig. 9.5: The cost change with ΔT_{min}.

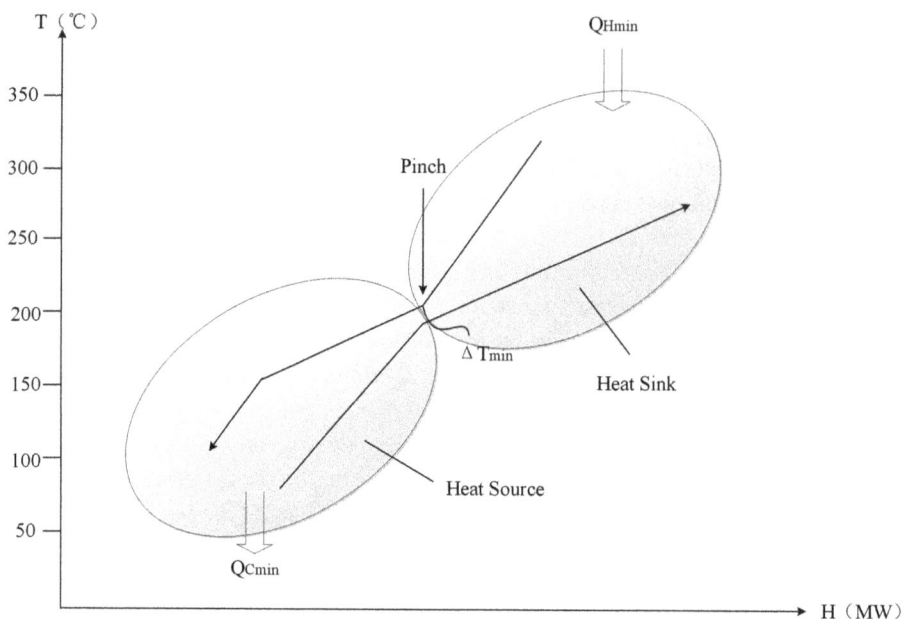

Fig. 9.6: Net heat source and heat sink divided by the pinch.

XH amount of heat needs to be imported from the hot utility and the same amount of heat needs to be exported to the cold utility. By contrast, if heat transfer occurs in the opposite direction (i.e., hot streams below the pinch exchange heat with cold streams above the pinch), it leads to a temperature difference less than ΔT_{min}. Even if the temperature difference is positive, the trade-off between energy and capital cost in the determination of ΔT_{min} suggests that individual exchangers should have a temperature difference no smaller than ΔT_{min}.

To summarize, to achieve the energy target set by the composite curves, the following three rules must be obeyed:
1. no cold utility used above the pinch
2. no hot utility used below the pinch
3. no process-to-process heat transfer across the pinch

9.2.2 Capital cost-related targets

A good HEN is not only energy efficient, but also cost efficient. The main contributions to the operation cost of a HEN are the hot and cold utility consumptions (i.e., the energy targets). The principal components that contribute to the capital cost are the number of units (matches between hot and cold streams) and heat exchange area.

Generally, the number of units has a bigger influence than heat exchange area on the capital cost of a HEN. The number of units of a HEN can be deduced from graph theory, given by Equation (9.2).

$$U_{min} = N + L - S \qquad (9.2)$$

where U is the number of units, N the number of streams including utilities, L the number of independent loops, and S the number of separate subsystems.

A loop is a path that begins and ends at the same stream, like 1–2 in Figure 9.7. In Figure 9.7, we can find three loops, 1–2, 2–3, and 1–3, but only two are independent.

To determine the minimal number of units, L should be zero and S should be the maximum. The assumption that L is zero is reasonable. However, $S > 1$ is unusual and not easy to predict, and so the assumption for S is that there is only one subsystem (i.e., $S = 1$). Therefore, the minimal number of units is given by the following:

$$U_{min} = N - 1 \qquad (9.3)$$

To determine the heat exchange area target, the composite curves including the utility streams can be used. The composite curves are divided into vertical enthalpy intervals as shown in Figure 9.8. Under the assumption of counter-current heat transfer, the area requirement for enthalpy interval i is given by Equation (9.4).

$$A_i = \frac{1}{\Delta T_{lmi}} \sum_j \frac{q_j}{h_j} \qquad (9.4)$$

where A_i is the heat exchange area required by interval i, ΔT_{lmi} the log mean temperature difference for interval i, q_j the heat duty of stream j in interval i, and h_j the heat transfer coefficient of stream j.

The network area target is the summation of areas for all enthalpy intervals, as follows:

$$\sum A_i = \sum_i \frac{1}{\Delta T_{lmi}} \sum_j \frac{q_j}{h_j} \qquad (9.5)$$

9.2.3 Synthesis of HENs

When synthesizing a HEN, a grid diagram showing only heat transfer operations is used. Hot streams are at the top, running left to right, and cold streams are at the

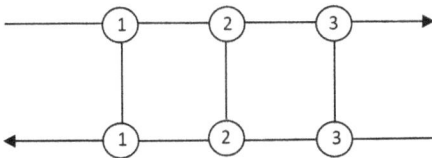

Fig. 9.7: A loop.

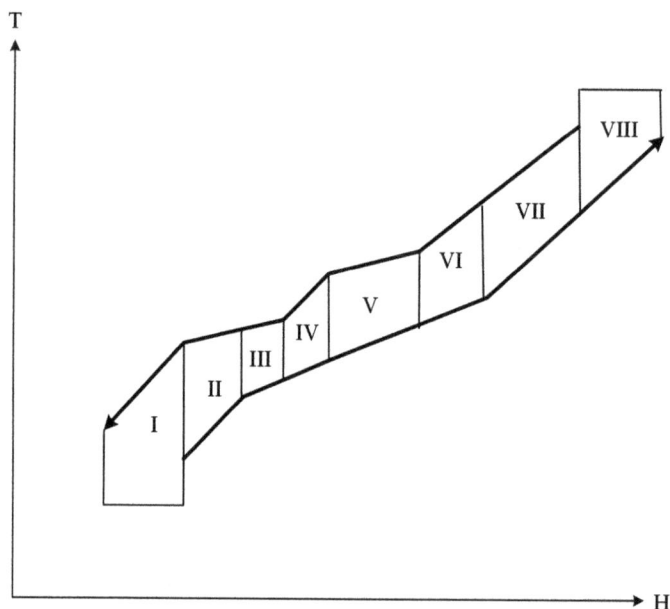

Fig. 9.8: Enthalpy intervals for determining the area target.

bottom, running right to left. A heat exchange match is represented by a vertical line joining two circles on the two streams being matched. An exchanger using hot utility (heater) is represented by a circle with an "H", and one using cold utility (cooler) is represented by a circle with a "C."

The pinch is the most constrained region of the HEN, because the temperature difference is minimal between all hot and cold streams and there is no heat transfer across the pinch. Considering that the pinch divides the process into two parts, the synthesis of a HEN is started at the pinch (i.e., from the pinch to the hotter side at the net heat sink, and from the pinch to the colder side at the net heat source). In a HEN design based on the pinch design method, matches between hot and cold streams should follow three rules:

1. The rule for number of streams

   Because no cold utility should be used above the pinch, each hot stream has to be cooled by cold streams (i.e., each hot stream has to be matched by one or more cold streams), so that above the pinch

$$N_H \leq N_C \tag{9.6}$$

   where N is the number of streams and subscripts H and C refer to hot and cold, respectively.

   Similarly, because no hot utility should be used below the pinch, the stream number below the pinch should be

$$N_H \geq N_C \tag{9.7}$$

It should be pointed out that this rule is not for the real system, but for the design. If a real system is not consistent with the rule, streams can be split to meet the rule. The rule is mainly for streams around the pinch. When away from the pinch, as long as the temperature difference is big enough, streams can be matched in sequence and the rule can be ignored.

2. The rule for heat capacity flowrates for each individual match

This rule is applied for the pinch match. At the pinch, the match starts with a temperature difference equal to ΔT_{min}. Moving away from the pinch, temperature differences must not decrease to ensure that the temperature difference is not less than ΔT_{min}. So, above the pinch

$$CP_H \leq CP_C \tag{9.8}$$

and below the pinch

$$CP_H \geq CP_C \tag{9.9}$$

If inequality 9.8 or 9.9 cannot be met at the pinch, a stream split can be considered.

3. The tick-off rule

The tick-off rule is used to keep the number of units to the minimum. To tick off a stream, the heat load of an individual unit is made as large as possible.

A network synthesized using the pinch design method, following the rule of no heat transfer across the pinch, has the minimum utility consumption and is called the initial network. However, because the network is designed separately for parts above and below the pinch, the number of units is normally more than the minimum of one. This means that there are some loops in the network. Next, we consider breaking the loops to reduce the number of units and, thus, reduce the capital cost of the network.

Heat duty can be shifted around a loop. If the duty of an exchanger is shifted to zero, the exchanger is removed and the loop is broken. However, although the change in heat duties around the loop maintains the network heat balance and the supply and target temperatures of the streams, the temperature differences of the exchangers in the loop are changed.

If the temperature differences of the exchangers are expected to be greater than ΔT_{min}, more utilities have to be used by shifting the heat duty along utility paths in a similar way as for loops. A utility path is a heat flow path from a hot utility to a cold utility, including all matches along the path.

9.2.4 HEN synthesis example

The data for hot and cold streams are given in Table 9.1. Composite curves obtained using this data are shown in Figure 9.9 for $\Delta T_{min} = 10\,°C$. From the composite curves, it can be seen that the pinch is at 90 °C for hot streams and 80 °C for cold streams, and the minimum hot and cold utilities are 20 kW and 60 kW, respectively.

Design of the initial network is started at the pinch. Above the pinch, there are two hot streams and two cold streams, which meet the rule for the number of streams. According to the rule for heat capacity flowrates, streams 2 and 3 exchange heat, as do streams 4 and 1. The heat load of each match is determined using the tick-off rule. The remaining heat demand of stream 1 is supplied by the hot utility. The network above the pinch is shown in Figure 9.10(a).

Below the pinch, there are two hot streams and one cold stream, which also meet the rule for number of streams. According to the rule for heat capacity flowrates, cold stream 3 can only match stream 1 at the pinch. This match can use up the heat load of stream 1 based on the tick-off rule. Stream 3 is away from the pinch; therefore, there is no need to apply the rule for heat capacity flowrates. The remaining heat load of stream 3 is heated by stream 2. The remaining cooling demand of stream 2 is supplied by the cold utility. The network below the pinch is shown in Figure 9.10(b).

Combining the networks above and below the pinch, the whole network can be obtained, as shown in Figure 9.10(c). The initial network can reach the energy target, but the number of units is one more than the minimum. This means that there is a loop in the network. From Figure 9.10(c), it can be seen that exchangers 2 and 4 form a loop. Next, we consider how to break the loop.

Tab. 9.1: Stream data.

No.	CP (kW/°C)	Supply temperature (°C)	Target temperature (°C)
1	3.0	170	60
2	1.5	150	30
3	2.0	20	135
4	4.0	80	140

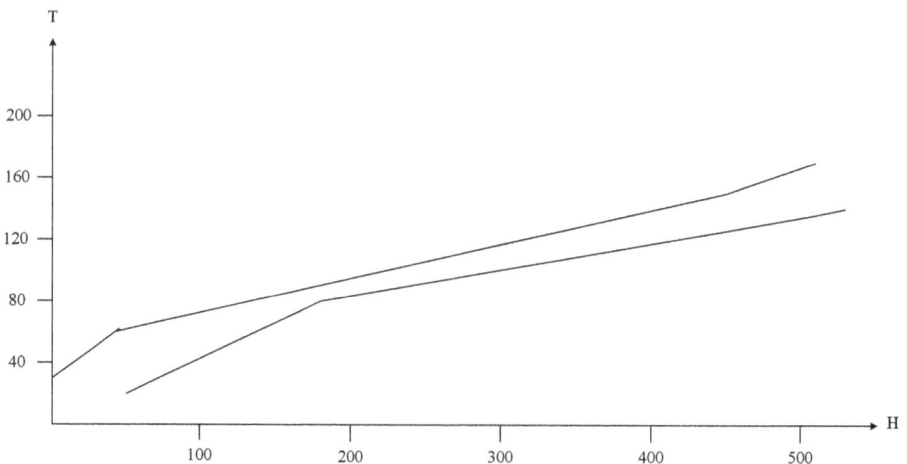

Fig. 9.9: Composite curves of the exercise.

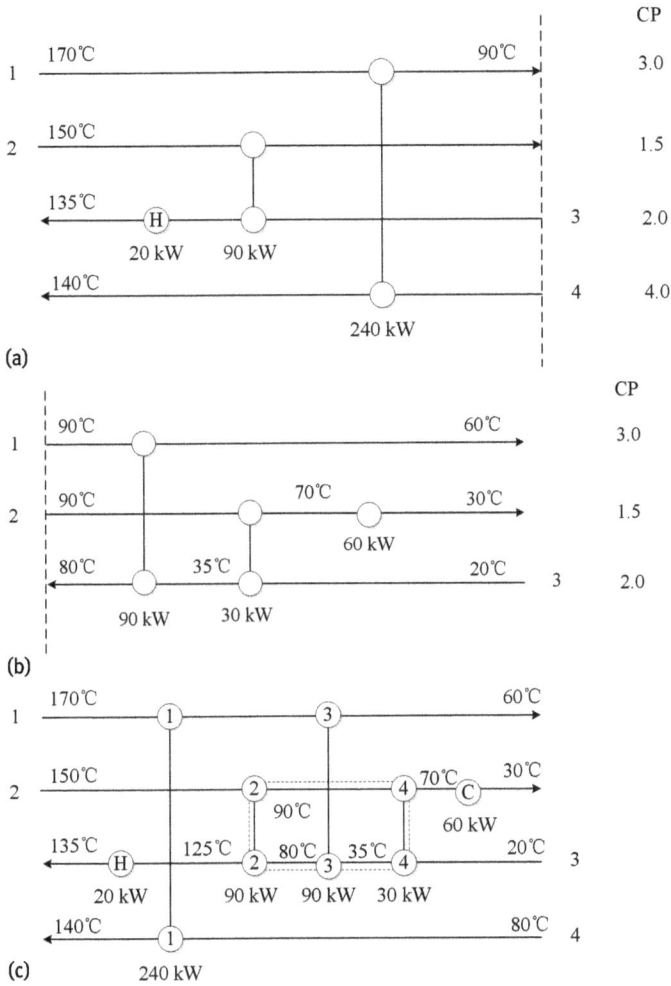

Fig. 9.10: The initial network: (a) network above the pinch, (b) network below the pinch, and (c) the whole network.

Heat can be shifted around a loop. In this loop, the heat duty of exchanger 4 is simply moved to exchanger 2, which means that exchanger 4 can be removed. At the same time, the duty of exchanger 2 becomes 120 kW, as shown in Figure 9.11(a). The change in heat duties around the loop maintains the network heat balance and does not change the supply and target temperatures of the streams. However, the temperatures around the loop change and, hence, the temperature differences of the exchangers in the loop change. Figure 9.11(a) shows that the temperature difference at the cold end of exchanger 2 + 4 is only 5 °C, which is less than ΔT_{min} (10 °C).

To maintain the temperature difference as 10 °C, more utilities have to be used by shifting the heat duty along the utility path. In this case, the utility path H → (2+4) →

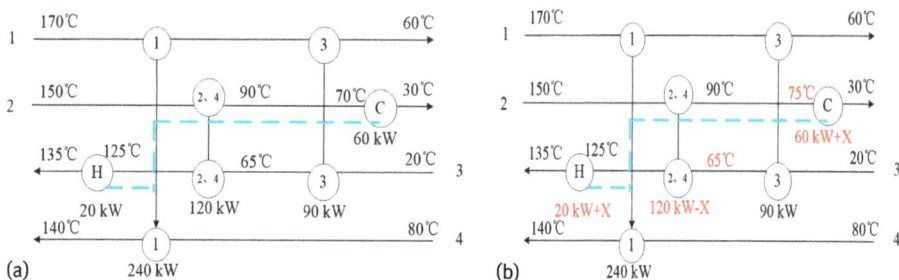

Fig. 9.11: Break the loop and maintain the minimum temperature difference: (a) before maintaining, (b) after maintaining.

C is the blue dashed line shown in Figure 9.11(b). In heater H, the load increases by X, the load of exchanger 2+4 decreases by X, and the load of cooler C also increases by X. To make the outlet temperature of exchanger 2 + 4 equal to 75 °C, X can be determined by the heat balance around cooler C as follows:

$$(75 - 30) \times 1.5 = 60 + X$$
$$X = 7.5\,\text{kW}$$

9.2.5 Retrofit of HENs

In this section, we consider retrofit of an existing HEN. The need to retrofit might arise from a desire to reduce the utility consumption of an existing network, increase throughput, modify the feed to the process, or modify the product specification.

An example of the retrofit of an existing network to reduce utility consumption is shown in Figure 9.12.

This is a simple process. The feed is heated in H1 from 30 to 120 °C and enters reactor R1. The two streams from R1 go to reactor R2 and column C1 without heating or cooling. The stream from R2 releases heat to the feed, drops in temperature from 180 to 80 °C, and then goes into C1. The top product of C1 goes into R1 after being heated from 60 to 100 °C. The bottom product of C1 is the final product of the process and is stored after being cooled from 130 to 40 °C, the heat of which is partly recovered by the top product of C1. The numbers in squares in Figure 9.12 represent the heat load of the corresponding heat exchangers. The real hot utility consumption is 102 kW (sum of the heat loads of H1 and H2), and the real cold utility consumption is 60 kW (the heat load of C). The retrofit process can be broken down into the following five steps:
1. Extract stream data
    In this example, the related data with the HEN can be extracted, as shown in Table 9.2.
2. Determine the value of ΔT_{min}

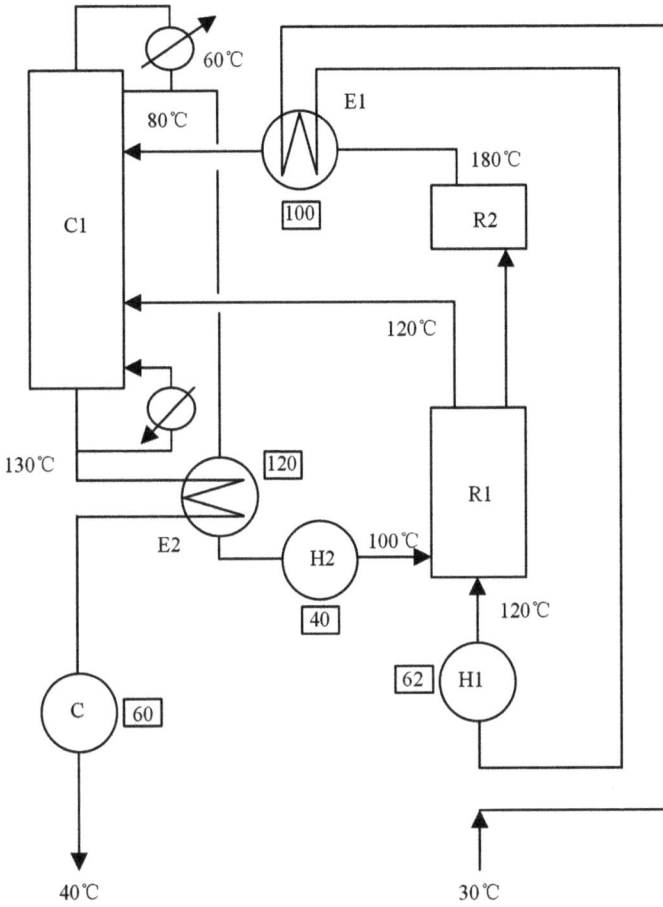

Fig. 9.12: An existing heat exchanger network.

Tab. 9.2: Stream data for the retrofit example.

No.	CP (kW/°C)	Supply temperature (°C)	Target temperature (°C)
1	1.8	30	120
2	4.0	60	100
3	1.0	180	80
4	2.0	130	40

In the existing network, the minimum temperature difference appears at the cold end of the match of streams 2 and 4, which is 10 °C. This value is reasonable, so in this example, ΔT_{min} is chosen as 10 °C.

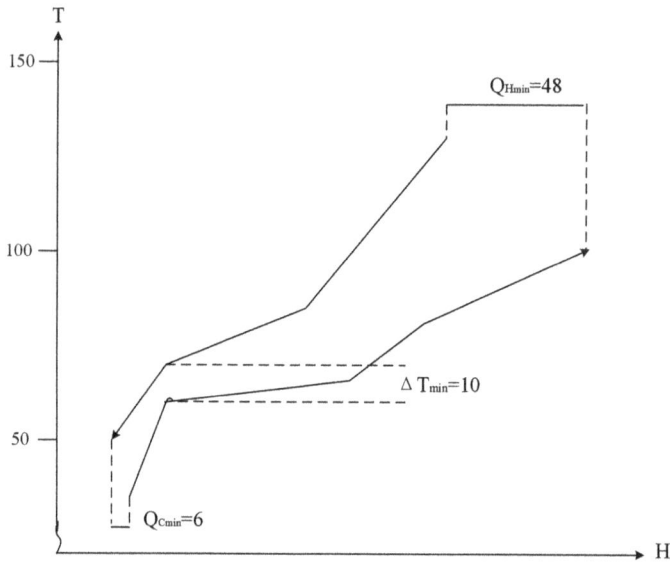

Fig. 9.13: $T-H$ diagram of the example.

3. Determine pinch location and the utility targets

Figure 9.13 shows the $T-H$ diagram for the example. It can be seen that the pinch locates at 70 °C for hot streams and 60 °C for cold streams. The minimum hot and cold utilities are 48 kW and 6 kW, respectively.

4. Analysis

According to the three rules in Section 9.2.1, all cross-pinch heat transfers are found. First, we check whether cold utility is used above the pinch. There is only one cooler (C) in the system, which cools stream 4 from 70 to 40 °C. This is below the pinch, so there is no cold utility used above the pinch. Then, we check whether hot utility is used below the pinch. There are two heaters in the system (H1 and H2). H1 heats stream 1 from 85.5 to 120 °C, and H2 heats stream 2 from 90 to 100 °C, both of which are above the pinch. Thus, there is no hot utility used below the pinch. Last, we check whether process-to-process heat transfer crosses the pinch. There are two process-to-process heat exchangers (E1 and E2). E2 is completely above the pinch because the hot stream (stream 4) changes from 130 to 70 °C, and the cold stream (stream 2) changes from 60 to 90 °C. However, E1 is set across the pinch, in which the hot stream (stream 3) changes from 180 to 80 °C to heat the cold stream (stream 1) from 30 to 85.5 °C. The heat transfer for heating the cold stream from 30 to 60 °C is across the pinch. The inappropriate heat transfer is 54 kW, which is the difference between the real hot (or cold) utility consumption and the minimum.

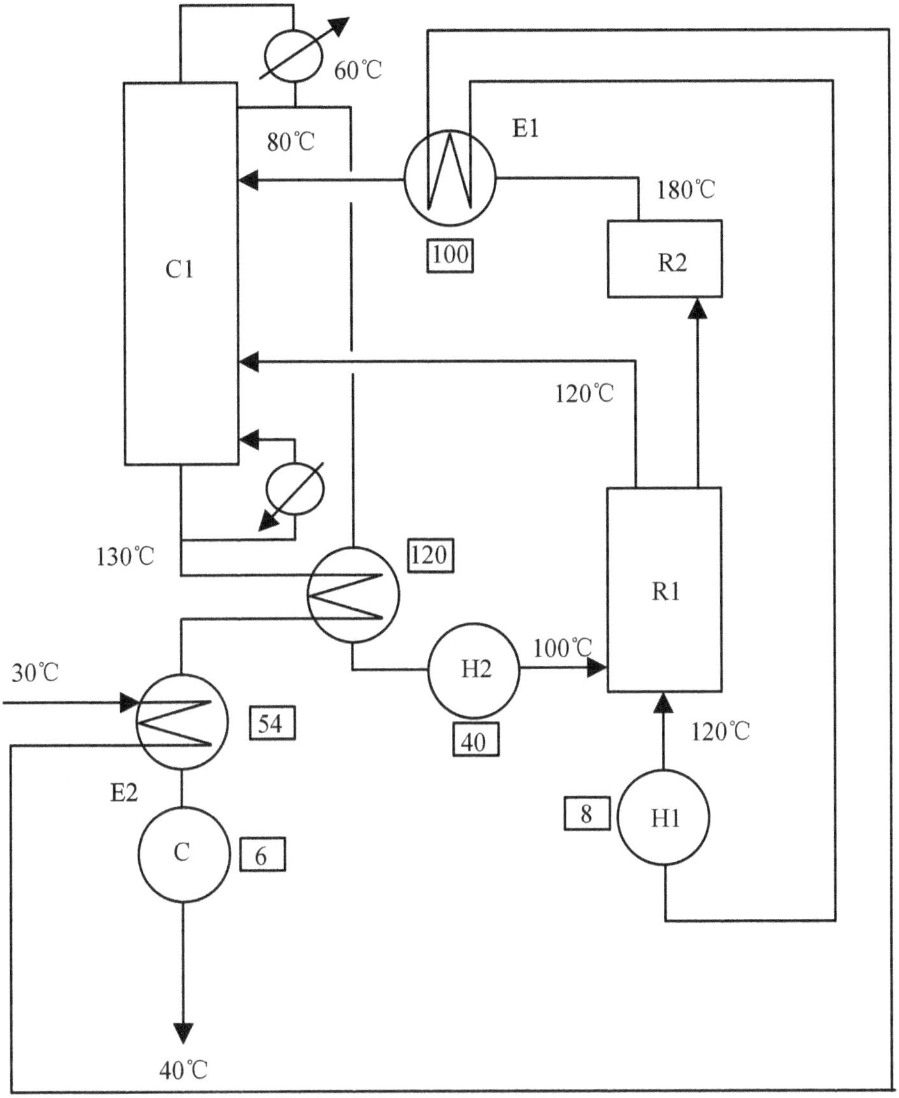

Fig. 9.14: The network after retrofit.

5. Eliminate cross-pinch heat transfers
   A retrofit is different from a new design, so the best design method is to evolve the network from the existing structure to identify the most crucial and cost-effective changes to the network structure. In this example, there is only one cross-pinch heat transfer. The heat load below the pinch of stream 1 should be supplied by a hot stream below the pinch. It can be seen from Figure 9.12 that stream 4 can be used. The network after retrofit is shown in Figure 9.14. It can be seen that the

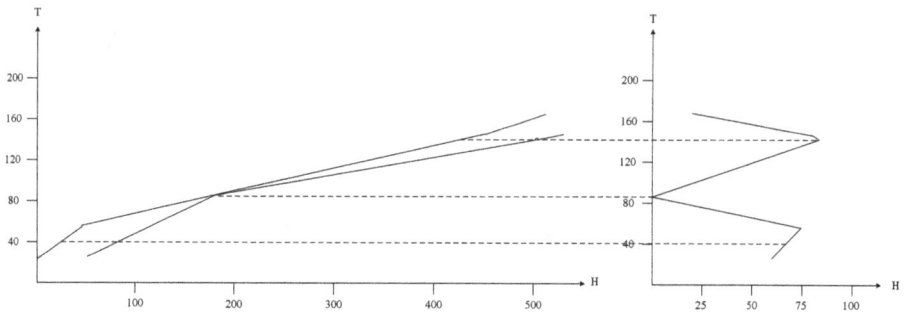

Fig. 9.15: Grand composite curve transformed from Figure 9.9.

total hot utility consumption drops from 102 to 48 kW, and the total cold utility consumption drops from 60 to 6. The increased heat recovery is 54.

9.3 Utility selection

After the HEN is synthesized, the heating and cooling duties not serviced by heat recovery must be provided by external utilities. To understand the interface between the process and the utility system, the grand composite curve is an appropriate tool for the selection of utilities [1, 2].

9.3.1 Grand composite curve

The grand composite curve can be obtained by transforming composite curves. The grand composite curve also has T and H coordinates, but the temperature plotted is shifted temperature and not actual temperature. Hot streams are represented $\Delta T_{min}/2$ colder and cold streams $\Delta T_{min}/2$ hotter than they are in practice. The hot composite curve is shifted downwards by $\Delta T_{min}/2$, and the cold composite curve is shifted upwards by $\Delta T_{min}/2$. The difference between the two shifted curves is the value of H in the grand composite curve under the corresponding shifted temperature. The point of zero heat flow in the grand composite curve is the pinch. The open "jaws" at the top and bottom are $Q_{H\,min}$ and $Q_{C\,min}$, respectively.

The grand composite curve transformed from Figure 9.9 is shown in Figure 9.15. Figure 9.16 gives a typical grand composite curve. The general trend of the grand composite curve is toward a positive direction. However, there are some curves that go toward a negative direction, as shown in Figure 9.16.

The shaded areas in Figure 9.16 are known as pockets. A pocket above the pinch means that a local heat source appears, the surplus of heat of which is usually provided to the cold streams in the pocket. Similarly, a pocket below the pinch means that

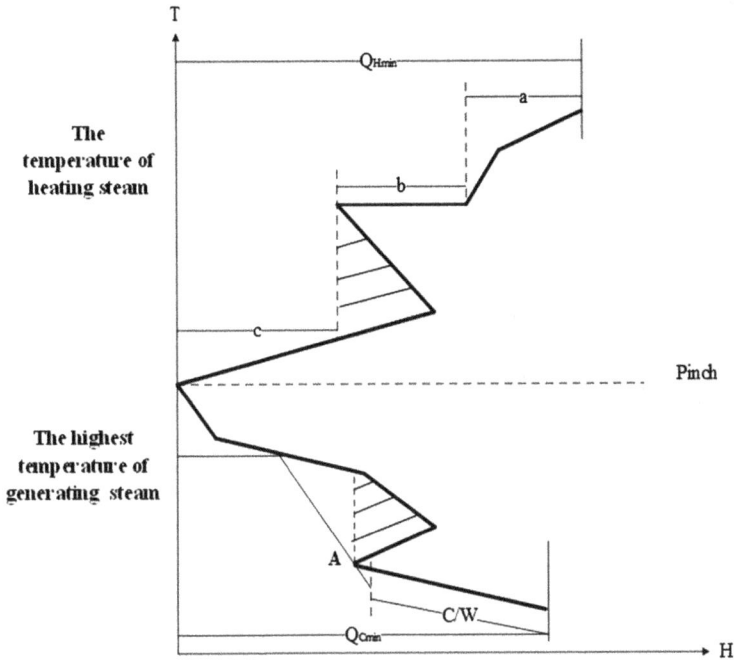

C/W——Cooling water line;
A——Steam preheating;
a——Line for high pressure steam;
b——Line for medium pressure steam;
c——Line for low pressure steam;

Fig. 9.16: Typical grand composite curve.

a local heat sink appears, the local heat deficit of which usually satisfies the cooling demand of the hot streams in the pocket. Thus, no external utilities are needed in the pockets.

When a pocket is big, the heat transfer process might cause energy degradation loss for the large temperature difference. In this case, recovering energy in the pocket can be considered as utilizing an energy cascade. Figure 9.17 is an example showing use of a lower quality utility to replace the higher quality process hot streams in the pocket to exchange heat with the cold streams in the pocket. The process hot streams are then used to generate higher quality utilities [4]. It is obvious that the energy performance of the process can be improved.

The profile of the grand composite curve represents residual heating and cooling demands after recovering heat within the shifted temperature intervals in the composite curves. Therefore, utility selection can be carried out according to the relation of heat flow and temperature presented by the grand composite curve.

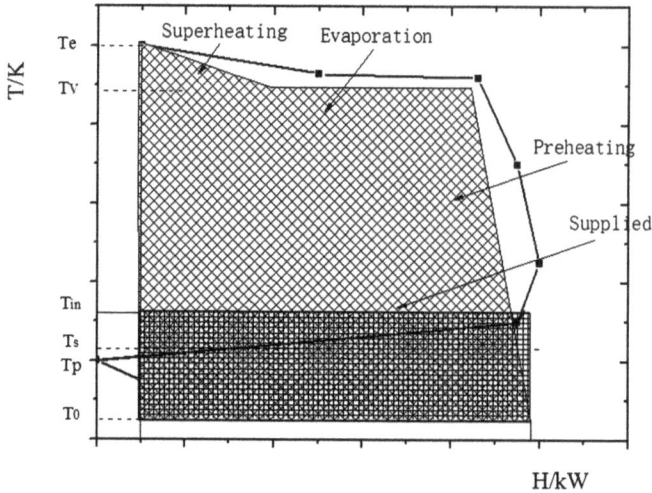

T_{in}——the temperature of supplied steam at lower pressure;
T_e——temperature of superheated steam at higher pressure;
T_v——evaporation temperature;
T_s——the initial temperature of preheating water;
T_p——temperature of pinch point;
T_0——environmental temperature;

Fig. 9.17: Grand composite curve after inclusion of energy media.

9.3.2 Utility selection

9.3.2.1 Hot utility selection

The most common hot utility is steam, which is usually available at several levels. High-temperature heating duties require furnace flue gas or a hot oil circuit.

If steam is used as the hot utility, most of the heat used is latent heat, so horizontal lines can be used to represent the corresponding temperatures. The heat load of steam at each level should be matched with the grand composite curve. Figure 9.16 shows the grand composite curve with three levels of steam (low, medium, and high pressure steam) used as the hot utility.

For furnace flue gas or hot oil, the curve presents a sloping profile because it is giving sensible heat rather than latent heat, as shown in Figure 9.18. In this case, the supply temperature and flowrate of the utility stream should be reasonably selected. A higher supply temperature, T_S, can both reduce the flowrate of the utility stream and increase the temperature difference in heaters, which decreases the heater area. However, T_S is constrained by the process conditions. Under a given T_S, with a lower flowrate, the temperature of the utility stream out of the process (T_r) is lower, resulting in less heat loss. This heat is that released by the utility stream in going from T_r to

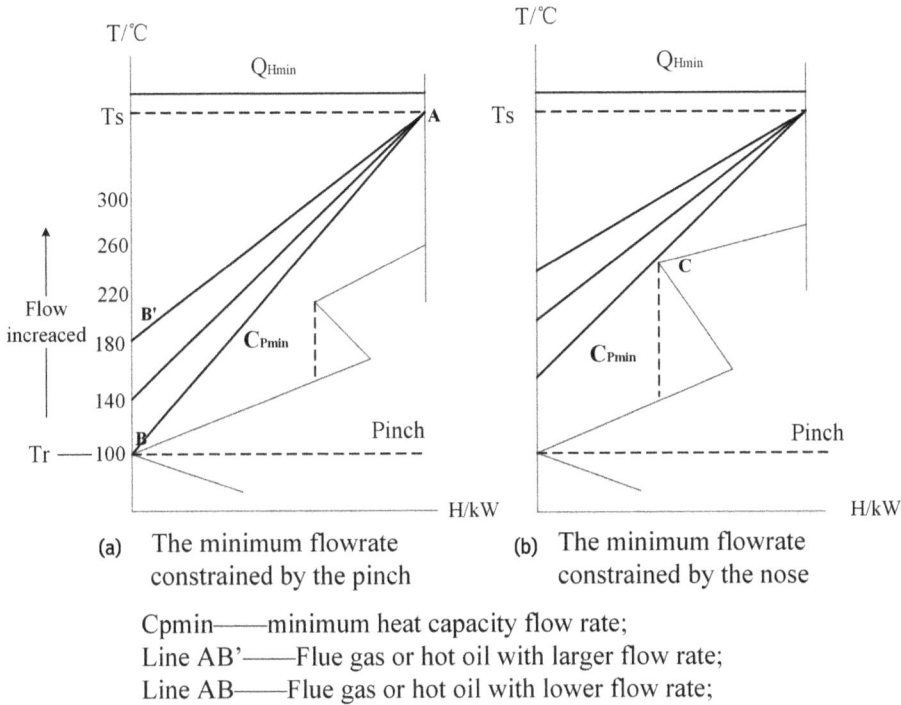

Cpmin————minimum heat capacity flow rate;
Line AB'————Flue gas or hot oil with larger flow rate;
Line AB————Flue gas or hot oil with lower flow rate;

Fig. 9.18: Flue gas or hot oil as hot utility.

the ambient temperature. However, for lower T_r, the temperature difference between the process and utility streams is lower, resulting in a higher capital cost. Thus, there is a trade-off between capital and energy costs. To attain the optimal energy performance, the minimum flowrate is determined, which is constrained either by the pinch, as shown in Figure 9.18a, or by a "nose," as shown by point C in Figure 9.18(b).

9.3.2.2 Cold utility selection

Cold utilities might be refrigeration, cooling water, air cooling, or even steam generation if heat needs to be rejected at high temperatures. The most usual cold utilities for common cooling (not at high or low temperatures) are environmental media (i.e., cooling water and air cooling).

Low-temperature cooling duties require refrigeration. The heat load of cold utilities at each level should be matched with the grand composite curve. Figure 9.19 shows the grand composite curve with one environmental medium and two levels of refrigeration.

If the rejected heat is at a high temperature, the heat can be used to generate steam. Figure 9.16 shows how to use the grand composite curve below the pinch to de-

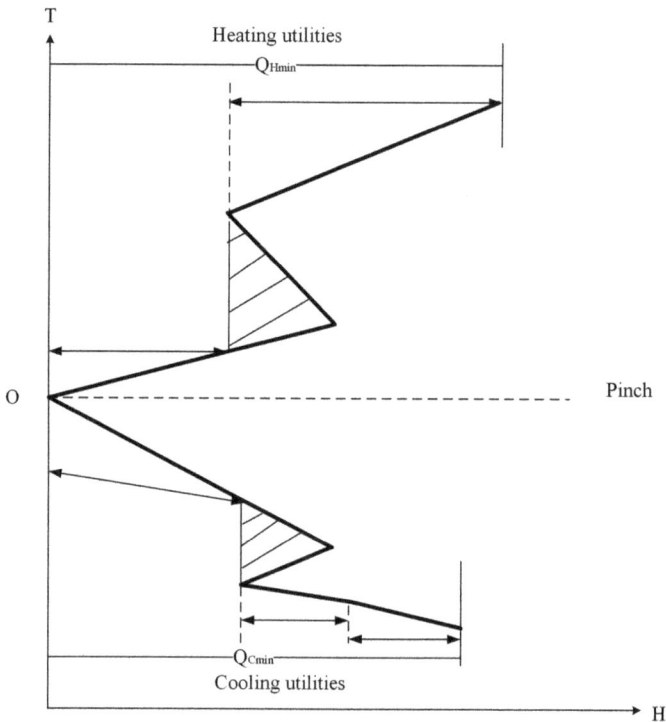

Fig. 9.19: Cold utility selection.

termine the evaporation temperature and heat load of the steam. It can be seen from Figure 9.16 that point A constrains the recovered heat by steam generation.

9.3.3 Combined heat and power generation

As the heat demand for utilities is at different temperatures, combined heat and power generation can improve energy efficiency. Such cogeneration can be carried out by heat engines, such as a steam turbine, gas turbine, or diesel engine. Figure 9.20 gives the energy balance around a heat engine.

Fundamentally, there are two possible ways to integrate a heat engine into a process: across or not across (above or below) the pinch. In Figure 9.21, the process is represented as a heat sink and a heat source (each of which is expressed as an ellipse) separated by the pinch.

Integration of a heat engine across the pinch as shown in Figure 9.21(a) is counterproductive. The process still requires $Q_{H\,min}$ and $Q_{C\,min}$, and the heat engine performs no better than if operated as stand-alone. There is no saving by integrating a heat engine across the pinch.

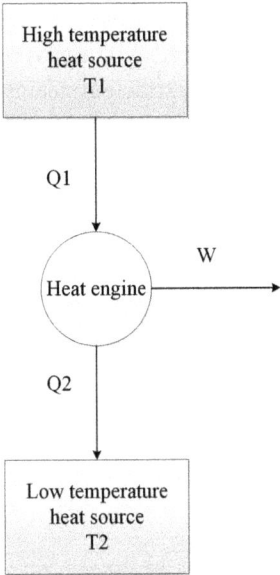

Fig. 9.20: Energy balance around a heat engine.

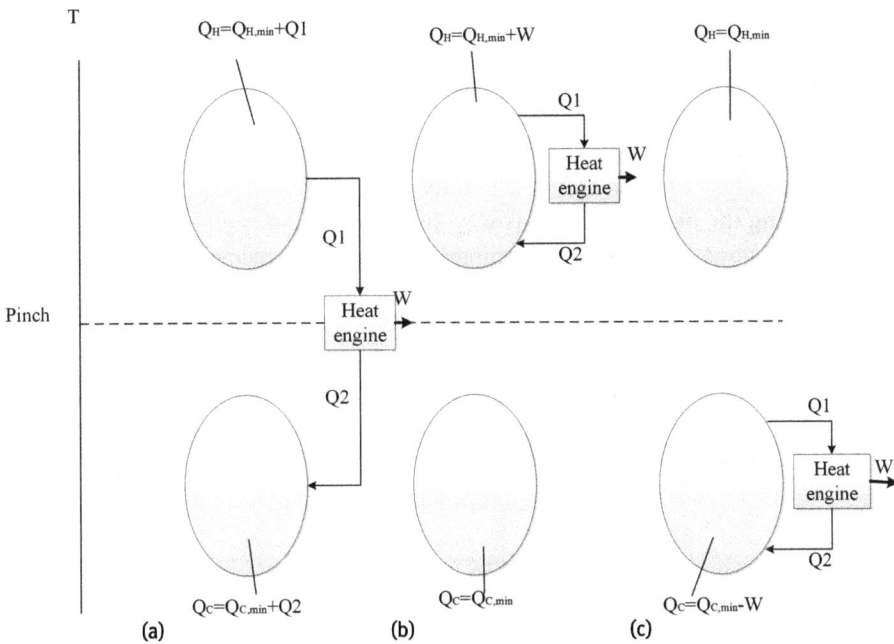

Fig. 9.21: Heat engine integrated into a process either (a) across, (b) above, or (c) below the pinch.

Figure 9.21(b) shows the heat engine integrated above the pinch. In this case, the rejected heat from the engine is used in the process. A typical example of this case is the use of back-pressure or extraction steam turbines, the exhaust or extraction steams of which are used as hot utilities in the process. The net effect in Figure 9.21(b) is the import of extra energy W from heat sources to produce power W. Because the process and heat engine act as one system, conversion of heat to power is at almost 100% efficiency. The heat loads of the exhaust or extraction steams can be determined by the H value at the corresponding temperatures on the grand composite curve. If the heat engine is a gas turbine, the turbine exhaust profile can be determined in the same way as the furnace flue gas, mentioned in Section 3.2.1.

A heat engine below the pinch, as shown in Figure 9.21(c), can definitely save energy, which is the situation with waste heat recovery for power generation.

In summary, to save energy, a heat engine should not be integrated across the pinch.

9.3.4 Integration of heat pumps

A heat pump is a device that takes low-temperature heat and upgrades it to a higher temperature to provide process heat. There are four types of heat pumps: compression heat pumps, absorption heat pumps, ejection heat pumps, and absorption heat transformers. The energy balance around a heat pump is shown in Figure 9.22.

There are two fundamental ways to integrate a heat pump with the process, across and not across the pinch, as shown in Figure 9.23.

Integration above the pinch is illustrated in Figure 9.23(a). This arrangement imports power or heat as W and saves hot utility as W. In other words, there is no saving by integrating the heat pump in this way. In this case, the capital cost for the heat pump is a waste of money and not economical. Integration below the pinch is shown in Figure 9.23(b). The result is worse economically because useful energy is turned into waste heat.

Integration across the pinch is illustrated in Figure 9.23(c). Because heat is pumped from a part of the process that is a net heat source to a part that is a net heat sink, this arrangement brings a genuine saving. Using the grand composite curve, the loads and temperatures of the cooling and heating duties of integrated heat pumps can be readily assessed. Thus, the appropriate placement of heat pumps is across the pinch.

The heat duties received and released by a heat pump should both match the grand composite curve and are less than the heat flow at the corresponding temperatures. Figure 9.24 gives the grand composite curves with and without a heat pump. If either heat duty is bigger than the heat flow at the corresponding temperature, the pinch point shifts, but no more energy can be saved. To achieve more hot and cold utility savings, more than one heat pump arrangement can be used [5].

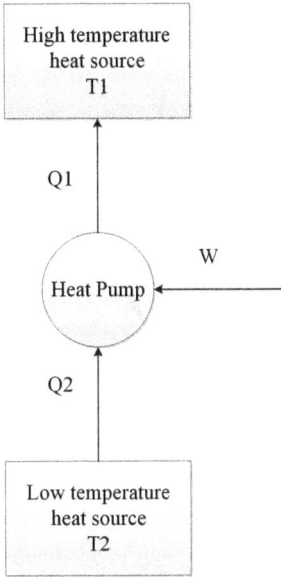

Fig. 9.22: Energy balance around a heat pump.

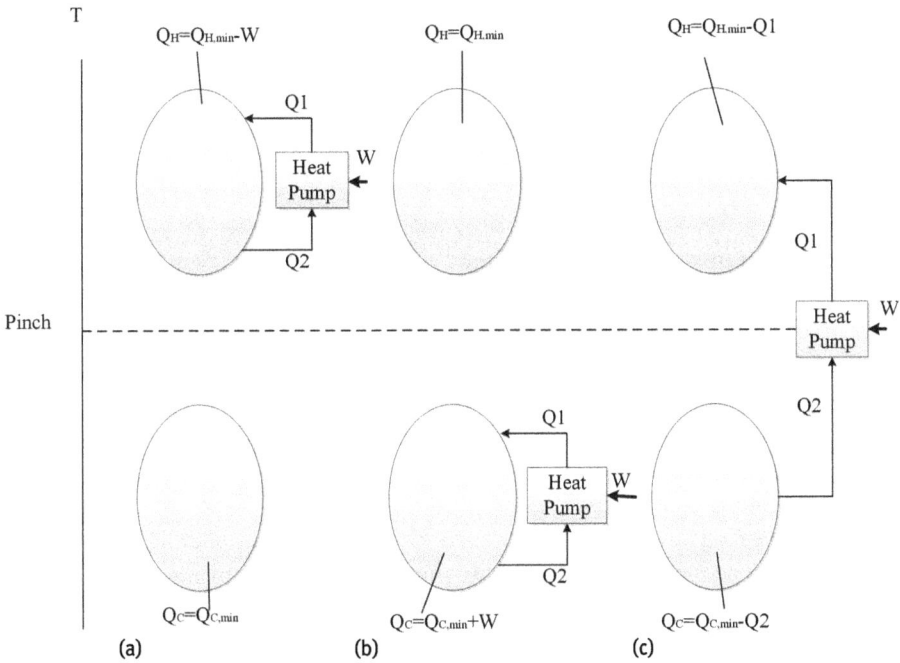

Fig. 9.23: Heat pump integrated into a process either (a) above, (b) below, or (c) across the pinch.

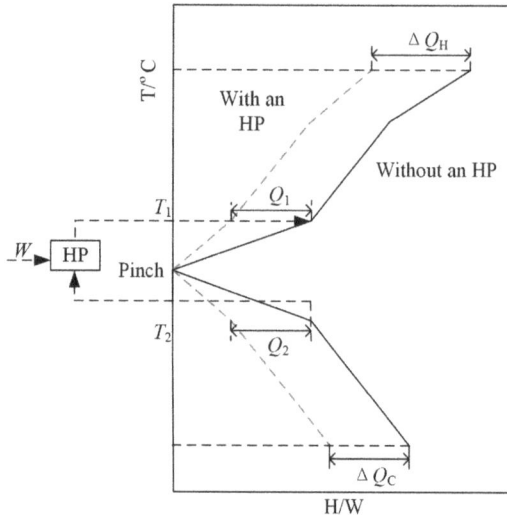

Fig. 9.24: Grand composite curves with and without a heat pump (HP).

9.4 Heat integration of reactors and distillation columns

The reactor and separation systems are at the center of a process. For energy efficiency, only separation systems that use energy as the separation agent (e.g., distillation, evaporation, and drying) are considered. In these separation systems, the distillation system is the most complicated so we focus on heat integration of distillation columns [1, 2].

9.4.1 Appropriate placement of reactors

First, we consider placement of reactors in different locations relative to the pinch. In Figure 9.25, the background process (which does not include the reaction heat) is represented as a heat sink and a heat source separated by the pinch.

Figure 9.25(a) shows the process with an exothermic reactor integrated above the pinch. The minimum hot utility can be reduced by the heat released by reaction (Q_{REA}). For comparison, Figure 9.25(b) shows an exothermic reactor integrated below the pinch. Because this part of the process is a heat source, the hot utility requirement cannot be reduced and the cold utility requirement increases by the value of Q_{REA}. Therefore, the appropriate placement for exothermic reactors is above the pinch.

Figure 9.26(a) shows an endothermic reactor integrated above the pinch. The endothermic reactor removes Q_{REA} from the process above the pinch, but an extra Q_{REA} must be imported from the hot utility to compensate. There is no benefit to integrating an endothermic reactor above the pinch.

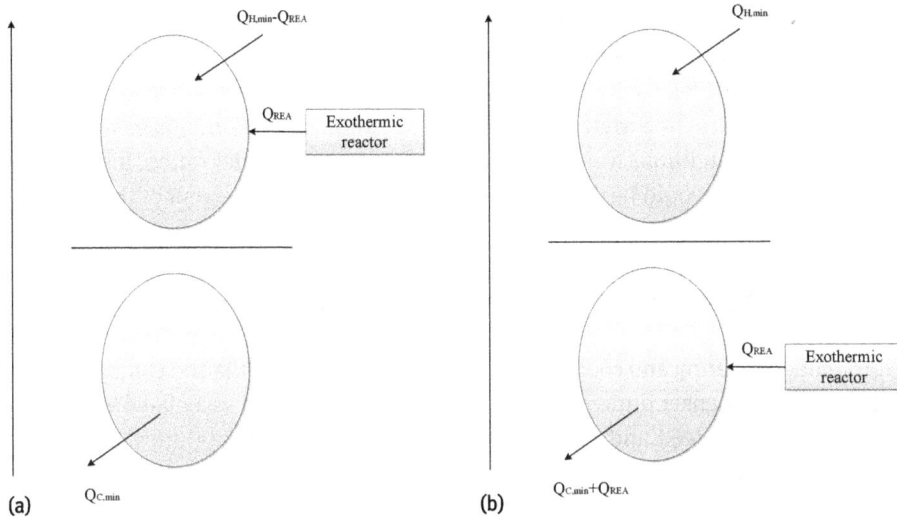

Fig. 9.25: Placement of an exothermic reactor (a) above and (b) below the pinch.

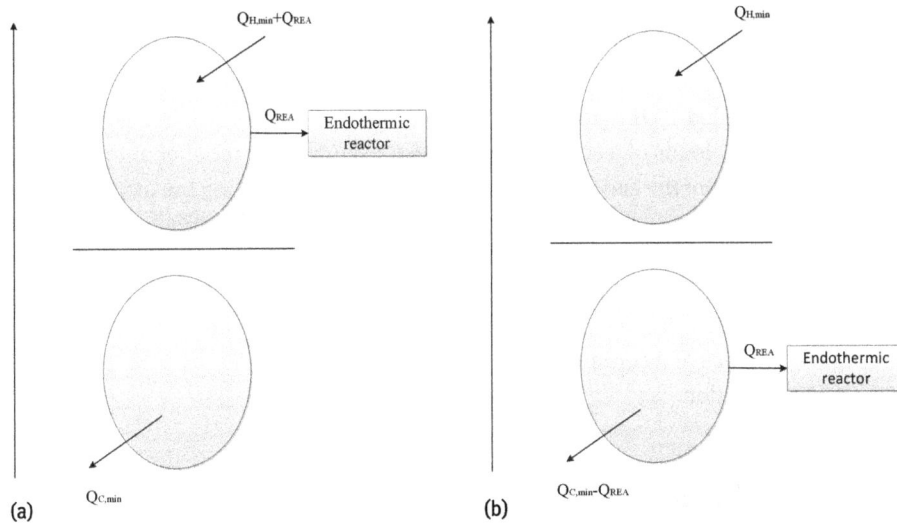

Fig. 9.26: Placement of an endothermic reactor (a) above and (b) below the pinch.

By contrast, Figure 9.26(b) shows an endothermic reactor integrated below the pinch. The reactor imports Q_{REA} from part of the process that needs to reject heat. Thus, integration of the reactor serves to reduce the cold utility consumption by Q_{REA}. There is also an overall reduction in the hot utility because, without integration, the process and reactor would require $Q_{H\,min} + Q_{REA}$ from the utility. Thus, the appropriate placement for endothermic reactors is below the pinch.

If a reactor is inappropriately placed, it is possible to change the process or the reactor conditions to correct this. Most often, however, the reactor conditions have been optimized for selectivity, catalyst performance, and so on, which, taken together with safety, construction material constraints, control, and so on, makes it unlikely that the reactor conditions would be changed to improve heat integration. Instead, the rest of the process would be changed to obtain appropriate placement of the reactor.

9.4.2 Heat integration characteristics of a single distillation column

The dominant heating and cooling duties associated with a distillation column are the reboiler and condenser duties. Although there are other duties associated with heating and cooling of feed and product streams, these sensible heat duties are usually much smaller than the latent heat in reboilers and condensers. Both reboiling and condensing processes normally take place over a range of temperatures. The heat to the reboiler is supplied at a temperature above the dew point of the vapor leaving the reboiler and the heat removed in the condenser is removed at a temperature lower than the bubble point of the liquid. Hence, for the sake of simplicity, both reboiling and condensing are assumed to take place at constant temperatures. A convenient representation of the column is a simple "box" representing the reboiler and condenser loads. The simplified $T-H$ diagram of a distillation column is shown in Figure 9.27.

The total heat loads of a column do not need to supply the reboiler at the bottom and be removed from the condenser at the top of the column. Along the upward direction in the stripping section of a column, the heating demand for vaporizing the light components gradually decreases. Similarly, along the downward direction in the rectifying section of a column, the cooling demand for condensing the heavy components decreases. Based on plate-by-plate calculation, a grand composite curve to show the heat characteristics of a column can be obtained, as shown in Figure 9.28. The pinch is at the feed location.

9.4.3 Heat integration of a distillation system

When a homogeneous multicomponent fluid mixture needs to be separated into a number of products by distillation, a number of columns (a distillation system) are normally used. If the condenser of a column can provide heat to the reboiler of another column, the heat in the two columns can be integrated. As shown in Figure 9.29, if the temperature of the condenser of column 1 is high enough, the condenser can provide heat to the reboiler of column 2, resulting in an energy-saving effect.

In Figure 9.30(a), the temperature of the condenser of column A is not high enough to provide heat to the reboiler of column B; similarly for columns B and C. In this case, to perform heat integration between columns, two methods can be considered. The

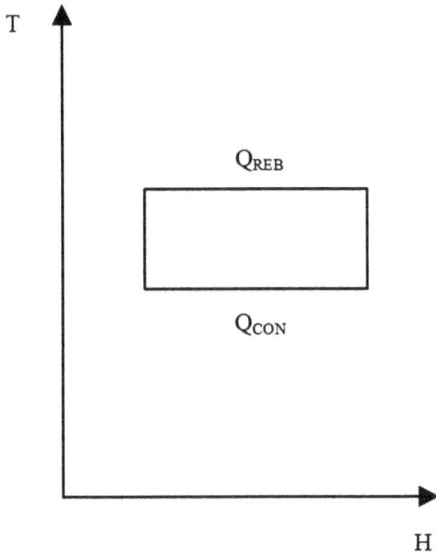

Fig. 9.27: Simplified *T–H* diagram of a distillation column (represented by the rectangle).

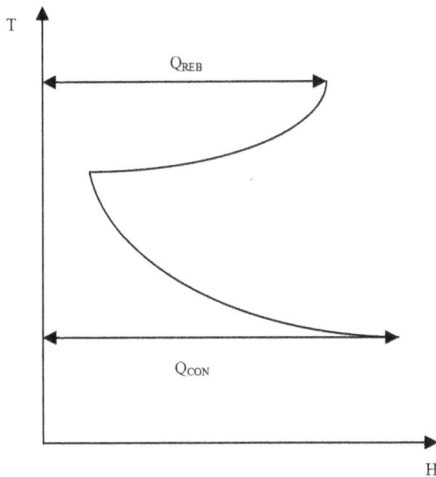

Fig. 9.28: Grand composite curve of a distillation column.

first method is to change the pressure of the columns, which changes the temperature distribution between columns. An increase in the pressure of the column used as heat provider, or a decrease in the pressure of the column used as heat receiver, might give sufficient temperature difference between the two columns, as shown in Figure 9.30(b). The second method is to use intermediate reboilers by considering the grand composite curves of the columns, as shown in Figure 9.30(c).

Heat integration of distillation columns can be used for distillation sequences (to separate a homogeneous multicomponent fluid mixture) and also for independent

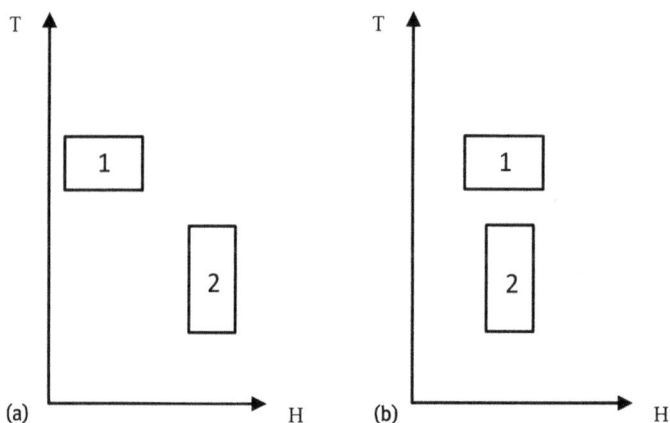

Fig. 9.29: Direct integrated columns. (a) Two columns without integrated, (b) integrated columns.

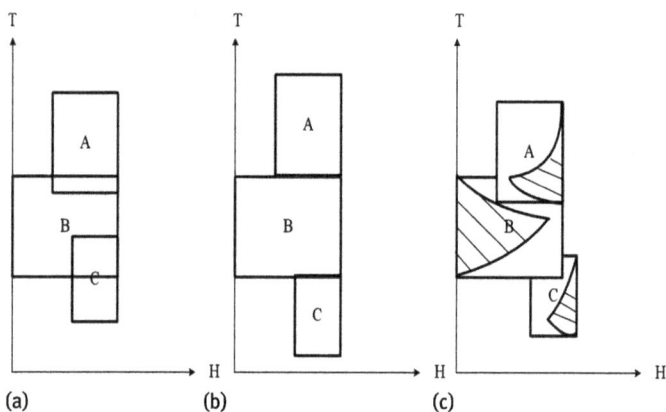

(a) The original temperature distribution
(b) Heat integration after adjusting pressures
(c) Heat integration by introducing intermediate reboilers

Fig. 9.30: Heat integration of distillation sequences.

columns. Such integration schemes can bring about a significant reduction in the energy requirement.

9.4.4 Appropriate placement of distillation column

We now consider the placement of a distillation column in different locations relative to the pinch of the background process (which does not include the reboiler and condenser). The column takes heat Q_{REB} into the reboiler at temperature T_{REB} and rejects heat Q_{CON} at a lower temperature T_{CON}. There, the column can be heat integrated with

T

$Q_H = Q_{H,min} + (Q_{REB} - Q_{CON})$

$Q_H = Q_{H,min}$

$Q_H = Q_{H,min} + Q_{REB}$

Q_{REB}

Column

Q_{CON}

Q_{REB}

Pinch

Column

Q_{REB}

Column

Q_{CON}

Q_{CON}

$Q_C = Q_{C,min}$

$Q_C = Q_{C,min} - (Q_{REB} - Q_{CON})$

$Q_C = Q_{C,min} + Q_{CON}$

(a)   Above the pinch   (b)   Blow the pinch   (c)   Across the pinch

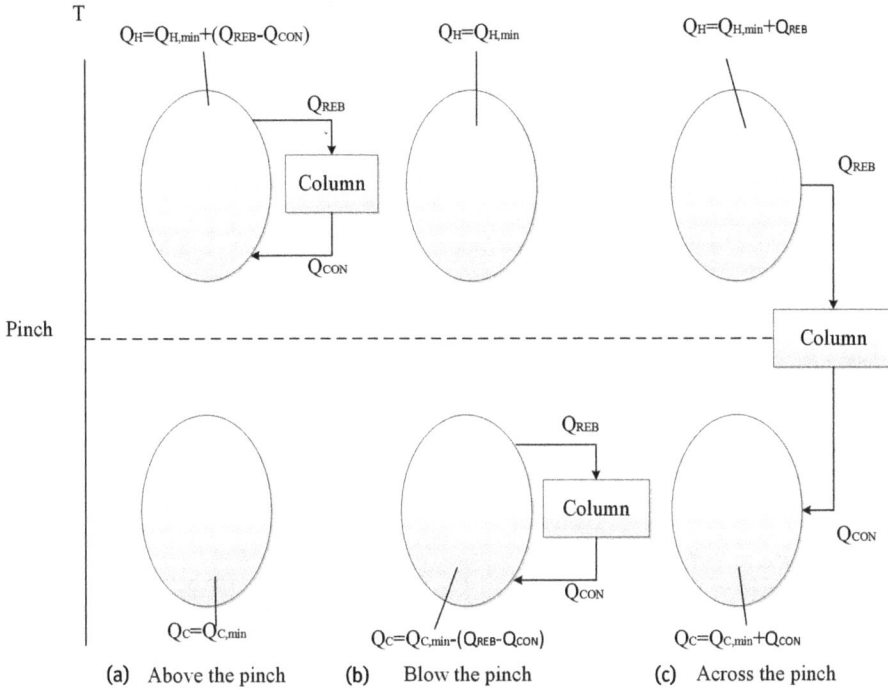

Fig. 9.31: Distillation column is integrated into a process.

the background process in two possible ways, that is, the reboiler and condenser can be integrated either across or not across the pinch.

Figure 9.31(a) shows a distillation column entirely above the pinch. The distillation column takes heat Q_{REB} from the process and returns Q_{CON} at a temperature above the pinch. The hot utility consumption changes by $Q_{REB} - Q_{CON}$. The cold utility consumption is unchanged. Usually, Q_{REB} and Q_{CON} have a similar magnitude. If $Q_{REB} \approx Q_{CON}$, then the hot utility consumption is $Q_{H\,min}$ and there is no additional hot utility required to run the column; it takes a "free ride" from the process. A column above the pinch means that the heat released from the condenser is supplied to heat some of the cold process streams above the pinch.

Heat integration below the pinch is illustrated in Figure 9.31(b). The hot utility is unchanged, but the cold utility consumption changes by $Q_{CON} - Q_{REB}$. Again, given $Q_{REB} \approx Q_{CON}$, the result is similar to heat integration above the pinch. A column below the pinch means that the reboiler is provided heat by some of the hot process streams below the pinch.

The arrangement for distillation across the pinch is shown in Figure 9.31(c). Heat Q_{REB} is taken into the reboiler above the pinch and heat Q_{CON} rejected from the condenser below the pinch. Because the process sink above the pinch requires at least $Q_{H\,min}$ to satisfy its enthalpy balance, the Q_{REB} removed by the reboiler must be com-

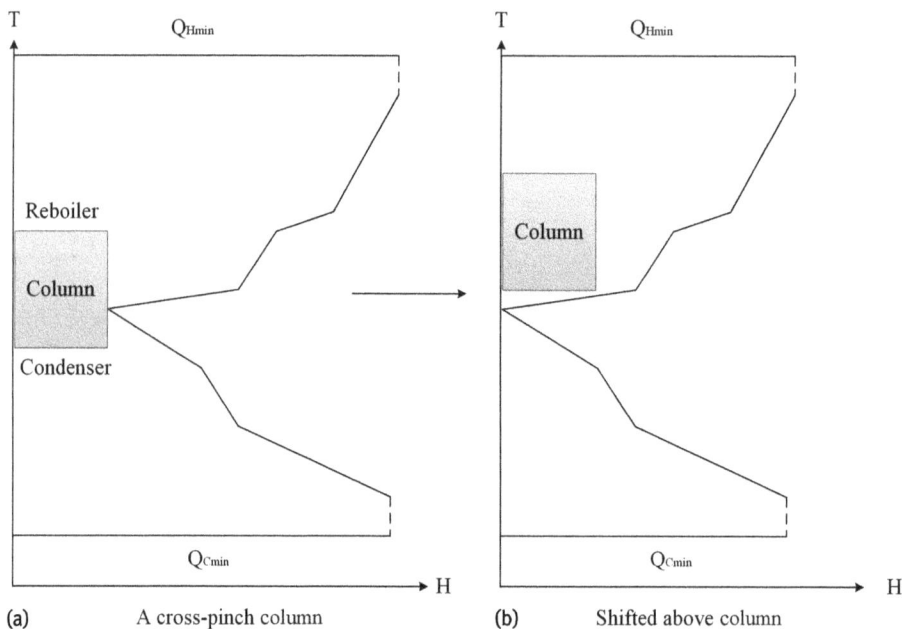

Fig. 9.32: Moving the column above the pinch.

pensated for by introducing an extra Q_{REB} from the hot utility. Below the pinch, the process needs to reject $Q_{C\,min}$ and an extra heat load Q_{CON} from the condenser needs to reject to the cold utility. Therefore, no savings are gained by the integration of a column across the pinch.

All these arguments can be summarized by a simple statement: the appropriate placement for separators is not across the pinch. Although the rule was originally stated with regard to distillation columns, it clearly applies to any separator that takes in heat at a higher temperature and rejects heat at a lower temperature.

9.4.5 Use of the grand composite curve for heat integration of distillation column

The appropriate placement rule can only be applied if the process has the capacity to provide or accept the required heat duties. A quantitative tool is needed to assess the source and sink capacities of any given background process. For this purpose, the grand composite curve can be used. A column should fit the grand composite curve of the background process.

If the reboiler and condenser duties of a column are on opposite sides of the pinch, as shown in Figures 9.32(a) and 9.33(a), the pressure of the column can be changed to achieve appropriate placement. In Figure 9.32(b), the pressure of the column is in-

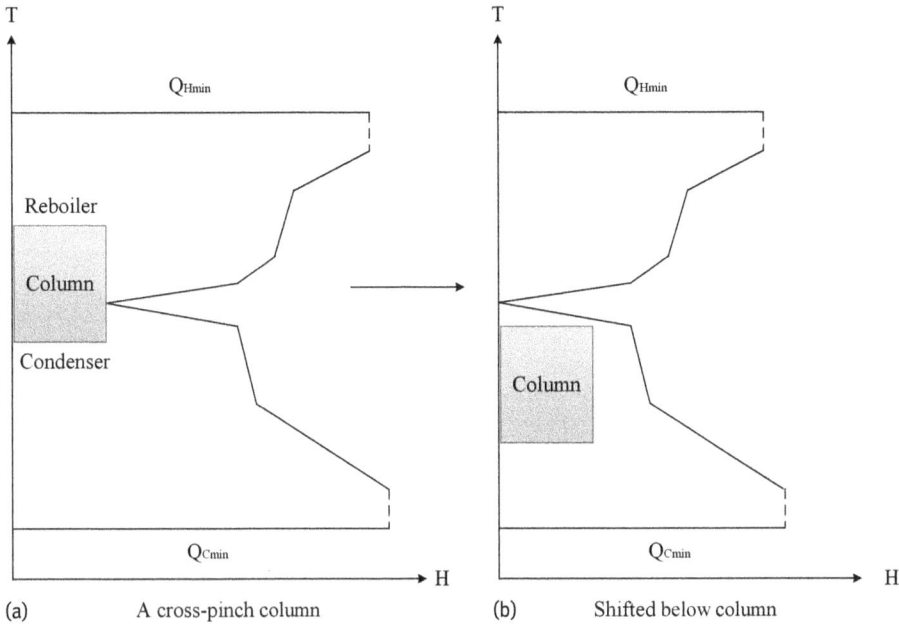

Fig. 9.33: Moving the column below the pinch.

creased to shift it to a location above the pinch, and in Figure 9.33(b), the pressure of the column is decreased to shift it to a location below the pinch.

Figure 9.34(a) shows that, although the reboiler and condenser duties are both below the pinch, the heat duties prevent a fit because the heat duties are larger. If the heat duty of the reboiler is only a little larger, decreasing the reflux ratio can be considered. If the heat duty of the reboiler is much larger, Figure 9.34 shows three possibilities. In Figure 9.34(a), part of the duty is accommodated, but less than the full reboiler and condenser duties. In Figure 9.34(b), the column is divided into two parts, one above the pinch, and the other below. In Figure 9.34(c), multieffect distillation is applied to decrease the heat duties and is placed either above or below the pinch.

Another design option that can be considered if a column does not fit into the grand composite curve is the use of an intermediate reboiler or condenser, as illustrated in Figure 9.35 for an intermediate condenser. The shape of the box is now altered, because the intermediate condenser changes the heat flow through the column, with some of the heat being rejected at a higher temperature in the intermediate condenser. The particular design shown in Figure 9.35 requires part of the heat rejected from the intermediate condenser to be passed to the process.

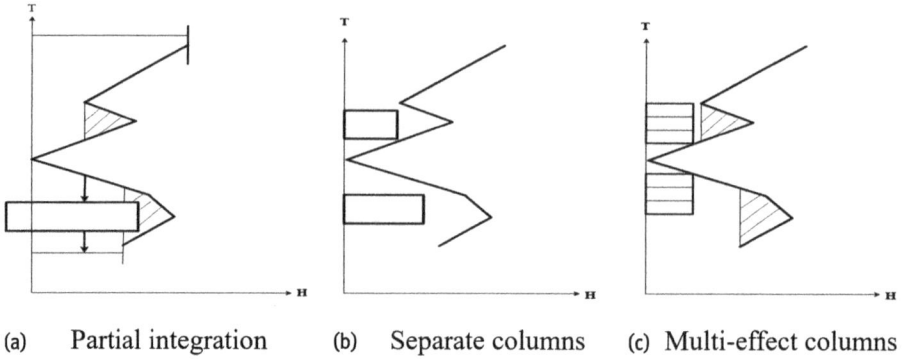

(a)   Partial integration   (b)   Separate columns   (c) Multi-effect columns

Fig. 9.34: Heat integration for larger duties.

9.5 Heat integration across plants

The discussion so far is mainly applicable for one plant. To improve energy efficiency further, the approach can be extended to site-wide applications (i.e., heat integration across plants).

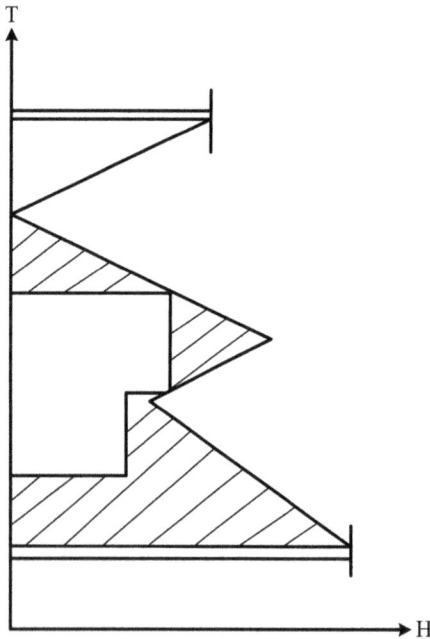

Fig. 9.35: Distillation column with intermediate condenser.

9.5.1 Total site profiles

The grand composite curve of a plant represents the heat available at every temperature level after all feasible intraplant heat integrations have been carried out [6]. The profile above the pinch represents a heat sink, whereas the profile below the pinch represents a heat source.

The site source profile is then constructed by combining the heat source information from all the available plants into a single profile, which is analogous to the hot composite curve for a single process. In a similar way, the site sink profile is formed by combining the heat sink information from all available plants, and is analogous to the cold composite curve for a single process. Together, the total site profiles give a simultaneous view of surplus heat and heat deficit for all the plants on the site, as shown in Figure 9.36.

By moving the sink profile toward the source profile, the amount of overlap of the profiles represents the amount of possible heat recovery Q_r. The position of the "total site pinch" indicates that heat recovery on the site is at a maximum, as shown in Figure 9.37.

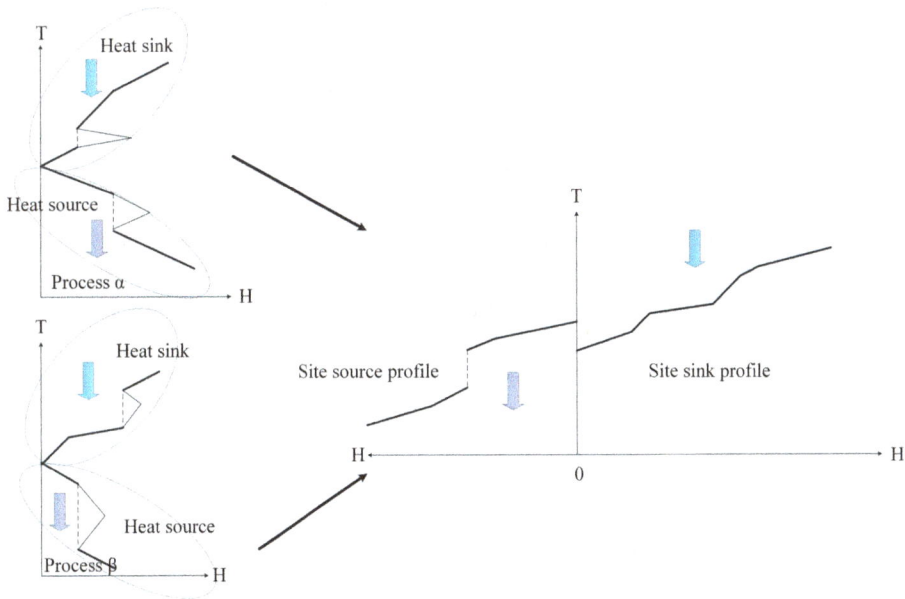

Fig. 9.36: Total site profiles.

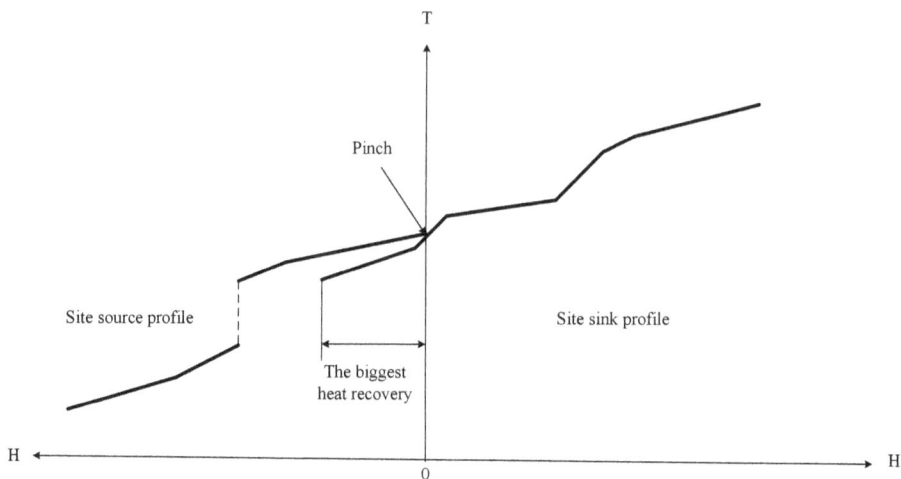

Fig. 9.37: Total site pinch.

9.5.2 Direct and indirect heat integration

Heat integration using process streams is defined as direct heat integration, and heat integration using an intermediate fluid is defined as indirect heat integration.

Direct and indirect heat integration feature different heat integration performances across plants. For direct heat integration, process streams exchange heat with each other directly, so that the heat integration process only experiences heat transfer once. This feature means that the heat recovery can be larger because of a smaller temperature difference during heat transfer between two plants. However, because process streams cannot be mixed with each other, when the number of process streams participating in heat integration is large, a number of heat transfer loops are required. Because of the large distance between plants, each loop requires the investment of pipeline. From the point of view of economics, if the number of process streams participating in heat integration is large, it is not beneficial to use the direct integration method.

By contrast, for indirect heat integration, heat is first transferred from hot streams to the intermediate fluid in one plant, and then transferred from the intermediate fluid to cold streams in another plant. Because of the multiple heat transfer, the total temperature difference has to be larger than for direct heat integration, resulting in lower energy recovery. Moreover, the number of heat exchangers is normally higher. The benefit of using indirect heat integration is that an intermediate fluid can be split and mixed, so that heat can be transported from one plant to another using one intermediate fluid loop. Pipeline investment can be reduced when the number of process streams participating in heat integration is high.

9.5.3 Direct heat integration

Considering direct heat integration [7], it is better to have a lower number of heat transfer loops. The following strategies can be used to decrease the number of loops:

1. If the pinch of a plant is close to the total site pinch, this plant can be integrated separately (i.e., streams are matched only in this plant). According to the rules of pinch approach, heat integration has no influence on energy savings for the total site, but can reduce the number of heat transfer loops.
2. When the pinch of a plant is much higher than the total site pinch, the hot streams below the plant pinch but above the total site pinch can be matched with cold streams in other plants. This plant can be treated as a heat source plant and only output surplus heat to other plants.
3. When the pinch of a plant is much lower than the total site pinch, the cold streams above the plant pinch can be matched with hot streams in other plants. This plant can be treated as a heat sink plant and only input heat from other plants.
4. Interplant transfer between two subsystems should be considered when there is a large difference between the two pinches of the plants. The hot streams below the pinch of the plant with higher pinch temperature can be matched with cold streams above the pinch of the plant with lower pinch temperature, so that the heating and cooling utilities of the two plants can be reduced at the same time. Interplant heat transfer between plants with similar pinch temperature do not need to be considered.
5. When any interplant heat transfer is needed to increase the feasibility of heat integration and give economic benefit, the relationship between the plants should be considered, such as the distance between plants and the parallel operation time of different plants.

9.5.4 Indirect heat integration

For indirect heat integration, steam and hot water are the most commonly used intermediate fluids. If steam is used as the intermediate fluid, some horizontal lines under the corresponding temperatures can be used to represent steam at different pressures, because most of the heat used is latent heat. Plants are served by their own utilities or by common central utilities. Figure 9.38 shows the location of utilities (steam mains) in the total site profile. From the site source profile, the steam generation of each level can be determined, as shown in Figure 9.38 (left). From the site sink profile, the steam requirement of each level can be determined, as shown in Figure 9.38 (left). This generated steam meets some of the heating requirements. The amount of heat recovery that can take place in the total site through the steam mains can be derived from the total site profile. By moving the sink profile toward the source profile (Figure 9.38 (middle), the amount of overlap of the profiles represents the amount of heat recovery that can

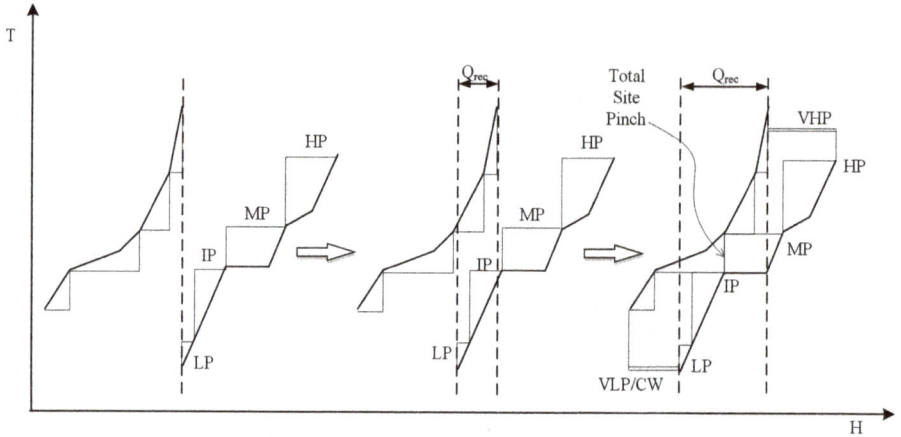

C/W——Cooling water; VLP——Very low pressure steam; LP——Low pressure steam;
IP——Intermediate pressure steam; MP——Medium pressure steam; HP——High pressure steam;
VHP——Very high pressure steam;

Fig. 9.38: Heat recovery targets when using steam as the intermediate fluid.

be obtained through the steam mains. The limit to heat recovery is reached when a sink profile steam main touches the source profile, as shown in Figure 9.38 (right) [6].

Steam is the most common intermediate fluid for recovering heat at high or medium temperatures. However, it cannot be used as an intermediate fluid for low-grade heat recovery. Hot water is often used for low-grade heat recovery.

The difference between hot water and steam when determining the energy target is that the line for hot water is oblique on the total site profiles whereas the line for steam is horizontal. Therefore, in heat integration across plants using hot water, it is important to determine the temperature and flow rate of the hot water in order to determine the energy target.

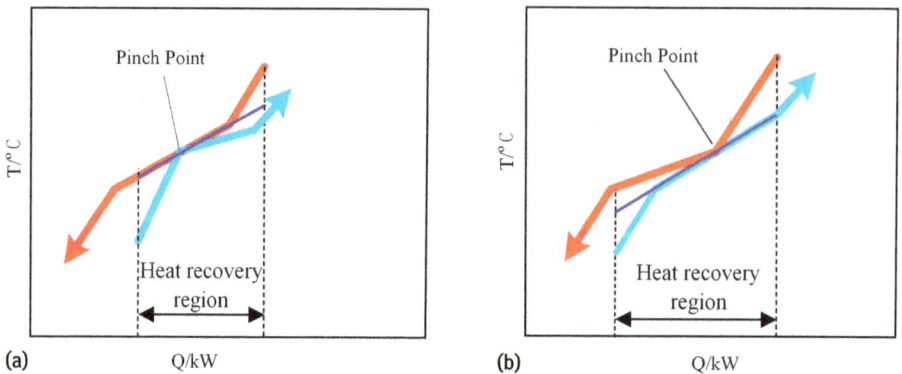

Fig. 9.39: Determining hot water flow rate from (a) the hot stream and (b) the cold stream.

Figure 9.39 shows how the flowrate of hot water can be determined. In total site profiles, the gradient of a line represents the heat capacity flowrate (*CP*), which is equal to the product of specific heat and flowrate. The flowrate of hot water can therefore be determined from the gradient of the line on a total site profile. In Figure 9.39, the thick curves indicate the total source profile (red line) and the total sink profile (blue line), while the thin line denotes the hot water line. The heat recovery region is defined as the region within which heat can be transferred from a heat source to a heat sink on the total site profile. Because no heat transfer occurs outside the heat recovery region, the hot water line should not extend beyond the boundaries of the region. Normally, a pinch point is formed by a convex point on one curve touching a straight line on the other. As shown in Figure 9.39(a), the pinch point is formed by the convex point on the heat sink profile touching a straight line on the heat source profile. It is obvious that maximum heat recovery can be achieved when the *CP* of the hot water line equals the *CP* of the straight line forming the pinch point [8].

9.6 An industrial case study

To make the above-mentioned synthesis approaches clear, we describe an industrial case study [3]. The whole system is divided into three subsystems: HEN subsystem, separation subsystem, and utility subsystem. All three subsystems are analyzed through the approaches introduced here. In addition, we consider the coordination between the three subsystems.

9.6.1 Procedure for energy integration for total site system retrofit

Energy integration for a total site system retrofit is not simply heat integration from one subsystem to another, because each step of the retrofit may affect the other subsystems. In the retrofit process, cycle analysis and calculation are needed to achieve the eventual balance.

Total site system retrofit energy integration analysis starts from the heat integration of each subsystem, especially the HEN, and is followed by consideration of coordination between subsystems. The concrete steps are shown in Figure 9.40. The pinch position of the HEN is determined first because the subsequent analysis and optimization is based on the pinch position. Then, the separation system is considered, including single columns and the column system. Next, coordination between the separation subsystem and the HEN subsystem is considered. The retrofit of the separation subsystem affects the HEN subsystem, causing a change in the HEN's matches, so that new heat integration for the HEN may be needed. The next step is coordination between the separation subsystem and the utility subsystem. The coordination between the HEN subsystem and the utility subsystem can also have an impact on the HEN; therefore,

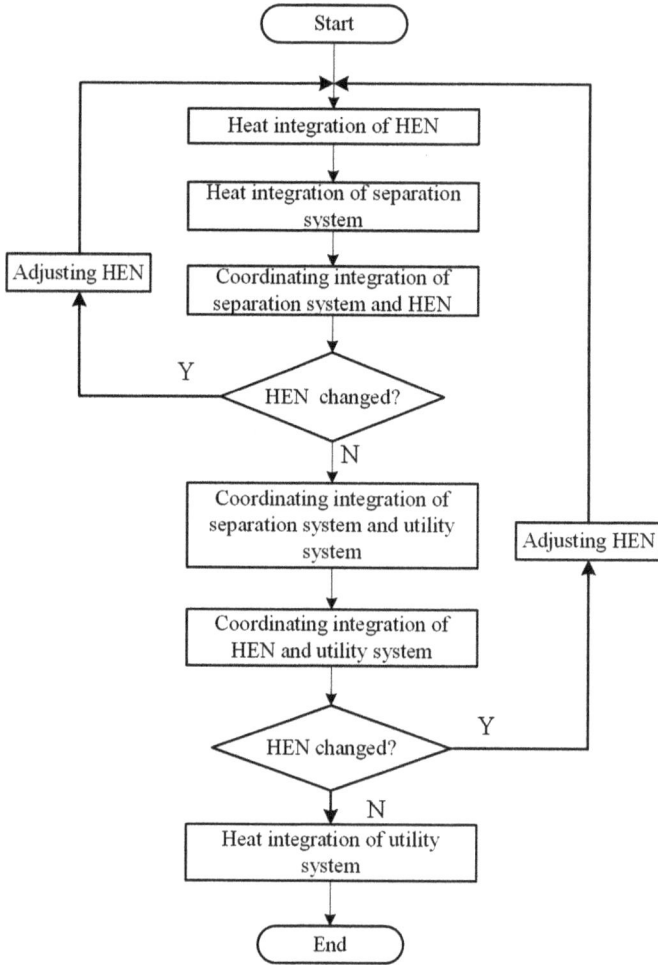

Fig. 9.40: Procedure of energy integration for total site system retrofit.

the heat integration of the HEN is considered again. After the retrofit, the supply and demand of different levels of steam are fixed. The last analysis is heat integration of the utility subsystem.

9.6.2 Basic data for the case study

The considered process is a caprolactam plant with six units, as shown in Figure 9.41. By extracting the flowrate, temperature, pressure, and component data for each stream through each heat exchanger, and simulating the heat duty of each heat exchanger using a commercial process simulation software (e.g., Aspen Plus), the stream data can be obtained, as shown in Table 9.3.

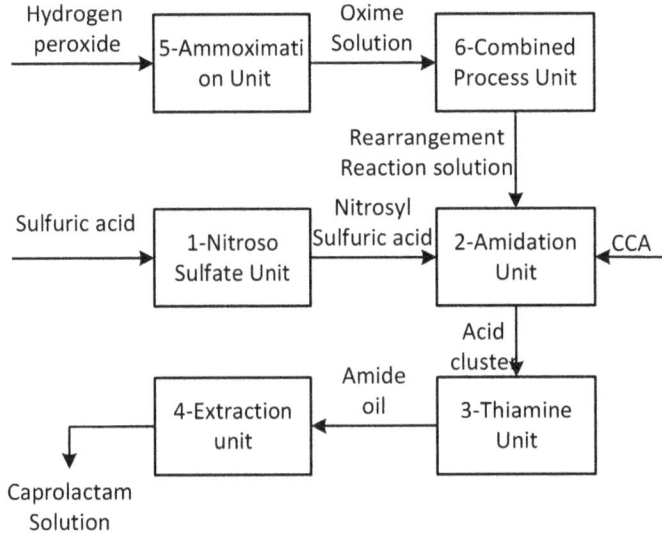

Fig. 9.41: Caprolactam production process diagram.

Tab. 9.3: Data for the hot streams in HEN

Unit no.	Stream no.	Heat exchanger on the stream	T_{in} (°C)	T_{out} (°C)	Duty (kW)
1	H1	E-102, E-103, E-104	870	168	5566.7
	H2	E-111	168	67	3313.4
	H3	E-107	67	35	353.3
	H4	E-108	35	11	570
	H5	E-110	52	44	172
2	H6	E-207	108	45	1557
	H7	E-208	108	45	1557
	H8	E-202, E-203	108	39	1694
	H9	E-204	71	48.5	5391
	H10	E-206	83.4	38	183
	H11	E-209	71	48.5	906
	H12	E-210	75.9	52.4	4760
3	H13	E-303	60.1	46	14,350
4	H14	E-401	105	66	1609.2
	H15	E-403, E-404	96	40	3367.8

Unit no.	Stream no.	Heat exchanger on the stream	T_{in} (°C)	T_{out} (°C)	Duty (kW)
	H16	E-406	97.5	82.5	1859
	H17	E-408	60	30	1326.8
5	H18	E-501	83	82	1019
	H19	E-505	80	50	1851
	H20	E-508	89	42	590
	H21	E-509	144	96	144.5
	H22	E-513	93.4	90	516
	H23	E-514	104	60	1285
	H24	E-516, E-517	105	45	1647

9.6.3 HEN subsystem integration

The temperature of the hot streams at the pinch point is 105.7 °C and that of the cold streams is 95.7 °C (ΔT_{min} = 10 °C). The energy saving potential is 6885 kW. The composite curve and the grand composite curve are shown in Figures 9.42 and 9.43, respectively.

After an initial retrofit (complying with the three principles of the pinch approach), six heat exchangers were added and 4946 kW hot and cold utility were saved.

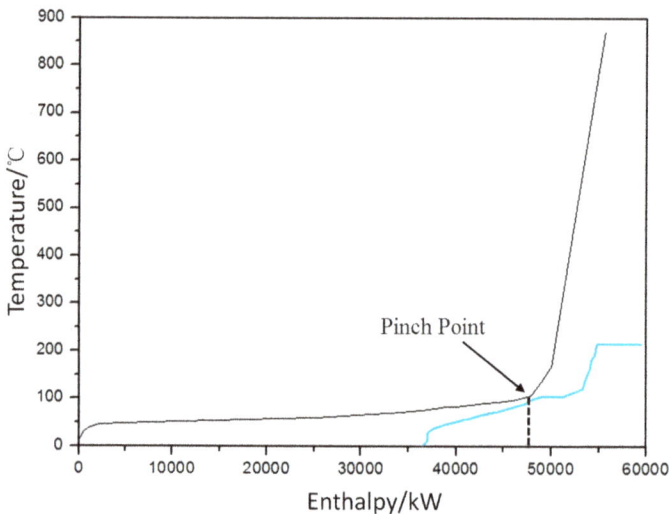

Fig. 9.42: Composite curve of the heat exchanger network.

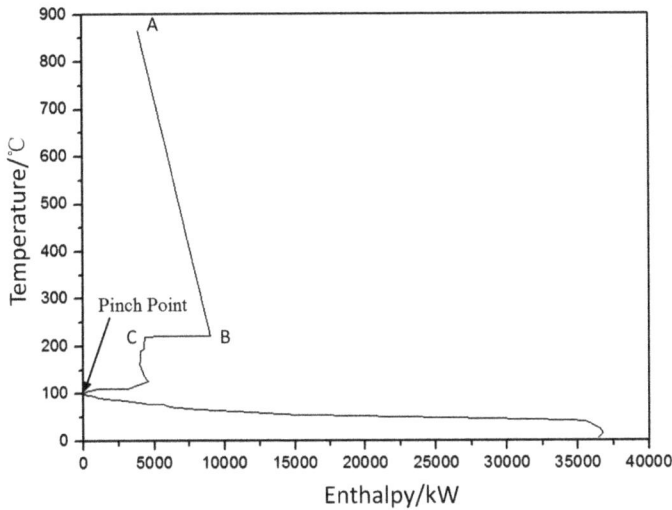

Fig. 9.43: Grand composite curve of the heat exchanger network.

9.6.4 Separation subsystem integration and its coordination with HEN subsystem

Five distillation columns in the separation subsystem are discussed. Their positions relative to the pinch point are shown in Table 9.4. From the reboiler and condenser temperatures, the columns that cross the pinch can be known. These columns are considered for moving to a position above or below the pinch point.

After simulation for pressure adjusting, only column T-401 can be moved above the pinch point. T-501 and T-502 still cross the pinch because they cannot be adjusted above or below the pinch point. For these columns, the reflux ratio is decreased to reduce their energy consumption.

After pressure adjusting, there are two columns above the pinch and one below. However, the heat duty released by the condensers of T-503 and T-401 is more than the demand of the HEN subsystem; therefore, reducing their reflux ratio is considered. The final positions of the columns in the grand composite curve are shown in Figure 9.44.

Tab. 9.4: Position between column and pinch point.

Column No.	Condenser temperature (°C)	Reboiler temperature (°C)	Relationship between column and pinch point
T-101	65	85	Below the pinch
T-401	97.5	104.3	Cross the pinch
T-501	75	118	Cross the pinch
T-502	54	124	Cross the pinch
T-503	115	159	Above the pinch

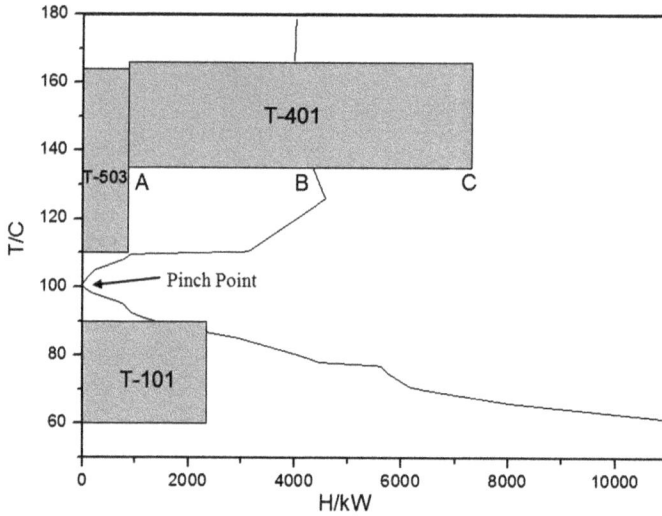

Fig. 9.44: Position of column on the grand composite curve.

9.6.5 Coordination of HEN subsystem and utility subsystem

In Figure 9.43, there is a huge "heat pocket" (A–B–C) in the curve, which exists in unit 1. The heat is recovered to generate high pressure steam (HPS; 225 °C, 1.7 MPa), as shown in Figure 9.45. However, some superhigh pressure steam (SHPS; 386 °C, 3.5 MPa) can be generated by using the heat between 436 °C and 870 °C at the mini-

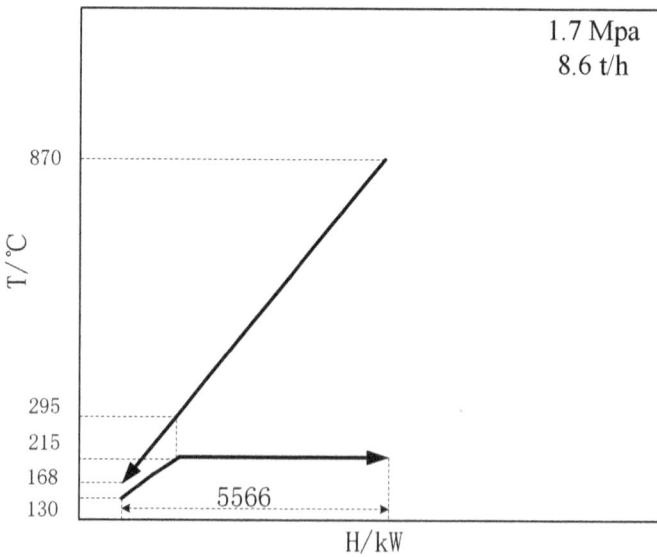

Fig. 9.45: Heat recovery situation of unit 1 before retrofit.

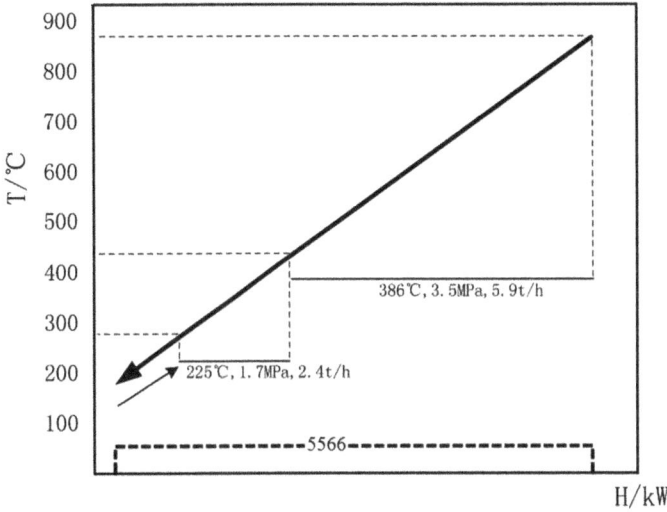

Fig. 9.46: Heat recovery situation of unit 1 after retrofit.

mum temperature difference of 50 °C. At the same time, the yield of the HPS is reduced, as shown in Figure 9.46. In this way, the flowrate of SHPS increases to 5.9 t/h and that of HPS decreases from 8.6 to 2.4 t/h.

At the same time, there are some improper usages of hot and cold utility in the HEN. In the retrofit process, it is preferable to replace high-quality utilities with low-quality utilities.

9.6.6 Steam subsystem retrofit

The changes in heat duty of different utilities as a result of the retrofit are shown in Table 9.5. These affect the conversion between different levels of steam. Changes to the flowrate of steam at different pressure levels are effected using valves to reduce the

Tab. 9.5: Heat duty changes of different utilities.

Utility	Change
SHPS	Production increased by 5.9 t/h
HPS	Production decreased by 6.2 t/h, usage decreased by 1.65 t/h
IPS	Usage increased by 15.76 t/h
MPS	Usage decreased by 6.9 t/h
LPS	Usage decreased by 29.66 t/h
Cooling water	Usage decreased by 492.49 t/h
Chilled water	Usage decreased by 128.45 t/h

Tab. 9.6: Flowrate through valves.

Pressure levels	Flowrate before retrofit (t/h)	Flowrate after retrofit (t/h)	Exergy loss after retrofit (MJ/h)
SHPS→HPS	32.3	34.21	8535
HPS→MPS	18.4	0	–
HPS→IPS	28.5	44.26	4339
MPS→LPS	20.6	9.1	980

Tab. 9.7: New turbines.

Turbine	Level change	Flowrate (t/h)	Work (kW)
T-1	SHP→HP	34.21	1324
T-2	HP→IP	44.26	1076
T-3	MP→LP	9.1	336
T-4	LP	18.16	1944

pressure and temperature, as shown in Table 9.6, which cause 13,854 MJ/h of exergy loss.

SHPS superhigh pressure steam, HPS high pressure steam, IPS intermediate pressure steam, MPS medium pressure steam, LPS low pressure steam

To decrease the exergy loss, cogeneration of power and heat is adopted. In this case, four steam turbines are added, as shown in Table 9.7. Turbine T-4 is a condensing steam turbine and the other three are back-pressure turbines. The total turbine work output is 4680 kW.

After a new balance of the steam subsystem, a total of 3.99 t/h SHPS can be saved and 4680 kW of power generated.

9.7 Concluding remarks

The energy efficiency of a process can be remarkably improved through process synthesis. Pinch analysis is a powerful tool for showing the energy performance in process synthesis. As a conceptual tool, represented by the composite curve and grand composite curve, pinch analysis can give insights into the process. The pinch point is a very important concept, and its relative position can indicate whether energy utilization is appropriate.

For HEN synthesis, to achieve the energy target set by the composite curves, there must be no cold utility used above the pinch, no hot utility used below the pinch, and no process-to-process heat transfer across the pinch, given that no individual heat exchanger has a temperature difference less than ΔT_{min}. The pinch design method is a step-by-step approach that allows the designer to interact with the design progress.

Using the grand composite curve, different utility scenarios with the corresponding duties can be screened quickly and conveniently. To save energy, a heat engine should not be integrated across the pinch. However, heat pumps should be placed across the pinch. The appropriate placement of reactors, as far as heat integration is concerned, is that exothermic reactors should be integrated above the pinch and endothermic reactors below the pinch.

The appropriate placement of heat-integrated distillation columns is not across the pinch.

The grand composite curve can be used as a quantitative tool to assess integration opportunities. If a distillation column cannot be integrated with the background process, or if the potential for integration is limited by the heat flows in the background process, then the distillation operation itself should be assessed and complex arrangements such as reflux ratio decrease, multieffect scheme, and intermediate reboiler or condenser considered.

The approach can be extended to site-wide heat integration across plants, including direct and indirect heat integration. The total site profiles can be constructed by combining the heat source information and heat sink information from all available plants into single profiles. Using the total site profiles, heat recovery targets can be determined.

As a graphical tool, application of pinch analysis is limited by the dimension of graphs. For complicated, multiobjective, or multiconstraint problems, superstructure-based mathematical programming can be used.

9.8 Bibliography

[1] Feng X, Wang Y. Energy saving principles and technologies of chemical processes. 4th edition, Beijing, Chemical Industry Press, 2015 (in Chinese).
[2] Smith RM. Chemical process: design and integration. Chichester, England: John Wiley & Sons, 2005.
[3] Feng X, Liang C. Strategy for Total Energy System Retrofit of a Chemical Plant, Chemical Engineering Transactions, 2013, 35, 145–150.
[4] Wang Y, Feng X, Cai Y, Zhu M. Improving process' efficiency by exploiting heat pockets in its heat exchange network, Energy, 2009, 34(11), 1925–1932.
[5] Yang M, Feng X, Chu KH. Graphical Analysis of the Integration of Heat Pumps in Chemical Process Systems, Industrial and Engineering Chemistry Research, 2013, 52(24), 8305–8310.
[6] Klemeš J, Dhole VR, Raissi K, Perry SR, Puigjaner L. Targeting and design methodology for reduction of fuel, power and CO2 on total sites. Applied Thermal Engineering, 1997, 17(8–10), 993–1003.
[7] Feng X, Pu J, Yang J, Chu KH. Energy recovery in petrochemical complexes through heat integration retrofit analysis, Applied Energy, 2011, 88(5), 1965–1982.
[8] Wang Y, Feng X, Chu KH. Trade-off between energy and distance related costs for different connection patterns in heat integration across plants. Applied Thermal Engineering, 2014, 70(1), 857–866.

Juan Gabriel Segovia-Hernández, Fernando Israel Gómez-Castro,
José Antonio Vázquez-Castillo, Gabriel Contreras-Zarazúa, and
Claudia Gutiérrez Antonio

10 Process synthesis and intensification by integration between process design and control

Abstract: Process chemical engineering must respond to the changing needs of industry to satisfy increasing social and market requirements, the need to save raw materials and energy, and environmental constraints. In the context of globalization and sustainability, process intensification is a means of meeting future demands in process chemical engineering. Process intensification refers to complex technologies that enable the replacement of large, expensive, and energy-intensive equipment or processes with smaller, less costly, and more efficient plants that minimize environmental impact, increase safety, and improve remote control and automation, or that combine multiple operations in a single apparatus or employ fewer devices. Operability and dynamic performance are important criteria in the selection of an adequate process, especially for intensified process where controllability is affected by the complexity of these processes. This chapter describes controllability analysis of several processes using closed-loop dynamic studies to analyze the effect of intensification on the dynamic behavior of the arrangements.

Keywords: Process intensification, control behavior, thermally coupling, reactive distillation

10.1 Introduction

The chemical industry is increasingly aware of the environmental and economic aspects of production processes. Under this scenario, there is a clear need to develop new processes that support these trends and represent a more profitable and sustainable option. It is well known that process synthesis applied in the early stages of development can produce suitable alternative designs [1].

Process synthesis addresses such concerns, commonly through rule-based heuristics, mass/energy integration, process optimization, and process intensification (PI). Recently, PI has attracted both industrial and academic interest because several reports have shown that it can produce economic and environmental improvements [2–4]. First, it is mandatory to understand the PI concept. Several authors have defined PI; however, this concept has evolved from simple to quite sophisticated approaches over time. The first PI concept reported was established at Imperial Chemical Industries during the 1970s [5], but this concept only considered economic aspects as the basic target. In 1995, while opening the 1st International Conference on Process In-

https://doi.org/10.1515/9783110465068-010

tensification in the Chemical Industry, Ramshaw, one of the pioneers in the field, defined PI as a strategy for both making dramatic reductions in the size of a chemical plant and reaching a given production objective [5]. While knowledge and research increase, the concept of PI also evolves. Numerous definitions have been proposed and a few are highlighted here. Lutze et al. [6] consider that a PI process must contain three different kinds of improvements that must be achieved through (1) integration of operations, (2) integration of functions, and (3) integration of phenomena. Ponce-Ortega et al. [7] defined PI as a process distinguished by five characteristics: reduced size of equipment, increased performance of the process, reduced equipment inventory, diminished use of utilities and raw materials, and increased efficiency of process equipment. Stankiewicz et al. [8] defined PI with the following all-encompassing definition: "The development of innovative apparatus and techniques that offer drastic improvements in chemical manufacturing and processing, substantially decreasing equipment volume, energy consumption, or waste formation, and ultimately leading to cheaper, safer, sustainable technologies and some research groups have focused their researches to propose a variety of intensified processes which have shown potentially to improve the established processes so far."

Initial application of PI technology dates back to the 1970s. Static mixers (widely used today) were one of the first PI inventions. Since then, a wide variety of novel/hybrid/intensified processes have been proposed using intensification design concepts.

PI is commonly associated with economic, process, and environmental benefits. The economic benefits are understandable because less operating costs and significant energy savings can be achieved. Furthermore, the size of equipment can be decreased and the costs related to pipework, layout engineering, etc. also decrease. From another point of view, the smaller equipment makes the intensified process easier to handle and more "transportable."

Regarding process benefits, several reports on reactive distillation columns claim higher selectivity, faster reactions, and improved product properties [9]. Moreover, because the chemical processes are commonly carried out under dangerous conditions, safety is a priority. Through PI, the inherent safety of a process is improved by using smaller processing units and thus reducing the volume of potentially hazardous chemicals. Another process advantage of using PI approaches is the increase in both heat and mass transfer rates, which would not be possible with conventional technologies [10].

PI technologies have a high impact in environmental issues. It is clear that using PI concepts can result in smaller equipment, so that less material is needed to build equipment, resulting in lower environmental impact. Furthermore, by reducing or eliminating undesired byproducts, the necessary energy to purify the target product is reduced and any discharge of those byproducts is also reduced. Thus, processes based on PI concepts include a clean technology philosophy.

As previously mentioned, processes based on PI can provide multiple benefits. Since PI concepts were introduced during the 1970s, a variety of intensified processes have been reported. Examples of such processes include microreactors and their utilization in specific industrial applications [11, 12], static mixers for intensifying mass transfer limited reactions [13, 14], hybrid separations involving melt crystallization and membrane utilization [15, 16], and reactive distillation for the production and separation of valuable chemicals [17, 18].

10.2 Process synthesis and integration of process design and control

Despite the proven benefits of intensified processes, several issues hinder more extended implementation of PI technologies in industry, as outlined next.

Resistance to new technologies: Most companies are risk-averse to trying new equipment or technologies because there is a lack of data on previous case studies to take into account as a base case for real operations. This understandable behavior produces a lack of reliable cost–benefit analyses for future cases. Furthermore, there is a lack of industrial suppliers for industrial applications or PI technologies [6]. Therefore, designing novel PI processes is expensive and technologically uncertain because the capital cost of a novel process is usually more expensive than that of traditional equipment.

Dynamic behavior: Because intensified processes claim an improvement in comparison with traditional technologies, the new processes probably lead to control problems.

The particularities of a given process: There are often important issues related to the features of a given process and, therefore, the possible implementation of PI in that process should be addressed. The part of the process that can be intensified to achieve a relevant improvement, the PI equipment that could be used, and other PI options for the process must be approached. However, there is no guarantee that the best PI option will be chosen if these aspects of the process are addressed in a cursory way. Thus, a suitable strategy must be utilized to approach these issues.

Process synthesis can help in finding the optimal intensified options for a process. In general, process synthesis involves the creation of multiple intensified options, evaluation of these options according to previously defined criteria, and identification of the best option. In this way, selection of the best PI option for a process can be reached by incorporating process synthesis and PI strategies in a given process. By using process synthesis approaches and PI implementation for generating intensified process alternatives, some relevant works have been developed. Particular attention has been devoted to intensified separation systems, including thermally coupled distillation columns, separation systems with membrane utilization, and reactive separation systems such as reactive distillation, reactive extraction, and reactive absorption.

The thermally coupled distillation column is a good example of intensified equipment. One of the most interesting fully thermally coupled columns is the divided-wall column (DWC), which has been proposed in several industries for achieving reductions in energy consumption. The DWC is an alternative to traditional separation configurations for targeting a sustainable separation operation [19]. A framework for developing multiple DWC intensified alternatives is the best way to obtain optimum selection of this equipment. In this context, Rong [20] proposed a systematic approach for generation of DWCs in a four-step procedure involving subspaces from simple separation columns, thermally coupled columns, thermodynamically equivalent columns, and DWCs. Caballero and Grossman [21] developed an approach based on mathematical programming for the design of fully thermally coupled sequences, including DWCs. More recently, Torres Ortega et al. [22] developed a stochastic optimization strategy for developing multiple intensified columns for the separation of a multicomponent mixture. The results of this work showed significant reductions in both energy consumption and capital cost of the intensified alternatives compared with conventional configurations.

Process synthesis has also been applied to the development of intensified separation systems. The work by Brunetti et al. [23] offered an approach for preliminary design of a membrane-integrated biogas separation system using polymeric membranes. The method defined the operation conditions and performance limits of the configuration in the separation system. Olán-Acosta et al. [24] developed a method for the synthesis of reactive liquid–liquid extraction that was based on reactive liquid–liquid equilibrium correlations and geometric concepts. Reactive absorption, a unit operation comprising the absorption of gases in liquid solutions with simultaneous chemical reactions in a single apparatus, was also developed using process synthesis. For example, Yildirim et al. [25] provided a complete review of recent implementation of reactive absorption, highlighting applications of this intensified technology designed to reduce environmental impact by treating and purifying harmful substances.

Reactive distillation columns (RDCs) are the best examples of PI. In reactive distillation, it is possible to perform the reaction and separation in a single unit of equipment and, thus, obtain reductions in capital cost and energy consumption. Additionally, improvements in the selectivity, conversion, and separation of the reaction mixture can be achieved. Combination of the reaction operation and the separation operation in an RDC makes the problem of finding the optimum configuration a very complex task; therefore, suitable strategies for the synthesis and design of this intensified equipment are necessary. Only a few works proposing methods of synthesis and design in reactive distillation have been published. For instance, Li et al. [26] proposed a generalized method for the synthesis and design of RDCs involving the sequential search of all the decision variables and thermodynamic properties of the feed mixture. The method showed enough robustness to deal with multiple reacting mixtures and to propose various RDCs. Amte et al. [27] presented a deterministic optimization strategy for the conceptual design of reactive distillation in order to maximize the selectivity. Pǎtruþ et al. [28] introduced a combination of cyclic distillation and reactive distillation in order to develop a rigorous mathematical algorithm for design of reactive distillation combined with cyclic distillation. The authors obtained significant advantages compared with conventional reactive distillation, in particular lower energy requirements and higher flexibility in operation. Carrera-Rodriguez et al. [29] developed a short-cut method for the design of multicomponent RDCs utilizing distillation lines in combination with algebraic component mass balances. The results provided a good predesign for posterior implementation of a rigorous method.

Process synthesis and PI for generating intensified process alternatives have proven to be very useful strategies for selecting the best intensified configuration of a process, specifically in design of thermally coupled columns and RDCs in steady state approaches, where energy savings and reductions in capital cost have been achieved. Examples are given in the works on thermally coupled columns by Errico et al. [30], Cossio-Vargas et al. [31], and Huang et al. [32] and in works on RDCs by Cheng et al. [33], Machado et al. [34], Huang et al. [35], Chen et al. [36], and López-Saucedo et al. [37]. However, the operability and dynamics of an intensified process are crucial issues and many research groups are now focusing on these topics. For example, there have been many studies on control aspects of RDCs and DWCs [38, 39].

Operability and dynamic performance are important criteria in the selection of an adequate process, especially in intensified processes where controllability can be affected by the more complex structure of these processes. Several works and studies have focused on the evaluation of operability and dynamic performance in close-loop control analysis. A common technique is the use of indicators such as the integral of absolute error (IAE) or the integral of square error (ISE), which are numerical criteria for evaluating the dynamics of a process.

The dynamic response depends strongly on controller parameters such as integral time and static gain; hence, many works focus on tuning the parameters of controllers. Zavala-Guzman et al. [40] reported a technique for tuning the controller parameters in a DWC by approaching the stable pole assignment, resulting in a very easy, fast, and trustworthy technique for tuning. Other tuning works were reported by Segovia-Hernandez et al. [41] and Ramirez-Marquez et al. [42], who used a control optimum strategy with the IAE criterion to determinate the controller parameters for intensified processes to purify biobutanol. The main objective of these works was to use IAE criterion to determine the optimal dynamic response (i.e, fastest stabilization of a perturbation). In the case of RDCs, one of the most important control issues is sensitivity analysis, because small changes in the operating conditions can strongly affect the reaction. Huang et al. [43], Lee et al. [44], and An et al. [45] reported different sensitivity analyses for reactive distillation processes. They introduced perturbations in the set points and in the feed with the objective of observing whether it was possible to return to operability conditions. These kind of studies are of special importance regarding implementation of a temperature control loop. Temperature control is more widely used than composition control because it is easier to implement. Ramirez-Marquez et al. [46] reported a control study on a multitasking reactive distillation column (MTRD) for the production of silanes. In this kind of process, the objective is to carry out several reactions, depending on the necessities of the production. Different MTRD arrangements were reported and control analyses carried out, with the goal of determining the strategy that provided the best dynamic performance.

Process synthesis and the integration of process design and control, conceived as a combined and integrated approach, can be a suitable strategy for dealing with current challenges in the chemical engineering industry. However, most developed studies involving process synthesis and the design and control of intensified process are conceived as sequential methodologies. For example, in the first stage, the process is synthesized and designed to achieve the aims of the intensified process (e.g., product specifications that meet market requirements). In the second stage, the control aspects are analyzed and solved. This sequential methodology may present problems such as dynamic constraint violations, process overdesign or underperformance, and no guarantee of robust performance [47]. Another disadvantage is related to the way in which process decisions influence the control operation of the process. In the realistic scenario of a competitive market, chemical processes must operate as flexibly as possible in order to adapt in an adequate way to changes in product specifications, consumer demand, and raw materials. In this context, utilization of appropriate strategies for integrating synthesis, design, and control allows suitable operation of the process by improving profitability through an increase in throughput production and yield of high value products, in addition to minimization of energy consumption.

Thus, development of a methodology that involves process synthesis and intensification by integration between process design and control is highly desirable. In this context, a recent study involving the integration of process synthesis with the design and control of distillation columns for the separation of multicomponent mixtures has been published [48]. The study involved multiobjective optimization in terms of total cost and open loop control evaluation of conventional separation columns and DWCs. The results of this study offered design options for the trade-off between cost and control. The intensified designs showed better control properties, lower energy consumption, and lower total cost than conventional distillations columns.

Process synthesis and the integration of process design and control can also be a sustainability-related strategy for achieving sustainability in the design of process systems, specifically in intensified processes. Unique dynamics and control challenges emerge from these processes, as more than one operation is generally performed in a single apparatus, resulting in increased interactions and strong nonlinear dynamics in the performance of intensified processes. This is particularly true for miniaturized systems such as microreactors and microfluidics devices and for devices such as microwave-assisted reactors and reactive separations that use alternative energy sources [49].

In this work, we have carried out a controllability analysis using the closed-loop dynamic performance for several intensified processes, including conventional and thermally coupled designs. Thermally coupled distillation and RDCs are typical examples of PI.

10.3 Closed-loop analysis

Rigorous dynamic simulations under closed-loop operation were carried out. It is important to mention that there were no significant differences between procedures for closed-loop simulations of intensified and nonintensified processes. The works by Segovia-Hernández et al. [50–52] provide useful insights on the implementation of procedures for closed-loop simulations of both kinds of processes. To begin, several aspects must be defined, such as the control loops for each system, the type of process controller, and the values of controller parameters. Several techniques, such as the relative gain array method, can be used to fix the loops for a control system. In the case of distillation columns, however, such loops are fairly well established and used successfully in practice, at least for conventional columns. A well-known structure is based on energy balance considerations, which yields the so-called LV control structure in which the reflux flowrate L and the vapor boil-up rate V (affected directly by the heat duty supplied to the reboiler) are used to control the compositions of distillate and bottom output [53]. The control loops for the integrated systems were chosen from extensions of the practical considerations observed for conventional distillation columns. The control objective was to preserve the output streams at their designed purity specifications.

Two control loops arise naturally from experience of the operation of conventional columns. The reflux flowrate was used for control of top product, whereas the reboiler heat duty was chosen for control of bottom product. It should be mentioned that such control loops have been used with satisfactory results in previous studies conducted by the author's research group on thermally coupled systems [50, 54–56]. The choice of controller was based on ample use of the proportional-integral mode for distillation systems in industrial practice. During selection of controller parameters, special care was taken to provide a common method for each of the sequences under comparison. A tuning procedure was used that involved minimization of the IAE for each loop of each scheme [57]. Therefore, for each loop, an initial value of the proportional gain was set. Values of the integral reset time were searched until a local optimum value of the IAE was obtained. The process was repeated for other values of the proportional gain. The selected set of controller parameters was the one that provided a global minimum value of the IAE. Although the tuning procedure was fairly elaborate, the control analysis was based on a commonly used method for tuning controller parameters.

The simulations involved solution of a rigorous tray-by-tray model of each sequence, together with the standard equations for the PI controllers for each control loop (with the parameters obtained through minimization of IAE criterion). The objective of the simulations was to find and compare the dynamic behavior of the systems under feedback control mode. Servo control was used to carry out the closed-loop analysis, whereby a step change was induced in the set point for each product composition under single-input, single-output (SISO) feedback control.

10.3.1 Case study 1: reactive distillation to produce diphenyl carbonate

The first case study involves one of the best examples of PI, reactive distillation implemented on a process to produce diphenyl carbonate (DPC) through a "green route." Traditional pathways to produce the polycarbonate precursor involve the highly toxic gas phosgene, whereas the alternative pathway consists of an esterification reaction between dimethyl carbonate (DMC) and phenyl acetate (PA), giving DPC and methyl acetate (MA) as products [58].

$$DMC + PA \rightleftharpoons DPC + 2MA$$

The advantages of using this route are (1) high equilibrium constant, (2) no formation of azeotropes, and (3) no side reactions. Thus, the process for purification of DPC is simplified because it does not require extractive distillation for separation of the azeotrope.

We studied two reactive distillation sequences for production of DPC using the pathway described, conventional reactive distillation (CRD) and thermally coupled reactive distillation (TCRD) sequences. Figure 10.1 shows the main features of the sequence designs for DPC production. It is worth noting the significant energy savings of TCRD compared with CRD. Both sequences consisted of two columns, the first being the RDC. The reagent with higher boiling point (PA) was fed near the top of the RDC and the reagent with lower boiling point (DMC) was fed near the bottom. The second column was the recovery column (RC) for the byproduct MA. Reagent DMC was returned from the bottom of the RC column to the RDC column. In all cases, the feed flows of PA and DMC were 10 kmol/h and 5.06 kmol/h, respectively. The purity required for DPC and MA was 99.5 mol%. The topology of the designs for CRD and TCRD were taken from a previous work by Cheng et al. [59].

A detailed explanation of the procedure for controllability analysis of the reactive configurations is given next. The TCRD sequence was selected as a representative example to illustrate the procedure. To aid comprehension, descriptions of the nine steps executed during the procedure are provided, together with illustrative figures.

(a)

(b)

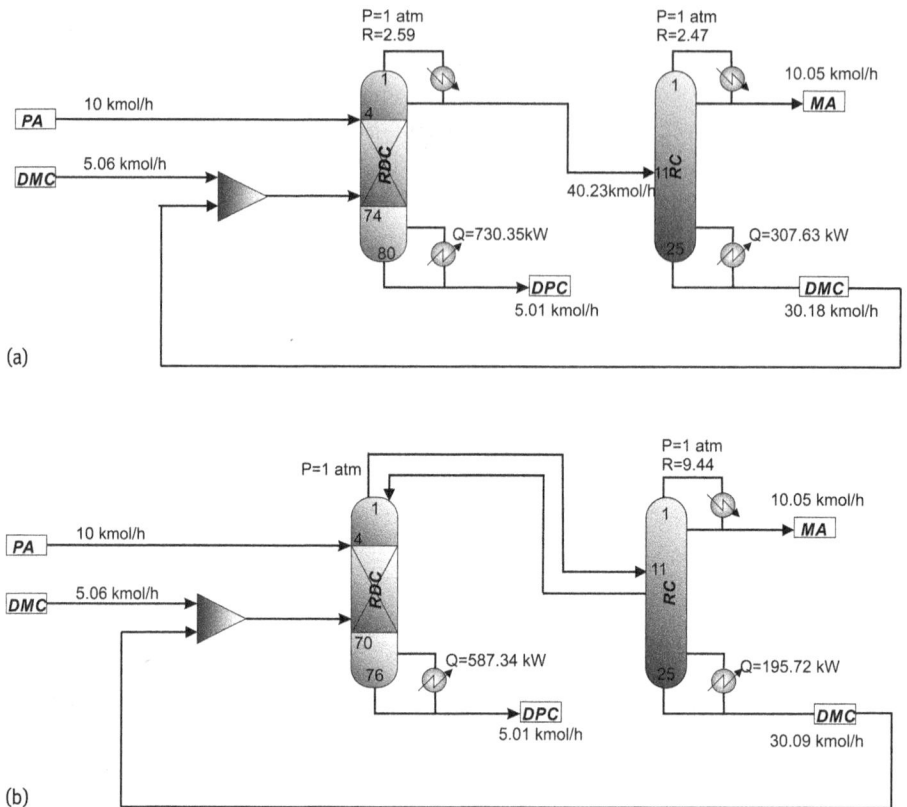

Fig. 10.1: Studied reactive distillation sequences: (a) conventional and (b) thermally coupled.

Step 1: initialization of the dynamic simulation

To perform the controllability analysis, the steady-state simulations first need to be exported to the Aspen Dynamics environment. Before exporting the simulation model, the equipment must be sized, so that the software can automatically calculate the hydraulics of the system using the provided data. The first step after exportation is initialization of the dynamic simulation, which indicates whether the simulation was exported with success. Initialize the simulation by clicking on the toolbar and changing the specification from *Dynamic* to *Initialization*, as shown in Figure 10.2. After the change, press the *Run* button, next to *Initialization*. If the initialization is successful, a window message appears: *The run has completed.*

Then, proportional-integral controllers are set at the product streams, using the manipulated variable/measured variable pairing shown in Table 10.1. To provide a detailed explanation of the placing of controllers in the TCRD, the controller in the DPC main product stream is selected as an illustrative example. The next steps explain how to place this controller, establish a performance criterion, run a dynamic simulation, and make a plot in Aspen Dynamics.

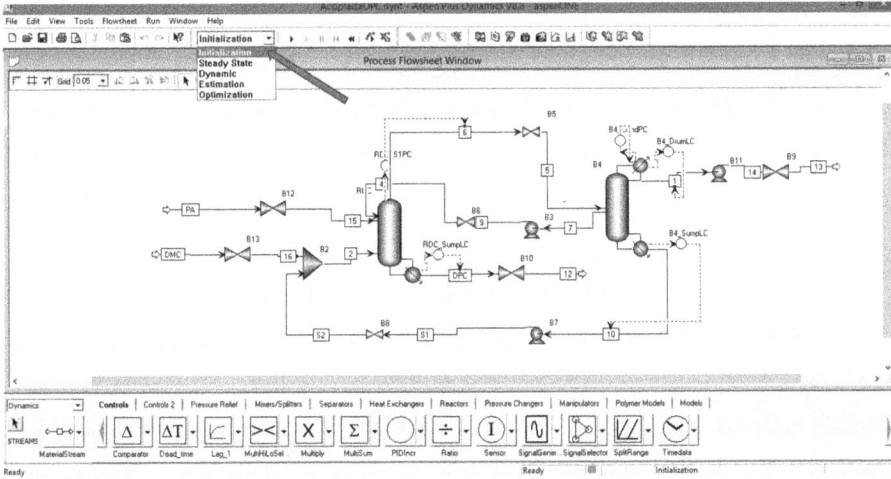

Fig. 10.2: Initializing the simulation by changing from *Dynamic* to *Initialization*.

Tab. 10.1: Control pairings for the reactive distillation sequences.

CRD		TCRD	
Manipulated variable	Measured variable	Manipulated variable	Measured variable
Reboiler duty of RDC	Composition of DPC	Reboiler duty of RDC	Composition of DPC
Reflux ratio of RC	Composition of MA	Reflux ratio of RC	Composition of MA

Step 2: placing the controller

After initialization, it is necessary to place the controller. For this, click on the *Controls* option and choose *PIDIncr*, as shown in Figure 10.3. The *PIDIncr* contains all kinds of controllers (proportional, proportional-integral, and proportional-integral-derivative). Once the controller is selected, click on the flowsheet window to place this controller.

Step 3: connecting the controller and the controlled variable

When the controller has been selected and placed, the next step is setting up the connection between the controller and the variable that needs to be monitored. To connect the controller, click the square that contains the *MaterialStream* (Figure 10.4) and select the stream with the name *ControlSignal*. This connects the controller with the material stream that contains the variable to control. In this case, the stream called *DPC* contains the variable, and so the *ControlSignal* stream is connected to the *DPC* stream.

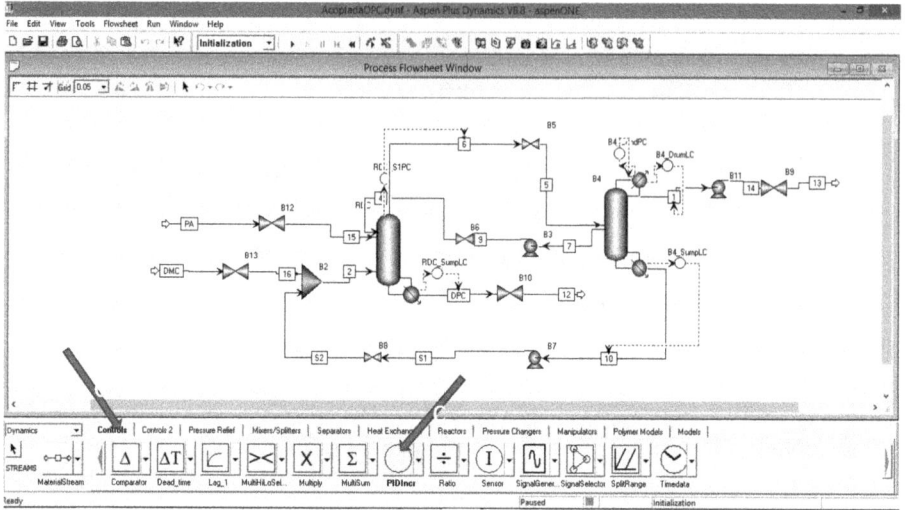

Fig. 10.3: The choice of *PIDIncr* in the *Controls* tab.

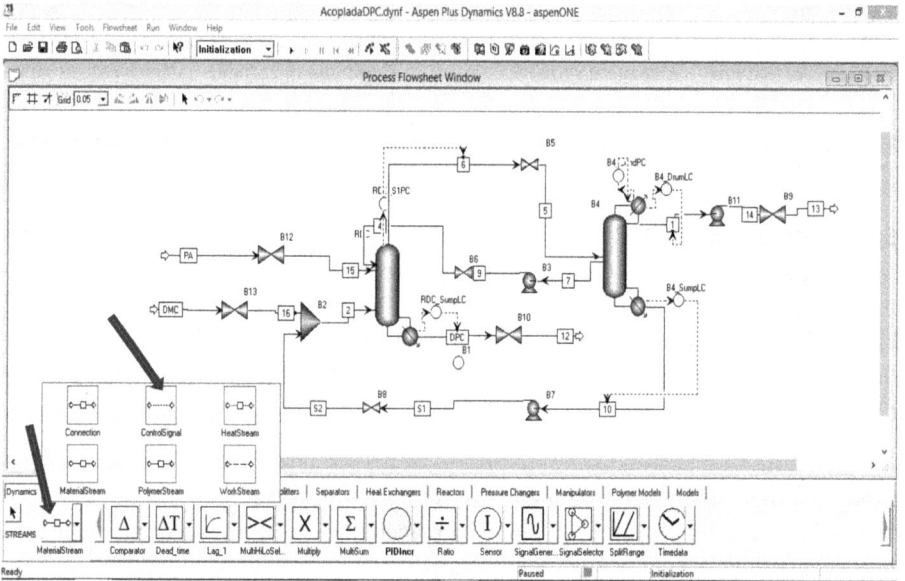

Fig. 10.4: Connecting the controller and the monitored variable.

When the *ControlSignal* stream is connected to the material stream, a window with all variables that contain the material stream appears. In this case study, control of the mole fraction of DPC is of special interest and is selected as the control variable (Figure 10.5).

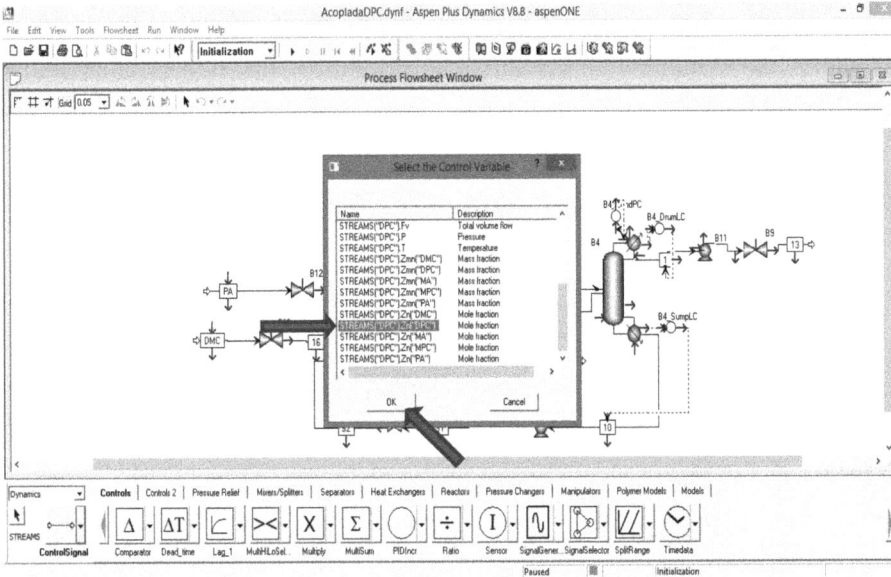

Fig. 10.5: Selection of the control variable as mole fraction of DPC.

Step 4: connecting the variable control to the input of controller

After selection of the control variable, the *ControlSignal* stream needs to be connected from the DPC material stream to the controller input. When the control signal is connected, a window with the options *Process variable* and *Remote setpoint* appears. Select *Process variable* because DPC mole fraction was selected in the last step and it is a variable from the process that needs to be controlled (Figure 10.6).

After the control input is connected, it is necessary to connect the controller output with the manipulated variable. Another control signal stream is required for this connection, and the option *Controller output* needs to be selected (Figure 10.7).

A traditional way to control of the bottom of a distillation column is with the reboiler duty. Therefore, it should be selected as a manipulated variable and the *Control Signal* stream connected to the reboiler duty. Finally, the *Specified reboiler duty* control variable is chosen (Figure 10.8).

Step 5: initialization of the controller

When the controller has been set as presented in step 2, initialization is necessary to establish the set point in the controller. For configuration, click in the controller. A new window appears, in which the user needs to click *Configure*, as shown in Figure 10.9. After choosing *Configure*, the window for this option will appear.

The *Configure* window contains different tabs. In the *Tuning* tab, select the *Reverse* controller action; next, click on *Initialize Values* to establish the initial values of the set

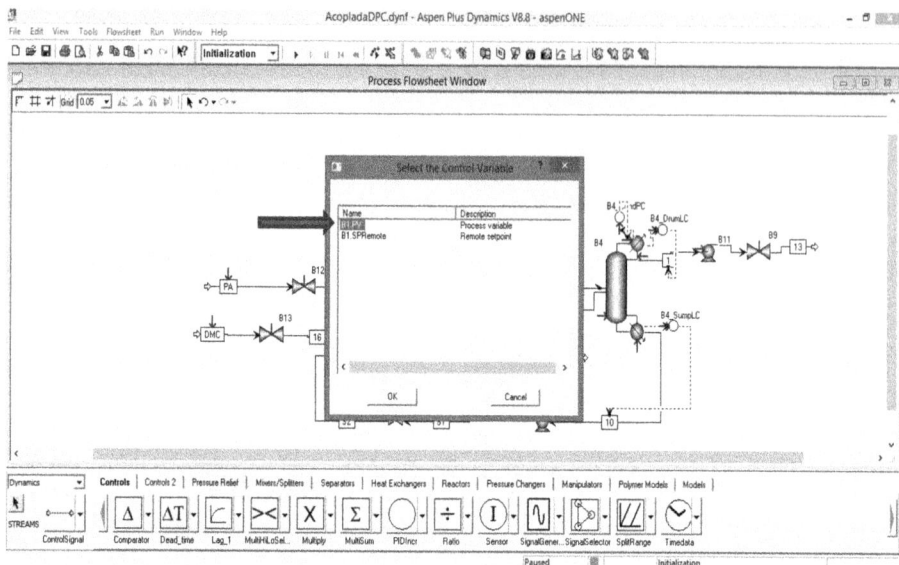

Fig. 10.6: Choice of the control variable using the *Process variable* option.

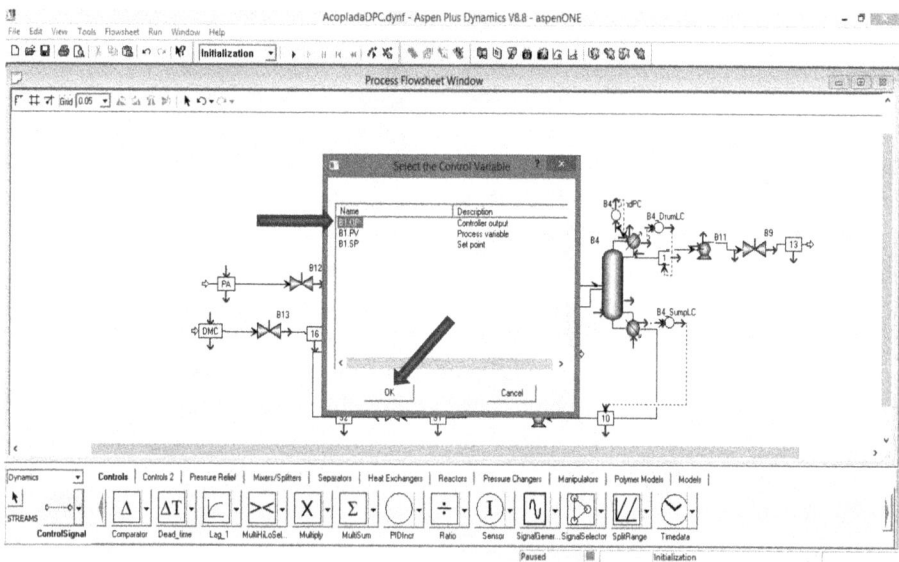

Fig. 10.7: Choice of the *Controller output*.

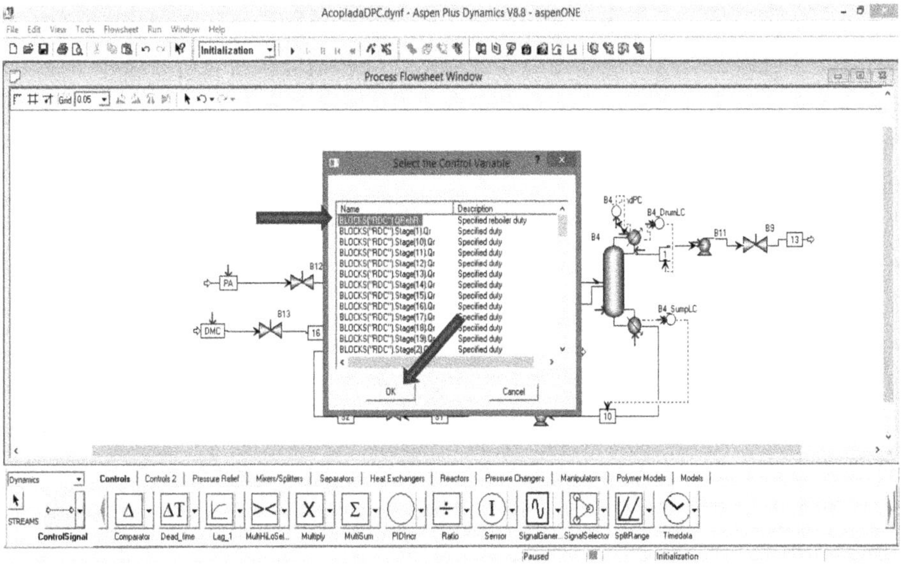

Fig. 10.8: Selecting the *Specified reboiler duty* control variable.

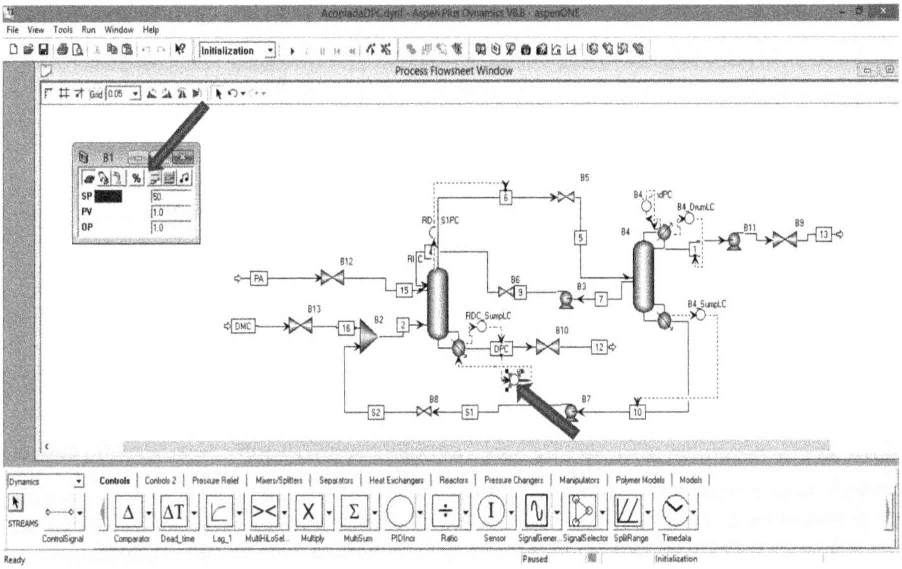

Fig. 10.9: Configuring the controller by selecting the *Configure* tab.

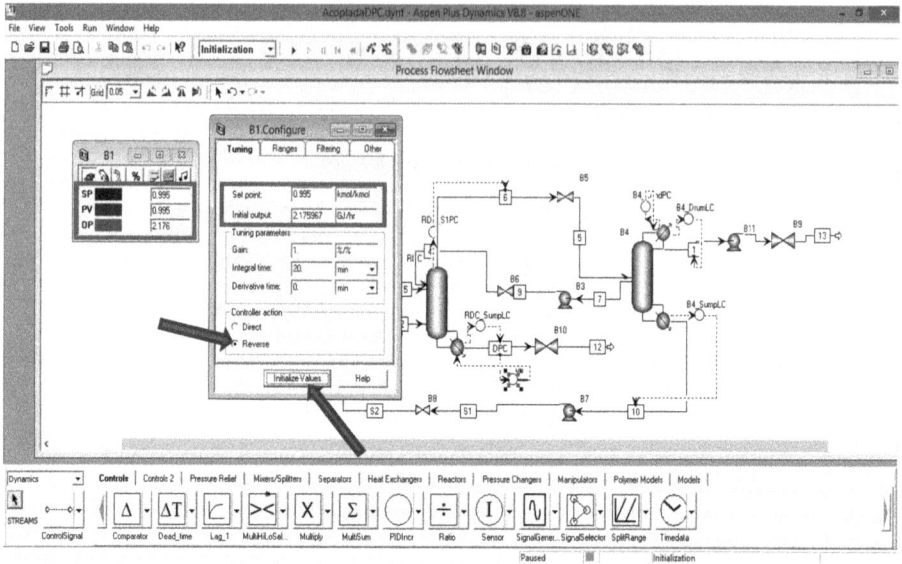

Fig. 10.10: Selecting the controller action of *Reverse* and initialization of the controller.

point. One way to corroborate the correct initialization of the controller is by checking the values locked in the indicated rectangles shown in Figure 10.10. These values must correspond to the values of the control variable and the manipulated variable, in this case, the mole fraction of DPC and reboiler duty, respectively.

Step 6: how to collocate a performance criterion
So far, assembly of the controller has been defined with respect to the control and manipulated variables. In control studies, it is very important to evaluate the performance of the process with specific parameters in the controller. The most common methods of evaluating this performance are the IAE and the ISE. The IAE is recommended for small errors, especially in the order of decimals, whereas the ISE is a better criterion for larger errors, in the order of units [57]. For more information about IAE and ISE criteria, the reader is advised to consult the literature on fundamentals of control. In this case study, the IAE criterion is selected because the errors in the mole fraction purity are considered small. To evaluate this criterion in Aspen Dynamics, click the lower tab *Controls 2* and choose the *IAE* option. After the selection of *IAE*, this block needs to be located in the main flowsheet window.

Similar to the procedure for location of the controller (step 4), the *IAE* block needs to be connected to the DPC stream by a control signal stream. This procedure is presented in Figures 10.11 and 10.12. First, the *ControlSignal* stream is connected to the DPC stream (Figure 10.11).

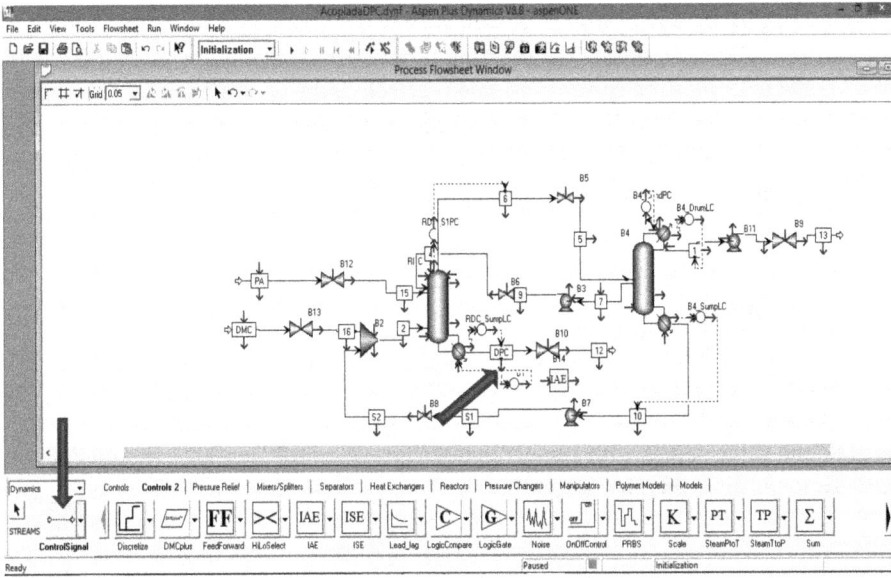

Fig. 10.11: Connecting the *ControlSignal* stream to the DPC stream.

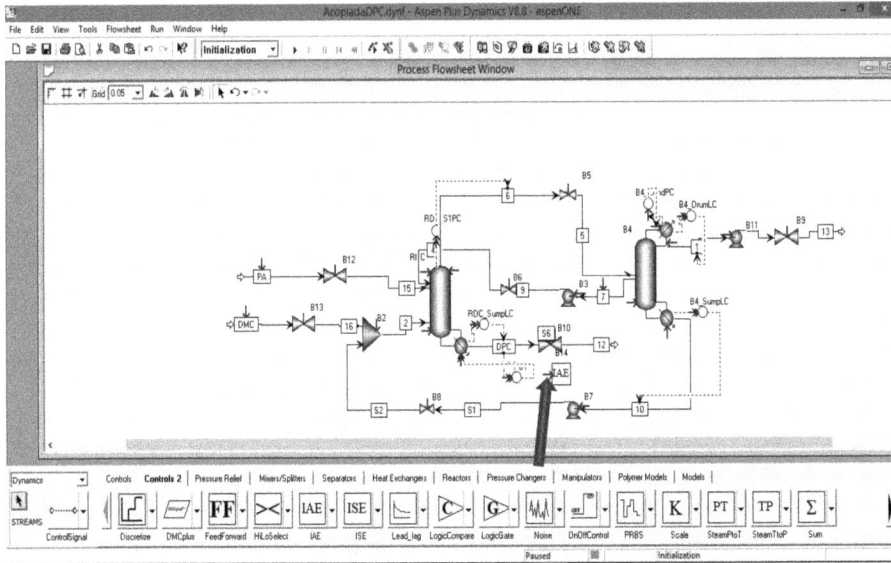

Fig. 10.12: Connecting the *ControlSignal* stream from the DPC material stream to the *IAE* square input.

After the *ControlSignal* stream is connected to the material DPC stream, a window with all variables that contain the material DPC stream appears, the mole fraction of DPC is selected. Once the control variable is selected, the *ControlSignal* stream needs to be connected from the DPC material stream to the *IAE* square input, as shown in Figure 10.12. After the control signal is connected, a window with the legend *Select the Control Variable* appears and the *Input signal* option is selected for the *IAE* square.

Step 7: initialization of performance criterion

After setting the performance criterion, it is necessary to initialize the criterion. The user needs to click on the *IAE* square. A new window appears, which contains the values of the *Input signal*, *Absolute error*, and *IAE*. After the window appears, the next step is to click on the button *Run*. After clicking this button, the values in the window must be the same values with respect to mole faction of DPC, and the legend *The run has completed* appears (Figure 10.13).

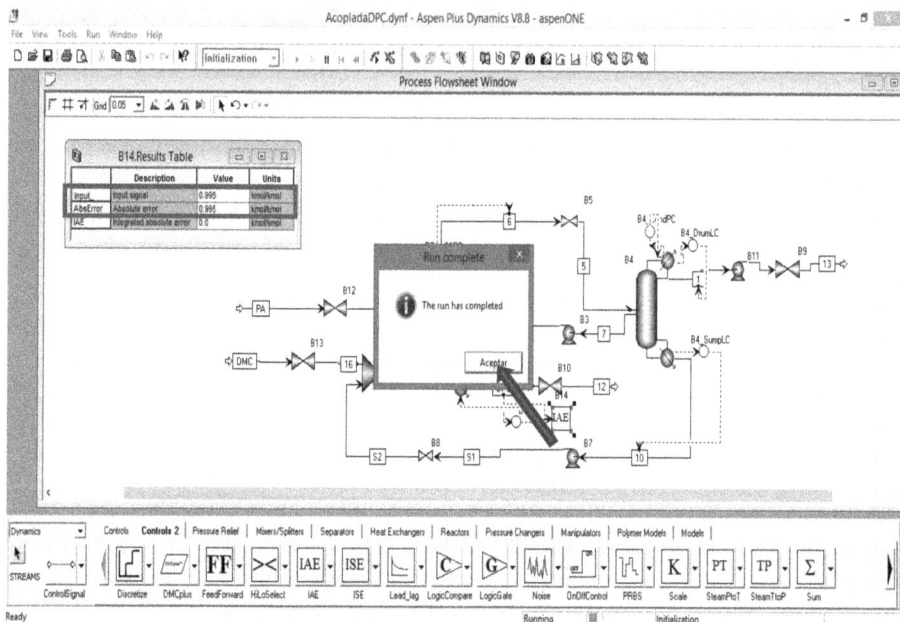

Fig. 10.13: Completed initialization of the performance criterion.

Step 8: how to start a dynamic simulation

In previous steps, the user has been working with configuration of the controller and the performance criterion. The next step is to realize a control study. The most common studies involve changes to the set point, which enable visualization of how the system responds to a perturbation in process inputs. To realize the study, change the simulation from *Initialization* to *Dynamic*. Then, click on the controller and change the set point. The change in the set point needs to be small in order to see whether the process can be stabilized. In this case, the perturbation is set to 1% below the nominal value.

It is important to mention that the change in set point must also be made for the performance criterion. Click with the right button of the mouse on the *IAE* square and select the option *Configure*. A new window with the option *Set point* appears and the user must set the new set point (Figure 10.14).

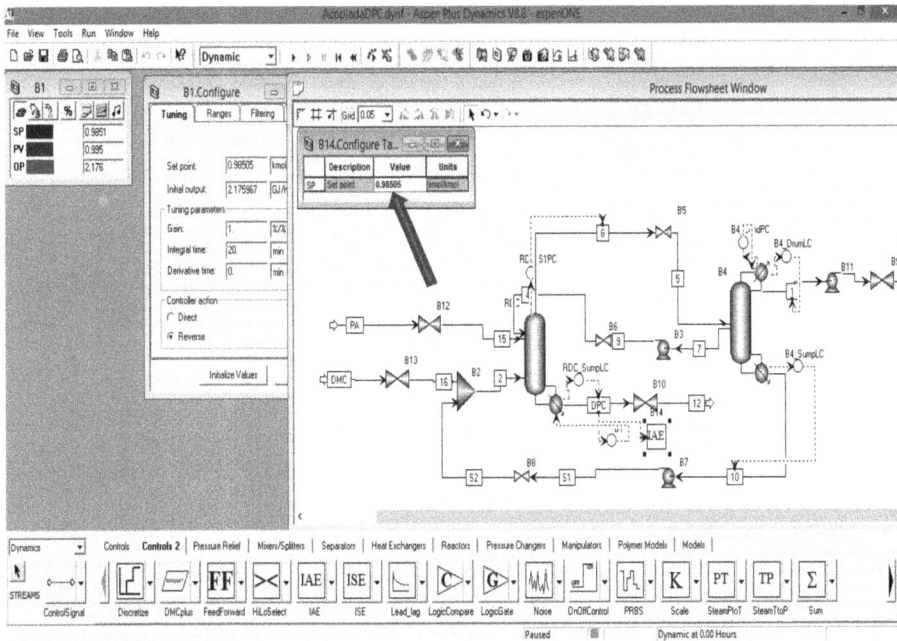

Fig. 10.14: Setting the new set point in the *IAE* performance criterion.

Step 9: how to plot in Aspen Dynamics

The last step is to make plots in Aspen Dynamics. The plots are very important because they allow visualization of the dynamic response of the process and can be a good visual criterion. To make a plot, click the option *New Form* on the toolbar, as shown in Figure 10.15. Then, click twice on the DPC stream to see the composition of the mole fraction of DPC. Click once and drag the mouse to the vertical axis in the *Plot*. Finally, click the *Run* button on the toolbar to start running the dynamic simulation.

Steps 1 to 9 provide the procedure for placing the DPC controller, establishing a performance criterion, running a dynamic simulation, and generating a plot in Aspen Dynamics for this controller. The procedure is similar for implementation of the MA controller in this study.

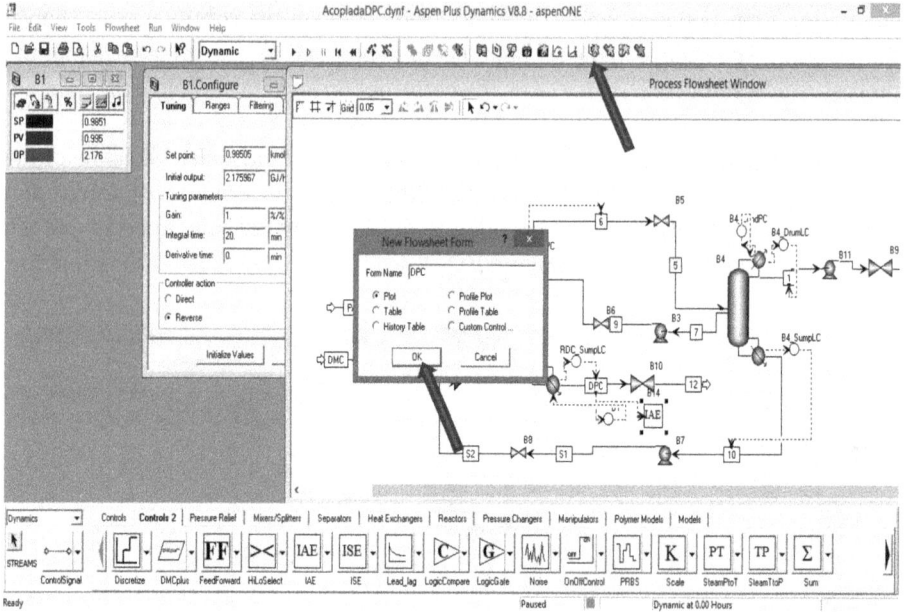

Fig. 10.15: Making a plot by clicking in the option *New Form*.

Results of the control analysis for the reactive sequences

The closed-loop control policy was performed under composition disturbance scenarios. This type of analysis is useful for investigating theoretical properties and dynamic behavior under feedback control. Additionally, the control analysis can reveal the best structures from a dynamic point of view and which schemes show better dynamic behavior. The control test was performed by first inducing a step change in the set point for each product composition under SISO feedback control at each output flowrate. For the closed-loop control policy, the analysis was based on proportional-integral composition controllers. The reason for using composition controllers is simply that 'back-off' from the purity specifications makes composition control simpler. This type of controller was chosen because of its wide use in industrial practice.

When a controller is used, the main issue is tuning the controller. In this study, a common strategy was used to compare and optimize the controller parameters. Because we considered PI controllers, the proportional gain (Kc) and reset times were tuned for each scheme studied; in addition, we evaluated dynamic performance using the IAE criterion. A key part of the dynamic analysis of each loop was the selection of control outputs and their respective manipulated variables. In this manner, structures based on energy balance considerations were used to control the distillate and bottom output compositions. The *LV* control structure uses the reflux flowrate L and the vapor boil-up rate V as the manipulated variables. In other words, we chose the corresponding reflux flowrate for the top of the column and the reboiler heat duty at the bottom of column as the manipulated variables.

In general, feedback control requires a model that describes the effect of the inputs (flows) on the outputs (product composition). This does not imply that the *LV* control structure is the preferred selection for control tests; the choice was made because *L* and *V* have a direct influence on composition and their effect is, consequently, only weakly dependent on the tuning of the level loops. It also natural to consider the column model in terms of *L* and *V* as manipulated inputs. This type of control loop has been applied with satisfying results in industry.

Summarizing, to tune each controller, an initial value of proportional gain was set, and a range of integral reset times was tested with this fixed value until a local optimum in the IAE value was obtained. This methodology was repeated with other proportional gain values until a global minimum in the IAE value was detected. It is important to stress that the strategy used here for tuning the controllers is widely described in the published literature as suitable for implementation in intensified systems. Papers by Segovia-Hernández et al. [41], Errico et al. [60], and Ramírez-Márquez et al. [61] attest to the usefulness of this strategy. Note that this procedure was conducted considering one control loop at a time until all control loops were considered. For dynamic analysis, individual set point changes of − 1.0% for product composition in the nominal state were implemented. We wished to demonstrate the dynamic properties of the intensified systems in each analyzed process under study and determine the best system. This is the reason to use the set point change for dynamic analysis of the intensified processes. Works by Segovia-Hernández et al. [62, 63] and Hernández et al. [64, 65] describe how set point changes have been used for dynamic analysis of intensified processes.

Figure 10.16 shows the best IAE values for different values of *K* and *t* for the TCRD sequence and how to obtain the best IAE values and the minimum IAE value.

Table 10.2 provides the best IAE values for the MA and DPC controllers in a TCRD sequence. The minimum value of IAE for the MA controller is 0.006223 for *K* = 125 and

Tab. 10.2: Best values of IAE for the MA and DPC controllers in TCRD.

MA controller			DPC controller		
K	*t* (min)	IAE	*K*	*t* (min)	IAE
125	37.5	0.006223*	125	1	0.000461
145	39	0.006471	155	1	0.000367
165	40	0.006635	185	1	0.000299
185	41	0.006801	200	1	0.000275
210	42	0.006966	230	1	0.000241
230	43	0.007131	250	1	0.000225*
* Minimum IAE value			* Minimum IAE value		

(a)

(b)

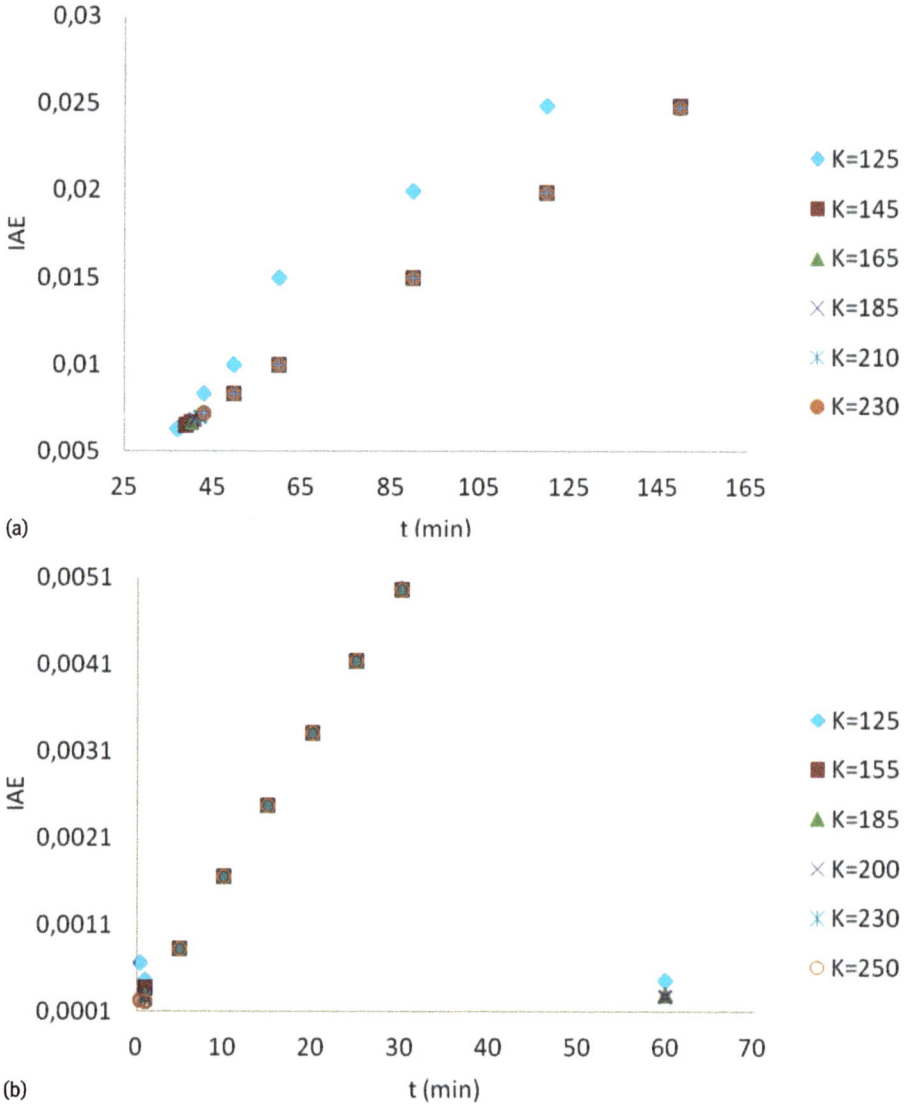

Fig. 10.16: IAE of controllers in thermally coupled reactive distillation: (a) MA controller and (b) DPC controller.

$t = 37.5$ min, whereas the minimum value of IAE for the DPC controller is 0.000225 for $K = 250$ and $t = 1$ min.

To obtain the minimum values of IAE for DPC and MA controllers in a CRD sequence, a similar procedure as for a TCRD sequence was carried out (i.e., by varying the gain K and integral time t). Table 10.3 provides the mimimum values of IAE for the controllers in CRD and TCRD sequences.

Tab. 10.3: Minimum values of IAE for the DPC and MA controllers in TCRD and CRD.

Sequence	DPC controller			MA controller		
	K	t (min)	IAE	K	t (min)	IAE
TCRD	250	1	0.000225	125	37.5	0.006223
CRD	250	9	0.001524	14	16	0.002892

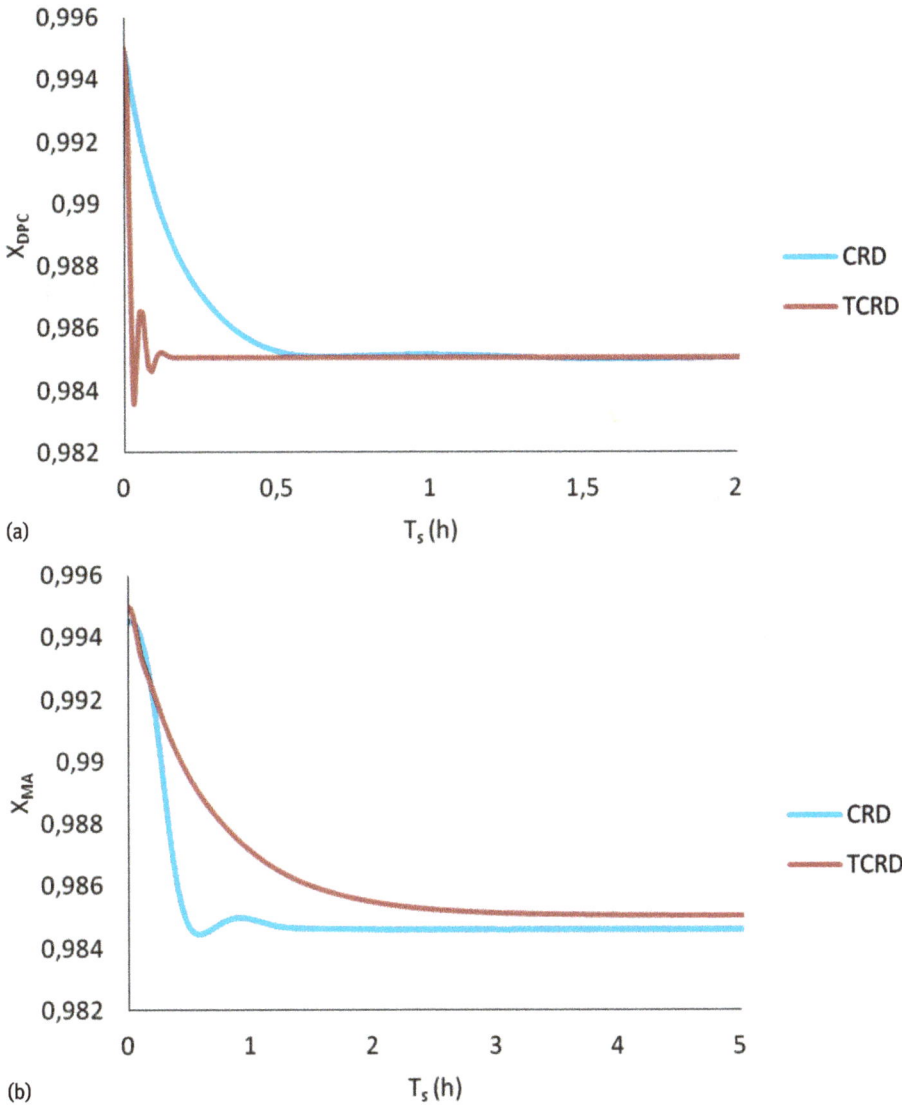

Fig. 10.17: Responses for DPC and MA components: (a) CRD with $K = 250$, $t = 9$; TCRD with $K = 250$, $t = 1$ and (b) CRD with $K = 14$, $t = 16$; TCRD with $K = 125$, $t = 37.5$

Figure 10.17 shows the responses of the MA and DPC components in both sequences for the minimum IAE values in each controller.

Prediction of the transient response of a process is very important because the effective control of a process must be known. As mentioned before, closed-loop simulations were performed by introducing a step change in the set point for the product composition under SISO feedback control. All simulations were performed in Aspen Dynamics, and PI controllers were considered. The parameters of those controllers were tuned to minimize the IAE as criterion. Using this tuning methodology, it is clear that the minimum IAE value is not guaranteed (unlike under a rigorous optimization strategy). However, this parametric methodology produced an IAE surface showing that the obtained IAE value was totally located in the zone where the minimum IAE values were situated.

The results from the individual servo test applied to CRD and TCRD designs are shown in Figure 10.17. Considering the DPC stream, the TCRD exhibited the best dynamic behavior because the settling time was lower than with CRD; additionally, the IAE value in Table 10.3 was lower for CRD. On the other hand, the conventional sequence exhibited longer settling times. Considering the MA stream, CRD showed the best dynamic response because the settling time was lower than for the TCRD configuration. This dynamic response was consistent with the low value of IAE shown in Table 10.3.

In general, the inclusion of some thermal coupling improved the dynamic behavior, as tested under a closed-loop control policy for the response of main product.

10.3.2 Case study 2: thermally coupled distillation columns for the separation of a multicomponent mixture

The second case study was the separation of a multicomponent hydrocarbon stream. The mixture used was from the work of Gutiérrez-Antonio et al. [66] and corresponds to a stream leaving a cracking reactor. The stream to be separated contained linear hydrocarbons in the proportions shown in Table 10.4. The hydrocarbons are grouped to form three products: light gases (L), middle-boiling components (M), and heavy components (H). It is worth noting that the heavy components represent the highest contribution to the feed composition. Feed flowrate was set to 0.51 kmol/h, and the objective of the separation train was recovery of 99% of the key components.

The three fractions were separated in two intensified distillation sequences: a thermally coupled direct sequence (TCDS) and a thermally coupled indirect sequence (TCIS), as shown in Figure 10.18(a, b). It has been reported that, depending on the feed composition and the nature of the components to be separated, such systems can reduce heat duty by up to 30% in comparison with conventional distillation trains [67].

The optimization methodology for such systems has been reported in previous works [66], and the designs with the lowest heat duty were taken for this study. The

Tab. 10.4: Feed composition of the mixture under analysis.

Component	Content (mol%)	Product ID
Propane (C3)	9.04	L
n-Butane (nC4)	6.86	
n-Pentane (nC5)	5.53	
n-Hexane (nC6)	4.63	
n-Heptane (nC7)	3.98	
n-Octane (nC8)	5.32	M
n-Nonane (nC9)	4.74	
n-Decane (nC10)	4.27	
n-Undecane (nC11)	3.89	
n-Dodecane (nC12)	3.57	
n-Tridecane (nC13)	3.29	
n-Tetradecane (nC14)	3.06	
n-Pentadecane (nC15)	0.26	
n-Hexadecane (nC16)	0.26	
n-Heptadecane (nC17)	20.63	H
n-Octadecane (nC18)	20.67	

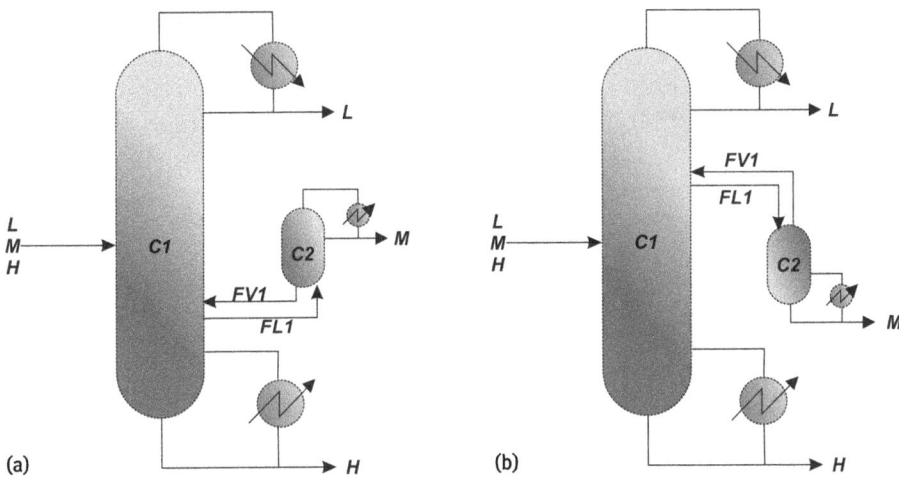

Fig. 10.18: Studied sequences: (a) thermally coupled direct sequence and (b) thermally coupled indirect sequence.

TCIS required about 20% more heat in the reboiler than the TCDS. The main characteristics of the analyzed designs are presented in Figure 10.18.

To perform the controllability analysis, the steady-state simulations were first exported to the Aspen Dynamics environment. The simulation was initialized to ensure that the steady-state simulation was correctly converted into a dynamic simulation. Then, proportional-integral controllers were set at the product streams, using the ma-

Tab. 10.5: Control pairings for the studied sequences.

TCDS		TCIS	
Manipulated variable	**Measured variable**	**Manipulated variable**	**Measured variable**
Reflux ratio C1	Composition of C3	Reflux ratio C1	Composition of C3
Reboiler duty C1	Composition of nC18	Reboiler duty C1	Composition of nC18

(a)

(b)

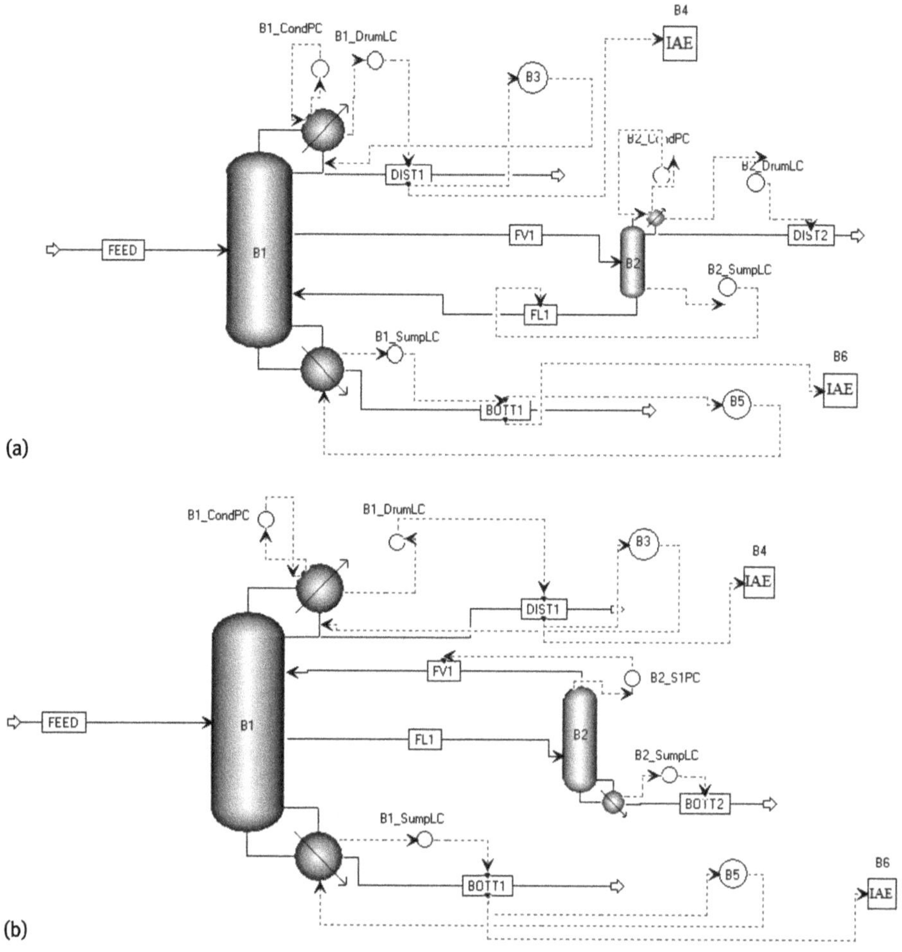

Fig. 10.19: Implementation of PI controllers in Aspen Dynamics: (a) thermally coupled direct sequence and (b) thermally coupled indirect sequence.

nipulated variable/measured variable pairing shown in Table 10.5. This control structure is one of the most commonly used and has been reported as a good alternative to one-point control [68].

Figure 10.19 shows the pairings implemented in the software Aspen Dynamics. Set point changes were applied to propane and n-octadecane because they had the highest contents in the corresponding stream and were therefore taken as representative components. For both studied systems, the manipulated variables remained the same because the streams to be perturbed were obtained at the main column, which had no changes in the distribution of components from one sequence to the other.

Results of the control analysis for thermally coupled distillation columns
Once the controllers were set in the simulation environment, the tuning procedure could start. Pairs of values for the gain K and the integral time t were tested, and the responses of the composition for the simulation time t_s registered. To demonstrate the results obtained at this step, Figure 10.20 shows the composition of propane when $K = 10$, for different values of the integral time. In the case of the TCDS, it can be seen in Figure 10.20(a) that low values of t caused an oscillatory response, with difficulties in stabilizing at the new set point. As the value of t increased, fewer oscillations occurred. On the other hand, for the TCIS, Figure 10.20(b) shows that a low value of t caused an overshoot, but then the response stabilized at the new set point. For $t > 10$, the overshoot disappeared but the response time increased. Nevertheless, for all cases, the response stabilized within 5 h, whereas in the TCDS case, the response showed oscillations for more than 50 h.

Figure 10.21 shows the responses for the composition of nC18 when $K = 50$. In the case of the TCDS, shown in Figure 10.21(a), a similar response was observed for each value of the integral time. Nevertheless, for $t > 150$, the response was slightly slower. Stabilization could be observed at around 25 h. For the TCIS, shown in Figure 10.21(b), an overshoot occurred at low values of t, but then the response quickly stabilized after about 3 h. Nevertheless, as the value of t increased, the response was unstable; for $t > 10$ it was not possible to reach a new stationary point. It can be observed that, for the same value of K, the TCDS required higher values for the integral time, but, in contrast to the TCIS, showed no oscillations in the responses.

It is clear that the responses of the composition can change in terms of the parameters of the controller. Thus, is not practical to evaluate the performance of the control system only by analysis of the responses. A better way to compare the different configurations is through criteria such as the IAE. Figure 10.22 shows the IAE for the C3 controller at various values of K and t. For the TCDS, Figure 10.22(a) shows that, at low values of K, increasing the value of the integral time caused an increase in the IAE. For $K > 10$, the opposite occurred. It is clear that high K and high t are required to achieve low values of the IAE. In the case of the TCIS, Figure 10.22(b) shows that IAE decreased when low values of integral time were used. It is also clear that the IAE decreased as K increased. Nevertheless, the difference was not important for $K > 30$.

(a)

(b)

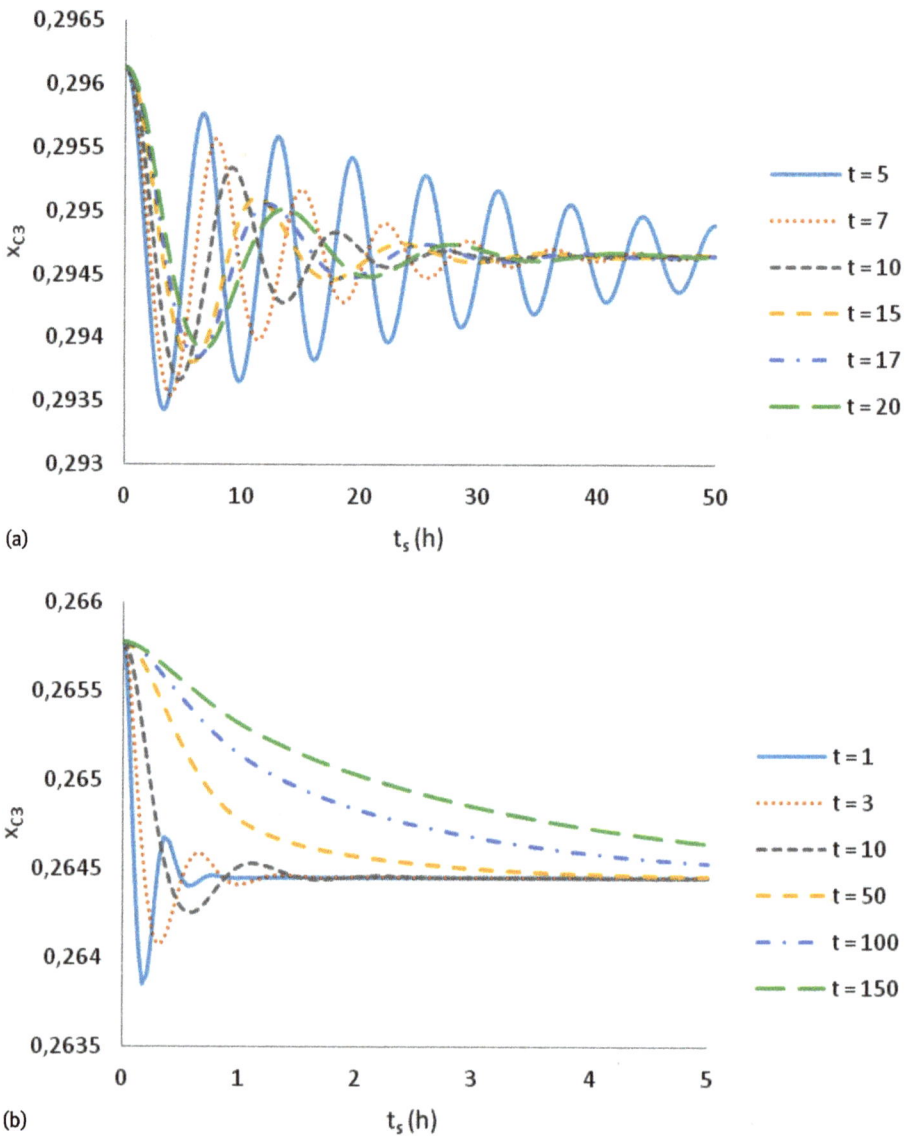

Fig. 10.20: Responses for the component C3 when $K = 10$: (a) thermally coupled direct sequence and (b) thermally coupled indirect sequence.

Changes in the IAE for the nC18 controller are presented in Figure 10.23. For the TCDS, Figure 10.23(a) shows that low values of IAE could be achieved for low values of K and low values of t. Figure 10.23(b) shows that, for the TCIS, high values of K were required to reduce the IAE. Nevertheless, the reduction was not significant for $K > 150$.

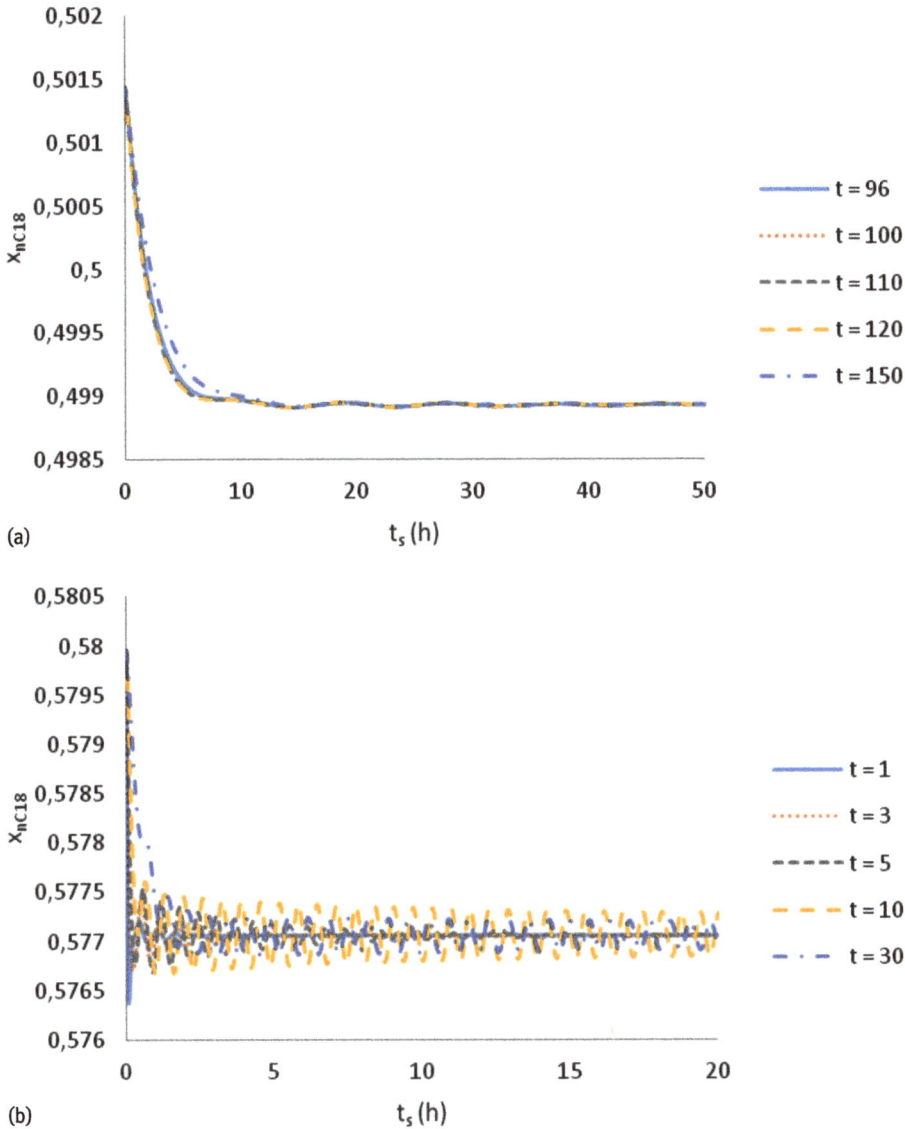

(a)

(b)

Fig. 10.21: Responses for the component nC18 when $K = 50$: (a) thermally coupled direct sequence and (b) thermally coupled indirect sequence.

Moreover, low values of t were preferable because they allowed further reductions of IAE.

To determine the most reliable configuration in terms of control, the best cases for different values of K are presented in Table 10.6 for the C3 and nC18 controllers. Table 10.6 demonstrates that, for any value of K, the TCIS always presented a lower

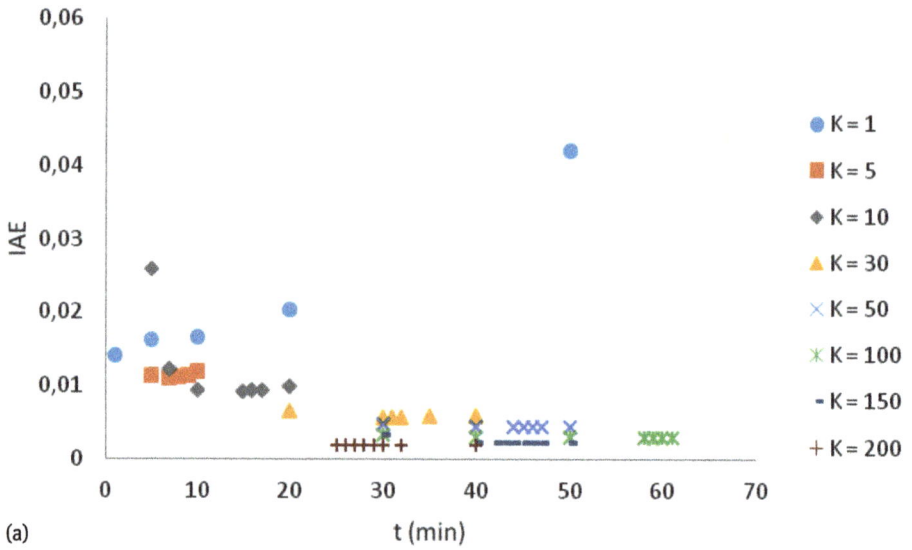

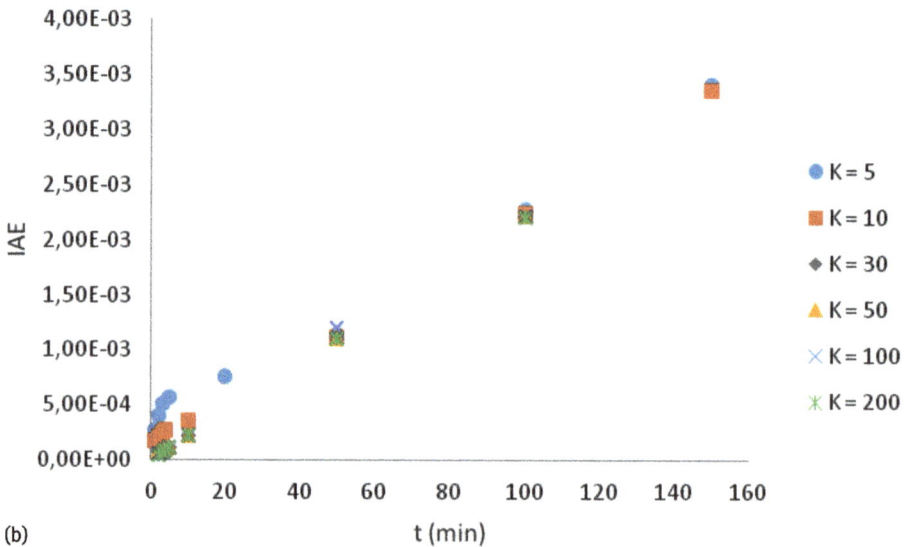

Fig. 10.22: IAE for the C3 controller: (a) thermally coupled direct sequence and (b) thermally coupled indirect sequence.

value of IAE than the TCDS. Moreover, the TCDS required high values of t to minimize the IAE, whereas the TCIS required low values t. It can also been observed that, for $K > 30$, the reduction in IAE for the TCIS was relatively small. Thus, it might not be necessary to use high values of K to obtain a proper response when the set point of C3 is modified.

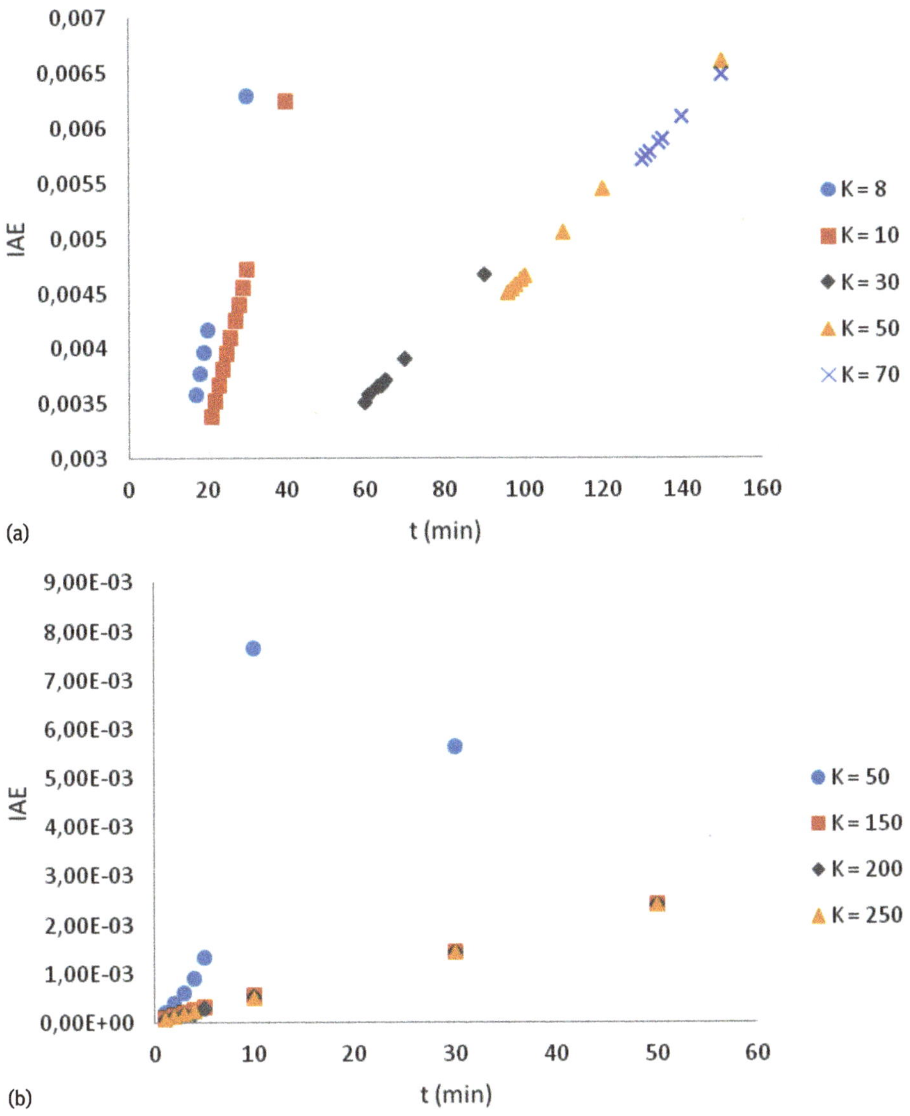

(a)

(b)

Fig. 10.23: IAE for the nC18 controller: (a) thermally coupled direct sequence and (b) thermally coupled indirect sequence.

Considering the nC18 controller, a minimal value of IAE could be observed for the TCDS when $K = 30$ and $t = 60$ min; higher values of K increased the IAE. In the case of the TCIS, the IAE was significantly lower for TCIS than for TCDS. Although increasing K helped to make the IAE smaller, the decrease was relatively small for $K > 150$. In general, for both controllers, the TCIS showed better values of IAE, but the nC18

Tab. 10.6: Best values of IAE for the C3 and nC18 controllers in TCDS and TCIS.

C3 controller						nC18 controller					
TCDS			TCIS			TCDS			TCIS		
K	t (min)	IAE	K	t (min)	IAE	K	t (min)	IAE	K	t (min)	IAE
5	7	0.01104	5	1	2.74×10^{-4}	8	17	0.00358	50	1	2.12×10^{-4}
10	15	0.00928	10	1	1.79×10^{-4}	10	21	0.00338	150	1	8.85×10^{-5}
30	31	0.00567	30	1	9.96×10^{-5}	30	60	0.00351	200	1	7.26×10^{-5}
50	44	0.00431	50	2	7.76×10^{-5}	50	97	0.00454	250	1	6.40×10^{-5}
100	58	0.00291	100	2	5.39×10^{-5}	70	130	0.00570	–	–	–
150	43	0.00229	200	2	4.35×10^{-5}	–	–	–	–	–	–
200	29	0.00184	250	2	4.35×10^{-5}	–	–	–	–	–	–
250	8	0.00127	–	–	–	–	–	–	–	–	–

controller required higher values of K. Nevertheless, it is important to recall that the TCIS had a higher energy requirement than the TCDS.

10.4 Conclusions

The application of PI to distillation sequences can result in important reductions in energy requirements and total annual costs. Nevertheless, it is also important to consider the operability of the obtained systems when establishing the benefits of intensification. In this work, several conventional and intensified distillation designs were evaluated through a closed-loop control policy. Considering all designs, the intensified designs exhibited better behavior under a feedback control test. Furthermore, the conventional designs exhibited worse dynamic behavior than intensified designs.

The relation between intensified design and control properties is not new. Indeed, a similar situation was reported by Segovia-Hernandez et al. [55], who describe in detail the good dynamic behavior when the configuration of the systems has been identified. Here, the cases of reactive distillation and the separation of multicomponent mixtures have been shown to follow similar trends to those previously reported for the separation of ternary mixtures in thermally coupled sequences [55].

10.5 Bibliography

[1]   Babi DK, Cruz MS, Gani, R. Fundamentals of Process Intensification: A Process Systems Engineering View. In: Process Intensification in Chemical Engineering, 2016, 7–33. Springer International Publishing.

[2]   US Department of Energy. Process intensification: technology assessments. Quadrennial Technology Review, 2015, pp 1–29.

[3]   Van Gerven T, Stankiewicz A. Structure, energy, synergy, time – The fundamentals of process intensification. Industrial & engineering chemistry research, 2009, 48(5), 2465–2474.

[4]   Nikačević NM, Huesman AE, Van den Hof PM, Stankiewicz AI. Opportunities and challenges for process control in process intensification. Chemical Engineering and Processing: Process Intensification, 2012, 52, 1–15.

[5]   Boodhoo K, Harvey A (eds). Process intensification technologies for green chemistry: engineering solutions for sustainable chemical processing. John Wiley & Sons, 2013.

[6]   Lutze P, Gani R, Woodley JM. Process intensification: a perspective on process synthesis. Chemical Engineering and Processing: Process Intensification, 2010, 49(6), 547–558.

[7]   Ponce-Ortega JM, Al-Thubaiti MM, El-Halwagi MM. Process intensification: new understanding and systematic approach. Chemical Engineering and Processing: Process Intensification, 2012, 53, 63–75.

[8]   Stankiewicz AI, Moulijn JA. Process intensification: transforming chemical engineering. Chemical Engineering Progress, 2000, 96(1), 22–34.

[9]   Luyben WL, Yu CC. Reactive distillation design and control. John Wiley & Sons, 2009.

[10] Hessel V. Novel process windows–gate to maximizing process intensification via flow chemistry. Chemical Engineering & Technology, 2009, 32(11), 1655–1681.

[11] Holvey CP, Roberge DM, Gottsponer M, Kockmann N, Macchi A. Pressure drop and mixing in single phase microreactors: Simplified designs of micromixers. Chemical Engineering and Processing: Process Intensification, 2011, 50(10), 1069–1075.

[12] Ganapathy H, Steinmayer S, Shooshtari A, Dessiatoun S, Ohadi MM, Alshehhi M. Process intensification characteristics of a microreactor absorber for enhanced CO_2 capture. Applied Energy, 2016, 162, 416–427.

[13] Al Taweel AM, Azizi F, Sirijeerachai G. Static mixers: Effective means for intensifying mass transfer limited reactions. Chemical Engineering and Processing: Process Intensification, 2013, 72, 51–62.

[14] Al Taweel AM, Li C, Gomaa HG, Yuet P. Intensifying mass transfer between immiscible liquids: using screen-type static mixers. Chemical Engineering Research and Design, 2007, 85(5), 760–765.

[15] Micovic J, Beierling T, Lutze P, Sadowski G, Górak A. Design of hybrid distillation/melt crystallisation processes for separation of close boiling mixtures. Chemical Engineering and Processing: Process Intensification, 2013, 67, 16–24.

[16] Pal P, Dekonda VC, Kumar R. Fermentative production of glutamic acid from renewable carbon source: Process intensification through membrane-integrated hybrid bio-reactor system. Chemical Engineering and Processing: Process Intensification, 2015, 92, 7–17.

[17] Patidar P, Mahajani SM. Esterification of fusel oil using reactive distillation. Part II: process alternatives. Industrial & Engineering Chemistry Research, 2013, 52(47), 16637–16647.

[18] Kolah AK, Asthana NS, Vu DT, Lira CT, Miller DJ. Triethyl citrate synthesis by reactive distillation. Industrial & Engineering Chemistry Research, 2008, 47(4), 1017–1025.

[19] Dejanović I, Matijašević L, Olujić Ž. Dividing wall column—a breakthrough towards sustainable distilling. Chemical Engineering and Processing: Process Intensification, 2010, 49(6), 559–580.

[20] Rong BG. Synthesis of dividing-wall columns (DWC) for multicomponent distillations—A systematic approach. Chemical Engineering Research and Design, 2011, 89(8), 1281–1294.

[21] Caballero JA, Grossmann IE. Optimal synthesis of thermally coupled distillation sequences using a novel MILP approach. Computers & Chemical Engineering, 2014, 61, 118–135.

[22] Torres-Ortega CE, Errico M, Rong BG. Design and optimization of modified non-sharp column configurations for quaternary distillations.Computers & Chemical Engineering, 2015, 74, 15–27.

[23] Brunetti A, Sun Y, Caravella A, Drioli E, Barbieri G. Process Intensification for greenhouse gas separation from biogas: More efficient process schemes based on membrane-integrated systems. International Journal of Greenhouse Gas Control, 2015, 35, 18–29.

[24] de los Ángeles Olán-Acosta M, Castrejón-González EO, Alvarado JFJ, Rico-Ramírez V. Approximate design method for reactive liquid extractors based on thermodynamic equilibrium correlations. Chemical Engineering Research and Design, 2016, 109, 443–454.

[25] Yildirim Ö, Kiss AA, Hüser N, Leßmann K, Kenig EY. Reactive absorption in chemical process industry: a review on current activities. Chemical engineering journal, 2012, 213, 371–391.

[26] Li P, Huang K, Lin Q. A generalized method for the synthesis and design of reactive distillation columns. Chemical Engineering Research and Design, 2012, 90(2), 173–184.

[27] Amte V, Nistala SH, Mahajani SM, Malik RK. Optimization based conceptual design of reactive distillation for selectivity engineering.Computers & Chemical Engineering, 2013, 48, 209–217.

[28] Pǎtruţ C, Bîldea CS, Kiss AA. Catalytic cyclic distillation–a novel process intensification approach in reactive separations. Chemical Engineering and Processing: Process Intensification, 2014, 81, 1–12.

[29] Carrera-Rodríguez M, Segovia-Hernández JG, Hernández-Escoto H, Hernández S, Bonilla-Petriciolet A. A note on an extended short-cut method for the design of multicomponent reactive distillation columns.Chemical Engineering Research and Design, 2014, 92(1), 1–12.

[30] Errico M, Tola G, Rong BG, Demurtas D, Turunen I. Energy saving and capital cost evaluation in distillation column sequences with a divided wall column. Chemical Engineering Research and Design, 2009, 87(12), 1649–1657.

[31] Cossio-Vargas E, Barroso-Munoz FO, Hernandez S, Segovia-Hernandez JG, Cano-Rodriguez MI. Thermally coupled distillation sequences: steady state simulation of the esterification of fatty organic acids.Chemical Engineering and Processing: Process Intensification, 2012, 62, 176–182.

[32] Huang S, Li W, Li Y, Ma J, Shen C, Xu C. Process Assessment of Distillation Using Intermediate Entrainer: Conventional Sequences to the Corresponding Dividing-Wall Columns. Industrial & Engineering Chemistry Research, 2016, 55(6), 1655–1666.

[33] Cheng JK, Lee HY, Huang HP, Yu CC. Optimal steady-state design of reactive distillation processes using simulated annealing. Journal of the Taiwan Institute of Chemical Engineers, 2009, 40(2), 188–196.

[34] Machado GD, Aranda DA, Castier M, Cabral VF, Cardozo-Filho L. Computer simulation of fatty acid esterification in reactive distillation columns. Industrial & Engineering Chemistry Research, 2011, 50(17), 10176–10184.

[35] Huang K, Chen H, Zhang L, Wang S, Liu W. Effective arrangement of an external recycle in reactive distillation columns. Industrial & Engineering Chemistry Research, 2014, 53(5), 1986–1998.

[36] Chen H, Zhang L, Huang K, Yuan Y, Zong X, Wang S, Liu L. Reactive distillation columns with two reactive sections: Feed splitting plus external recycle. Chemical Engineering and Processing: Process Intensification, 2016, 108, 189–196.

[37] Lopez-Saucedo ES, Grossmann IE, Segovia-Hernandez JG, Hernández S. Rigorous modeling, simulation and optimization of a conventional and nonconventional batch reactive distillation column: A comparative study of dynamic optimization approaches. Chemical Engineering Research and Design, 2016, 111, 83–99.

[38] Lu S, Lei Z, Wu J, Yang B. Dynamic control analysis for manufacturing ethanol fuel via reactive distillation. Chemical Engineering and Processing: Process Intensification, 2011, 50(11), 1128–1136.

[39] Rewagad RR, Kiss AA. Dynamic optimization of a dividing-wall column using model predictive control. Chemical Engineering Science, 2012, 68(1), 132–142.

[40] Zavala-Guzmán AM, Hernández-Escoto H, Hernández S, Segovia-Hernández JG. Conventional proportional–integral (PI) control of dividing wall distillation columns: systematic tuning. Industrial & Engineering Chemistry Research, 2012, 51(33), 10869–10880.

[41] Segovia-Hernandez JG, Vazquez-Ojeda M, Gómez-Castro FI, Ramírez-Márquez C, Errico M, Tronci S, Rong BG. Process control analysis for intensified bioethanol separation systems. Chemical Engineering and Processing: Process Intensification, 2014, 75, 119–125.

[42] Ramírez-Márquez C, Segovia-Hernández JG, Hernández S, Errico M, Rong BG. Dynamic behavior of alternative separation processes for ethanol dehydration by extractive distillation. Industrial & Engineering Chemistry Research, 2013, 52(49), 17554–17561.

[43] Huang K, Wang SJ. Design and control of a methyl tertiary butyl ether (MTBE) decomposition reactive distillation column. Industrial & engineering chemistry research, 2007, 46(8), 2508–2519.

[44] Lee HY, Lai IK, Huang HP, Chien IL. Design and control of thermally coupled reactive distillation for the production of isopropyl acetate.Industrial & Engineering Chemistry Research, 2012, 51(36), 11753–11763.

[45] An D, Cai W, Xia M, Zhang X, Wang F. Design and control of reactive dividing-wall column for the production of methyl acetate. Chemical Engineering and Processing: Process Intensification, 2015, 92, 45–60.

[46] Ramírez-Márquez C, Sánchez-Ramírez E, Quiroz-Ramírez JJ, Gómez-Castro FI, Ramírez-Corona N, Cervantes-Jauregui JA, Segovia-Hernández JG. Dynamic behavior of a multi-tasking reactive distillation column for production of silane, dichlorosilane and monochlorosilane. Chemical Engineering and Processing: Process Intensification, 2016, 108, 125–138.

[47] Seferlis P, Georgiadis MC (eds). The integration of process design and control, vol. 17. Elsevier, 2004.

[48] Vázquez-Castillo JA, Segovia-Hernández JG, Ponce-Ortega JM. Multiobjective optimization approach for integrating design and control in multicomponent distillation sequences. industrial & Engineering Chemistry Research, 2015, 54(49), 12320–12330.

[49] Daoutidis P, Zachar M, Jogwar SS. Sustainability and process control: A survey and perspective. Journal of Process Control, 2016, 44, 184–206.

[50] Segovia-Hernández JG, Hernández S, Rico-Ramírez V, Jiménez A. A comparison of the feedback control behavior between thermally coupled and conventional distillation schemes. Computers & chemical engineering, 2004, 28(5), 811–819.

[51] Segovia-Hernández JG, Hernández S, Jiménez A. Analysis of dynamic properties of alternative sequences to the Petlyuk column. Computers & Chemical Engineering, 2005, 29(6), 1389–1399.

[52] Segovia-Hernández JG, Hernández S, Jiménez A, Femat R. Dynamic behavior and control of the Petlyuk scheme via a proportional-integral controller with disturbance estimation (PII2). Chemical and biochemical engineering quarterly, 2005, 19(3), 243–253.

[53] Häggblom KE, Waller KV. Control structures, consistency, and transformations. In Practical distillation control, pp 192–228. Springer US, 1992.

[54] Jiménez A, Hernández S, Montoy FA, Zavala-García M. Analysis of control properties of conventional and nonconventional distillation sequences. Industrial & engineering chemistry research, 2001, 40(17), 3757–3761.

[55] Segovia-Hernández JG, Hernández S, Jiménez A. Control behaviour of thermally coupled distillation sequences. Chemical Engineering Research and Design, 2002, 80(7), 783–789.

[56] Segovia-Hernández JG, Hernández S, Jiménez A. Análisis dinámico de secuencias de destilación térmicamente acopladas. Información Tecnológica, 2002, 13(2), 103.

[57] Stephanopoulos G. Chemical process control: an introduction to theory and practice, 1984.

[58] Tuinstra H, Rand CL. U.S. Patent No. 5,349,102. Washington, DC: U.S. Patent and Trademark Office, 1994.

[59] Cheng K, Wang SJ, Wong DS. Steady-state design of thermally coupled reactive distillation process for the synthesis of diphenyl carbonate. Computers & Chemical Engineering, 2013, 52, 262–271.

[60] Errico M, Ramírez-Márquez C, Torres Ortega CE, Rong BG, Segovia-Hernandez JG. Design and control of an alternative distillation sequence for bioethanol purification. Journal of Chemical Technology and Biotechnology, 2015, 90(12), 2180–2185.

[61] Ramírez-Márquez C, Cabrera-Ruiz J, Segovia-Hernández JG, Hernández S, Errico M, Rong BG. Dynamic behavior of the intensified alternative configurations for quaternary distillation. Chemical Engineering and Processing: Process Intensification, 2016, 108, 151–163.

[62] Segovia-Hernández JG, Hernández-Vargas EA, Márquez-Munoz JA. Control properties of thermally coupled distillation sequences for different operating conditions. Computers & Chemical Engineering, 2007, 31(7), 867–874.

[63] Segovia-Hernández JG, Hernández S, Femat R, Jiménez A. Control of thermally coupled distillation arrangements with dynamic estimation of load disturbances. Industrial & Engineering Chemistry Research, 2007, 46(2), 546–558.

[64] Hernández S, Sandoval-Vergara R, Barroso-Muñoz FO, Murrieta-Dueñas R, Hernández-Escoto H, Segovia-Hernández JG, Rico-Ramirez V. Reactive dividing wall distillation columns: simulation and implementation in a pilot plant. Chemical Engineering and Processing: Process Intensification, 2009, 48(1), 250–258.

[65] Hernandez S, Segovia-Hernandez JG, Juarez-Trujillo L, Estrada-Pacheco JE, Maya-Yescas R. Design study of the control of a reactive thermally coupled distillation sequence for the esterification of fatty organic acids. Chemical Engineering Communications, 2010, 198(1), 1–18.

[66] Gutiérrez-Antonio C, Gómez-Castro FI, Hernández S, Briones-Ramírez A Intensification of a hydrotreating process to produce biojet fuel using thermally coupled distillation. Chemical Engineering and Processing: Process Intensification, 2015, 88, 29–36.

[67] Tedder DW, Rudd DF. Parametric studies in industrial distillation: Part I. Design comparisons. AIChE Journal, 1978, 24(2), 303–315.

[68] Skogestad S, Lundström P, Jacobsen EW. Selecting the best distillation control configuration. AIChE Journal, 1990, 36(5), 753–764.

Index

2-ethoxy-2-methylbutane 232
3D printing 225, 238
3D-printed distillation column 226

A
ABE mixture 192
advantages of biodiesel 246
advantages of bioethanol 246
agricultural residues 243
agronomy 304
alternative space 8
analysis 377
analysis-dominated method 24
analytic and synthetic methods 23
animal manure 243
antisolvent crystallization 315
application of microscale distillation 231
approximation of a DWC system 281
Artemisia annua 309
artemisinin 309
artificial neural network 311
Aspen 378
Aspen Plus 203
Aspen Plus V8.8 270
Aspen's Economic Analyzer V8.8 270
axiomatic approach 299
azeotropes 48
azeotropic behavior 232
azeotropic mixtures 194

B
background process 352
basic concepts and elements 39
basic elements 2
basic elements and concepts 3
basic facets of process intensification 9
basic facets of process synthesis 4
basic hardware elements 20, 49
basic scientific methods 23
basic software elements 18, 48
bioalcohols 194
biobutanol 192, 199
bioethanol 189
biofuels consumption 244
biogenic carbon 245
biomass 243

biomass matrix 320
biomass supply chain 244
biosynthetic pathways 292
branch and bound 267

C
capital cost 373
case studies 143
– biodiesel production 163
– fatty esters synthesis 169
– methyl acetate production 173
catalytic distillation 161, 162
cavitation 125
chemometrics 303
closed heat integration 55
closed-loop 376
close-loop control 374
column internals 162
column sections 267
column section's functionality 268
column sections' recombination 267
cometabolites 296
complex configurations 207
complex distillation flowsheets 42
component mass balance 272
composite curve 327
composite membranes 191
composition 389
conceptual design of DWCs 91
conceptual process design 4, 13
conflict-based method 298
control loop 375
controllability 377
controller 378, 379
convergence approach 7
cooling crystallization 316
cosolvency 318
counterflow extractors 196
criterion 92, 384
critical quality attributes 303

D
data and information translation 11
deactivation 127
design (working) process 22
design of experiments 299

https://doi.org/10.1515/9783110465068-011

dielectric constant 114
dietary supplements 294
differential evolution 204
diffusivity efficiency 186
dihydroartemisinic acid 314
dipole moment 113
direct heat integration 358
disadvantages of biodiesel 246
disadvantages of bioethanol 246
distillation 252, 371
distillation configurations 42
distillation intensification 34
distillation sequences 351
distillation systems 3
distinct distillation subspaces 44
distinct features 57
distinct separation sequence (DSS) 50
distinct subspaces 51
divergence approach 8
diverse scopes of study 26
divided wall columns 268
dividing-wall column (DWC) 34
dividing-wall columns (DWC) 90
dominant criteria 45
dynamic behavior 372
dynamic performance 375
dynamics 375

E

emergence of the technical system 21
energy consumption 373
energy crisis 183
energy crops 243
energy efficiency 348
energy integration 261
energy targets 328
energy world consumption 241
environmental conditions 296
environmental medium 343
epoxidation 117
equilibrium 309
equilibrium and non-equilibrium stages 147
errors 384
esterification 255
ethanol-water azeotrope 252
evaluation indicators 11
evaluation of flowsheet 308
evolutionary design approach 297
external thermal coupling 103

extract 195
extraction 310
extractive distillation 191, 197, 199, 253

F

FAEE 254
FAME 254
fatty acid 254
features of a technical system 17
feedback 388
fermentation 250
fermentation broth 251
FFA 256
fixed cost investment 273
flash column chromatography 314
flowering plants 292
forest products 243
fuel properties 247
fully sloppy split 46
functional parts and components 18

G

gain 388
GAMS 274
gasoline gallon equivalent 270
general elements of the methodology 106
general procedure 29
general systematic procedure 83
generalized systematic procedure 97
GHG emissions 245
global optimal 276
global optimum performance 2
glycerol 253
good 266
grand composite curve 340
grid diagram 331

H

hardware elements 19
hardware system 49
heat capacity flowrates 333
heat engines 344
heat exchange area 331
heat exchanger network 325
heat integration between columns 350
heat integration of utilities 326
heat integration principle 58
heat pipe 221
heat pump 346

heater area 342
heat-integrated thermally coupled
    configurations 57
heterogeneous catalysts 131
heuristic approach 299
heuristics 9, 306
high performance liquid chromatography 321
homogeneous azeotropic mixtures 189
hybrid configuration 199, 200
hybrid distillation/pervaporation system 189
hybrid flowsheet 185, 202, 207
hydrogenation 129
hydrolysis 250

I
impurities 317
independent loops 331
indirect heat integration 358
industry 389
inhibitory compounds 250
initial process flowsheet 305
integral 395
integral of absolute error 374
integral of square error 374
integration 258
Integration strategies 258
intended individual splits 50
intensification 138, 262
intensification factor 129
intensified 372
intensified alternative 206, 207
intensified distillation system 37, 43, 74
intensified simple column configurations 74
intensified structures 203
intermediate fluid 358
intermediate reboiler or condenser 355
intermediate transport submixtures 66
internal dividing walls 37
internal thermal coupling 103
interplant transfer 359
intrinsic thermodynamic inefficiency 53

L
levels of intensification 277
life-cycle assessment 204
lignocellulosic materials 247
liquid hot water pretreatment 276
liquid–liquid extraction 195

liquid–liquid extraction-assisted distillation
    199
liquid–liquid extractor 198
loss factor 114
low-grade heat recovery 360

M
maceration 310
manipulated 395
manipulated variable 381
manufacturing of phytochemicals 299
market survey 293
mass balances 236
mass communication 55
mass integration 260
mass transfer 138, 147, 229, 230
mass-separation agent 198
mathematical modeling 308
maximum heat recovery 327
mean end analysis method 298
mechanically press 255
mechanism 43, 54, 63, 69, 78, 83, 93
medicinal plants 294
membrane selectivity 186
metal foam 221–224, 229
methodology 1, 12, 22, 39, 63, 105, 297
microplant for impurity accumulation 236
microwave irradiation 111
miniaturized pilots 213
minimum 389
minimum flowrate 343
minimum temperature difference (ΔT_{min}) 327
MINLP 265
mixed integer nonlinear programming 298
mixer-settlers 196
mixture 392
mobile phase 317
modeling 138
mole fraction 381
movable column section 68, 69
multicomponent distillation 41
multieffect distillation 355
multi-objective optimization 204
multivariate data 303
multivariate tools 301

N
natural products 291
net heat sink 329

net heat source 329
nonequilibrium 147
nonproduct column sections 75
nonsharp separation sequences 42
nonsharp sequence configurations (NSSCs) 54
nonsharp split 45
novel equipment and devices 10
NREL 279
NRTL 270
number of units 331

O
object system 22
objective result of the PS or PI work 14
one-way transport side-rectifier 76
one-way transport side-stripper 76
onion diagram 325
online gas chromatograph 235
open heat integration 55
optimization 33, 259, 308
ordinary distillation 44
original thermally coupled configuration 62, 64

P
packed bed reactor 128
paclitaxel 296
PARAFAC 311
parallel scheme 188
parameters 392
Pareto-optimal solutions 205, 207
partial oxidation 135
partially sloppy split 46
PAT framework 303
performance 386
permeate 190, 193
pervaporation 185, 192
pervaporation unit 193
pervaporation-assisted fermentation 194
pharmaceutical production 129
pharmacognosy 304
phenomena-driven design 298
PI work 12
pilot plant sizes 214
pinch point 329
pinch position 326
pinch technology 183
placement of reactors 348
Plasmodium falciparum 309
plots 387

pockets 340
polymorphs of artemisinin 315
positive balance 245
post-distillation arrangements 190
post-distillation schemes 188
practical synthesis 7
predistillation scheme 188
prefractionator 97
pretreatment 250
primary metabolites 291
problem scope and task 4
process alternatives 7
process analysis 26
process analytical chemistry 300
process analytical technology 300
process analyzers 301
process data 297
process development 214–216, 219, 236, 239
process flowsheet 304
process integration 184
process intensification 183, 201, 370
process intensification (PI) 1, 41
process intensification applications 10
process intensification evaluation indicators 11
process intensification methods and techniques
    10
process intensification problems and tasks 9
process optimization 184
process synthesis 184, 201, 297, 370
process synthesis (PS) 1, 41, 263
process synthesis applications 6
process synthesis evaluation indicators 6
process synthesis methods and techniques 5
process synthesis problems and tasks 5
process system engineering 183
process technical system 17
product column sections 75
product type 45
properties and attributes 31
proportional gain 376
PS work 12
pure synthesis 8

Q
quality by design 301

R
raffinate 195
raw material 296

reaction 377
reaction kinetics 147
reaction rate 150
reactive 371
reactive distillation
– applications 163
– benefits and limitations 146
– configuration 162
– control 152
– design 152
– dividing-wall column 161
– equipment 161
– feasibility evaluation 145, 154, 155
– modeling 147
– PID controller 160
– process dynamics 160
– technical evaluation 145
– working rinciple 144
reactive DWC 161
reboiler types 220
reflux ratio 355
reflux ratio control 219
relative volatility 44
residue curve map 149
response 387
response surface methodology 311
retentate 190
rigorous simulations 279

S
scientific method 22
secondary metabolites 291
separation 392
separation and dehydration 251
separation train 267
sequences 390
set point 381
sharp separation sequences 42
sharp split 45
short-rotation energy crops 245
simple column configuration 35
simple column configuration (SCC) 52, 54
simple column sequences 267
simple distillation column 46
simulation 387
single-section side columns 75, 83
small-scale distillation 217
social synergy 244
software and hardware elements 35, 38

software and hardware systems 2, 16, 47
software elements 18
software system 47
solubility of artemisinin 317
sonochemistry 124
specific energy demand 200
specific procedures 30
specific technical state 31
SSCF 251
standardization 294
starch 135
state of a technical system 31
states of the technical system 35
steady-state and dynamic properties 28
structural degrees of freedom 68
structure 394
subject designer/investigator 22
subjective endeavor of the PS or PI work 14
subspace 43
supercritical 257
superheating 116
superstructure synthesis problem 265
supply temperature 342
surface area 137
synalysis method 24
synthesis 372
synthesis methods 90
synthesis procedure 200, 202, 203
synthesis-dominated method 23, 105
system concept 15
systematic methodology 3, 43, 103
systematic precedure 78
systematic procedure 28, 34, 57, 69
systematic synthesis procedure 105
systems analysis 15
systems synthesis 15

T
technical system 1, 30
technical system concept 16
technological routes 265
technology whole 32
temperature control 238
temperature profile 217, 221, 222
temperature–enthalpy (T–H) diagram 326
thermal coupling 63, 202
thermal coupling mechanism 62
thermally coupled 373
thermally coupled alternatives 205

thermally coupled configuration 36
thermally coupled sequences 267
thermodynamic data 319
thermodynamic system 15
thermodynamically equivalent configurations
    206
thermodynamically equivalent sequences 268
thermodynamically equivalent side-column
    structure 71, 91
thermodynamically equivalent structure 36, 62
three rules 330
time 395
total annual cost 197
total costs of manufacturing 271
total site profiles 357
traditional distillation configurations 62
traditional medicinal systems 290
transducer 136
transesterification 256
triglycerides 255

tuning procedure 376
two-way transport side-rectifier 76
two-way transport side-stripper 76

U
ultrasonic cavitation 124
uncertainty and randomness 33
uses for glycerol 258
utility path 333
utility system 340

V
vapor permeation 191
variable 378
vertical walls 281

W
working process 38

Y
yield of artemisinin 319

www.ingramcontent.com/pod-product-compliance
Lightning Source LLC
Chambersburg PA
CBHW080647220326
41598CB00033B/5132